AF410814

Innovation in Structural Analysis and Dynamics for Constructions

Innovation in Structural Analysis and Dynamics for Constructions

Editors

Chiara Bedon
Flavio Stochino
Mislav Stepinac

Basel • Beijing • Wuhan • Barcelona • Belgrade • Novi Sad • Cluj • Manchester

Editors
Chiara Bedon
Department of Engineering
and Architecture
University of Trieste
Trieste, Italy

Flavio Stochino
Department of Civil,
Environmental Engineering
and Architecture
University of Cagliari
Cagliari, Italy

Mislav Stepinac
Faculty of Civil Engineering
University of Zagreb
Zagreb, Italy

Editorial Office
MDPI
St. Alban-Anlage 66
4052 Basel, Switzerland

This is a reprint of articles from the Topical Collection published online in the open access journal *Buildings* (ISSN 2075-5309) (available at: https://www.mdpi.com/journal/buildings/topical_collections/Innovat_Struct_Analys_Dynami_Construct).

For citation purposes, cite each article independently as indicated on the article page online and as indicated below:

Lastname, A.A.; Lastname, B.B. Article Title. *Journal Name* **Year**, *Volume Number*, Page Range.

ISBN 978-3-0365-8570-3 (Hbk)
ISBN 978-3-0365-8571-0 (PDF)
doi.org/10.3390/books978-3-0365-8571-0

Cover image courtesy of Jordan Opel

Contents

About the Editors

Chiara Bedon

Chiara is Associate Professor of Structural Engineering at University of Trieste (Italy), Department of Engineering and Architecture, where she is instructor for the courses of 'Structural Analysis', 'Steel Structures' and 'Structural Design and Rehabilitation'. She received a PhD degree in 2012, and since 2009 she has been involved in international research projects and networks, as an expert member, national delegate, task leader or advisor (JRC-ERNCIP, NATO-SPS, EU-COST, etc.). Editorial Board member for several ISI journals. Top Scientist for research impact since 2019 (Stanford std cit indicators), she is listed as Top Italian Scientists (TIS) since 2021 and Top Italian Women Scientist (TIWS), for the Engineering - Civil field. Core Group Member for the ongoing EU-COST Action CA20139 "HELEN" - Holistic design of taller timber buildings (2021-2025).

Flavio Stochino

Flavio is Assistant Professor of Theory and Design of Structures at University of Cagliari (Italy), Department of Civil Environmental Engineering and Architecture. He obtained is Ph.D. degree in Structural Engineering at University of Cagliari in 2013. Then, he moved to the University of Sassari (Italy), working on computational mechanics problems as a Post Doc researcher until 2016. In that year, he became Dresden Junior Fellow at TU Dresden (Germany) where he worked on some advanced computational techniques for fracture mechanics. In 2017, he won the PhD ITalents call, funded by CRUI and Confindustria. Actually, he is involved in several research projects, editorial board members and committees.

Mislav Stepinac

Mislav is Assistant Professor of Structural Engineering at the University of Zagreb (Croatia), Faculty of Civil Engineering. He is involved in teaching activities as instructor for the courses of 'Earthquake Engineering', 'Existing Masonry Structures', 'Precast Reinforced Concrete Structures' and 'Concrete and Masonry Structures'. As the principal investigator for the "ARES" project on the assessment and rehabilitation of existing structures, he is involved in the development of contemporary methods for the structural analysis of masonry and timber constructions. Core Group Member for the ongoing EU-COST Action CA20139 "HELEN" - Holistic design of taller timber buildings (2021-2025).

Preface

The structural analysis and dynamic assessment for materials, components and systems is a rather challenging issue. As guest editors for this Topical Collection, we hope that it could stimulate further research and innovative development in the field of structural analysis and innovation for building and infrastructure applications.

In this regard, we express our gratitude to all authors who contributed to the collection, as well as to the reviewers, who provided their valuable feedback. We also thank the editorial staff, especially Emma Fang, and the publisher, for the technical support.

We hope that readers will find this Topical Collection informative and inspiring for future innovative developments.

Chiara Bedon, Flavio Stochino, and Mislav Stepinac
Editors

Article

Passive Control of Ultra-Span Twin-Box Girder Suspension Bridges under Vortex-Induced Vibration Using Tuned Mass Dampers: A Sensitivity Analysis

Seyed Hossein Hosseini Lavassani [1,*], Denise-Penelope N. Kontoni [2,3,*], Hamed Alizadeh [1] and Vahidreza Gharehbaghi [4]

1   Department of Civil Engineering, Faculty of Engineering, Kharazmi University, Tehran 15719-14911, Iran; hamed.alizade.eng@gmail.com
2   Department of Civil Engineering, School of Engineering, University of the Peloponnese, GR-26334 Patras, Greece
3   School of Science and Technology, Hellenic Open University, GR-26335 Patras, Greece
4   School of Civil Engineering, University of Kansas, Lawrence, KS 66045, USA; vahidrqa@ku.edu
*   Correspondence: lavasani@khu.ac.ir (S.H.H.L.); kontoni@uop.gr (D.-P.N.K.)

Abstract: Suspension bridges' in-plane extended configuration makes them vulnerable to wind-induced vibrations. Vortex shedding is a kind of aerodynamic phenomenon causing a bridge to vibrate in vertical and torsional modes. Vortex-induced vibrations disturb the bridge's serviceability limit, which is not favorable, and in the long run, they can cause fatigue damage. In this condition, vibration control strategies seem to be essential. In this paper, the performance of a tuned mass damper (TMD) is investigated under the torsional vortex phenomenon for an ultra-span streamlined twin-box girder suspension bridge. In this regard, the sensitivity of TMD parameters was addressed according to the torsional responses of the suspension bridge, and the reached appropriate ranges are compared with the outputs provided by genetic algorithm. The results indicated that the installation of three TMDs could control all the vulnerable modes and reduce the torsional rotation by up to 34%.

Keywords: ultra-span suspension bridge; vortex-induced vibrations; tuned mass damper (TMD); finite element method (FEM); sensitivity analysis; genetic algorithm

Citation: Hosseini Lavassani, S.H.; Kontoni, D.-P.N.; Alizadeh, H.; Gharehbaghi, V. Passive Control of Ultra-Span Twin-Box Girder Suspension Bridges under Vortex-Induced Vibration Using Tuned Mass Dampers: A Sensitivity Analysis. *Buildings* **2023**, *13*, 1279. https://doi.org/10.3390/buildings13051279

Academic Editors: Chiara Bedon, Flavio Stochino and Mislav Stepinac

Received: 10 April 2023
Revised: 10 May 2023
Accepted: 12 May 2023
Published: 14 May 2023

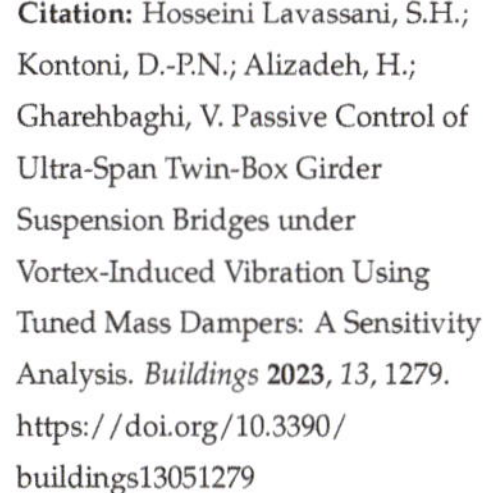

1. Introduction

Within the variety of bridge designs, suspension bridges are known for their capacity to span vast distances, often being the top choice for connecting distant points through extended structural spans. Consequently, the significance of wind-driven oscillations for these bridges increases as their span lengthens. These bridges can vibrate independently in four primary modes—vertical, longitudinal, lateral, and torsional—or as a blend of these modes [1–3]. Aerodynamic phenomena, such as vortex-shedding, galloping, buffeting, static divergence, and flutter, are caused by the interaction between airflow and a broad structure like a bridge deck. Vortex shedding is the primary cause of structural vibrations in the direction perpendicular to the airflow, although other velocity-dependent forces become relevant when the induced motion is substantial [4].

Recently, vortex-induced vibration (VIV) has been detected on numerous bridges globally, including Japan's Trans-Tokyo Bay Bridge [5], Russia's Volgograd Bridge [6], China's Xihoumen Bridge, Yingwuzhou Yangtze River Bridge, and Humen Bridge [7], as well as Denmark's Great Belt Bridge. It is essential to acknowledge that vortex shedding-induced oscillations are predominantly resonant, self-regulating, narrowly focused, and usually occur at low wind speeds, impacting driving safety, contributing to long-term fatigue damage, and reducing the average bridge life span [8–10].

If it is determined that a structure does not meet aerodynamic requirements, there are three strategies for bridge control under VIV as follows [11–13]:

- Prevent vortex shedding by refining the bridge deck's aerodynamic design, utilizing distinct streamlined box girders.
- Modify the bridge's dynamic properties by changing its structural layout.
- Enhance stability by incorporating additional attachments such as spoilers, guide vanes, or flapping plates.

The tuned mass damper (TMD) system, a passive mechanical control approach, falls under the third strategy and is frequently employed in suspension bridges to manage vibrations resulting from various load types. Three primary parameters make up a TMD, specifically, mass ratio, tuning frequency, and damping ratio [14,15]. Originally conceived by Frahm [16], the TMD was a spring-mass system that could only dampen vibrations of a single frequency because of its inability to retain excess energy. Den Hartog [17] later proposed a set of optimal TMD design formulas based on fixed-point theory and included a viscous damper into the system, making it the most commonly used setup.

A novel passive system was tested in a wind tunnel and shown to be effective by Kwon et al. [18], who used a TMD mechanism to activate a plate and adjust the airflow on the deck. Gu et al. [19] proposed a new lever-type TMD to counteract wind-induced vibrations, testing their suggested control system on the Yichang suspension bridge and demonstrating greater efficiency than a passive TMD. Pourzeynali and Datta [20] examined the effects of TMD parameters on accelerating the rate of flutter, finding that TMD did not introduce any instability. Chen and Kareem [21] studied TMD's effectiveness in controlling a bridge's flutter, including an optimal TMD design strategy based on the negative damping of systems.

A unique control strategy for dampening modal coupling effects and reducing resonant vibrations via TMDs was proposed by Chen and Cai [22]. The effectiveness of TMD in increasing the gallop speed of flexible suspension bridges without modifying their forms was studied by Abdel-Rohman and John [23]. Domaneschi et al. [24] studied how TMD can be used to manage buffeting on suspension bridges. For the Vincent Thomas suspension bridge, Alizadeh et al. [25] conducted a sensitivity analysis of flutter velocity concerning the gyration radius and placement of TMDs. Kontoni and Farghaly [26] used tuned mass dampers to mitigate soil-structure interaction effects on a cable-stayed bridge's seismic response. Finally, Mansouri et al. [27] investigated how far-fault earthquake duration, severity, and size affected the seismic response of RC bridges retrofitted with seismic bearings.

Patil and Jangid [28] and Bandivadekar and Jangid [29] reported that the performance of TMD depends on tuning frequency and optimum damping ratio. Researchers have made numerous efforts to address the mistuning issue of single TMD (STMD), such as implementing nonlinear TMD [30], adaptive-passive TMD [31], and semi-active TMD [32]. Additionally, a new idea involving multiple-tuned mass dampers (MTMD), which consists of an array of sub-TMDs with differing frequencies, was put forth by researchers [33,34].

In all types of vibrations, the mass ratio was introduced as the most important factor, causing the large static displacement in the single and multi-TMD configuration [35]. While in torsional motion, the distribution of mass block should be accounted for instead of its weight, increasing the mass does not necessarily improve the efficacy. In fact, the optimum distribution of mass should simultaneously provide the best performance and lightest mass.

In this paper, the effects of TMD parameters on response reduction under torsional VIV are addressed by sensitivity analysis. In this regard, an ultra-span streamlined twin-box girder suspension bridge is chosen for the case study. The VIV analysis will be done on certain vulnerable modes. Then, TMDs are placed along the span according to the shape of the considered modes. The variation of four main parameters called mass ratio, distance factor, damping ratio, and frequency ratio are investigated in the reduction in VIV. The appropriate range is provided by interpreting the related curvatures, and then the results are compared with the outputs of the genetic algorithm. Finally, the most important points will be represented.

2. Aerodynamic Phenomena

Airflow, approaching the deck, causes two independent static and dynamic responses. The evaluation of static response is simple, while conditions are too sophisticated in computing the dynamic responses. Totally, the following three main reasons make dynamic responses:

1. Oncoming airflow inherently includes turbulence, i.e., it is fluctuating in time and space.
2. The flow separates at the body's sharp edges because of the friction that causes further turbulence and vortex development on the surface. As a result, there is vortex shedding in the body's wake since the flow past the body is unstable, with a variable fraction alternating from side to side.
3. The flexible body will be vibrated by fluctuating forces. In this condition, the interaction between the oscillating body and alternation flow will cause additional forces.

In wind engineering, the first item is named buffeting, the second one is known as vortex shedding, and finally, the third part expresses the motion inducted or self-excited forces. Commonly, VIV will have occurred in low velocities, the buffeting will be seen in higher ones, and the motion-induced forces affect the structure in too-high velocities. It is worth noting that this division is only for convenience. There are some other phenomena disturbing the stability of the bridge, such as static divergence, galloping, motion stability limit in torsion, and flutter [36]. As per mentioned contents, the aerodynamic phenomena can be categorized according to the instability and serviceability limits, as shown in Figure 1.

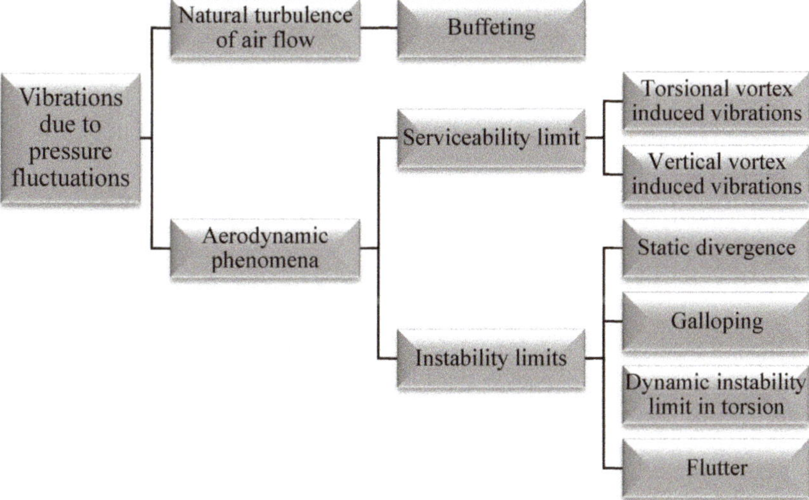

Figure 1. Aerodynamic phenomena of suspension bridges.

3. Vortex Shedding Vibration

Vortex shedding causes bridges to vibrate in vertical and torsion degrees of freedom. VIV will be remarkable around the natural frequencies of bridge modes; it can be considered as a narrow-band frequency process. In fact, when the corresponding frequency of vortices is equal (or nearly equal) to the bridge modes frequency, the vibrations will be seen in larger amplitude. Thus, there is a certain velocity for each mode that resonance occurs in that.

$$V_{R_i} = f_i D / St \tag{1}$$

where f_i, D, and St are the mode frequency, depth of deck, and Strouhal number, respectively.

As VIV is the narrow band process, it is rational to evaluate the bridge response in certain modes. Thus, lower modes should be more accurately studied. As mentioned, the total response is divided into the static and dynamic parts, such that the mean value of the dynamic response is zero. In this regard, the auto spectral density of response ($S_r(\omega)$) and its variance (σ_r^2) can be written as follows:

$$\left.\begin{array}{c} S_{r_n}(\omega) = \phi_n^2(x_r) \cdot \left|\hat{H}_{\eta_n}(\omega)\right|^2 \cdot S_{\hat{Q}_n}(\omega) \\[2mm] \sigma_{r_n}^2 = \int_0^\infty S_{r_n}(\omega)d\omega \\[2mm] n = \begin{cases} z \\ \theta \end{cases} \end{array}\right\}, \qquad (2)$$

where z and θ signify the vertical and torsional degrees of freedom, respectively. In addition, ω, ϕ_n and x_r denote the frequency, mode shape, and position in which the computation is done in there, respectively. The normalized frequency response function ($\hat{H}_{\eta_n}(\omega)$) and normalized auto spectral density of wind load ($S_{\hat{Q}_n}(\omega)$) are as follows:

$$\left.\begin{array}{c} \hat{H}_{\eta_n}(\omega) = \left[1 - \left(\frac{\omega}{\omega_n}\right)^2 + 2i\cdot(\zeta_n - \zeta_{ae_n})\cdot\frac{\omega}{\omega_n}\right]^{-1} \\[3mm] S_{\hat{Q}_n}(\omega) = 2\lambda D \cdot \dfrac{S_{qn}(\omega)\cdot \int_{Lexp} \phi_n^2(x)dx}{\left(\omega_n^2 \tilde{M}_n\right)^2} \end{array}\right\} \qquad (3)$$

where L_{exp} and λ are the length of the bridge experiencing the vortex inducted loads and the non–dimensional coherence length scale of vortices, respectively. In addition, $\tilde{M}_n$ is modal mass. The aerodynamic damping (ζ_{ae}) can be evaluated by the following relation:

$$\begin{array}{c} \zeta_{ae_z} = \dfrac{\tilde{C}_{ae_{zz}}}{2\omega_z M_z} = \dfrac{\rho B^2 H_1^*}{4\tilde{m}_z} \cdot \dfrac{\int_{Lexp} \phi_z^2 dx}{\int_L \phi_z^2 dx} \\[3mm] \zeta_{ae_\theta} = \dfrac{\tilde{C}_{ae_{\theta\theta}}}{2\omega_\theta M_\theta} = \dfrac{\rho B^4 A_2^*}{4\tilde{m}_\theta} \cdot \dfrac{\int_{Lexp} \phi_\theta^2 dx}{\int_L \phi_\theta^2 dx} \end{array} \qquad (4)$$

where $\tilde{m}_n$ is the modally equivalent and evenly distributed mass:

$$\tilde{m}_n = \frac{\tilde{M}_n}{\int_L \phi_n^2 dxn} \qquad (5)$$

In addition, the auto spectral density of the load may be evaluated by the following relation:

$$\begin{bmatrix} S_{qz}(\omega) \\ S_{q\theta}(\omega) \end{bmatrix} = \frac{\left(\frac{1}{2}\rho V^2\right)^2}{\sqrt{\pi}\cdot\omega_n} \cdot \begin{bmatrix} \dfrac{(B\cdot\hat{\sigma}_{qz})^2}{b_z} \cdot \exp\left\{-\left(\dfrac{1-\frac{\omega}{\omega_z}}{b_z}\right)^2\right\} \\[4mm] \dfrac{(B^2\cdot\hat{\sigma}_{q\theta})^2}{b_\theta} \cdot \exp\left\{-\left(\dfrac{1-\frac{\omega}{\omega_\theta}}{b_\theta}\right)^2\right\} \end{bmatrix} \qquad (6)$$

where ρ, V, B, and b_n are the air density, wind velocity, width of the deck, and bandwidth parameters. In addition, the standard deviation of load can be written as follows:

$$\sigma_{qz} = \frac{1}{2}\rho V^2 B \hat{\sigma}_{qz} \quad \text{and} \quad \sigma_{q\theta} = \frac{1}{2}\rho V^2 B^2 \hat{\sigma}_{q\theta} \qquad (7)$$

4. The Considered Bridge

The Halsafjorden suspension bridge, planned by the Norwegian Public Roads Administration (NPRA), aims to connect the west side of the fjord to Akvik on the east. Concrete towers, standing 265 m tall, support the main span on both sides, with cables extending down to rock anchor blocks across a 410-m side span. The cables have a sag of 205 m, resulting in a sag ratio of 0.1. The main cables possess a tensile strength of 1860 MPa, and each has a constructive steel area of A = 375 cm^2. Hangers are spaced 30 m apart.

The deck features a twin box girder design, selected for its advantageous aerodynamic characteristics. Each box girder has a height of 2.5 m, a width of 11 m, and an area of 0.4430 m^2, while the gap between them measures 10 m. Transverse stiffened steel girders, with a height of 2.5 m, a width of 1.5 m, and a constructive steel area of 0.132 m^2, connect the two box girders. These transverse girders align with the hangers and are positioned at 30-m intervals along the bridge span. The bridge's configuration and cross-section can be seen in Figures 2 and 3, while Table 1 presents the bridge's structural and geometrical property specifications.

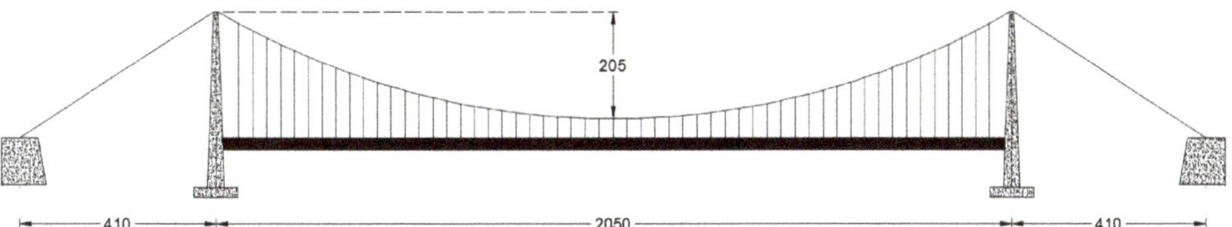

Figure 2. In-plane view of the considered bridge (Dimensions in m).

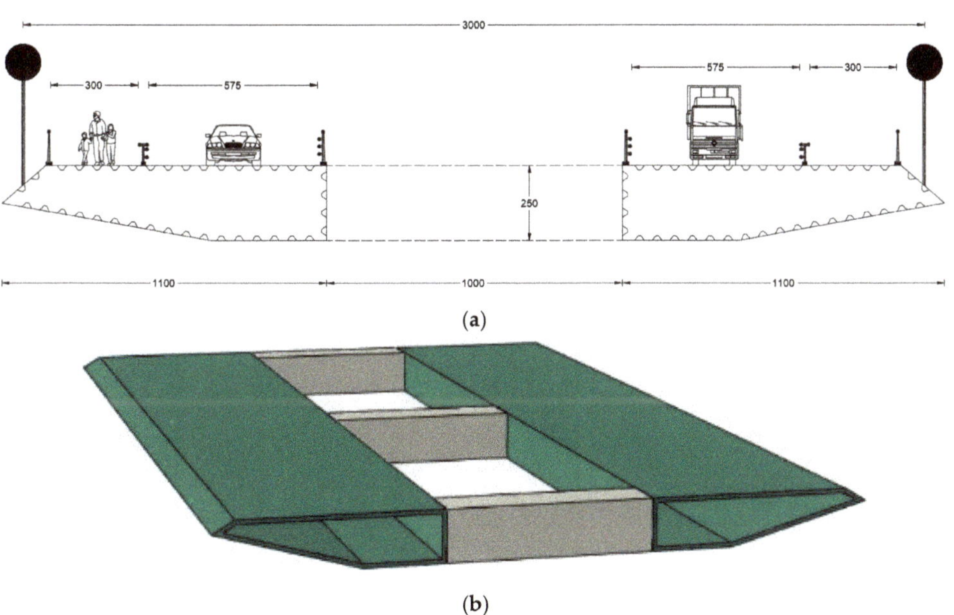

(a)

(b)

Figure 3. Cross-section of the considered bridge, (**a**) Geometry of the deck (Dimensions in m), and (**b**) Configuration of the deck.

Table 1. Structural and geometrical specifications of considered bridge.

No.	Parameters	Value
1	Length of span, L (m)	2050
2	Virtual length of cable, L_E (m)	3160
3	Width of deck, b (m)	32
4	Depth of deck, d (m)	2.5
5	Gyration radius, r (m)	0.62
6	Cross section of main cable, A_c (cm^2)	375
7	Modulus of elasticity of girder, E_g $\left(\frac{N}{m}\right)$	2.00×10^{11}
8	Shear modulus of elasticity of girder, G_g $\left(\frac{N}{m}\right)$	8.0×10^{10}
9	Warping constant, Γ (m^6)	1.712
10	Second moment of inertia, I (m^4)	0.4414
11	Modulus of elasticity of main cable, E_c $\left(\frac{N}{m}\right)$	2.00×10^{11}
12	Horizontal tension of main cable, H_w (kN)	248,050
13	Dead weight of main cable, w $\left(\frac{kg}{m}\right)$	9825

To assess the bridge's dynamic properties, the FEM is employed to calculate the structural properties matrices. The bridge is divided into 100 specific finite elements for this purpose. Each finite element consists of two nodes at its endpoints, with multi degrees of freedom including bending rotation, torsional rotation, vertical displacement, and warping as illustrated in Figure 4. Lavasani et al. [14] provide a detailed explanation of the computation process, and the dynamic specifications of the bridge are presented in Table 2. In Table 2, TS and TA denote the torsional symmetric and torsional antisymmetric mode shape of the bridge, respectively. To investigate the VIV, the flutter derivatives of the bridge are needed, as depicted in Figure 5.

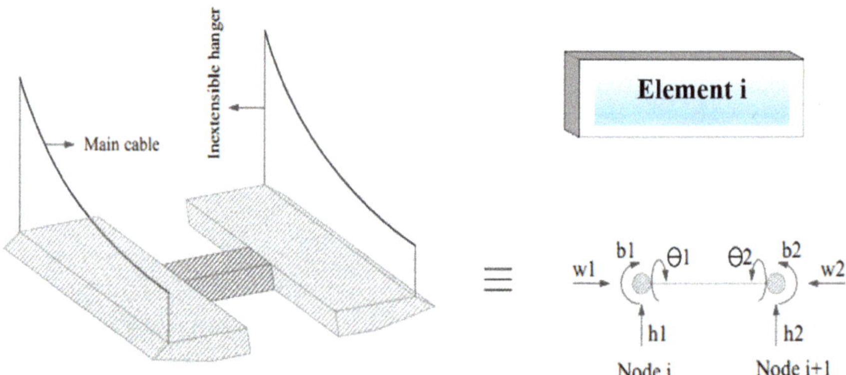

Figure 4. Finite element of the suspension bridge model.

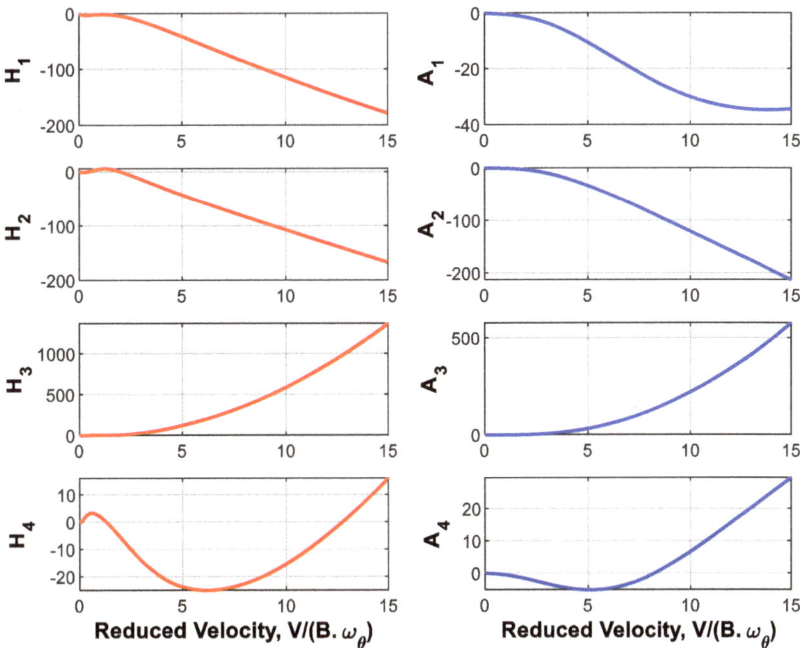

Figure 5. Flutter derivatives of the considered bridge.

Table 2. Mode shapes, frequencies, and periods of considered bridge.

Mode Shape		$\omega_i\left(\frac{rad}{s}\right)$	$T_i\ (s)$
TA1		1.0043	6.28
TS1		1.1072	5.71
TS2		1.5815	3.98
TA2		2.0086	3.14
TS3		2.5218	2.50
TA3		3.0131	2.09

TMD Device

Tuned mass dampers have been widely employed in numerous studies for vibration control [26,36–43], with their straightforward design process and low maintenance costs being their key attributes. On the other hand, their downsides include high sensitivity to mistuning, significant space requirements for installation, and an inability to adapt their parameters to dynamic applied loads. Nonetheless, TMDs are often the first choice for mitigating vibrations in structures, both in research and practical applications. Typically, TMDs are configured to target a specific mode (usually the lower one) to dissipate incoming dynamic energy through their dampers [37]. Adjusting parameters such as mass ratio, damping ratio, and tuning frequency is crucial during the design process. Consequently, these parameters must be set to optimal values, which can be determined through an optimization process.

A too-heavy mass ratio causes considerable static deformations and changes the natural dynamic specifications. On the other hand, a light mass block does not have enough efficacy. The damping ratio should be adjusted to values, allowing free movement of the mass block. The tuning frequency may be named as the most sensitive parameter of it. As a simple way to overcome this problem, instead of tuning the TMD to a certain mode, it is rational to tune it to a ratio of a certain value that may involve the other vulnerable close modes. In this regard, the prone modes should be determined, and then an optimal ratio can be provided. The demonstration of attachment of TMD to the considered bridge is demonstrated in Figure 6. According to Figure 6, two masses are attached to the bottom of the section on the outside.

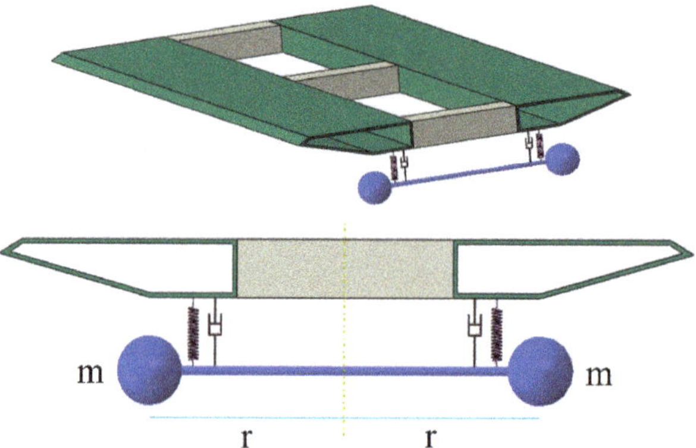

Figure 6. Details of TMD installation.

5. Numerical Analysis

In this paper, the torsional VIV of the considered bridge is controlled by TMD. In this regard, some parameters are used for evaluating the wind-corresponding functions, provided in Table 3. To reach a comparative view, the uncontrolled responses should be evaluated. Regarding the narrow band frequency process, the resonance velocity of each mode and its corresponding response are required. In terms of the considered bridge, computations indicate that the first six modes expose high amplitude vibrations compared to other ones, shown in Figures 7–9. In fact, the higher modes are not vulnerable to VIV.

Table 3. Wind properties.

ρ (kg/m^3)	St (−)	$\hat{\sigma}_{q_\theta}$ (−)	b_θ (−)
1.25	0.1	0.3	0.2

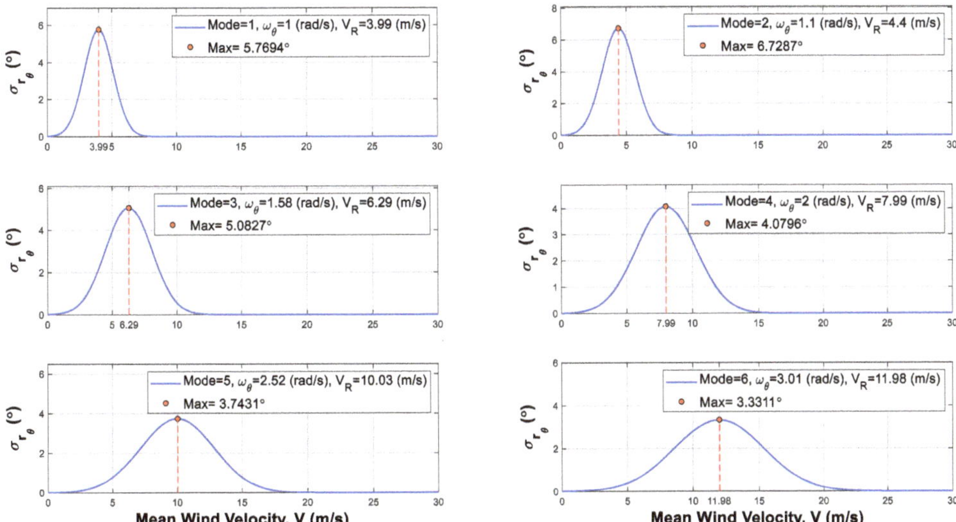

Figure 7. The standard deviation of six initial modes.

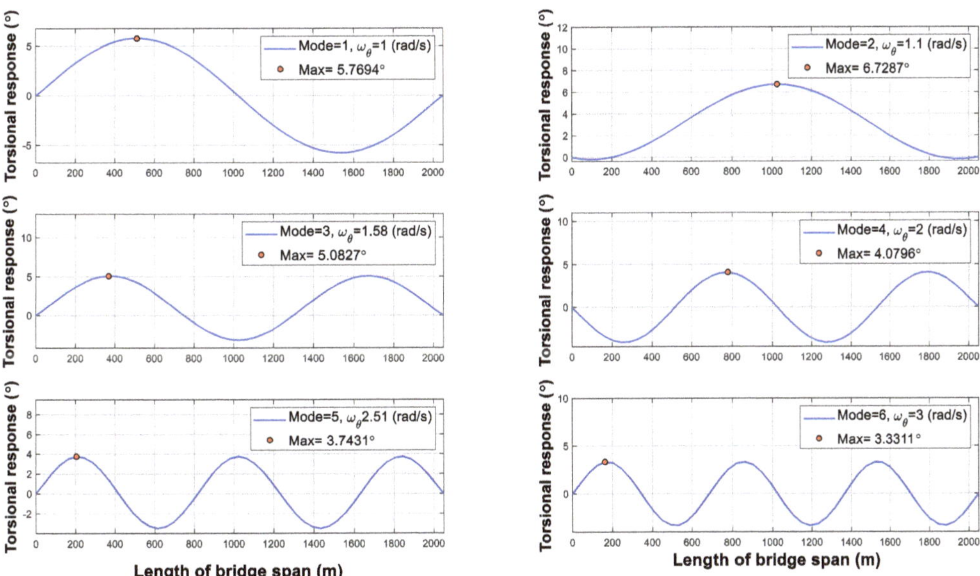

Figure 8. Total response of the bridge.

Figure 9 shows that the first and second modes experience higher amplitude vibrations, disturbing the serviceability limit. Thus, the mode shapes of these modes determine the placement of TMDs. According to Figure 7, $\frac{L}{4}$, $\frac{L}{2}$ and $\frac{3L}{4}$ are the most important points. Mentioned points also undergo almost high amplitudes in the other modes. Thus, the analysis is conducted according to two following conditions:

(a)   One TMD is placed at the middle point.

(b)   Two TMDs are placed at the $\frac{L}{4}$ and $\frac{3L}{4}$ points.

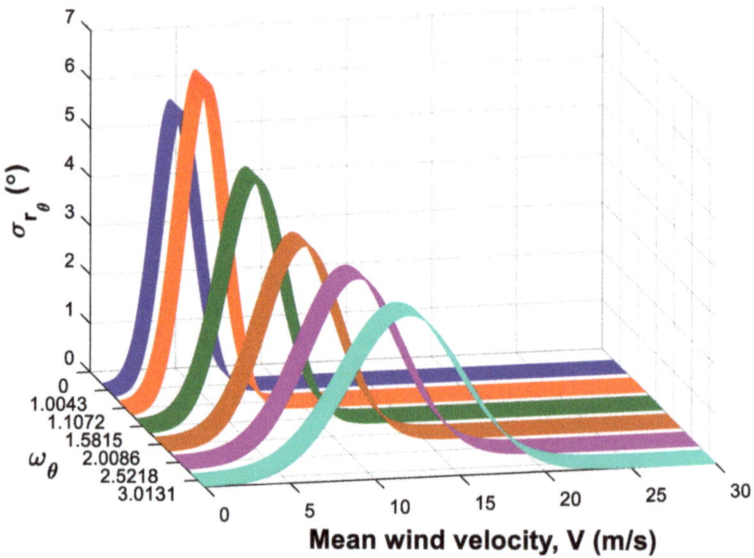

Figure 9. Order and magnitude of six initial modes along the wind velocity.

In each case, its parameters behavior is addressed by sensitivity analysis, and the related curvatures will be interpreted.

5.1. Mass Polar Moment of Inertia

The mass polar moment of inertia involves two primary parameters: mass ratio and the distance between the mass and the rotation axis, which determine the distribution of mass around the rotation axis. In this context, this distance is referred to as the distance factor (r). The optimization process aims to achieve the lightest mass with an appropriate distribution around the rotation axis. For this purpose, r is assumed to be 1, and the sensitivity of the mass ratio is examined based on the standard deviation of the midpoint (L/2) and the point with the maximum torsional response across the entire span. It is important to note that adding a TMD may alter the location of the maximum response. Figure 10 depicts the relevant curvatures for the mass ratio.

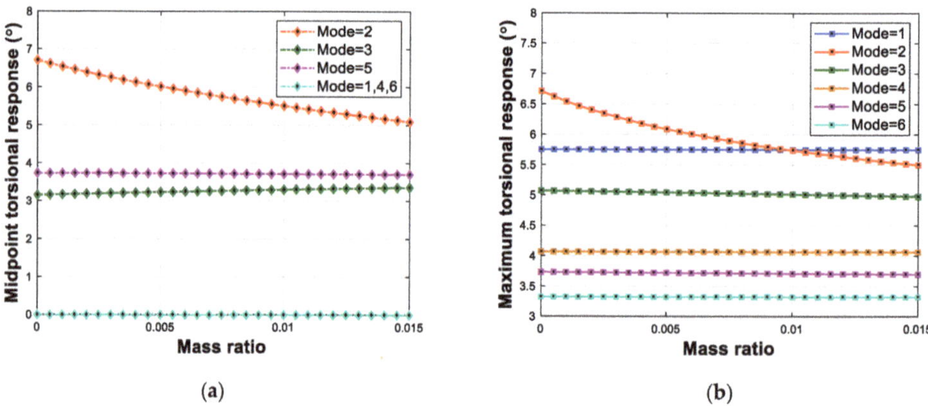

(a)

(b)

Figure 10. Variation of the maximum torsion of (**a**) $\frac{L}{2}$ point and (**b**) whole span, in relation to the mass ratio, in the one TMD condition.

According to Figure 10, the maximum corresponding response of uncontrolled VIV in the middle point and whole span is the same as and equal to 6.72 degrees. In Figure 10a, by increasing the mass ratio, the middle point responses of the second mode were decreased by an appropriate slope, while the other modes' corresponding responses have stayed unchangeable. In Figure 10b, the maximum response of the whole bridge in the second mode has been decreased by an increment of mass ratio. Comparing Figure 10a,b indicates that the local and total performance of TMD is 25.5 (6.72 was decreased to 5 degrees) and 18 (6.72 was decreased to 5.5 degrees) percent, respectively. Hence, the local performance of TMD is more suitable than its total performance. In addition, another distinguishing feature of the Figure is that although TMD cannot reduce VIV-related responses of other modes, it does not increase them.

Now, the computations are carried on by investigating the effects of the distance factor. In this regard, Figure 11 demonstrates the related curvatures of it.

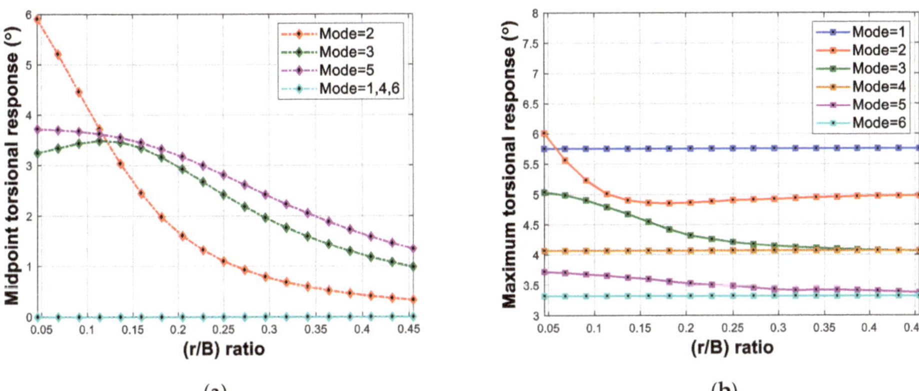

(a) (b)

Figure 11. Variation of the maximum torsion of (**a**) L/2 point and (**b**) whole span, in relation to the distance factor, in the one TMD condition.

In the horizontal axis of Figure 11, $\frac{r}{B}$, signifying distance factor to deck width ratio, is utilized rather than r. This ratio may provide more practical considerations. Figure 11a shows the maximum response of the middle point in the second mode, 6 degrees, is decreasing by an increment of $\frac{r}{B}$. The gradient of the second mode curvatures has been decreased by an increment of $\frac{r}{B}$, revealing there is a certain value that after it the increment of $\frac{r}{B}$ is not too effective. Here, the mentioned value is 0.25. Mode three curvature is similar to mode five, and a little ascending has been seen in the initial part of it. Figure 11b shows that the appropriate range of $\frac{r}{B}$ is 0.1 up to 0.2. Like the mass ratio parameter, the local performance of TMD ($6 \rightarrow 0.4 \Rightarrow 93\%$) is too better than the total performance of it ($6 \rightarrow 5 \Rightarrow 16\%$).

Comparing Figures 10 and 11 discloses that the $\frac{r}{B}$ parameter can reduce the other modes' responses, or vice versa to mass ratio that cannot change the other modes' responses.

Figures 12 and 13 demonstrate the sensitivity of responses in relation to the variation of mass ratio and distance factor for the two TMDs condition.

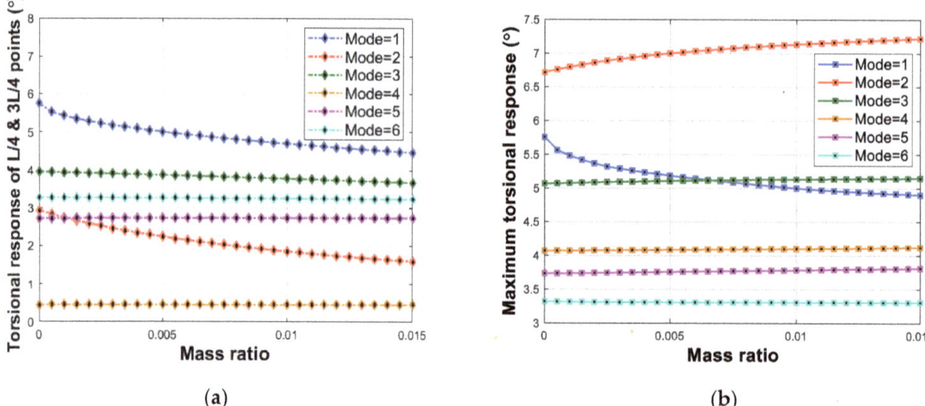

(a)

(b)

Figure 12. Variation of the maximum torsion of (**a**) $\frac{L}{4}$ and $\frac{3L}{4}$ points and (**b**) whole span, in relation to the mass ratio, in the two TMDs condition.

Figure 12a displays the main role of mass ratio in response reduction of first and second modes corresponding responses. Figure 12b shows that by increasing the mass ratio, the second-mode responses are intensified, while first-mode ones are decreased. Hence, two TMDs, located at the $\frac{L}{4}$ and $\frac{3L}{4}$ points, provide positive local performance in all considered modes; however, their total one is not positive for all modes.

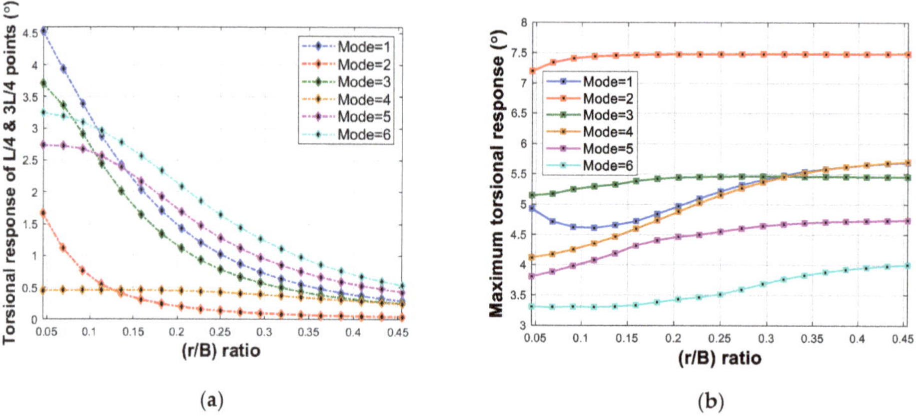

(a)

(b)

Figure 13. Variation of the maximum torsion of (**a**) $\frac{L}{4}$ and points and (**b**) whole span, in relation to the distance factor, in the two TMDs condition.

Figure 13a indicates that the increment of the distance factor considerably improves the local performance of TMD for all modes. Figure 13b shows that increasing the distance factor first reduces the responses; however, after $\frac{r}{B} = 0.2$, the responses of other modes are intensified.

5.2. Damping Ratio

In order to investigate the sensitivity of VIV against the damping ratio, it changes in the 0 up to 20% range, demonstrated in Figure 14. In addition, the mass ratio and distance factor are considered 1% and 0.2, respectively, for both one and two TMD conditions.

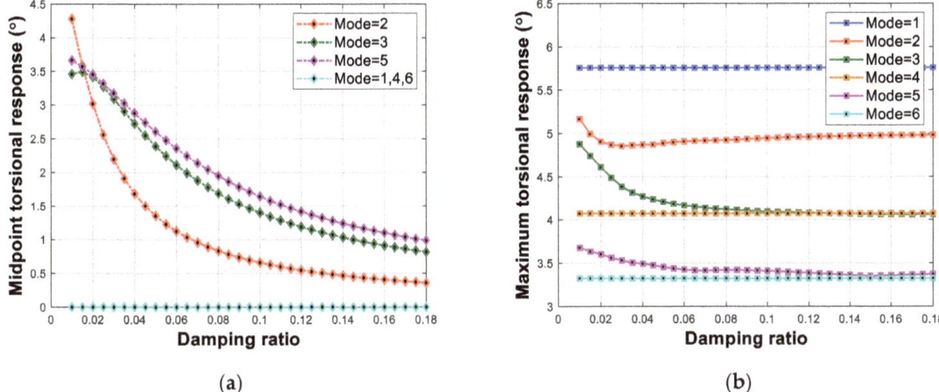

Figure 14. Variation of the maximum torsion of (**a**) $\frac{L}{2}$ point and (**b**) whole span, in relation to the damping ratio, in the one TMD condition.

Figure 14a discloses that by increasing the damping ratio the $\frac{L}{2}$ point corresponding responses in modes two, three, and five are considerably reduced. Of course, the second mode response reduction gradient is too high compared to other ones in the 0 up to 10 percent range. Figure 14b indicates that the effect of the damping ratio on the maximum response of the whole bridge is too low compared to any certain point like $\frac{L}{2}$. Especially, for mode two, the local performance of TMD is 90% $(4.3 \rightarrow 0.4)$, while the total one is 7.6% $(5.2 \rightarrow 4.8)$. Thus, the damping ratio parameter is effective in the local performance of TMD, and its total efficacy is nuance.

Figure 15 demonstrates the sensitivity of responses in relation to the variation of damping ratio for the two TMDs condition.

Like the one TMD condition, the damping ratio renders suitable local performance, and its appropriate range is 0 up to 10 percent.

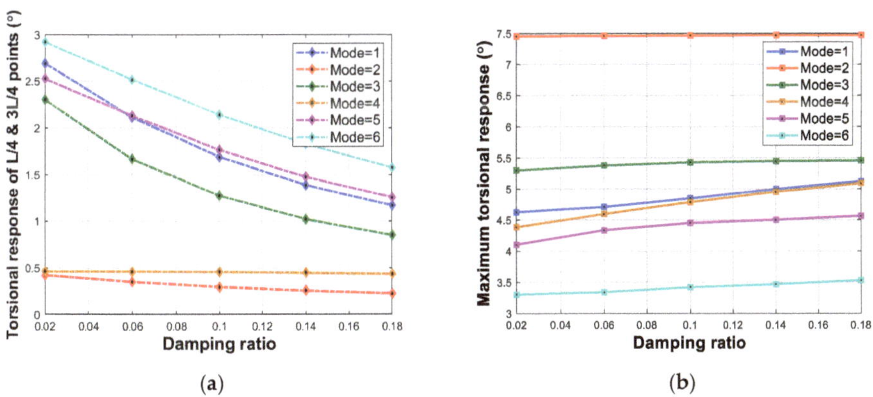

Figure 15. Variation of the maximum torsion of (**a**) $\frac{L}{4}$ and $\frac{3L}{4}$ points and (**b**) whole span, in relation to the damping ratio, in the two TMDs condition.

5.3. Frequency Ratio

According to mentioned contents of the former sections, the mistuning of TMD can considerably decline the efficacy of it. In this regard, the responses should be addressed in a range around a certain mode frequency. Figure 16 shows the VIV corresponding response in the 0.7 up to 1.4 variations of the frequency ratio. The mass ratio, the r/B, and the damping ratio are 1%, 0.2, and 10%, respectively.

Figure 16a reveals that the corresponding responses of $\frac{L}{2}$ point are declined by the increment of the frequency ratio. The final part of the second mode's corresponding curvature indicates that when the tuning frequency deviates considerably from the frequency of the modes, unfavorable effects will emerge. Figure 16a,b display that after $f = 1.3\omega_2$, the maximum response of the whole span and $\frac{L}{2}$ point will be increased.

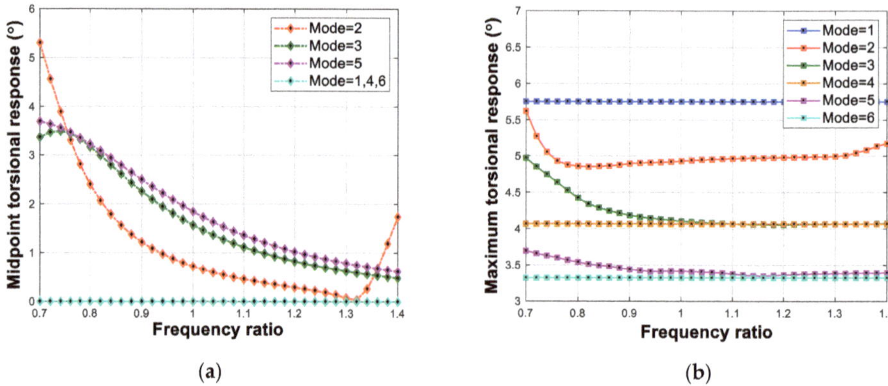

(a) (b)

Figure 16. Variation of the maximum torsion of (a) $\frac{L}{2}$ point and (b) whole span, in relation to the frequency ratio, in the one TMD condition.

Figure 17 demonstrates the sensitivity of responses in relation to the variation of frequency ratio for the two TMDs condition. It is worth noting that the mass ratio, $\frac{r}{B}$, and damping ratio are considered 1%, 0.2, and 10%, respectively.

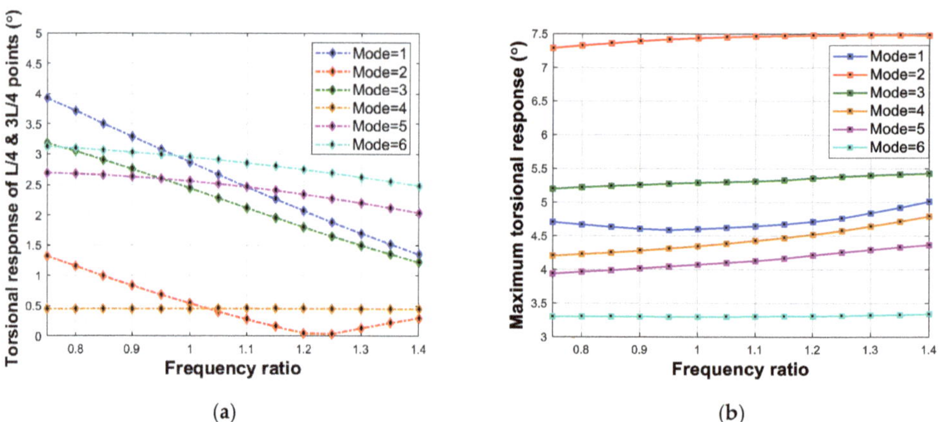

(a) (b)

Figure 17. Variation of the maximum torsion of (a) $\frac{L}{4}$ and $\frac{3L}{4}$ points and (b) whole span, in relation to the frequency ratio, in the two TMDs condition.

A similar statement of the one TMD condition can be expressed for the two TMDs. It is worth noting that one TMD is tuned to the second mode, and two TMDs are tuned to the first mode.

5.4. Combined Condition

There, TMDs whose appropriate parameters were achieved using genetic algorithm and sensitivity analysis were attached to $\frac{L}{4}$, $\frac{L}{2}$ and $\frac{3L}{4}$ points to reduce the VIV responses are provided in Tables 4 and 5.

Table 4. Optimum parameter of TMD reached by genetic algorithm.

TMDs	Mass Ratio (%)	Distance Factor (*m*)	Frequency Ratio	Damping Ratio (%)
Middle one	0.6	5.21	1.2	11.35
Side ones	1.35	2.5	1.25	12.65

Table 5. Optimum parameter of TMD reached by sensitivity analysis.

TMDs	Mass Ratio (%) [0.5, 1.5]	Distance Factor (*m*) [4.04, 5.4]	Frequency Ratio [1, 1.25]	Damping Ratio (%) [5, 12.5]
Middle one	1	4.8	1.25	10
Side ones	1	4.8	1.25	10

The middle point TMD parameters are adjusted according to the second mode, and two other ones obey the first mode specifications. Figures 18 and 19 show the standard deviation of the critical point of each mode and the total response of the bridge in uncontrolled and controlled conditions, respectively.

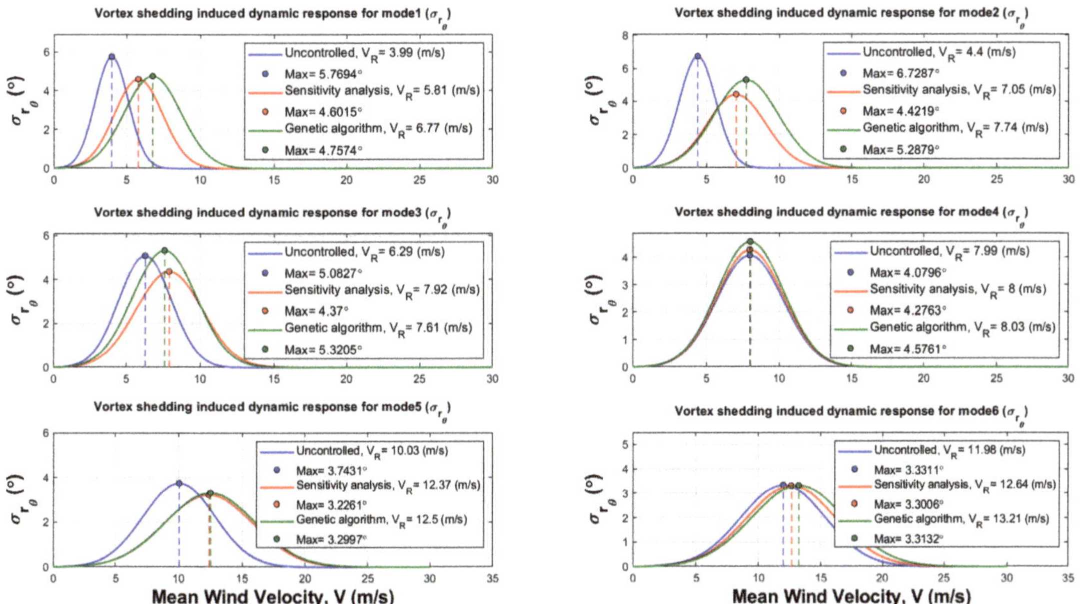

Figure 18. Standard deviation in uncontrolled and controlled conditions.

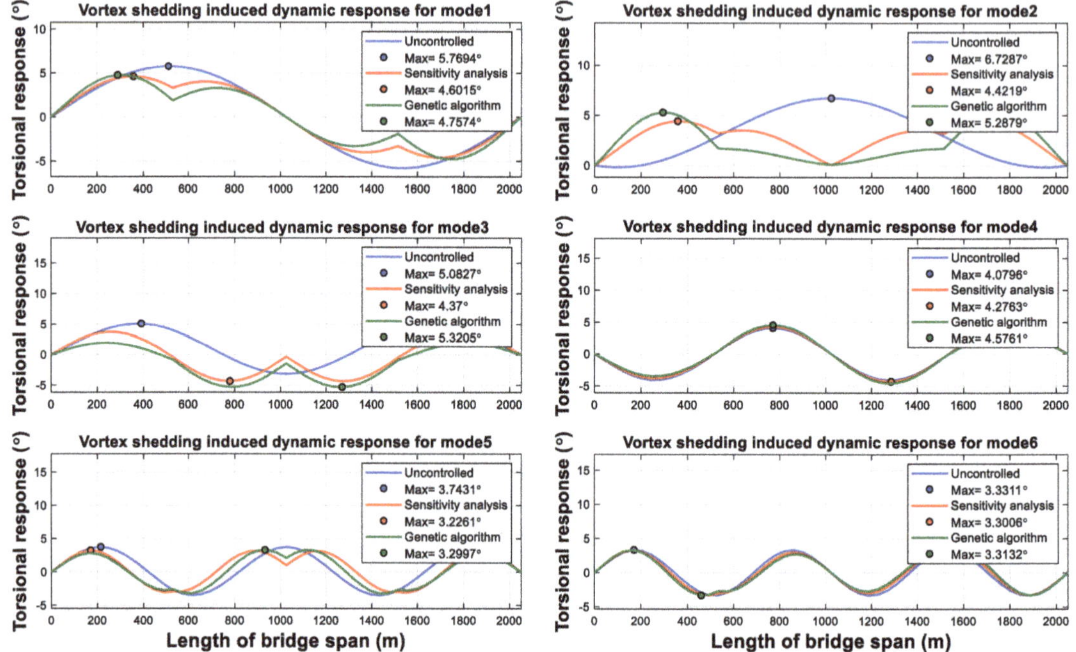

Figure 19. Total response of bridge in uncontrolled and controlled conditions.

Figure 18 indicates that TMD can successfully reduce the standard deviation of critical points. There is a little increment in mode four, which is negligible. Thus, the local performance of TMD is better in lower modes and has not had a negative role in the other modes. In addition, the parameters achieved by sensitivity analysis provide better performance than genetic algorithm ones.

For addressing the performance of TMD in response to reduction during the time, the following relation can be utilized:

$$S_x(\omega_j) = \omega_j^2 / (2\Delta\omega_j) \tag{8}$$

$$x(t) = \sum_{j=1}^{N} c_j \cos(\omega_j t + \psi_j) \tag{9}$$

$$c_j = \left[2S_x(\omega_j)\,\Delta\omega_j\right]^{1/2} \tag{10}$$

where Ψ_j is arbitrary phase angles between zero and 2π, $\Delta\omega$ is the frequency segment, and ω is the middle point of each segment.

Figures 20 and 21 show the time history of torsional responses. Figures 20 and 21 show that TMDs successfully mitigate the VIV in considered modes.

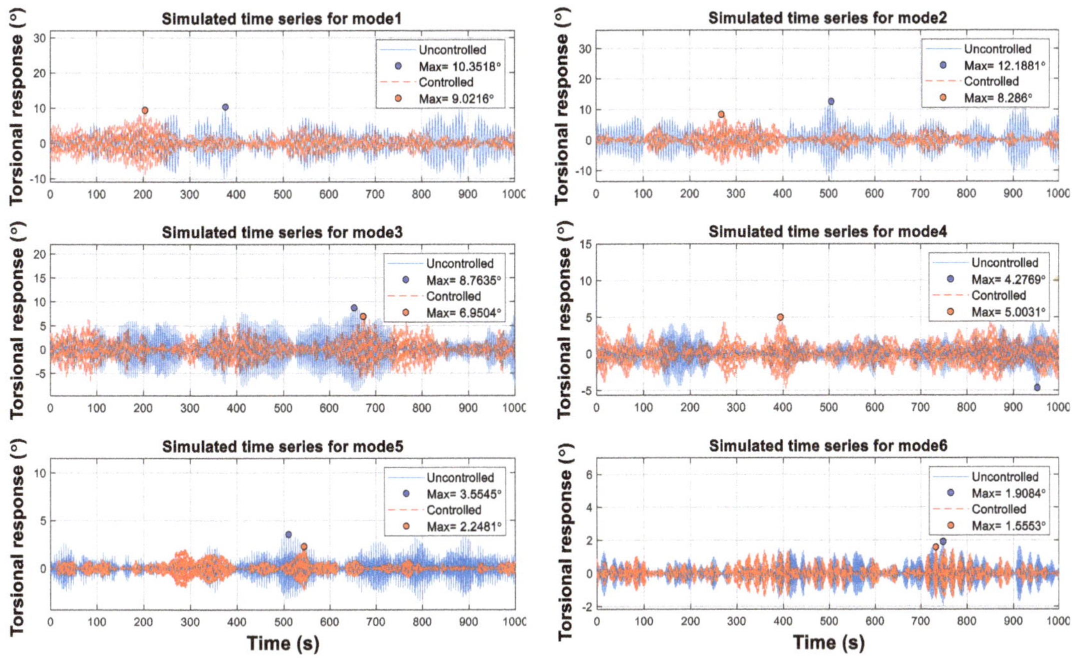

Figure 20. Time history of torsional response at the 4 $\frac{m}{s}$ velocity at each mode critical point.

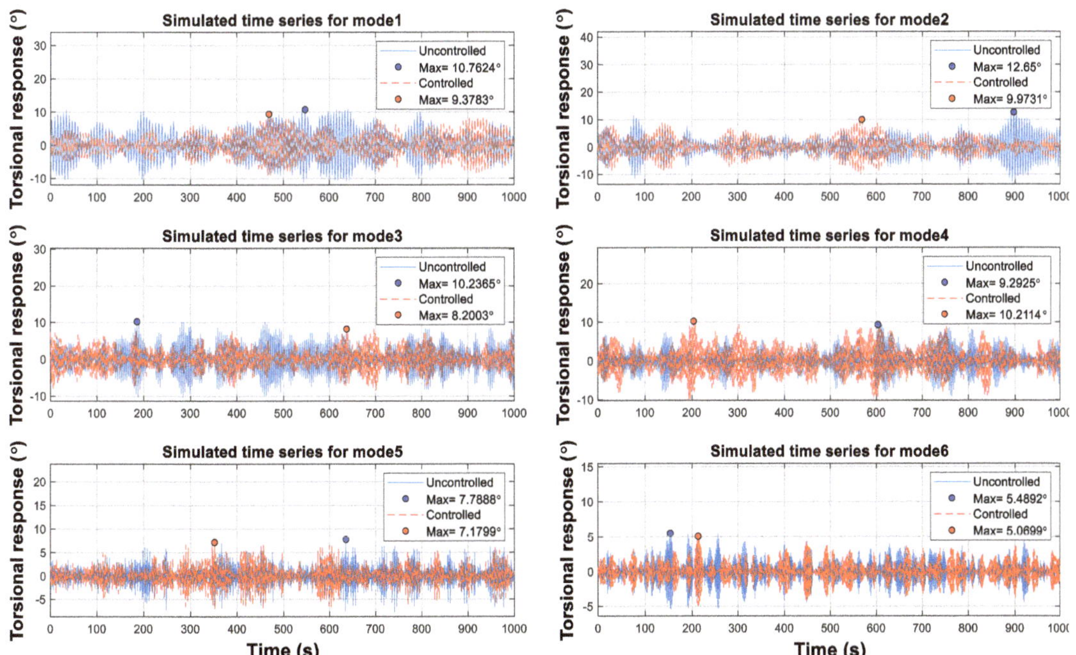

Figure 21. Time history of torsional response at the resonance velocity of each mode at its critical point.

6. Conclusions

In this paper, the torsional VIV of an ultra-span twin box girder suspension bridge, with a 2050 m span length, was controlled by TMDs. Firstly, VIV analysis of the bridge was done to determine the vulnerable modes that were the six initial ones. The parameters of TMD named mass ratio, distance factor, damping ratio, and frequency ratio were investigated under VIV to achieve their optimum ranges. A sensitivity analysis was conducted according to each mode critical point and the maximum response along the span. After computing the appropriate parameters by sensitivity analysis and genetic algorithm, the torsional VIV was controlled and comparisons were made. The most important results are as follows:

The mass polar moment of inertia should be optimized in two independent phases, including mass ratio and distribution of mass around the torsion axis. Increment of mass ratio and $\frac{r}{B}$ improve the local and total performance of TMD, and increment of $\frac{r}{B}$ has a stronger impression on the local performance in comparison to the total one. The $\frac{r}{B}$ has the most integral role in reducing the vulnerable modes in comparison to the mass ratio. The appropriate range of mass ratio and $\frac{r}{B}$ are 0.5 up to 1.5 percent and 0.15 up to 0.2, respectively.

For improving the local performance, by moving away from the middle point, the mass blocks become lighter, and their distance increase. The TMD, located at the middle point, has the most weight, and the distance of its mass block is the minimum. The better total performance needs heavier mass blocks placed closer to each other.

The damping ratio has perfectly local performance, and its total efficacy is nuance. Its optimum range is 0 up to 10 percent.

The increment of frequency ratio increases the local and total efficacy of TMD such that the local one is more sensible. The optimum range of it is 1 up to 1.3 times to the considered mode. A higher value of 1.3 causes unfavorable results.

For the considered bridge, attaching three TMDs to $\frac{L}{4}$, $\frac{L}{2}$ and $\frac{3L}{4}$ points, comprising a 1% mass ratio, 0.15 $\frac{r}{B}$, 10% damping ratio, and 1.25 frequency ratio, decreased the maximum response, $6.72°$, to $4.42°$, indicating a 34% response reduction. The considered TMDs suitably mitigated the vibration amplitude during the resonance velocity of vulnerable modes preventing the occurrence of fatigue damage in the long term.

Author Contributions: Conceptualization, S.H.H.L. and D.-P.N.K.; methodology, S.H.H.L. and D.-P.N.K.; software, S.H.H.L., D.-P.N.K., H.A. and V.G.; validation, S.H.H.L., D.-P.N.K., H.A. and V.G.; formal analysis, S.H.H.L. and D.-P.N.K.; investigation, S.H.H.L., D.-P.N.K., H.A. and V.G.; resources, S.H.H.L. and D.-P.N.K.; data curation, S.H.H.L., D.-P.N.K., H.A. and V.G.; writing—original draft preparation, S.H.H.L., D.-P.N.K., H.A. and V.G.; writing—review and editing, S.H.H.L. and D.-P.N.K.; visualization, S.H.H.L., D.-P.N.K., H.A. and V.G.; supervision, S.H.H.L. and D.-P.N.K.; project administration, D.-P.N.K. All authors have read and agreed to the published version of the manuscript.

Funding: This research received no external funding.

Institutional Review Board Statement: Not applicable.

Informed Consent Statement: Not applicable.

Data Availability Statement: Data are available within the article.

Conflicts of Interest: The authors declare no conflict of interest.

References

1. Alizadeh, H.; Lavassani, S.H.H. Flutter Control of Long Span Suspension Bridges in Time Domain Using Optimized TMD. *Int. J. Steel Struct.* **2021**, *21*, 731–742. [CrossRef]
2. Alizadeh, H.; Hosseini Lavassani, S.H. The effect of soil-structure interaction on longitudinal seismic responses of suspension bridges controlled by optimal TMD. *J. Struct. Constr. Eng.* **2022**, *8*, 439–458. [CrossRef]
3. Hosseini Lavassani, S.H.; Alizadeh, H.; Doroudi, R.; Homami, P. Vibration control of suspension bridge due to vertical ground motions. *Adv. Struct. Eng.* **2020**, *23*, 2626–2641. [CrossRef]

4. Basu, R.; Vickery, B. Across-wind vibrations of structure of circular cross-section. Part II. Development of a mathematical model for full-scale application. *J. Wind Eng. Ind. Aerodyn.* **1983**, *12*, 75–97. [CrossRef]
5. Fujino, Y.; Yoshida, Y. Wind-induced vibration and control of Trans-Tokyo Bay crossing bridge. *J. Struct. Eng.* **2002**, *128*, 1012–1025. [CrossRef]
6. Weber, F.; Maślanka, M. Frequency and damping adaptation of a TMD with controlled MR damper. *Smart Mater. Struct.* **2012**, *21*, 055011. [CrossRef]
7. Gao, D.; Deng, Z.; Yang, W.; Chen, W. Review of the excitation mechanism and aerodynamic flow control of vortex-induced vibration of the main girder for long-span bridges: A vortex-dynamics approach. *J. Fluids Struct.* **2021**, *105*, 103348. [CrossRef]
8. Park, J.; Kim, S.; Kim, H.-K. Effect of gap distance on vortex-induced vibration in two parallel cable-stayed bridges. *J. Wind Eng. Ind. Aerodyn.* **2017**, *162*, 35–44. [CrossRef]
9. Wang, W.; Wang, X.; Hua, X.; Song, G.; Chen, Z. Vibration control of vortex-induced vibrations of a bridge deck by a single-side pounding tuned mass damper. *Eng. Struct.* **2018**, *173*, 61–75. [CrossRef]
10. Xue, Z.; Han, B.; Zhang, H.; Xin, D.; Zhan, J.; Wang, R. External suction-blowing method for controlling vortex-induced vibration of a bridge. *J. Wind Eng. Ind. Aerodyn.* **2021**, *215*, 104661. [CrossRef]
11. Li, K.; Ge, Y.; Guo, Z.; Zhao, L. Theoretical framework of feedback aerodynamic control of flutter oscillation for long-span suspension bridges by the twin-winglet system. *J. Wind Eng. Ind. Aerodyn.* **2015**, *145*, 166–177. [CrossRef]
12. Bai, H.; Ji, N.; Xu, G.; Li, J. An alternative aerodynamic mitigation measure for improving bridge flutter and vortex induced vibration (VIV) stability: Sealed traffic barrier. *J. Wind Eng. Ind. Aerodyn.* **2020**, *206*, 104302. [CrossRef]
13. Zhou, R.; Ge, Y.; Liu, Q.; Yang, Y.; Zhang, L. Experimental and numerical studies of wind-resistance performance of twin-box girder bridges with various grid plates. *Thin-Walled Struct.* **2021**, *166*, 108088. [CrossRef]
14. Hosseini Lavassani, S.H.; Alizadeh, H.; Homami, P. Optimizing tuned mass damper parameters to mitigate the torsional vibration of a suspension bridge under pulse-type ground motion: A sensitivity analysis. *J. Vib. Control* **2020**, *26*, 1054–1067. [CrossRef]
15. Elias, S.; Matsagar, V. Research developments in vibration control of structures using passive tuned mass dampers. *Annu. Rev. Control.* **2017**, *44*, 129–156. [CrossRef]
16. Frahm, H. Device for Damping Vibrations of Bodies. US Patent US989958A, 30 October 1909.
17. Den Hartog, J.; Ormondroyd, J. Theory of the dynamic vibration absorber. *ASME J. Appl. Mech.* **1928**, *50*, 11–22.
18. Kwon, S.-D.; Jung, M.S.; Chang, S.-P. A new passive aerodynamic control method for bridge flutter. *J. Wind Eng. Ind. Aerodyn.* **2000**, *86*, 187–202. [CrossRef]
19. Gu, M.; Chen, S.; Chang, C. Control of wind-induced vibrations of long-span bridges by semi-active lever-type TMD. *J. Wind Eng. Ind. Aerodyn.* **2002**, *90*, 111–126. [CrossRef]
20. Pourzeynali, S.; Datta, T. Control of flutter of suspension bridge deck using TMD. *Wind Struct.* **2002**, *5*, 407–422. [CrossRef]
21. Chen, X.; Kareem, A. Efficacy of tuned mass dampers for bridge flutter control. *J. Struct. Eng.* **2003**, *129*, 1291–1300. [CrossRef]
22. Chen, S.; Cai, C. Coupled vibration control with tuned mass damper for long-span bridges. *J. Sound Vib.* **2004**, *1*, 449–459. [CrossRef]
23. Abdel-Rohman, M.; John, M.J. Control of wind-induced nonlinear oscillations in suspension bridges using a semi-active tuned mass damper. *J. Vib. Control* **2006**, *12*, 1049–1080. [CrossRef]
24. Domaneschi, M.; Martinelli, L.; Po, E. Control of wind buffeting vibrations in a suspension bridge by TMD: Hybridization and robustness issues. *Comput. Struct.* **2015**, *155*, 3–17. [CrossRef]
25. Alizadeh, H.; Lavassani, S.H.H.; Pourzeynali, S. Flutter Instability Control in Suspension Bridge by TMD. In Proceedings of the 11th International Congress on Civil Engineering, Tehran, Iran, 8–10 May 2018.
26. Kontoni, D.-P.N.; Farghaly, A.A. Mitigation of the seismic response of a cable-stayed bridge with soil-structure-interaction effect using tuned mass dampers. *Struct. Eng. Mech.* **2019**, *69*, 699–712. [CrossRef]
27. Mansouri, S.; Kontoni, D.-P.N.; Pouraminian, M. The effects of the duration, intensity and magnitude of far-fault earthquakes on the seismic response of RC bridges retrofitted with seismic bearings. *Adv. Bridge Eng.* **2022**, *3*, 19. [CrossRef]
28. Patil, V.B.; Jangid, R.S. Optimum multiple tuned mass dampers for the wind excited benchmark building. *J. Civ. Eng. Manag.* **2011**, *17*, 540–557. [CrossRef]
29. Bandivadekar, T.; Jangid, R. Optimization of multiple tuned mass dampers for vibration control of system under external excitation. *J. Vib. Control* **2013**, *19*, 1854–1871. [CrossRef]
30. Wang, Y.; Feng, C.; Chen, S. Damping effects of linear and nonlinear tuned mass dampers on nonlinear hinged-hinged beam. *J. Sound Vib.* **2018**, *430*, 150–173. [CrossRef]
31. Shi, W.; Wang, L.; Lu, Z.; Wang, H. Experimental and numerical study on adaptive-passive variable mass tuned mass damper. *J. Sound Vib.* **2019**, *452*, 97–111. [CrossRef]
32. Dai, J.; Xu, Z.-D.; Gai, P.-P.; Xu, Y.W. Mitigation of vortex-induced vibration in bridges using semi active tuned mass dampers. *J. Bridge Eng.* **2021**, *26*, 05021003. [CrossRef]
33. Ubertini, F.; Comanducci, G.; Laflamme, S. A parametric study on reliability-based tuned-mass damper design against bridge flutter. *J. Vib. Control* **2015**, *23*, 1518–1534. [CrossRef]
34. Pisal, A.Y.; Jangid, R.S. Vibration control of bridge subjected to multiaxle vehicle using multiple tuned mass friction dampers. *Int. J. Adv. Struct. Eng.* **2016**, *8*, 213–227. [CrossRef]

35. Tao, T.; Wang, H.; Yao, C.; He, X. Parametric sensitivity analysis on the buffeting control of a long-span tripletower suspension bridge with MTMD. *Appl. Sci.* **2017**, *7*, 395. [CrossRef]
36. Strømmen, E. *Theory of Bridge Aerodynamics*; Springer Science & Business Media: Berlin/Heidelberg, Germany, 2010. [CrossRef]
37. Amini, F.; Doroudi, R. Control of a building complex with magneto-rheological dampers and tuned mass damper. *Struct. Eng. Mech.* **2010**, *36*, 181–195. [CrossRef]
38. Bortoluzzi, D.; Casciati, S.; Elia, L.; Faravelli, L. Design of a TMD solution to mitigate wind-induced local vibrations in an existing timber footbridge. *Smart Struct. Syst.* **2015**, *16*, 459–478. [CrossRef]
39. Debbarma, R.; Das, D. Vibration control of building using multiple tuned mass dampers considering real earthquake time history. *Int. J. Civ. Environ. Eng.* **2016**, *10*, 694–704.
40. Kontoni, D.-P.N.; Farghaly, A.A. Enhancing the earthquake resistance of RC and steel high-rise buildings by bracings, shear walls and TMDs considering SSI. *Asian J. Civ. Eng.* **2023**. [CrossRef]
41. Kontoni, D.-P.N.; Farghaly, A.A. The effect of base isolation and tuned mass dampers on the seismic response of RC high-rise buildings considering soil-structure interaction. *Earthq. Struct.* **2019**, *17*, 425–434. [CrossRef]
42. Kontoni, D.-P.N.; Farghaly, A.A. TMD effectiveness for steel high-rise building subjected to wind or earthquake including soil-structure interaction. *Wind Struct.* **2020**, *30*, 423–432. [CrossRef]
43. Hosseini Lavassani, S.H.; Shangapour, S.; Homami, P.; Gharehbaghi, V.; Farsangi, E.N.; Yang, T.Y. An innovative methodology for hybrid vibration control (MR + TMD) of buildings under seismic excitations. *Soil Dyn. Earthq. Eng.* **2022**, *155*, 107175. [CrossRef]

 *buildings*

Perspective

Role of In-Field Experimental Diagnostic Analysis for the Derivation of Residual Capacity Indexes in Existing Pedestrian Glass Systems

Chiara Bedon *, Salvatore Noè, Marco Fasan and Claudio Amadio

Department of Engineering and Architecture, University of Trieste, Via Valerio 6/1, 34127 Trieste, Italy; noe@units.it (S.N.); mfasan@units.it (M.F.); amadio@units.it (C.A.)
* Correspondence: chiara.bedon@dia.units.it; Tel.: +39-(04)-05583837

Abstract: The use of simplified tools in support of the mechanical performance assessment of pedestrian structures is strongly attractive for designers due to practical efficiency, as well as for researchers in terms of innovation and the assessment of new proposals. On the side of design, the vibration serviceability requires that specific comfort levels for pedestrians are satisfied by taking into account conventional performance indicators and the class of use, or the structural typology for pedestrian systems' object of analysis. A major issue, in this context, is represented by long-term performance of systems (especially pedestrian) that are based on innovative or sensitive materials and possibly affected by degradation or even damage, and thus potentially unsafe. Consequently, it is clear that, especially for in-service structures, the availability of standardized non-destructive protocols for a reliable (and possibly rapid) structural assessment can represent an efficient support for diagnostics. This perspective paper poses the attention on the residual capacity quantification of laminated glass (LG) pedestrian structures, and on the assessment of experimental and/or numerical tools for their analysis. To this aim, three modular units belonging to two different indoor, in-service pedestrian systems are taken into account like pilot studies. On the practical side, as shown, a primary role is assigned to Operational Modal Analysis (OMA) procedures, which are used on-site, to quantify their structural performance based on vibration response, including damage detection and inverse characterization of materials degradation. As shown, based on earlier detailed validation, it is proven that a rapid structural assessment can be based on a single triaxial Micro Electro-Mechanical System (MEMS) accelerometer, which can be used to derive relevant capacity measures and indicators. To develop possible general recommendations of technical interest for in-service LG pedestrian systems, the so-calculated experimental performance indicators are assessed towards various traditional design procedures and literature approaches of classical use for structural diagnostic purposes, which are presently extended to the structural typology of LG systems.

Keywords: laminated glass (LG); pedestrian systems; walk-induced vibrations; glass fracture; non-destructive in-field experiments; Finite Element (FE) numerical modelling; damage measure; residual capacity

Citation: Bedon, C.; Noè, S.; Fasan, M.; Amadio, C. Role of In-Field Experimental Diagnostic Analysis for the Derivation of Residual Capacity Indexes in Existing Pedestrian Glass Systems. *Buildings* **2023**, *13*, 754. https://doi.org/10.3390/buildings13030754

Academic Editor: Hugo Rodrigues

Received: 22 February 2023
Revised: 10 March 2023
Accepted: 11 March 2023
Published: 13 March 2023

1. Introduction

In the civil and structural engineering fields, monitoring and diagnostic tools have a primary role for safety level preservation. To minimize the number of possible injuries in case of structural damage, the availability of simplified and standardized operational protocols for in-field diagnostics is particularly advantageous. In general, the primary objective of a possibly rapid but especially robust/sound structural safety assessment for a given construction/building system is to quickly inspect and quantitatively evaluate its residual load-bearing capacity under ordinary or even exceptional design actions, and, consequently, to address whether the presence/initiation/propagation of any damage can represent a major issue and possible risk for the safety of customers. As such, this task is particularly challenging, particularly in case of accidental damage or extreme events on buildings ([1,2], etc.).

In the framework of a residual capacity assessment of traditional existing buildings (reinforced concrete-framed structures, masonry constructions, etc.), two primary concerns need to be properly taken into account, namely consisting of (i) a quick evaluation and preliminary detection of "structural" and "non-structural" components; and (ii) the sound individuation of any visual sign of damage, and thus, in the consequent detection of possible "unsafe" members (Figure 1). For "structural" components in reinforced concrete buildings, for example, visual damage can be recognized in the form of typical failure evidence like cracks, spalling, etc., which are well-known conditions associated to possible risk. In-field diagnostic inspections, with eventual localized measurements of few key parameters, can suggest the opportunity and the urgency of a rapid retrofit intervention, or at least the convenience of an additional structural health monitoring process for a given time interval [3] before any kind of retrofit planning. Ad-hoc protocols for in-field diagnostic campaigns, in this regard, may be required and may appear particularly convenient, especially for those buildings and constructions in which, due to severe damage, the accessibility or the practical execution of in-field measurements could be extremely difficult or unsafe [4]. For cultural heritage diagnostics considerations, such as ancient masonry structures or monumental buildings, specific uncertainties could require additional appropriate assessments in problem solving [5]. Overall, based on the above considerations, it is clear that the availability of systematic, robust monitoring and diagnostics protocols, as well as the availability of simplified operational methods for in-field interventions, assumes an increasingly strategic role [6,7].

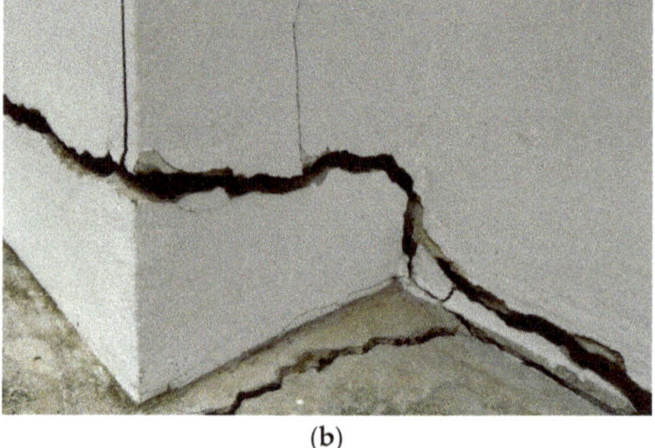

(**a**) (**b**)

Figure 1. Examples of damage in (**a**) masonry or (**b**) concrete structures (figures reproduced with permission from Unsplash).

Typical monitoring tools of buildings can be efficiently adapted to bridge structures and even pedestrian walkways, which have a direct interaction with occupants and thus, are possibly associated with human discomfort and even risks of falling in cases of structural damage. The use of accelerometers and vibration-based damage detection for bridges, in particular, is notoriously characterized by several advantages compared to traditional methods of non-destructive evaluation, as well as a reduced cost thanks to recent technologies [8,9]. For traditional bridge structures, recent studies show that Virtual Reality (VR) environments, in the same way as innovative image-based structural health monitoring strategies, can also integrate classical monitoring tools and instruments [10,11]. In addition, maximum predictive details and data can be achieved from the development of integrated smart sensors based on the application of GPS receivers, accelerometers, and smartphones for bridges [12]. As a common rule of the multitude of research studies and industrial applications, there is the general need of engineering knowledge and monitoring of performance indicators for those structural parameters—to address in terms of conventional performance indicators—which have a primary role in safety issues with a predictivity capacity that can anticipate possible severe damage. This primary goal, which is especially challenging for ancient structures, can take advantage of robust procedures of analysis, which are in most cases integrated by experimental, analytical, and numerical methods, as well as reference performance limits to address and satisfy [13–15].

Such a basic need of deep engineering knowledge and damage prevention for safety maximization, in general terms, can also be rationally expected for innovative constructional solutions and/or materials, which, for example, are subjected to unfavourable operational conditions or —compared to traditional solutions—a lack of sufficiently deep engineering knowledge in term of residual capacity assessments. In the field of laminated glass (LG) elements, for example, one of the major open challenges is represented by post-fracture behaviour characterization and optimization. This aspect is particularly urgent (but presently solved by an overdesign of new LG members), especially for load-bearing components characterized by a prevailing human-structure interaction (i.e., floors, partitions, etc.) or even unfavourable operational conditions (Figure 2a,b).

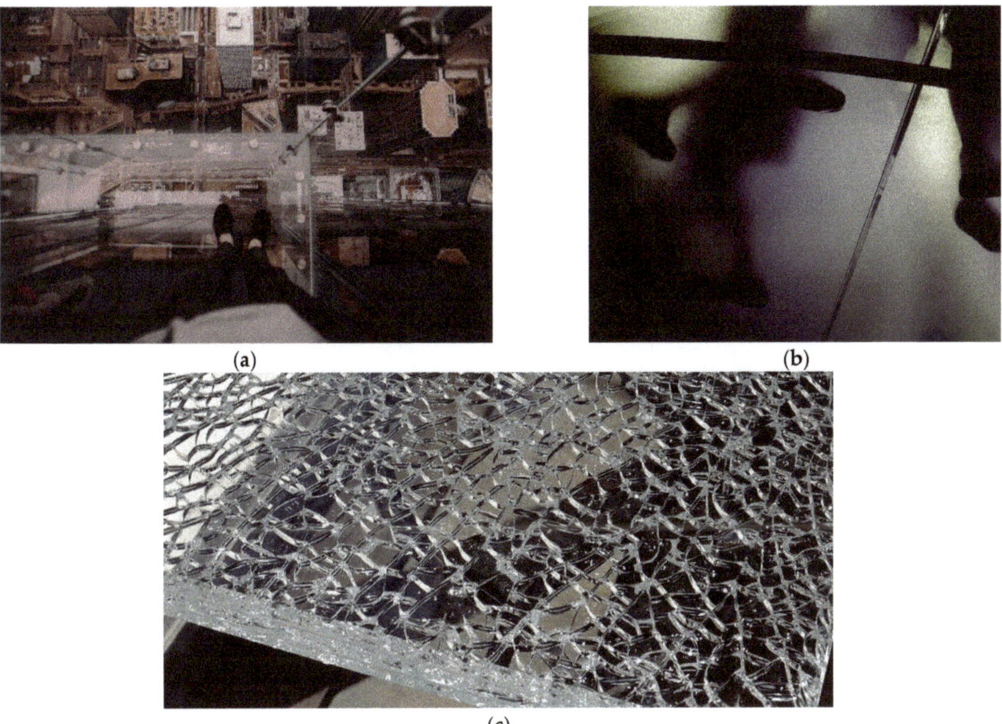

Figure 2. Examples of (**a**,**b**) glass pedestrian systems and (**c**) LG fracture (figures reproduced with permission from Unsplash).

Typical examples are pedestrian LG systems under long-term effects and subjected to random walks [16,17] or balustrades with a deliberate fracture [18]. Due to intrinsic material properties and structural design assumptions, even partial evidence of glass fracture (i.e., Figure 2c) or material degradation can, in fact, suggest urgent maintenance and retrofit interventions [19], or even the replacement of original components by higher robustness elements. In this regard, it is recognized that the visual detection of glass fracture, even partial, for an existing structural glass system still represents a "no-return" condition for the structure itself, and thus, the origin of possible short-term risks for customers. On the other side, glass fracture itself does not correspond to immediate structural collapse and thus, could offer a minimum post-fracture resistance to facilitate a fast retrofit intervention. In some cases, the progressive degradation of mechanical parameters can be detected early by interlayer discolouring, which can possibly represent an efficient indicator of incoming severe deterioration phenomena. However, on the other side, it is hardly quantifiable [20]. A quantitative in-field measure of key performance indicators for improved control and risk minimization of post-fracture performances and residual capacities of glass structures, in this regard, should unavoidably take advantage from a dedicated, easy-to-apply and optimised experimental protocol. Apart from risk minimization in case of mechanical capacity degradation or damage, the present approach fulfils various serviceability aspects for the whole lifetime of a given glass structure.

In the same way of damage and collapse prevention, additional attention to design procedures is, in fact, also given to comfort maximization for those customers that take advantage of structural glass functionality (i.e., against vibrations, etc.) during daily activities. Overall, it is clear that the satisfaction of appropriate comfort and safety levels for in-service LG structures is necessarily correlated to a robust engineering knowledge of real-time performances and mechanical properties of stand-alone or assembled components. The higher is the direct interaction of customers with glass structures (i.e., pedestrian systems, balustrades, etc.) and the need of dedicated diagnostic protocols for risk minimization, comfort optimization, and thus, functionality preservation. The intrinsic material and geometrical features of typical use for LG solutions, more in detail, suggest the need of dedicated assessment methods and the consequent instruction for the correlation of evidence discovered in experimental outcomes of "Current" load-bearing capacities to reliable performance/safety indicators. Huang et al. [21], for example, proposed a rapid safety assessment of curtain wall panels based on their remote vibration frequency measurement. Starting from modal analysis results, the study showed that the first order inherent frequency of linearly restrained curtain wall panels is expected to decrease with an increase of sealant failure and degradation.

In this paper, in accordance with the investigation earlier reported in [22] and based on the above considerations, the in-field experimental derivation of possibly efficient and easy-to-use structural performance indicators for safety and residual capacity assessments of in-service LG pedestrian systems is explored. The feasibility and potential of procedural steps as in Section 2, in support of structural health diagnostics, are addressed by taking into account three different case-study applications. Limits and open gaps are also discussed in Sections 3–5 about the elaboration of a robust methodology of general applicability.

2. Research Methods

2.1. Constituent Materials

Talking about diagnostics in glass constructional members is particularly challenging for several reasons. On one side, the engineering knowledge for existing in-service structures is limited, because monitoring data and diagnostic programs are still very few [16,17]. Lack of experimental data to support the interpretation of long-term or accidental behaviours for these special structures is a first practical obstacle.

In parallel, material intrinsic features and their sensitivity to several external and operational aspects represent an additional source of uncertainty for diagnostics [23,24], which can be still addressed by experimental tools and monitoring programs. However, this necessitates (as it happens for different structural typologies) a wide set of experimental data, case studies, and real applications.

Structurally speaking, the working assumption for design is, in fact, that glass material behaves as linear elastic material with brittle behaviour in tension, while the glass layers are bonded by viscoelastic interlayers (see Table 1 [23,24]). Moreover, in practical applications for in-service LG systems, all the above aspects are further mutually affected by the final destination these structural systems have in buildings (i.e., type of loading, etc.), by external ambient conditions (see also Sections 3–5) and by the occurrence of possible degradation in materials.

Table 1. Summary of typical mechanical properties for constituent LG materials. * = depending on time loading, temperature, humidity, etc.

Material	Elastic Modulus [MPa]	Poisson' Ratio [-]	Density [kg/m^3]	Behaviour
Glass	70,000	0.23	2500	Brittle elastic in tension
PVB (interlayer)	Variable *	0.45	1000	Viscoelastic

2.2. Procedural Steps for Residual Capacity Assessment

Based on above considerations, technical issues, and lack of specific regulations for in-service LG structural assessment, it is clear that the availability of a standardized procedural protocol is a primary need [22]. In this paper, the attention is thus focused on the elaboration and practical application of a possible diagnostic methodology in which a primary role is assigned to in-field testing, but experimental evidence is then properly integrated (Figure 3). To note that—as a first pilot experimental and numerical assessment—the specific technology of structural LG pedestrian systems is taken into account.

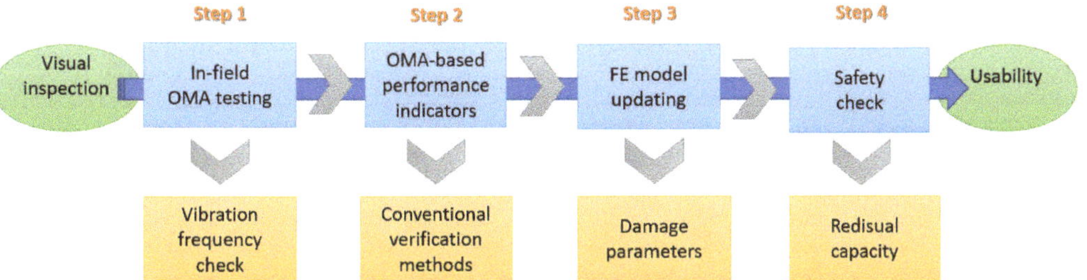

Figure 3. Proposed procedural steps for safety and residual capacity assessment of existing, in-service LG pedestrian systems.

Differing from "ad hoc" laboratory protocols and testing configurations, the experimental assessment of in-service LG systems is notoriously affected by several technical challenges and restrictions, which suggest the use of a minimum number of instruments and the maximization of experimental outputs (Step 1 in Figure 3). To this aim, a preliminary visual inspection is also recommended (Figure 3).

Once the most relevant performance indicators are extrapolated from in-field testing, specific limits and indicators are required for safety assessment (Step 2 in Figure 3). To note, LG systems may offer relatively "high vibration frequency", according to definitions from ISO 10137:2007 [25], but they still suffer from marked sensitivity to vibrations or even damage. In this context, the preliminary knowledge of the real vibration frequency for an in-service structural system (and its possible sensitivity and modification under external loads) notoriously represents a powerful indicator for structural diagnostics [16,17,19]. At the same time, the vibration frequency itself is limited in interpretation as single parameter for monitoring purposes. A more refined analysis of experimental evidence, based on post-processing elaboration of basic in-field data, is thus recommended to derive possible quantification of damage parameters under real-time operational conditions.

A more detailed interpretation of experimental outputs can take further advantage of the support of dedicated Finite Element (FE) numerical models in order to characterize basic equivalent material properties for the constituent LG components (Step 3 in Figure 3). Most importantly, these FE models—once validated in experimental data—can support a more realistic quantification of long-term effects for the examined LG systems and thus a rationale measure of residual capacity parameters, which are of primary interest for safety purposes in structural systems interacting with occupants (Step 4 in Figure 3). All these steps are applied to three different case-study systems and discussed in terms of practical convenience, impact, present uncertainties, and future developments.

2.3. Operational Modal Analysis (OMA) Testing for In-Service LG Systems

Operational Modal Analysis (OMA) for structures and building components—Step 1 in Figure 3—is known to represent a robust and efficient technique, able to offer a multitude of material and damage parameters (Figure 4a). Major benefits of OMA techniques are related to possible application in various structural components without the need of destructive interventions and service interruptions [26]. The optimal setup definition is a critical step to capture relevant dynamic mechanical parameters, especially for complex assemblies. OMA is, in fact, very attractive because tests are generally cheap and fast, and they do not usually interfere with the normal use of the structure. Moreover, the identified modal parameters are representative of the actual behaviour of the structure under in-service conditions since they refer to realistic levels of vibration in the structure and not to artificially generated vibrations [27]. Successful experimental research studies can be found in the literature for a multitude of civil engineering applications, most of them consisting in towers and minarets [28,29], bridges [30,31], or special structures [32,33]. For glass structures, practical OMA evidence can also be coupled to different customers behaviours and reactions for comfort analysis [34,35].

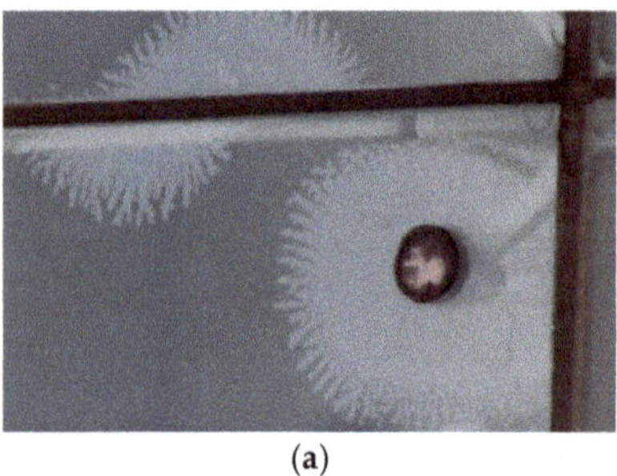
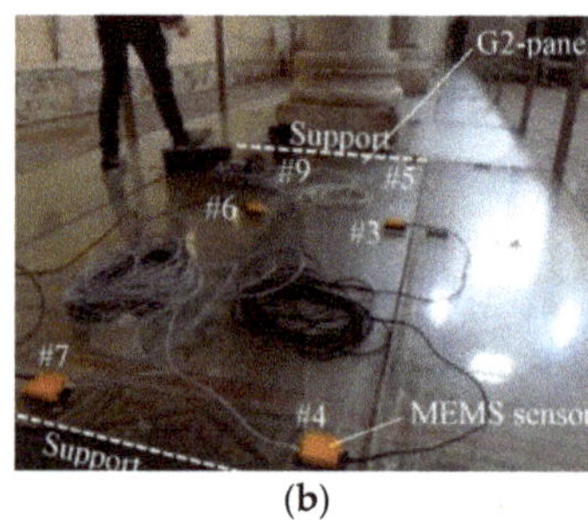

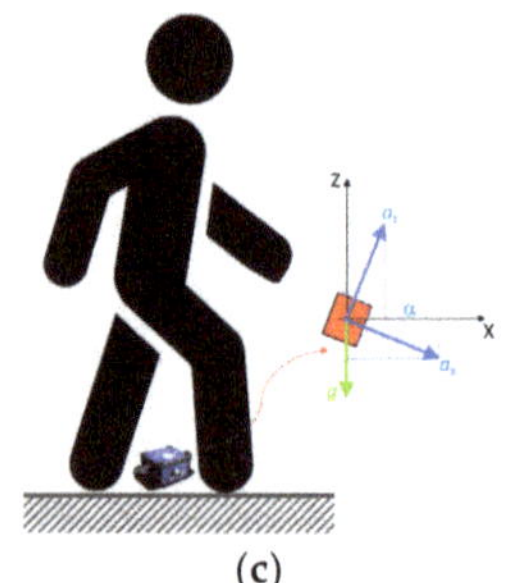

(a)   (b)   (c)

Figure 4. Application of Operational Modal Analysis to in-service LG pedestrian systems: (**a**) example of delamination in LG (figure reproduced from [36] under the terms and conditions of CC-BY license agreement); (**b**) overview of past experimental study with multiple MEMS sensors (figure reproduced from [19] under the terms and conditions of CC-BY license agreement); and (**c**) schematic representation of in-field assessment based on single MEMS setup.

In the framework of existing LG systems, the typical size of panels and structural mass parameters can be efficiently addressed based on experimental tools and protocols for OMA testing and low-level imposed vibrations. In [16,17], for example, multiple Micro Electro-Mechanical System (MEMS) accelerometers, resulting from prototypes validated in [9], have been used for diagnostic purposes of a suspension glass walkway (Figure 4b). A total of six MEMS sensors were placed on the investigated modules to capture the typical beam-like bending vibration response. Partial glass fracture and damage effects under random walks were explored in [19]. The effect of psychological discomfort for customers asked to move in the context of glass structures, including LG pedestrian systems, was addressed in [34,35]. As a matter of fact, the primary concern for customers may be represented by uncertainty of actual safety levels of glass pedestrian systems.

Following earlier experiences, especially [16,17], the present study shows recent trends of in-field experimental diagnostic tools for LGs. To facilitate an efficient in-field investigation, the acquisition system is optimized in number of sensors and recorded data to maximize the interpretation of in-field experimental outcomes based on a single MEMS sensor (Figure 4c). This kind of working assumption corresponds to a diagnostic protocol that could be extremely advantageous for those situations affected (as it is for in-service systems in general) by major operational and technical limitations and restrictions for testing (i.e., due to normal service use of the structure object of study). In addition, the proposed approach is advantageous for all those situations affected by major technical issues as it is for emergency conditions and thus where it is not possible to take advantage of time, resources, instruments, and integrated tools, which are of typical use of more sophisticated experimental protocols. On the side of in-field experimental testing, it is, in any case, important to note that the setup arrangement or other material issues could make particularly challenging the inverse detection of structural parameters, as discussed in the following.

2.4. Vibration Frequency Estimation for In-Service LG Systems

As far as OMA techniques are applied to an in-service LG system, and, in particular, to a LG pedestrian structure, the first performance indicator to experimentally address is represented by a preliminary—but still possibly meaningful for early damage detection—vibration frequency estimate (Step 1 in Figure 3). The latter is notoriously able to take into account several influencing parameters that, for LG systems, have intrinsic modifications and trends compared to other constructional materials and components, such as possible modification of materials (including degradation events like in Figure 4a), but also important effects of interaction with occupants (see, for example, Figure 4b,c). The intrinsic advantage of frequency analysis based on in-field testing is thus a more realistic measurement of "actual" performances compared to, for example, analytical calculations. A beam-like LG member, regardless of its constituent layers, can be, in fact, theoretically assimilated to a slender Euler–Bernoulli beam, which is characterized in out-of-plane bending and vibration performances by an equivalent, monolithic $b \times h$ section, and its response is governed by:

$$\frac{\partial^2}{\partial x^2} EJ(x) \left(\frac{\partial^2 v(x,t)}{\partial x^2} \right) + \rho A \frac{\partial^2 v(x,t)}{\partial t^2} = 0 \tag{1}$$

where $v(x,t)$ is the vertical displacement, at the abscissa x and time instant t, E and ρ are the modulus of elasticity and density of glass material (Table 1), J is the second moment of area, and A is the cross-section. However, as also shown in [36], Equation (1) supports a rational vibration frequency estimation only under the ideal assumptions of perfect restraints (i.e., simply support or clamp) and rigid bonding of the constituent LG layers:

$$f_1 = \frac{\omega_1}{2\pi} = \frac{1}{2\pi} \sqrt{\frac{\beta_1^4 E}{12m} h^3} \tag{2}$$

where β_n is given in Table 2 and m is the mass per unit of length.

Table 2. Reference wavenumbers β_n for monolithic beams with simply supported ideal end restraints and bending span L_{ef}.

Mode Order n		
1	2	3
π/L_{ef}	$2\pi/L_{ef}$	$3\pi/L_{ef}$

For LG sections, in support of Equation (2), major benefits and accuracy of estimates can be obtained from the use of an equivalent monolithic glass thickness $h_{ef} = h$ [37,38]. This assumption is of utmost importance to include—even in simplified way—possible viscous effects of bonding interlayers and thus capture the corresponding frequency shift between the lower "*abs*" bound (weak bond of glass layers) and an upper "*full*" limit (rigid bond):

$$f_{1,abs} \leq f_1 \leq f_{1,full} \tag{3}$$

However, multiple intrinsic limits affect the usability of Equation (2) for "real" LG structural members and, consequently, the correlation in Equation (3) can also suffer for major sensitivity. Among others, Equation (2) lacks, in fact, major effects due to real, non-ideal boundaries. For the specific case of LG pedestrian systems, the occupant's mass is disregarded. In addition, Equation (2) neglects the progressive modification of shear stiffness for the bonding viscoelastic interlayer with ageing and time loading [39,40] (see Figure 5) or even possible delamination phenomena [36], as well as unsymmetrical of flexibility parameters of restraints [41].

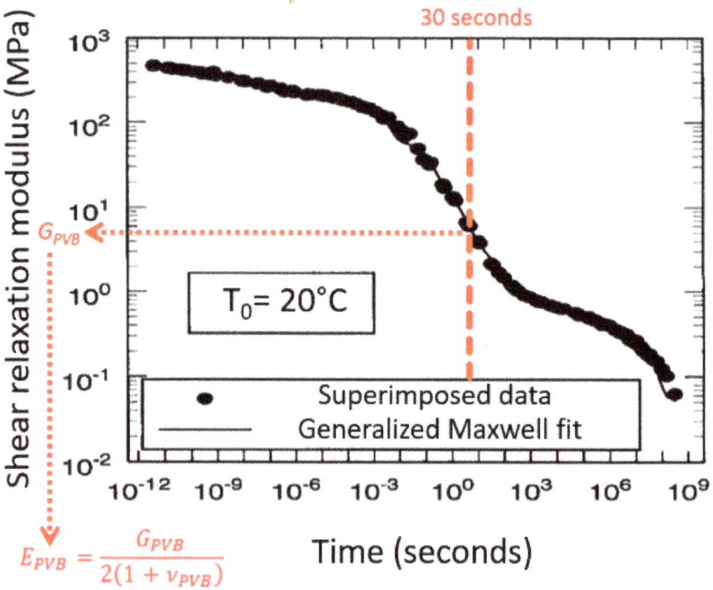

Figure 5. Example of practical derivation of equivalent shear relaxation modulus for PVB interlayer based on experimental master-curve (original figure adapted from [40] under permission from John Wiley and Sons©, Copyright license number 5327091377077, June 2022).

To note, vibration frequency changes, according to several literature applications for load-bearing cantilever of beam elements made of various constructional materials, is a relevant parameter, especially for damage severity detection and quantification (i.e., presence of cracks, and detection of their depth/size/location). Moreover, many other factors and boundary conditions may affect the vibration frequency for glass elements, as a major effect of their intrinsic material properties and needs for restraints. The stiffness and mechanical efficiency of fixing systems, in this case, may be efficiently addressed by frequency estimates (see for example Figure 6), as well as for LG components, which are still in the uncracked stage.

A major challenge for diagnostic purposes in LG structures is thus represented, according to Step 2 in Figure 3, by the experimental derivation and interpretation of relevant performance indicators. These are inclusive of vibration frequencies and, especially, their assessment towards reference performance indicators, which presently are not available for LG structures [34]. For the current study, all these elaborations are proposed in Section 4, by taking into account three different case-study systems.

(a)   (b)

Figure 6. Example of accident in a 100 m high bridge in China due to wind loading: (**a**) initial configuration for the intact bridge and (**b**) post-accident configuration.

2.5. Finite Element (FE) Model Updating for In-Service LG Systems

For SHM and diagnostic applications, the use of Finite Element (FE) numerical methods is known to represent a strategic method in support of mechanical analysis and inverse detection of unknown parameters. As such, it is expected that FE model updating can also efficiently support the residual capacity assessment of in-service LG structures (Step 3 in Figure 3). Moreover, specific details that are intrinsic of LG structures should be necessarily taken into account.

Most practical examples and case-study applications from literature are focused on bridge structures or historical buildings affected by structural damage after earthquakes [27]. The combined use of Genetic Algorithms and Artificial Intelligence tools can be also extremely efficient in model updating and optimal calibration of input parameters [42–45], including complex geometries and damage scenarios [46,47].

In the present study, FE model updating is primarily used to detect and quantify damage severity (including bonding degradation), and to address its effects on the performance of the examined in-service LG systems. On the practical side, it is clear that such a methodology requires detailed knowledge of geometrical properties for structural and non-structural components, which have a primary role in dynamic response assessment, especially for LG systems (i.e., [41]). For the herein discussed FE model updating and fitting procedure, more in detail, the target performance indicator is represented by the experimentally derived fundamental vibration frequency of the examined systems under random walks, which is known to represent a first strategic parameter for SHM purposes [25]. In this regard, it is worth to note that LG pedestrian systems are generally characterized by the use of repeated modular units, often limited in size, with relatively small structural mass and high or low vibration frequency. Accordingly, relevant dynamic effects could be also expected from the interaction of structures with occupants [17,48,49].

2.6. Final Verification Check and Residual Capacity Quantification

The final procedural diagnostic step for residual capacity quantification (see Step 4 in Figure 3) is finalized when the residual load-bearing capacity and thus the need of any maintenance intervention can be univocally quantified. There are no doubts that a robust engineering knowledge of "Current" capacity for a given existing structural system and its response to ordinary design actions is crucial for both safety and comfort purposes.

However, such a quantitative consideration represents a major challenge in the overall procedure in Figure 3, and the response is mutually affected by a multitude of working steps and operational assumptions, both for experiments and integrated models. Among others, a major issue can be represented by comparison and quantitative assessment of experimental evidence towards an "initial", or "Time 0", set of performance trends, which, in most cases, are unknown. Thus, it is clear that such a series of considerations may necessitate the support of "ad-hoc" numerical tools (for material properties sensitivity assessment, etc.), which could implicitly involve additional uncertainties in calibration.

Most importantly, once the "Current" performance is known, a specific reference level is needed for comparative purposes (for example, in terms of stress analysis or deflection amplitude). This crucial assessment step could be solved based on conventional verification procedures for stress verification (at the Ultimate Limit State—ULS) and deformation limit prevention (at the Serviceability Limit State—SLS) by taking into account the reference parameters conventionally in use for the "Time 0" structural design of LG structural systems.

As a final result, the major, initial issue or residual capacity could be thus clarified. Is the "Current" system able to ensure a sufficient functionality against ordinary loads? In addition, how much of the mechanical degradation/damage of materials affect its original load-bearing capacity? In the present study, following Figure 3, such an assessment and verification procedure is extrapolated for the selected case-study systems based on the Italian CNR-DT 210/2013 recommendations for design of structural glass elements [24]. To this aim, experimental and numerical evidence are first discussed in Sections 3–5, based on Steps 2–4 of Figure 3.

3. Practical Applications for Selected LG Pedestrian Systems

3.1. Geometrical and Mechanical Properties

The procedural steps, as shown in Figure 3 are applied to in-service case-study systems that were accessible for in-filed testing and/or practical interest for the present investigation. All the examined LG slabs are characterized by linearly restrained edges along two sides only (i.e., beam-like setup) and a triple LG section (i.e., three glass layers bonded by PVB). All the samples, moreover, are part of indoor pedestrian systems located in Friuli Venezia Giulia Region (Italy) and installed in the context of two historical churches where they are used to allow visibility of underground Roman age manufacts. Major geometrical details of practical interest for mechanical characterization are summarized in Table 3, while selected photos are collected in Figure 7.

Table 3. Summary of geometrical features for the examined LG pedestrian systems (beam-like simply supported setup).

Specimen	Dimensions [m]	Span [m]	Total Thickness [mm]	Cross-Section	Mass [kg]	R_M Equation (4)	λ Equation (5)
SM#1–LGU	0.51 × 2.80	0.51	27.52	8/10/8 + 0.76 PVB	93	1.16	68
SM#2–LGU	1.35 × 2.65	2.65	37.52 + 6	12/12/12 + 0.76 PVB + AN cover	320	4	245
SM#3–LGF	1.35 × 2.65	2.65	37.52 + 6	12/12/12 + 0.76 PVB + AN cover	320	4	245

The first examined system, SM#1–LGU, is located in San Giorgio di Nogaro (Udine). The in-field experiments were carried out in December 2020, $\approx$15 years apart from its original construction, on a reference modular unit characterized by 8/10/8 mm thick, fully tempered (FT) glass layers (0.76 mm PVB bonds). The dimensions of modules were L = 2.80 m in length by B = 0.51 m in width, and LG panels were linearly restrained at the edges by hollow box steel members (60 mm × 100 mm in section, 5 mm in thickness), specifically properly arranged to create a grid for LG modules. To note, the original pedestrian system object of experiments was retrofitted, starting from Spring 2021, and replaced by newly designed LG components with similar geometrical and mechanical properties (Figure 7a).

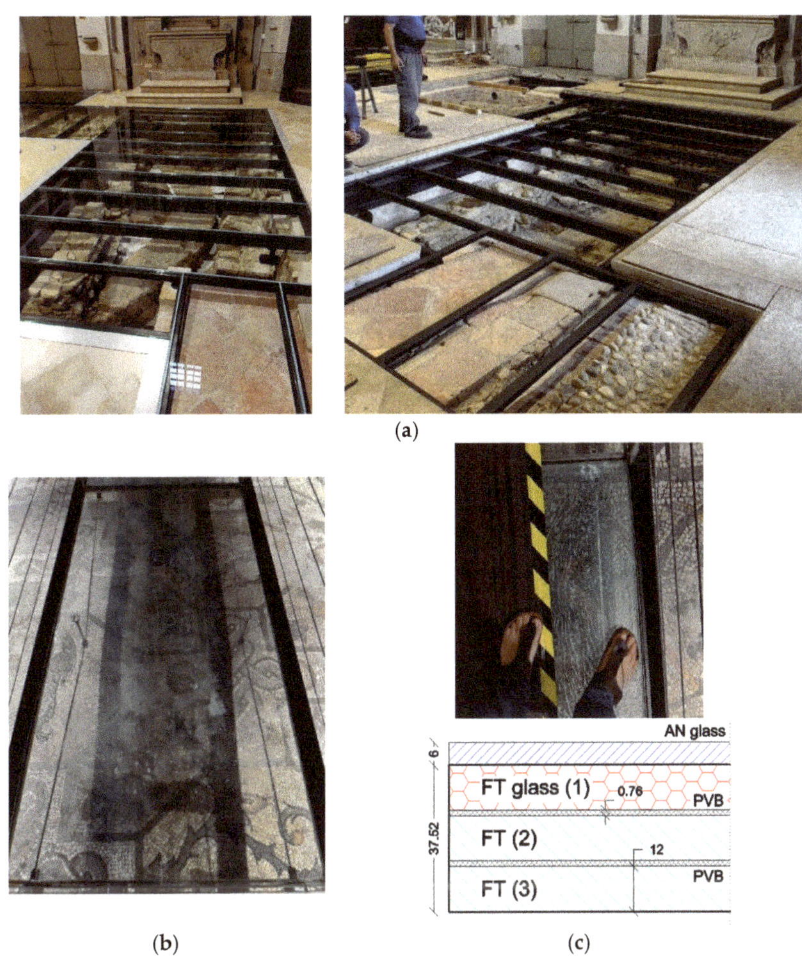

Figure 7. Overview of selected modular units for the examined LG pedestrian systems: (**a**) SM#1–LGU layout (pictures taken during retrofit interventions in Spring 2021, Courtesy of Seretti Vetroarchitetture); (**b**) top view of SM#2–LGU modular unit (reproduced from [19] under the terms and conditions of CC-BY license); and (**c**) schematic representation of glass fracture detail for SM#3–LGF panel.

The second and third examined LG samples are presently located in Aquileia (Udine) and are part of the suspension walkway investigated in [16,17]. Each module is characterized by dimensions of $L = 2.65$ m in length by $B = 1.35$ m in width (Figure 7b). To note, an additional sacrificial, protective layer of annealed (AN) glass (6 mm in thickness) was positioned on the top surface of LG panels (Table 2). As for the SM#1–LGU system, the case-study examples in Aquileia were subjected to in-field testing ≈ 15 years apart from its original construction. At the time of the experiments, moreover, the difference of two selected modular systems was represented by the presence of intact glass layers (SM#2–LGU), or by the presence of (accidental) partial fracture for one of the constituent glass layers in the resisting LG section (SM#3–LGF, which was replaced after testing, see Figure 7c).

In terms of the preliminary mechanical characterization of selected systems, some comparative properties are also summarized in Table 3 in terms of mass and slenderness. In particular, the mass ratio R_M is given by:

$$R_M = \frac{M_{structure}}{M_{occupant}} \tag{4}$$

while the geometric slenderness is calculated as:

$$\lambda = \frac{L_{ef}}{\bar{\rho}} \tag{5}$$

on the base of the effective bending span L_{ef} (from Table 3) and on the radius of gyration given by:

$$\bar{\rho} = \sqrt{\frac{J}{A}} \tag{6}$$

To note, a single occupant (adult volunteer, $M = 80$ kg) was invited to take part in the experimental measurements on the three different in-service LG slabs. In addition, the bonding contribution of PVB foils was preliminary disregarded.

It is worth noting in Table 2 that R_M from Equation (4) has major effects on the assessment of human-structure interaction phenomena. Compared to other constructional typologies, for LG pedestrian solutions, it is typically small [17,19]. A mutual interaction of several aspects should be necessarily taken into account. Such a condition may, in fact, negatively affect—compared to other structural typologies—both the mechanical performance in terms of stress peaks and deflections under ordinary loads, and the corresponding comfort level of pedestrian under human-induced vibrations.

As shown in Table 3, the geometric slenderness from Equation (5) is also strongly affected by typical geometrical features of LG pedestrian systems and differs from other constructional typologies.

3.2. Finite Element Model Updating

The numerical analysis of selected slabs was carried out in ABAQUS [50]. Element features, mesh size, and features and material properties were calibrated to optimize the computational efficiency of simulations, as well as by taking into account past modelling efforts for similar structural systems [16,19]. Linear elastic material laws were taken into account for glass and PVB (Table 1), as well as metal sub-components [24]. The geometrical description of constituent components, as in Figure 8, was based also on visual inspections and technical drawings.

Careful attention in modelling should be paid, case by case, especially for those structural and non-structural details that have a primary role in glass applications. As a basic step towards the frequency assessment with the in-field experimental output, a linear modal frequency analysis procedure was taken into account. Such a modelling choice was adopted to capture the vibration frequency and (especially in case of damage) fit the unknown material properties in terms of degradation features and damage severity, towards the experimentally derived vibration frequencies for the selected samples. For general applications on LG systems, most of the attention should be given to the reproduction of geometrical and mechanical feature of real restraints, given that they have a primary role in vibration and bending performances. To this aim, a preliminary visual inspection for the in-service system object of study could also facilitate detecting possible anomalies or even initial damage (if any).

For the present practical application, Figure 8 shows an example of assembled modules, with evidence of vertical (out-of-plane) displacement contour plots of fundamental modal shape of their first vibration mode.

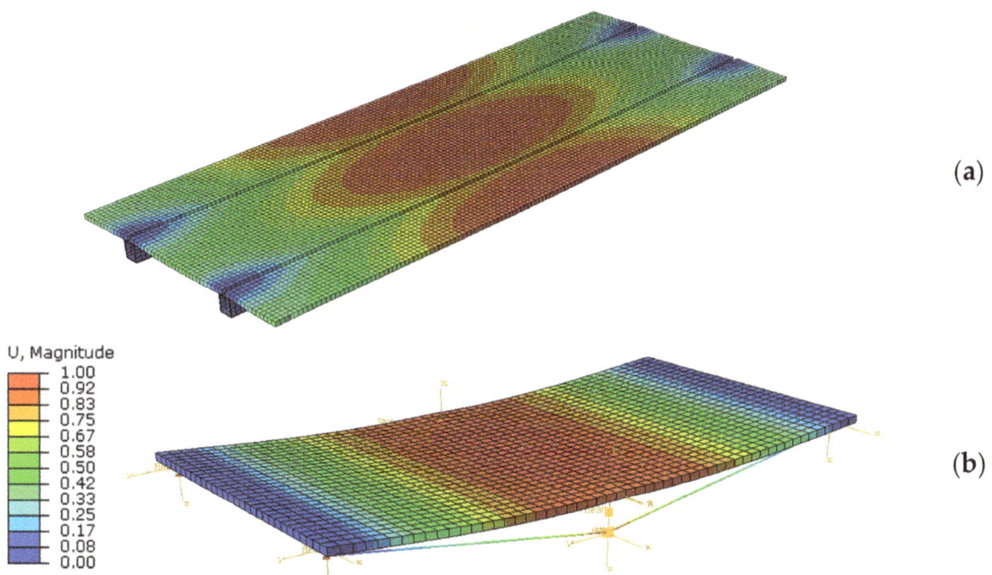

Figure 8. Qualitative comparison of fundamental modal shapes (out-of-scale) based on linear modal frequency response analysis of (**a**) SM#1–LGU module and (**b**) SM#2–LGU or SM#3–LGF modules (ABAQUS).

4. Diagnostic Investigation and Assessment of Experimental Performance Indicators

4.1. Vibration Frequency

As previously discussed, the basic assumption of the present application is that for the experiments herein reported, a single triaxial MEMS accelerometer like in [19] was used for all the examined systems. Output-only test data were collected under the effects of normal walks or in-place jumps.

The detailed experimental methods are also discussed in [19]. The diagnostic investigation was based on the experimental analysis from different test repetitions on the selected modules, with a total of nine test trials for the SM#1–LGU system, 14 repetitions for the SM#2–LGU module, and 18 for the SM#3–LGF system. The typical experimental records were collected, for all the case-study systems, in terms of vertical acceleration time histories similar to Figure 9, with the sensor placed in the centre of each slab. To note, case by case, the experimental setup and the detailing of sensor setup should be preliminary addressed to capture relevant data.

Based on experimental evidence like in Figure 9, the results in Figure 10 show selected examples of corresponding FFT signals. At a preliminary stage, a major difference in the vibration response of examined modules can be easily perceived, for example. Compared to existing design standards to prevent severe vibration issues in pedestrian systems, it is possible to see that the measured FFT peaks are associated to vibration frequencies that are significantly higher than the recommended minimum value of 5 Hz [25,34]. Moreover, typical LGs are characterized by intrinsic features that assign them a particular dynamic behaviour compared to other systems [16,17].

For the presently examined LG systems, the fundamental frequency was experimentally quantified in 30 Hz (±0.39) for the SM#1–LGU system. For the SM#2–LGU and SM#3–LGF slabs with similar geometrical and mechanical properties, but different damage severity, it resulted in 15.05 Hz (±0.2) and 13.8 Hz (±0.21), thus up to −8.3% vibration frequency decrease due to glass fracture [19]. Often, no interventions and verifications are required for traditional slabs and floors with a fundamental frequency higher than 5–8 Hz

(ISO 10137:2007 [25]). On the other side, it was also shown in [34] that existing reference indicators and comfort assessment procedures cannot be directly applied to LG systems.

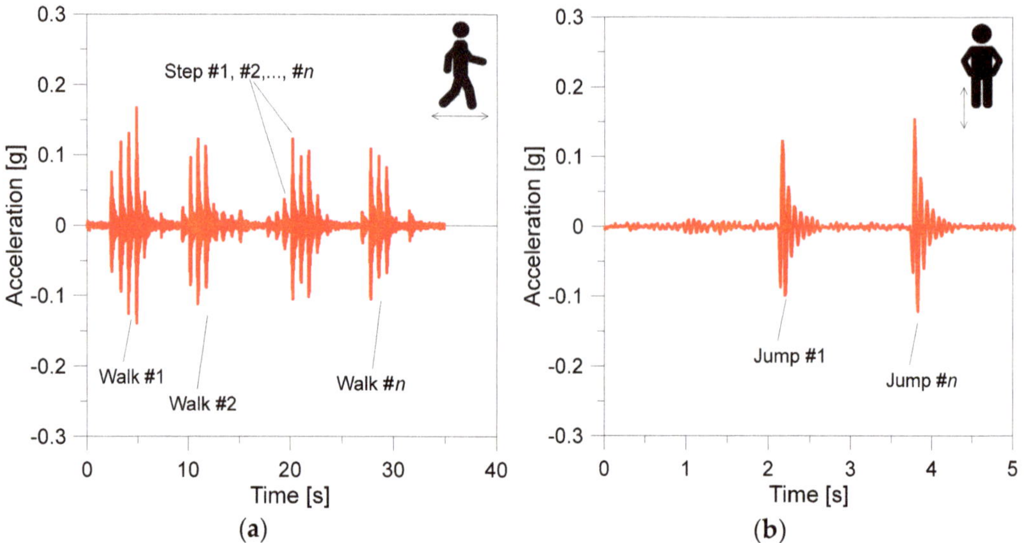

Figure 9. Example of in-field acquisition (selection) of acceleration records (vertical component) un-der (**a**) linear walking path or (**b**) on-site jump.

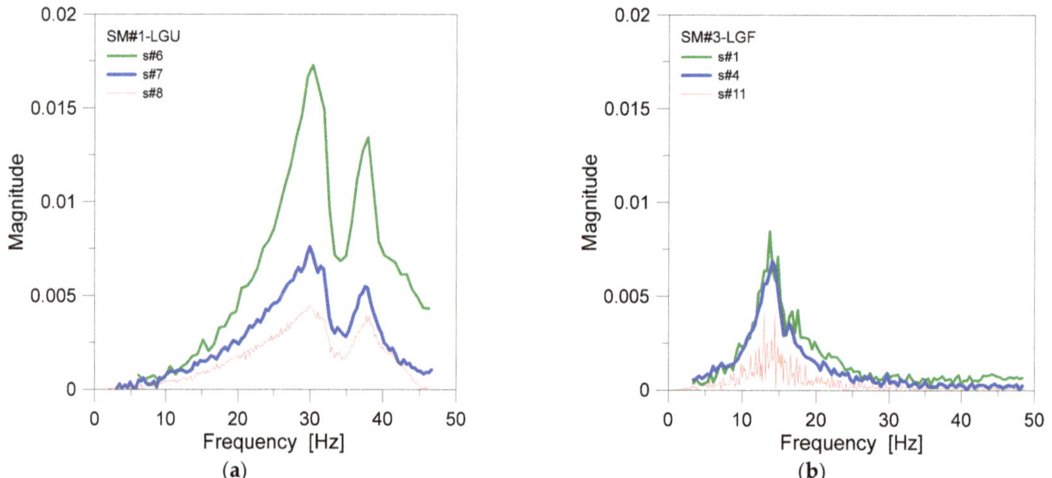

Figure 10. Frequency domain response of the examined systems: example of selected experimental signals for the (**a**) SM#1–LGU and (**b**) SM#3–LGF modules.

In this regard, the availability of intermittent in-field monitoring records for in-service LG structures would support the characterization of their vibration frequency (from "Time 0" of original installation apart) and also facilitate any kind of diagnostic analysis and maintenance plan.

4.2. Damping

FFT signals, as in Figure 11, can be also addressed in terms of diagnostic parameters. In the present study, data were further analysed in the post-processing stage because of the utmost importance for simplified damping estimates [51]. Even under uncertainty, among others, the half-power bandwidth method is the most representative and widely used approach for damping estimation due to its simplicity in implementation. The analysis of available experimental signals in the frequency domain gives, in fact:

$$\zeta = \frac{1}{2Q} \tag{7}$$

with:

$$Q = \frac{f_{max}}{f_{m,2} - f_{m,1}} \tag{8}$$

where f_{max} is the resonant frequency and $f_{m,1}$, $f_{m,2}$ are the frequencies at the left-hand and right-hand sides of f_{max}, respectively. To note, for the presently examined in-service systems, the average damping was quantified in 6.78% for the SM#1–LGU module, 7.25% for SM#2–LGU, and 8.95% for SM#3–LGF.

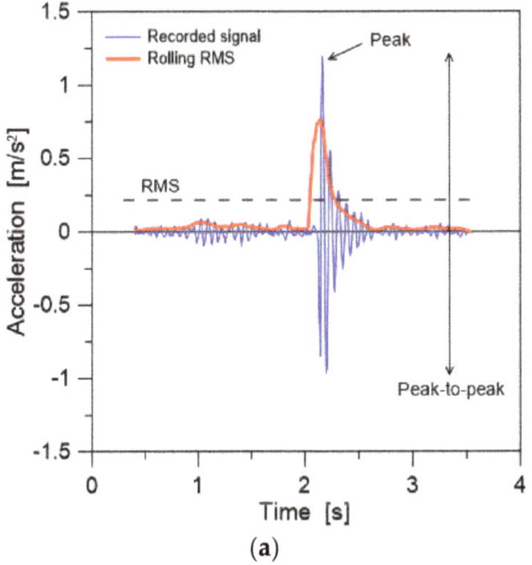

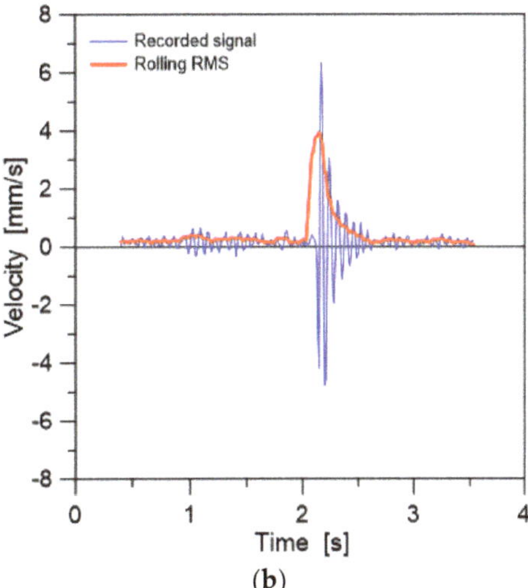

Figure 11. Qualitative example of typical (**a**) acceleration and (**b**) velocity experimental signal due to a single footfall of the involved volunteer.

The above values, whilst not fully exhaustive for diagnostic purposes, still offer the opportunity to capture relevant modifications in time (i.e., under repeated monitoring steps), as well as towards LG systems of literature (if any). According to past literature findings, damping for monolithic glass in the uncracked stage is, in fact, generally very small and can be typically expected in around 1.5% [41], but for LG sections, it can be relatively higher, and there is experimental evidence of damping terms up to 5–6% [52,53] or even higher [54]. For the specific analysis of SM#2–LGU and SM#3–LGF systems, as in the present study, additional damping contributions can be justified from the role of partially soft/flexible restraints, which are interposed to the metal substructure [41]. Moreover, the most important outcome is in the variation of SM#2–LGU-to-SM#3–LGF estimates, which was calculated in a +23.5% for the SM#3–LGF module. In this sense, damping evidence confirms higher contributions for the SM#3–LGF system affected by glass fracture and suggests the definition of possible threshold limits in support of monitoring and diagnostic analyses (especially when damping evidences are integrated to other performance indicators).

4.3. Vibration Assessment Based on Existing Conventional Approaches

An additional quantitative assessment of structural behaviour and capacity for a given in-service systems can be extrapolated from a comparative analysis and evolution of classical vibration parameters under ordinary loads. The vibration assessment of herein examined LG systems was further carried out in terms of structural safety and comfort, maximum vertical acceleration, peak-to-peak acceleration, RMS acceleration (Equation (9)), or rolling RMS acceleration, respectively (Equation (10)):

$$a_{RMS} = \sqrt{\frac{1}{T} \int_0^T a(t)^2 dt} \tag{9}$$

$$a_{RMS}(t) = \sqrt{\frac{\sum_0^n a(t)^2}{n}} \tag{10}$$

with T (in seconds) the total duration of each signal and n the number of recorded data in a time interval of 0.5 s.

Finally, the rolling RMS velocity was also taken into account, since it is traditionally representative of robust feedback for floor vibrations:

$$v_{RMS}(t) = \sqrt{\frac{\sum_0^n v(t)^2}{n}} \tag{11}$$

An example can be seen in Figure 11, as obtained from a single footfall of the involved volunteer (SM#3–LGF). To note, multiple design standards are available in support of the vibration serviceability assessment of pedestrian systems, but no specific rules and recommendations are available for glass structures. In the present application, for example, the reference limits were derived from ISO 10137:2007 [25] and compared to available experimental indicators based on post-processed signals.

As far as the acceleration values are considered, typical results take the form as in Figure 12, where data are grouped for SM#1–LGU and SM#2–LGU or SM#3–LGF systems, respectively. Regardless of the constructional details, and even possible damage, it is possible to see that the dynamic response of the three different LG systems under random normal walks is associated to absolute vertical acceleration peaks, which exceed the limit values for "indoor footbridges". The mean peak was, in fact, measured in 2.921 m/s^2 for the SM#1–LGU system, 1.315 m/s^2 for SM#2–LGU, and 1.512 m/s^2 for SM#3–LGF.

The measured RMS acceleration values from Equation (10) are also compared in Figure 12 for quantitative comparison with the "ISO baseline curve". Based on [25], the limit RMS acceleration is extrapolated on the base of the corresponding input vibration frequency of the system object of study, as well as its destination of use.

The typical recommended values of threshold baseline Multiplying Factors (MF) are summarized in Table 4 from [25]. Starting from the value of $a_{ISO,baseline}$ = 0.005 m/s^2 for vibration frequencies up to 8 Hz, Table 3 indicates that the top limit acceleration:

$$a_{RMS} = MF \times a_{ISO,baseline} \tag{12}$$

should not be exceeded, where:

$$a_{ISO,baseline} = a_{ISO,baseline}(f_1) \tag{13}$$

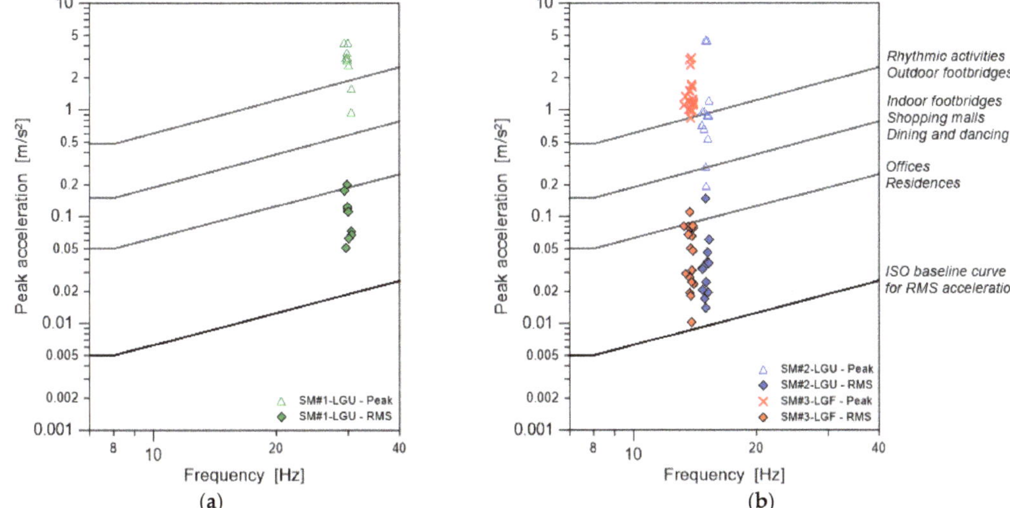

Figure 12. Reference ISO 10137:2007 [25] acceleration curves for RMS and peak acceleration on floors. Application of recommended limits to the presently examined (**a**) SM#1–LGU and (**b**) SM#2–LGU, SM#3–LGF systems.

Table 4. Reference limit multiplying factors (MF) and experimentally derived MF values (Equation (14)) for continuous and intermittent vibrations, based on ISO 10137:2007 [25] (*) from Equation (9) and experimental signals; (**) from Equation (13) and ISO baseline curve.

Floor	Time	Recommended MF Limit	Experimentally Calculated MF (Average) from Equation (14)
Critical working area	Day & Night	1	
Residential	Day	2 to 4	0.109 */0.02 ** = 5.47 for SM#1–LGU
	Night	1.4	0.040 */0.007 ** = 5.71 for SM#2–LGU
Quite office, Open plan	Day & Night	2	0.051 */0.006 ** = 8.50 for SM#3–LGF
General office	Day & Night	4	
Workshop	Day & Night	8	

For the present application example, the a_{RMS} values were first calculated based on experimental signals and Equation (9). The $a_{ISO,baseline}$ amplitude was then expressed from Equation (13), based on available experimental frequencies. The "Current" MF was hence quantified as:

$$MF = \frac{a_{RMS}}{a_{ISO,baseline}} \tag{14}$$

As shown in Table 4, the experimental MF values for SM#1–LGU and SM#2–LGU systems are mostly comparable (around six), but, indeed, they exceed the recommended thresholds from [25], for all floor types and destinations. To note, in particular, the highest MF value was calculated especially for the SM#3–LGF system with partial glass fracture and hence higher sensitivity to normal walks (MF = 8.50, +49%). The comparison towards the uncracked SM#2–LGU system (with identical size and geometrical features but intact glass) is a further confirmation of severe MF variation, which could be possibly considered as an additional meaningful parameter for damage detection.

In terms of RMS velocity from Equation (11), based on [55,56], the attention could be focused on limit values and ranges, which are proposed to detect critical serviceability configurations, as a function of the dynamic response of floors and on their destination of use.

For the present application, for example, the average peak of rolling RMS velocity was calculated in 7.11 mm/s for the SM#1–LGU system, 3.61 mm/s for the SM#2–LGU system, and 4.44 mm/s for the SM#3–LGF system. To note, from [55,56], the examined rolling RMS velocity values should be accounted as Class "E" of comfort and may, consequently, result in preferably suitable vibration performances for "industrial or sport" only. The present experimental evidence is, in fact, out of range for the recommended limits in case of other common destinations of use, especially the "residential or office" class.

However, such a comparison and quantitative classification gives a first suggestion for possible monitoring and mitigation interventions only, rather than efficiently supporting a concise diagnostic analysis.

4.4. Direct Structural Assessment Based on In-Field Performance Indicators

OMA techniques are particularly simple to apply in systems under normal operational conditions, so the critical stage is represented by collection of a sufficient number of signals for data interpretation and diagnostic analysis.

In Figure 13a, the vertical acceleration peak is shown as a function of the rolling RMS value, where each dot corresponds to a test configuration. A rather linear correlation for all experiments can also generally be noted, as suggested by the reported R-square correlation coefficient. Moreover, the SM#3–LGF system with glass fracture has the lowest correlation of experimental results (R-square = 0.93), which is relatively low compared to intact systems. While such a comparison should be extended to multiple slab units with different damage scenarios, such a correlation could be a parameter of practical feedback for diagnostic purposes. In addition, a correlation coefficient tending to 1 could represent an optimum for a safety check and integrity assessment. At this stage, moreover, the minimum number of experimental repetitions and configurations required for "robust" feedback is still uncertain to be generalized.

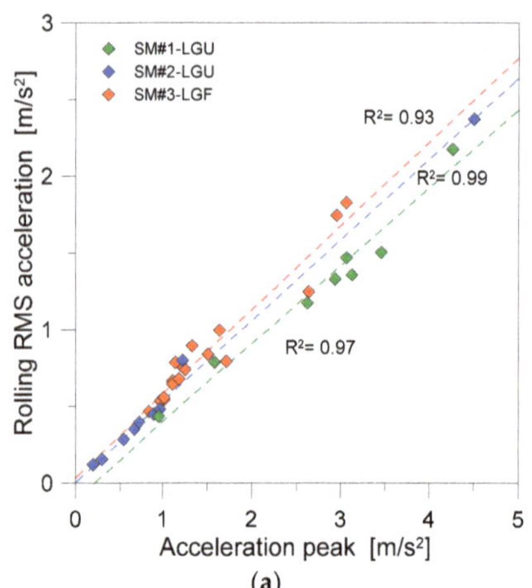

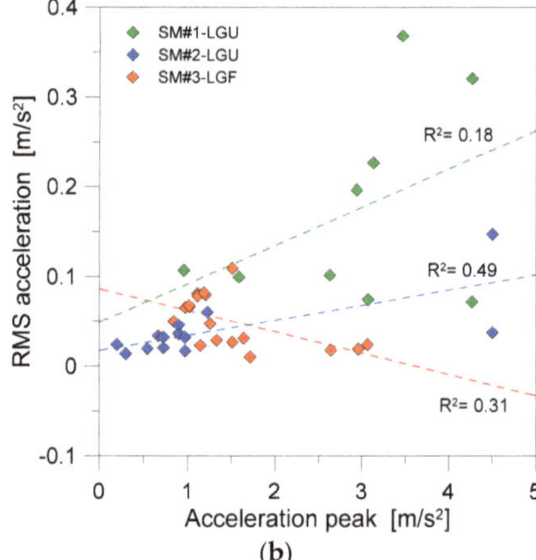

Figure 13. Correlation analysis of experimental performance indicators for the examined slab units: (**a**) rolling RMS acceleration as a function of the vertical acceleration peak; and (**b**) RMS acceleration as a function of the vertical acceleration peak.

In Figure 13b, the RMS acceleration value is indeed shown as a function of the vertical acceleration peak. Differing from Figure 13a, a major scatter from the linear regression method can be noticed for the SM#3–LGF sample affected by glass fracture, which has a relatively low R-square correlation coefficient but also a substantially different trend of experimental dots compared to the intact SM#1–LGU and SM#2–LGU samples. The potential of such a kind of comparative analysis should be elaborated further, including multiple LG systems and testing configurations. To note, the trend of vibration frequency with acceleration peak can also reveal major structural modifications and damage severity, as well as particular discomfort under ordinary loads, as it was for the example discussed in [16,17].

4.5. Material Characterization Based on In-Field Experiments and Finite Element Updating

Apart from in-field experimental measures and derived performance indicators on the side of structural performance diagnostics, a more refined and advanced protocol for safety assessment (as in Figure 3) necessarily requires the use of FE numerical models able to capture the geometrical and mechanical features of the real systems object of study and thus integrate and extend experimental evidence by model updating.

For example, based on FE systems like in Figure 8, one could pose the attention on the inverse characterization of the equivalent shear stiffness for the bonding PVB foils, or, even (in case of glass fracture) on the calibration of an equivalent, reduced E_{crack} modulus for the fractured glass layer (as it is for the SM#3–LGF system). Typical comparative results are proposed in Figure 14. To note, the other input material properties were kept fixed in $E = 70$ GPa for intact glass, $\nu = 0.23$ and $\rho = 2500$ kg/m^3, with $\nu_{PVB} = 0.49$ and $\rho_{PVB} = 1000$ kg/m^3 for PVB foils (Table 1).

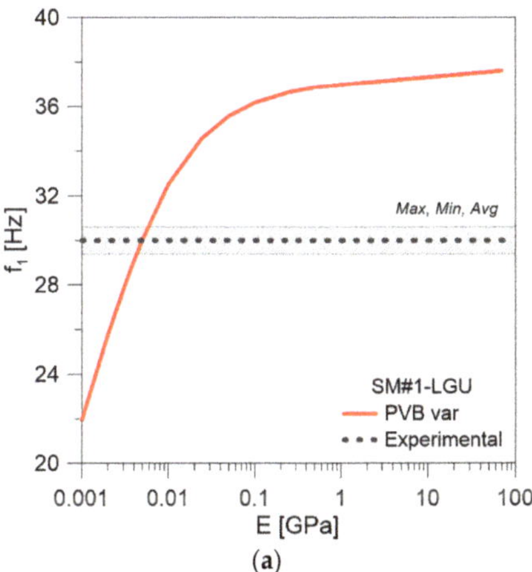

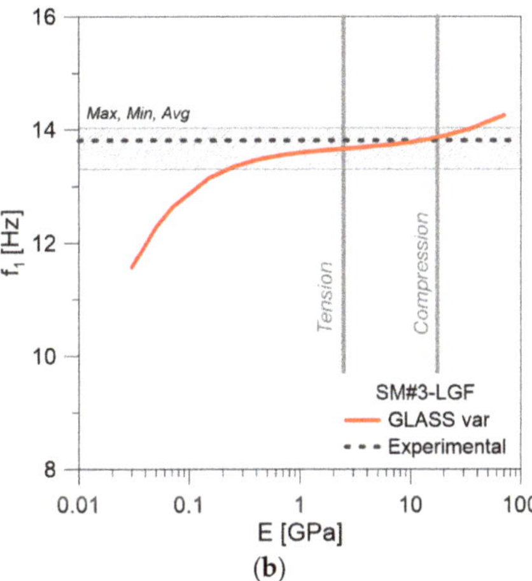

Figure 14. Material characterization based on FE model updating: (**a**) derivation of equivalent interlayer stiffness for SM#1–LGU system; and (**b**) equivalent modulus for fractured glass layer (SM#3–LGF system).

Figure 14a shows the effect of PVB stiffness on the fundamental frequency of the SM#1–LGU module. Compared to the average experimental vibration frequency, it can be seen that the best match is found for E_{PVB} modulus for PVB foils in the order of ≈4–5 MPa. Interestingly, the so-calculated equivalent modulus for PVB is in the same order of magnitude of the study reported in [16,19].

In Figure 14b, curve-fitting based on model updating is proposed for the fractured SM#3–LGF system in which the degradation of PVB foils acts in combination with glass fracture. The parametric FE analysis shows that the best match of experimentally derived frequency is in the order of $E_{crack} \approx 15$ GPa for the fractured glass layer, and this finding is in close correlation with the compressive fractured glass modulus calibrated in [18]. Further support from FE model updating could derive also from analysis of local effects in vibrational and dynamic terms, such as, for example, deriving from special fixing systems of typical use in LG applications [34].

5. Residual Structural Capacity Assessment

5.1. Quantification of Mechanical Degradation and Load-Bearing Capacity Loss

The final sub-stage of Step 4 in Figure 3 is of utmost importance for safety assessment as it is associated to the final residual capacity quantification. On the other side, such a procedural stage necessarily requires a robust engineering characterization of the LG system object of study (based on in-field testing) as well as an accurate mechanical characterization (based on integration from FE models like in Figure 7). In the present application example, the analysis of stress peaks and deflections in glass components was in fact carried out based on previous experimental–numerical evidence.

More in detail, the FE models, such as in Figure 7, were adapted to quasi-static, non-linear incremental mechanical analysis where the input material properties were kept fix as in the preliminary frequency analysis (and Table 1), but the attention was focused on the quantification of structural behaviours under ordinary design actions. The selected LG pedestrian systems were, in fact, investigated under the effects of self-weight and a distributed accidental vertical load ($Q_k = 4 \text{ kN/m}^2$ its characteristic value). Furthermore, the analysis was carried out for the pedestrian modules as in the "Current" situation and at "Time 0", that is, with a short-term elastic modulus for PVB (i.e., Figure 5) and ideally intact glass layers.

The so-collected numerical results are summarized in Table 5, where the percentages scatter between the initial design stage, "Time 0", and the present situation; "Current" structural performances are also reported. In the analysis of performance indicators, moreover, maximum stress peaks are calculated:

- At the mid-span of short edges in free bending for the SM#1–LGU system.
- In the region of mechanical pint supports for SM#2–LGU and SM#3–LGF systems.

Table 5. Structural performance analysis for the examined modular units (with single occupant and $M = 80$ kg), with evidence of "Time 0" and "Current" behaviours.

Sample	Vibration Frequency [Hz]			ULS Stress [MPa]			SLS Deflection [mm]		
	Time 0	Current	Δ [%]	Time 0	Current	Δ [%]	Time 0	Current	Δ [%]
SM#1–LGU	34.6	30	−13.3	4.31	5.27	+22.2	1.16	1.76	+51.7
SM#2–LGU	21.2	15.05	−28.3	17.28 (point-fixing) 6.29 (centre)	20.01 7.69	+15.8 +18.2	4.11	6.10	+48.4
SM#3–LGF	21.2	13.8	−34.9	17.28 (point-fixing) 6.29 (centre)	26.55 4.29	+53.64 −31.8	4.11	6.28	+52.8

From Table 5, some useful parameters can be easily derived to quantify the current capacity due to both long-term effects and unfavourable operational conditions. In terms of vibration frequency of the occupied modules, for example, it can be seen that the SM#1–LGU system with minimum geometric slenderness and relatively short bending span is less affected by mechanical degradation of PVB foils compared to the others. In contrary, the SM#3–LGF system affected by the additional fracture of glass shows the maximum frequency decrease.

As far as the principal ULS stress peaks in glass that are taken into account, Table 5 shows a rather balanced variation for SM#1–LGU and SM#2–LGU solutions. Conversely, the partial glass fracture in SM#3–LGF system gives clear evidence of more pronounced stiffness degradation as a direct result of the fractured glass layer in compression. The analysis of SLS deflections, finally, shows comparable modifications for the three examined systems. In this sense, another important aspect to note is that the analysis of deflections only may not be sufficiently exhaustive to capture the actual damage severity for a given LG system.

More robust feedback could be indeed derived from the frequency analysis and from the stress peak analysis in glass, by comparing the single experimental output data with recommended limit values of the literature (i.e., critical frequency range, etc.), or even by comparison of stress peaks in glass with the nominal material strength. Damping predictions from in-field experiments, finally, could be rather simple to measure and compare, as well as meaningful for damage quantification, but necessarily needing multiple comparative data and multiple similar systems for efficient and robust comparative analysis.

5.2. Safety Check against Ordinary Mechanical Loads

As far as the design parameters from [24], other design standards are taken into account. For example, the potential and efficiency of parametric results as in Table 5 can be further exploited in terms of residual capacity towards specific design conditions and limitations by standards. For newly designed LG systems, technical recommendations are, in fact, available to preserve appropriate safety levels and functionality under normal service conditions.

For LG plates with two linearly supported edges, as for the examined modules, it is recommended in [24] at the SLS deflection check that the limit value u_{lim} should not be exceeded, where:

$$u_{lim} = min \begin{cases} 50 \text{ mm} \\ \frac{L_{min}}{100} \end{cases} \tag{15}$$

with L_{min} the minimum edge size (u_{lim} = 5.1 mm for SM#1–LGU and u_{lim} = 13.5 mm for SM#2–LGU and SM#3–LGF).

Regarding the maximum ULS stress values for the verification in glass layers, the comparison is carried out towards the design strength defined as in [24], where it is assumed that:

$$f_{g;d} = f_{g;d,b} + f_{g;d,p} = \frac{k_{mod}k_{ed}k_{sf}\lambda_{gA}\lambda_{gl}f_{g;k}}{R_M\gamma_M} + \frac{k'_{ed}k_v\left(f_{b;k} - f_{g;k}\right)}{R_{M;v}\gamma_{M;v}} \tag{16}$$

with $f_{g;d,p} > 0$ for pre-stressed glass and:

$$k_{mod} = 0.585 \cdot t_L^{-1/16} \tag{17}$$

while the other coefficients and safety factors are defined in [24] to account for a multitude of production and loading/boundary condition features. The resistance verification notoriously requires that the stress effects of a given design action do not exceed the capacity of the system, that is:

$$\sigma_{max} \leq f_{g;d} \tag{18}$$

Further, for safety purposes, the ULS stress analysis is a primary verification check. For the present analysis, the application of Equation (18) to the examined in-service systems in ULS design conditions, with k_{mod} = 0.78 for temporary transient of pedestrians, resulted in strength values in the order of ≈60 MPa at the edges and ≈74 MPa in the centre of LG panels.

The so-derived ULS and SLS performance values are reported in Table 6 for the case-study applications in the form of maximum stress-to-strength ratio (ULS, from Equation (18)) and maximum deformation-to-deflection limit ratio (SLS, from Equation (15)).

From a practical point of view, such an assessment reveals the "real" capacity loss of a given LG system thanks to accurate mechanical calibration and to multiple performance indicators for "Time 0" and "Current" analyses. For the SM#2–LGU system, for example, the "Current" SLS deflection in Table 5 is associated to major safety and comfort, while a less-pronounced modification of performance can be seen in terms of ULS stress peaks. For the SM#3–LGF system with glass fracture, it is possible to see a rather unform safety check for ULS stress and SLS deformation values. It is worth noting that the "Current" stress condition, which is markedly increased in severity compared to SM#2–LGU system, further confirms the severity of the damage.

Table 6. Safety check for ULS stress and SLS deformation accounting for in-field experimentally derived degradation phenomena for the examined LG modular units.

	ULS Stress Ratio			SLS Deflection Ratio		
Sample	Time 0	Current	SAFE ($\leq$1)	Time 0	Current	SAFE ($\leq$1)
SM#1–LGU	0.071	0.087	Yes	0.22	0.35	Yes
SM#2–LGU	0.29	0.34	Yes	0.31	0.45	Yes
SM#3–LGF	0.29	0.45	Yes	0.31	0.47	Yes

Overall, it is important to note in Table 6 that the examined systems were still "safe" for occupants, as determined through current performance indicators and technical recommendations of ordinary use and structural design of "new" glass members. However, at the same time, the explored systems also give evidence of marked loss of structural capacity in their "Current" situation, compared to "Time 0". Such a procedure could be used to derive and quantify a robust and concise damage index of existing LG structural members and thus could represent an efficient performance indicator for monitoring purposes in early damage detection or early retrofit interventions, especially for those in-service LG systems where it is not possible to establish continuous or intermittent in-field experimental monitoring protocols.

6. Summary and Conclusions

A detailed engineering knowledge of current mechanical properties and residual load-bearing capacity levels for in-service laminated glass (LG) structures is of utmost importance in building management, especially for those structural systems that are characterized by direct interaction with occupants (i.e., pedestrian systems, balustrades, etc.). While, in certain conditions, damage can be visually detected (especially major glass cracks), there are several practical situations in which the in-service structure could be potentially unsafe for customers (due, for example, to unfavourable loading conditions or unfavourable ambient conditions facilitating the material degradation) without marked evidence of visual defects. In this sense, a harmonized protocol to (possibly rapidly) efficiently assess the "Current" structural safety level of a LG system compared to its "Time 0" performance, and thus to estimate the residual capacity is crucial for risk minimization and comfort/functionality optimization.

In this perspective paper, the attention was focused on a possible procedural protocol to generally apply to different LG pedestrian systems based on in-field experiments, analysis, and the assessment of experimental evidence and integrated model updating for structural estimates. To this aim, three different case-study systems belonging to two different (indoor) structures constructed in Italy were explored.

In terms of vibration assessment, the practical application study showed that:

- The estimation of the fundamental vibration frequency is a first relevant but not exhaustive step for quantitative characterization of safety levels in existing glass structures. LG pedestrian systems are often characterized by relatively high fundamental frequency but often have relatively small mass compared to occupants.
- Fast, intermittent in-field experimental measures based on OMA techniques could allow for the collection of a set of meaningful comparative data for an efficient check of mechanical features and modifications in a given existing structure. However, the experimental testing conditions should be possibly planned to reproduce the "normal" service configurations for the examined systems.
- For the proposed procedural steps and the methodology application to three different LG systems, the in-field testing highlighted the availability of multiple performance indicators, but also the need of robust engineering knowledge for their interpretation.
- Similarly, the availability of in-field experimental measures proved to be meaningful, especially when combined to refined Finite Element numerical models able to

indirectly quantify the long-term/damage effects, in terms of "Current" state-of-art condition and capacity in comparison to "Time 0" design performances.

On the other side, it was also shown that:

- Most of the existing conventional methods for vibration serviceability purposes are not specifically adaptable to LG systems.
- Damping estimates can represent an experimental output of simple calculation but are still often affected by a multitude of various influencing parameters and possibly characterized by a high sensitivity under test repetitions.

In terms of structural checks of "Current" performances, finally, it was shown that:

- Due to intrinsic material properties and structural design assumptions, long-term phenomena and material degradation can induce severe modifications compared to "Time 0" conditions. After installation, it is hence important to monitor the evolution of basic parameters (and thus the residual capacity) over time.
- Overall, such a kind of intermittent diagnostic approach can facilitate the early detection of unfavourable configurations and hence promptly prevent maintenance interventions before any kind of severe damage could take place.

Author Contributions: C.B.: conceptualization, methodology, formal analysis, investigation, resources, project administration, writing-original draft preparation, writing-review and editing; S.N., M.F. and C.A.: methodology, investigation, writing-original draft preparation, writing-review and editing. All authors have read and agreed to the published version of the manuscript.

Funding: This research received no external funding.

Data Availability Statement: Data will be shared upon request.

Acknowledgments: Seretti Vetroarchitetture and So.Co.Ba. are acknowledged for facilitating the in-field experimental investigation on case-study systems.

Conflicts of Interest: The authors declare no conflict of interest.

References

1. Baggio, C.; Bernardini, A.; Colozza, R.; Corazza, L.; Della Bella, M.; Di Pasquale, G.; Dolce, M.; Goretti, A.; Martinelli, A.; Orsini, G.; et al. *Field Manual for Post-Earthquake Damage and Safety Assessment and Short Term Countermeasures (AeDES)*; Report EUR 22868 EN.; Pinto, A.V., Taucer, F., Eds.; Joint Research Centre—Institute for the Protection and Security of the Citizen: Brussels, Belgium, 2007; ISSN 1018-5593.
2. Ministry of Business, Innovation and Employment (MBIE). *Field Guide: Rapid Post Disaster Building Usability Assessment—Earthquake*, 1st ed.; MBIE: Wellington, New Zealand, 2014; ISBN 978-0-478-41794-4 (Print), 978-0-478-41797-5 (Online).
3. Harirchian, E.; Lahmer, T.; Buddhiraju, S.; Mohammad, K.; Mosavi, A. Earthquake Safety Assessment of Buildings through Rapid Visual Screening. *Buildings* **2020**, *10*, 51. [CrossRef]
4. Stepinac, M.; Kisicek, T.; Renić, T.; Hafner, I.; Bedon, C. Methods for the Assessment of Critical Properties in Existing Masonry Structures under Seismic Loads—The ARES Project. *Appl. Sci.* **2020**, *10*, 1576. [CrossRef]
5. Lorenzoni, F.; Casarin, F.; Caldon, M.; Islami, K.; Modena, C. Uncertainty quantification in structural health monitoring: Applications on cultural heritage buildings. *Mech. Syst. Signal Process.* **2016**, *66*, 268–281. [CrossRef]
6. Masciotta, M.-G.; Ramos, L.F.; Lourenco, P.B. The importance of structural monitoring as a diagnosis and control tool in the restoration process of heritage structures: A case study in Portugal. *J. Cult. Herit.* **2017**, *27*, 36–47. [CrossRef]
7. Russo, S. Simplified procedure for structural integrity's evaluation of monuments in constrained context: The case of a Buddhist Temple in Bagan (Myanmar). *J. Cult. Herit.* **2017**, *27*, 48–59. [CrossRef]
8. Ali, A.; Sandhu, T.Y.; Usman, M. Ambient Vibration Testing of a Pedestrian Bridge Using Low-Cost Accelerometers for SHM Applications. *Smart Cities* **2019**, *2*, 20–30. [CrossRef]
9. Bedon, C.; Bergamo, E.; Izzi, M.; Noè, S. Prototyping and validation of MEMS accelerometers for structural health monitoring—The case study of the Pietratagliata cable-stayed bridge. *J. Sens. Actuator Netw.* **2018**, *7*, 30. [CrossRef]
10. Luleci, F.; Li, L.; Chi, J.; Reiners, D.; Cruz-Neira, C.; Necati Catbas, F. Structural Health Monitoring of a Foot Bridge in Virtual Reality Environment. *Procedia Struct. Integr.* **2022**, *37*, 65–72. [CrossRef]
11. Payawal, J.M.G.; Kim, D.-K. Image-Based Structural Health Monitoring: A Systematic Review. *Appl. Sci.* **2023**, *13*, 968. [CrossRef]
12. Guzman-Acevedo, G.M.; Vazquez-Becerra, G.E.; Millan-Almaraz, J.R.; Rodriguez-Lozoya, H.E.; Reyes-Salazar, A.; Gaxiola-Camacho, J.R.; Martinez-Felix, C.A. GPS, Accelerometer, and Smartphone Fused Smart Sensor for SHM on Real-Scale Bridges. *Adv. Civ. Eng.* **2019**, *2019*, 6429430. [CrossRef]

13. Buffarini, G.; Clemente, P.; Giovinazzi, S.; Ormando, C.; Scafati, T. Structural assessment of the pedestrian bridge accessing Civita di Bagnoregio, Italy. *J. Civ. Struct. Health Monit.* **2022**, 1–18. [CrossRef]
14. Norouzzadeh Tochaei, E.; Fang, Z.; Taylor, T.; Babanajad, S.; Ansari, F. Structural monitoring and remaining fatigue life estimation of typical welded crack details in the Manhattan Bridge. *Eng. Struct.* **2021**, *231*, 111760. [CrossRef]
15. Clemente, P. Monitoring and evaluation of bridges: Lessons from the Polcevera Viaduct collapse in Italy. *J. Civ. Struct. Health Monit.* **2020**, *10*, 177–182. [CrossRef]
16. Bedon, C. Diagnostic analysis and dynamic identification of a glass suspension footbridge via on-site vibration experiments and FE numerical modelling. *Compos. Struct.* **2019**, *216*, 366–378. [CrossRef]
17. Bedon, C. Experimental investigation on vibration sensitivity of an indoor glass footbridge to walking conditions. *J. Build. Eng.* **2020**, *29*, 101195. [CrossRef]
18. Kozłowski, M. Experimental and numerical assessment of structural behaviour of glass balustrade subjected to soft body impact. *Compos. Struct.* **2019**, *229*, 111380. [CrossRef]
19. Bedon, C.; Noè, S. Post-Breakage Vibration Frequency Analysis of In-Service Pedestrian Laminated Glass Modular Units. *Vibration* **2021**, *4*, 836–852. [CrossRef]
20. El-Sisi, A.; Newberry, M.; Knight, J.; Salim, H.; Nawar, M. Static and high strain rate behavior of aged virgin PVB. *J. Polym. Res.* **2022**, *29*, 39. [CrossRef]
21. Huang, Z.; Xie, M.; Du, J.Z.Y.M.; Song, H.-K. Rapid evaluation of safety-state in hidden-frame supported glass curtain walls using remote vibration measurements. *J. Build. Eng.* **2018**, *19*, 91–97. [CrossRef]
22. Bedon, C.; Noè, S. Rapid Safety Assessment and Experimental Derivation of Damage Indexes for In-Service Glass Slabs. In Proceedings of the Challenging Glass Conference Proceedings, Ghent, Belgium, 23–24 June 2022; Volume 8. [CrossRef]
23. EN 572–2:2004; Glass in Buildings—Basic Soda Lime Silicate Glass Products. CEN: Brussels, Belgium, 2004.
24. CNR-DT 210/2013; Istruzioni Per la Progettazione, L'esecuzione ed il Controllo di Costruzioni con Elementi Strutturali di Vetro. National Research Council of Italy (CNR): Rome, Italy, 2013. (In Italian)
25. ISO 10137:2007; Bases for Design of Structures -Serviceability of Buildings and Walkways against Vibrations. International Organisation for Standardization (ISO): Geneve, Switzerland, 2007.
26. Dimarogonas, A.D. Vibration of cracked structures: A state of the art review. *Eng. Fract. Mech.* **1996**, *55*, 831–857. [CrossRef]
27. Limongelli, M.P.; Manoach, E.; Quqa, S.; Giordano, P.F.; Bhowmik, B.; Pakrashi, V.; Cigada, A. Vibration Response-Based Damage Detection. In *Structural Health Monitoring Damage Detection Systems for Aerospace*; Springer Aerospace Technology: Cham, Switzerland, 2021.
28. Dogangun, A.; Acar, R.; Sezen, H.; Livaoglu, R. Investigation of dynamic response of masonry minaret structures. *Bull. Earthq. Eng.* **2008**, *6*, 505–517. [CrossRef]
29. Gentile, C.; Saisi, A. Ambient vibration testing of historic masonry towers for structural identification and damage assessment. *Constr. Build. Mater.* **2007**, *21*, 1311–1321. [CrossRef]
30. Ubertini, F.; Gentile, C.; Materazzi, A.L. Automated modal identification in operational conditions and its application to bridges. *Eng. Struct.* **2013**, *46*, 264–278. [CrossRef]
31. Zhu, X.; Cao, M.; Ostachowicz, W.; Xu, W. Damage Identification in Bridges by Processing Dynamic Responses to Moving Loads: Features and Evaluation. *Sensors* **2019**, *19*, 463. [CrossRef]
32. Gauron, O.; Boivin, Y.; Ambroise, S.; Saidou Sanda, A.; Bernier, C.; Paultre, P.; Proulx, J.; Roberge, M.; Roth, S.-N. Forced-vibration tests and numerical modeling of the Daniel-Johnson multiple-Arch Dam. *J. Perform. Constr. Facil.* **2018**, *32*, 04017137. [CrossRef]
33. Pereira, S.; Magalhães, F.; Cunha, Á.; Moutinho, C.; Pacheco, J. Modal identification of concrete dams under natural excitation. *J. Civ. Struct. Health Monit.* **2021**, *11*, 465–484. [CrossRef]
34. Bedon, C.; Fasan, M. Reliability of field experiments, analytical methods and pedestrian's perception scales for the vibration serviceability assessment of an in-service glass walkway. *Appl. Sci.* **2019**, *9*, 1936. [CrossRef]
35. Bedon, C.; Mattei, S. Facial Expression-Based Experimental Analysis of Human Reactions and Psychological Comfort on Glass Structures in Buildings. *Buildings* **2021**, *11*, 204. [CrossRef]
36. Bedon, C. Issues on the Vibration Analysis of In-Service Laminated Glass Structures: Analytical, Experimental and Numerical Investigations on Delaminated Beams. *Appl. Sci.* **2019**, *9*, 3928. [CrossRef]
37. Zemanova, A.; Zeman, J.; Janda, T.; Schmidt, J.; Sejnoha, M. On modal analysis of laminated glass: Usability of simplified methods and Enhanced Effective Thickness. *Compos. Part B Eng.* **2018**, *151*, 92–105. [CrossRef]
38. Galuppi, L.; Royer-Carfagni, G. Effective thickness of laminated glass beams: New expression via a variational approach. *Eng. Struct.* **2012**, *38*, 53–67. [CrossRef]
39. Andreozzi, L.; Briccoli Bati, S.; Fagone, M.; Ranocchiai, G.; Zulli, F. Weathering action on thermo-viscoelastic properties of polymer interlayers for laminated glass. *Constr. Build. Mater.* **2015**, *98*, 757–766. [CrossRef]
40. Bennison, S.J.; Jagota, A.; Smith, C.A. Fracture of Glass/Poly(vinyl butyral) (Butacite®) Laminates in Biaxial Flexure. *J. Am. Ceram. Soc.* **1999**, *82*, 1761–1770. [CrossRef]
41. Bedon, C.; Fasan, M.; Amadio, C. Vibration analysis and dynamic characterization of structural glass element with different restraints based on operational modal analysis. *Buildings* **2019**, *9*, 13. [CrossRef]
42. Levin, R.I.; Lieven, N.A.J. Dynamic finite element model updating using simulated annealing and genetic algorithms. *Mech. Syst. Signal Process.* **1998**, *12*, 91–120. [CrossRef]

43. Chisari, C.; Bedon, C.; Amadio, C. Dynamic and static identification of base-isolated bridges using Genetic Algorithms. *Eng. Struct.* **2015**, *102*, 80–92. [CrossRef]
44. Standoli, G.; Salachoris, G.P.; Masciotta, M.G.; Clementi, F. Modal-based FE model updating via genetic algorithms: Exploiting artificial intelligence to build realistic numerical models of historical structures. *Constr. Build. Mater.* **2021**, *303*, 124393. [CrossRef]
45. Sung, H.; Chang, S.; Cho, M. Efficient Model Updating Method for System Identification Using a Convolutional Neural Network. *AIAA J.* **2021**, *59*, 3480–3489. [CrossRef]
46. Teughels, A.; De Roeck, G. Damage detection and parameter identification by finite element model updating. *Arch. Comput. Methods Eng.* **2005**, *12*, 123–164. [CrossRef]
47. Zapico, J.L.; González, M.P. Numerical simulation of a method for seismic damage identification in buildings. *Eng. Struct.* **2005**, *28*, 255–263. [CrossRef]
48. Busca, G.; Cappellini, A.; Manzoni, S.; Tarabini, M.; Vanali, M. Quantification of changes in modal parameters due to the presence of passive people on a slender structure. *J. Sound Vib.* **2014**, *333*, 5641–5652. [CrossRef]
49. Cao, M.S.; Sha, G.G.; Gao, Y.F.; Ostachowicz, W. Structural damage identification using damping: A compendium of uses and features. *Smart Mater. Struct.* **2016**, *26*, 043001. [CrossRef]
50. Simulia, D.S. *ABAQUS Computer Software*; Dassault Systémes: Providence, RI, USA, 2021.
51. Clough, R.W.; Penzien, J. *Dynamics of Structures*; McGraw-Hill: New York, NY, USA, 1993; ISBN 0-07-011394-7.
52. Larcher, M.; Manara, G. *Influence of Air Damping on Structures Especially Glass*; JRC 57330 Technical Note; European Commission, Joint Research Centre—Institute for the Protection and Security of the Citizen: Ispra, Italy, 2010.
53. Ramos, A.; Pelayo, F.; Lamela, M.J.; Canteli, A.F.; Huerta, C.; Acios, A.P. Evaluation of damping properties of structural glass panes under impact loading. In *COST Action TU0905 Mid-term Conference on Structural Glass*; CRC Press: Boca Raton, FL, USA, 2013; ISBN 978-1-138-00044-5.
54. Lenci, S.; Consolini, L.; Clementi, F. On the experimental determination of dynamical properties of laminated glass. *Ann. Solid Struct. Mech.* **2015**, *7*, 27–43. [CrossRef]
55. Feldmann, M.; Heinemeyer, C.; Butz, C.; Caetano, E.; Cunha, A.; Galanti, F.; Goldack, A.; Hechler, O.; Hicks, S.; Keil, A.; et al. *Design of Floor Structures for Human Induced Vibrations*; EUR 24084 EN; Publications Office of the European Union: Luxembourg, 2009; p. JRC55118.
56. Sedlacek, G.; Heinemeyer, C.; Butz, C.; Völling, B.; Waarts, P.; Duin, F.; Hicks, S.; Devine, P.; Demarco, T. *Generalisation of Criteria for Floor Vibrations for Industrial, Office, Residential and Public Building and Gymnastic Halls*; Report number: EUR-21972-EN; European Commission: Luxembourg, 2006.

Article

Structural Identification from Operational Modal Analysis: The Case of Steel Structures

Flavio Stochino [1,*], Alessandro Attoli [2], Michele Serra [3], Alberto Napoli [4], Daniel Meloni [1] and Fausto Mistretta [1]

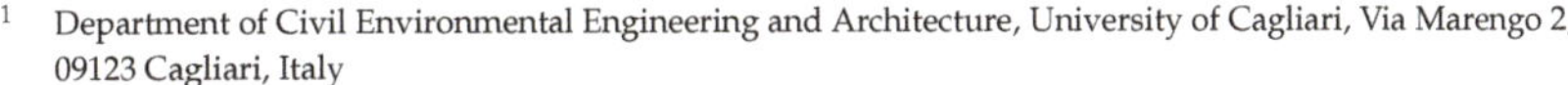

[1] Department of Civil Environmental Engineering and Architecture, University of Cagliari, Via Marengo 2, 09123 Cagliari, Italy
[2] INAF–OAC Osservatorio Astronomico di Cagliari, 09047 Selargius, Italy
[3] Secured Solutions srl, Via dell'Artigianato 11, 09122 Cagliari, Italy
[4] Freelance Engineer
* Correspondence: fstochino@unica.it

Abstract: In the case of old existing structures where the cultural value is very high, structural health analyses and investigations would be better performed without damages or service interruptions. Thus, modal analysis aimed at identifying eigenfrequencies and eigenmodes represents a very effective strategy to identify structural characteristics. In this paper, an innovative strategy to identify structural parameters exploiting the modal information obtained from operational modal analysis is proposed. The importance of the structural modeling in the problem formulation is highlighted. In the case of a simply supported beam, it was possible to assess the beam steel elastic modulus, while in the case of a cantilever beam, some constraint characteristics have been evaluated as well. In the steel frame case, the focus was on the constraint conditions of the structure determining the flexural stiffness of the springs representing the column base constraints. The method performances are promising for applications in larger structures such as bridges and buildings.

Keywords: modal analysis; structural identification; simulated annealing; operational modal analysis; steel

Citation: Stochino, F.; Attoli, A.; Serra, M.; Napoli, A.; Meloni, D.; Mistretta, F. Structural Identification from Operational Modal Analysis: The Case of Steel Structures. *Buildings* **2023**, *13*, 548. https://doi.org/10.3390/buildings13020548

Academic Editors: Emanuele Brunesi, Ling Yu and Hugo Rodrigues

Received: 9 December 2022
Revised: 17 January 2023
Accepted: 11 February 2023
Published: 17 February 2023

1. Introduction

A large part of European building constructions have exceeded their service life or require careful structural health analyses. This is particularly important in Italy where historical monuments are spread around the whole country and the main infrastructures were built in the 1950s and 1960s [1,2]. In this specific case and in case of monumental buildings where the cultural value is very high [3], structural health analyses and investigations should be performed while preserving the existing structures.

Thus, modal analysis aimed at identifying eigenfrequencies and eigenmodes represents a very effective strategy to identify structural characteristics. In particular, operational modal analysis (OMA) is a modal parameters identification technique based on vibration data collected when the structure is in service conditions [4]. Practically, OMA considers random environmental vibrations and cyclic loads on the structure as unknown sources of excitation. The easiest OMA technique is the peak picking (PP) method [5,6]. It is a frequency domain technique in which the natural frequencies are pointed as peaks in the power spectrum. Basic assumptions are that damping is low and the modes are well-separated. More complex methods such as frequency domain decomposition (FDD) [7], time domain decomposition (TDD) [8], and stochastic subspace identification (SSI) [9] are out of the scopes of this paper.

An emerging strategy to measure the modal properties without adding accelerometers to the system is 3D laser vibrometry, see [10–12], which has the main advantage in taking the measures from a distance. This can be useful in case of large structures that are not easily reachable.

The developments of computational mechanics allow to accurately model the structural behavior considering many parameters that often are not easily known. Indeed, a direct measure of these parameters would require high costs and would not always be compatible with the preservation of historical and monumental buildings.

For this reason, it is often more effective to formulate an inverse problem in which the eigenfrequencies measured through OMA represent a benchmark and the model parameters can be tuned in order to obtain numerical eigenfrequencies similar to the benchmark ones. Actually, this approach requires a good optimization strategy and presents quite a high computational cost. Usually, this optimization problem is represented by target functions that are neither continuous nor monotonic; consequently, the use of a heuristic algorithm such as simulated annealing [13–15], genetic algorithms [16,17], differential evolution [18], ant colony [19], and particle swarm [20] becomes mandatory.

Recently, [21] reported on an inverse problem for the structural identification of floor diaphragms using a perturbation approach. The study [22] presented an identification problem for a reinforced concrete, tall building based on model updating and experimental modal analysis. In [23], the authors showed the dynamic identification of the Baptistery of San Giovanni in Firenze (Italy) based on OMA and frequency domain decomposition.

With specific attention to steel structures, an interesting structural identification was performed for a steel footbridge in [24], while [25] presented an experimental modal analysis for a steel arch bridge. More recently, [26] presented an image-based operational modal analysis aimed at damage detection in a steel frame, and [27] reported on truss steel bridge damage identification through experimental modal analysis.

This paper presents a strategy to identify structural parameters exploiting the modal information obtained from OMA and developing an inverse problem in which the outcomes of experimental modal analyses are considered a known input while the structural parameters of a mathematical model representing the real structure are unknown. The use of the simulated annealing algorithm to minimize the difference between experimental eigenfrequencies and those obtained by the model represents a novelty in the literature. The importance of the structural modeling in the problem formulation is highlighted. This methodology has been applied to three different steel structures characterized by increasing complexity.

After this brief introduction, the paper is organized as follows: Section 2 presents a general description of the simulated annealing algorithm necessary for the proposed methodology, while the methodologies for structural identification are shown in Section 3. Finally, some conclusive remarks are drawn in Section 4.

2. Simulated Annealing

Initially introduced as a generic heuristic technique for discrete optimization, simulated annealing (SA) has become a widely used tool to tackle optimization problems in a wide range of application areas such as business, medicine, and engineering [13–15].

Also known as a local search algorithm, the SA does not find the optimal solution but provides an approximate solution very close to the optimal one. This is performed through a stochastic and iterative procedure in which the local search starts from an initial current configuration and the choice of the next configuration is randomly generated (randomized exploration).

The methodological process is inspired by the behavior of fluids when subjected to processing involving controlled cooling, as in the production of large crystals. Indeed, if cooling occurs rapidly, the crystal lattice may be affected by defects such as cracking or fracture. The annealing process, on the other hand, involves gradual cooling, bringing the structure of the crystal to an optimal and stable configuration according to the principle of minimum potential energy.

For each temperature value T belonging to a defined range, the solid can reach thermal equilibrium in which the probability of being in a state with energy E is defined by the Boltzmann distribution [28]:

$$Pr\{\mathbf{E} = E\} = \frac{1}{Z(T)} e^{-E/k_B\,T} \tag{1}$$

where $Z(T)$ is a normalization factor (also called partition function), which depends on the temperature T; k_B is the Boltzmann constant; and $e^{-E/k_B\,T}$ is the Boltzmann factor.

The stochastic component of the method lies in the application of the Monte Carlo method to generate successive configurations through small random perturbations starting from a current configuration with energy E_i.

According to the Metropolis Criterion [29], the energy difference:

$$\Delta E = E_j - E_i \tag{2}$$

between the perturbed j-th configuration E_j and the current i-th configuration E_i can be:

$\Delta E \leq 0$, then the j-th configuration replaces the previous one as it has a lower potential energy:

$\Delta E > 0$, then the j-th configuration is accepted with probability $e^{-\Delta E/k_B\,T}$.

The algorithm, therefore, does not exclude a priori the analysis of worst-case solutions but admits them with a decreasing probability as the temperature decreases. Indeed, the exponential function is governed by the relationship between the change in energy of two configurations and the temperature. The probability that the j-th worst solution will be confirmed decreases as ΔE increases. In addition, the temperature, T, is high in the early stages of the algorithm and low in the final stages.

Therefore, since as temperature decreases, the Boltzmann distribution concentrates on the lower energy states, at the end of the cooling process only the lower energy states are likely to occur.

One of the most important features of the SA is the robustness of the algorithm, due to the possibility of easily dealing with nonlinear problems with many variables, despite the presence of strongly discontinuous functions. However, it has some disadvantages:

-   it is not possible to know whether the solution found, which is a local minimum, coincides with the global minimum or how different the two values are;
-   the quality of the local minimum obtained depends on the initial configuration chosen, but no criterion establishes a way of selecting a starting point that allows good solutions to be obtained;
-   it may require very long computational times that cannot be predicted a priori.

Switching from the physical problems to combinatorial optimization problems [30], the energy becomes the cost function C, the temperature is a control parameter c, the particle configurations become the values of the problem variables and the search for the lowest energy state becomes the search for the solution that minimizes the cost function.

In order to ensure that the algorithm performs the calculation cycles, it is necessary to define the initial value of the control parameter c_0, the final value of the control parameter c_f that stops the algorithm, and the decrement law of the control parameter. The choice of c_0 must be made in such a way that, in the initial phase of the algorithm, all configurations must be approved, while on reaching c_f, any deterioration of the solution must not be accepted. Furthermore, the law of variation of the control parameter must be chosen considering that a higher cooling rate corresponds to a higher number of iterations to reach a new equilibrium state: this naturally influences the computational effort. If the control parameter is initially set equal to $c_0 = 0$, the SA operates similarly to a local minimum search algorithm.

A Matlab^{TM} 2013 version of SA is available in the literature [31], which initially requires parameters to be defined. Table 1 shows the input value adopted in this work.

Table 1. Parameters used in the optimization process.

Parameter	Entity
c_0	1
c_f	10^{-8}
k_B	1
N	150
Variation law of c	$c_{k+1} = \left(c_f/c_0\right)^{1/N} \cdot c_k$

3. Structural Identification

3.1. Proposed Method

The structural identification strategy proposed in this work is based on the following steps:

- Execution of operational modal analysis (OMA) focused on the extraction of modal eigenfrequencies from output only experimental data.
- Development of theoretical (analytical and numerical) structural models for dentification purpose.
- Identification of unknown structural parameters $(x_1, x_2, \ldots x_j)$ that minimize the following error function.

$$e = \sqrt{\sum_i^n \left(\frac{f_{is} - f_{in}(x_1, x_2, \ldots x_j)}{f_{is}}\right)^2} \tag{3}$$

where f_{is} is the i-th experimental eigenfrequency and f_{in} is the corresponding theoretical one obtained from a model depending on the set of chosen unknown parameters $(x_1, x_2, \ldots x_j)$. The latter represent properly chosen unknown characteristics of the structure such as structural stiffness, material mechanical property, boundary stiffness, etc. The optimization of Equation (3) can be performed using the SA algorithm. In this way, it is possible to identify the unknown parameters that are the target of the structural identification problem. In this procedure, the experimental eigenfrequencies represent the benchmark necessary for the structural identification strategy.

3.2. Experimental Apparatus

The proposed methodology will be tested using the experimental modal data obtained from steel beams and a steel frame built in the Materials Strength Laboratory of the University of Cagliari.

The experimental apparatus was composed of:

- Accelerometer sensor PCB 393C characterized by a sensitivity of 101.9 mV/(m/s^2) and a frequency range ($\pm 5\%$): 0.025 to 800 Hz, see Figure 1.
- Data acquisition hardware: Dymas 24. It is composed of a central unit with a CPU that has the task of managing the acquisition process, synchronization of the sensors, and data storage. The system is characterized by 30 channels managed by 5 acquisition cards. Each card has an internal memory of 1 GB and is equipped with a DSP (on-board) processor that autonomously manages the functions of digitization and amplification of the inputs and filters the signals. Each card is connected to the CPU by USB connections, see Figure 1.
- Dedicated software: DymaSoft^{TM} 3.6.3 for the connection and configuration of the system and VibroSoft^{TM} 3.2.19 aimed at displaying and processing the recorded data.

(a)

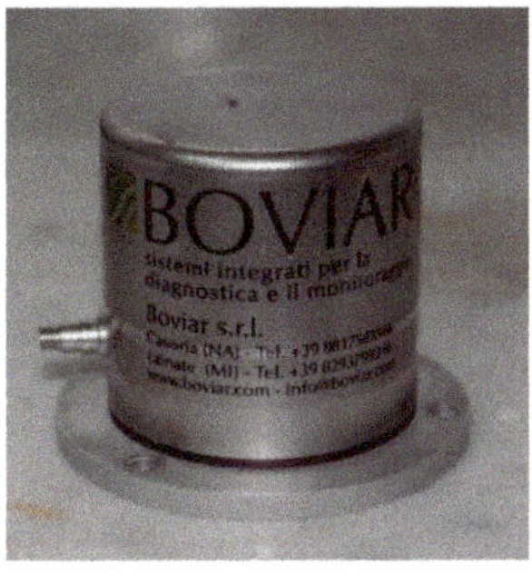

(b)

Figure 1. (**a**) Data acquisition hardware Dymas 24. It is composed of 30 channels (1) managed by 5 acquisition cards (2), an internal battery, (3) and a central unit with CPU that has the task of managing the acquisition process, synchronization of the sensors, and data storage (4); (**b**) PCB 393 C accelerometer.

Given the beam characteristics, in order to have a harmonic force with variable frequencies, we built a home-made vibrodyne, see Figure 2 and Table 2. It was made from a tin container in which a steel rotation axis with an eccentric mass was inserted. A potentiometer was adopted to control the rotation speed. The dynamic forces applied by this system can be expressed as:

$$\vec{F}(t) = M_r \cdot \omega_r^2 \sin\left(\omega_r^2\, t\right) \tag{4}$$

where M_r represents the rotating mass and ω_r^2 is its rotation frequency. The experimental modal analysis tests were developed using the hypothesis of white noise in the loading vibration. For this reason, the vibrodyne was used at different rotation speeds that were randomly varied by the users. This was conducted in order to perform a pure OMA without knowing the dynamic loading characteristics.

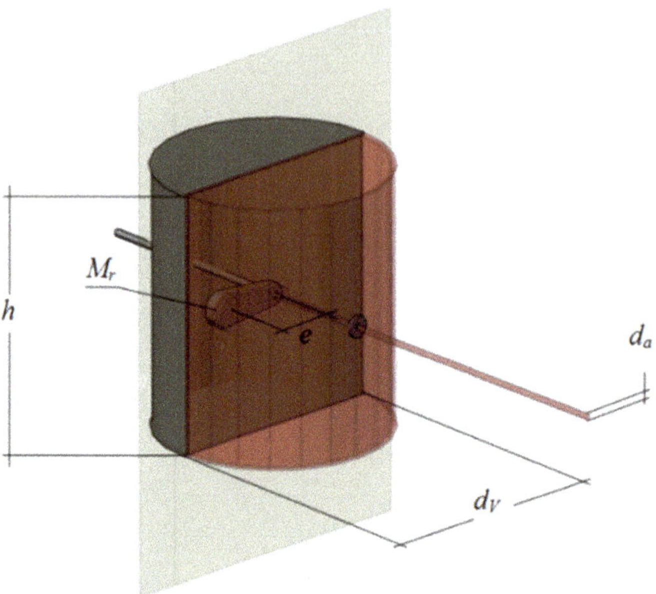

Figure 2. Vibrodyne scheme.

Table 2. Vibrodyne dimensions.

h	120 mm	mm
d_V	110 mm	mm
d_a	3 mm	mm
M_r	50 gr	gr
e	20 mm	mm

The experimental tests were developed considering first a steel beam characterized by different boundary conditions and then a more complex steel frame.

3.3. Steel Beams

The experimental modal analysis was developed while considering two beams with different boundary conditions. In the first case, the beam was simply supported, see Section 3.3.1, and in the second, it was clamped on one side and free on the other to analyze a cantilever beam, see Section 3.3.2.

3.3.1. Simply Supported Beam

The simply supported beam's geometrical characteristics are reported in Figure 3. Its density is 7746.90 kg/m^3, the steel longitudinal elastic modulus $E_{s\text{-}real}$ is 203 GPa, while dimensions are $L = 1499.00$ mm, $b = 118.82$ mm, and $h = 6.39$ mm.

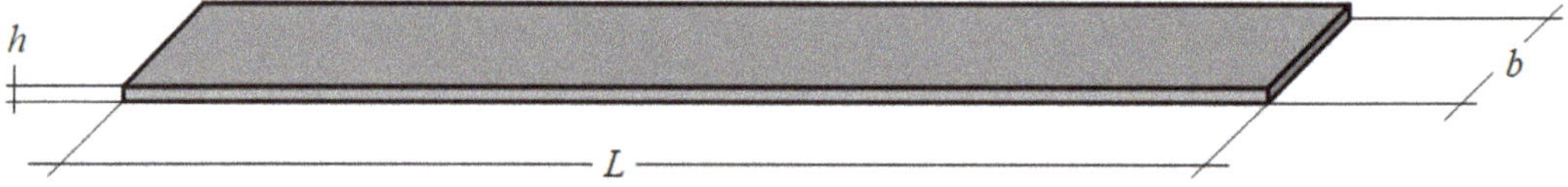

Figure 3. Steel beam geometrical characteristics.

Figure 3 presents the experimental set up. The positions of the vibrodyne and accelerometer were chosen in order to avoid the nodes of the beam eigenmodes, i.e., the sections that do not undergo any displacement during the vibrations in natural modes. The accelerometer is located 435 mm from the right side and symmetrically to the vibrodyne which was placed 435 mm from the left side, see Figure 4.

Figure 4. Simply supported beam experimental test scheme. The vibrodyne is on the left, while the accelerometer is on the right.

The vibrodyne was used by varying randomly the rotation speed. In this way, it was possible to record several samples with different harmonic force. In order to have a statistical consistency, the test was repeated 6 times with a sampling frequency equal to 500 Hz.

Actually, since an OMA approach was adopted, the random rotation speed of the vibrodyne was approximated as a white noise.

The acceleration response time histories were obtained for each case and the first 5 flexural modal frequencies were identified (see Table 3 and Figure 5) using the fast Fourier transform [32] and the peak picking method [5]. As is well-known, see [33], this technique is based on the low damping and well-separated modes hypotheses that can be assumed for this case.

Table 3. Experimental eigenfrequencies of the simply supported beam and the corresponding standard deviation (SD).

Mode	$f_{i,s}$ [Hz]	SD [Hz]
1	6.00	0.03
2	23.23	0.16
3	57.52	0.38
4	100.12	0.18
5	142.93	0.12

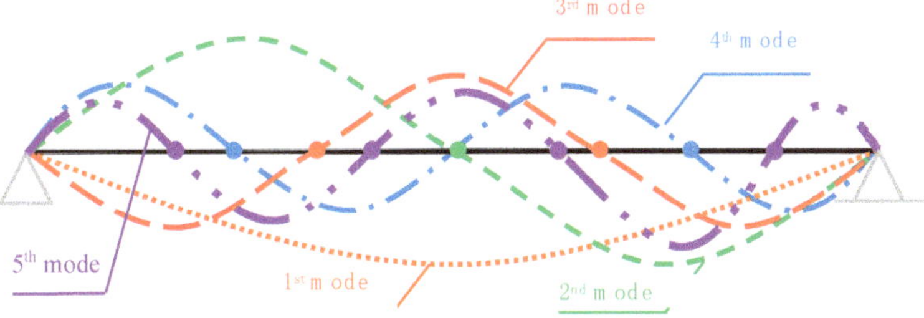

Figure 5. First 5 beam eigenmode shapes.

The matrix of the cross-spectral density (spectral matrix) presents on its diagonal terms the real valued autospectral densities and it is defined as:

$$G(f) = E_{xp}\left[A(f)A^H(f)\right] \tag{5}$$

where $A(f)$ is a vector containing the acceleration responses in the frequency domain and $A^H(f)$ is the complex conjugate transpose matrix, while E_{xp} represents the expected value. In the Peak Picking approach, in the neighbourhood of an eigenfrequency f_r the spectral matrix is approximated by:

$$G(f_r) \approx \alpha_r \Phi_r \Phi_r^H \tag{6}$$

where α_r is a parameter depending on the damping ratio, considering eigenfrequency, excitation spectra, and modal participation factor, see [33,34]. Φ_r is the mode shape vector corresponding to frequency f_r. In this paper, the beam eigenfrequencies were identified from the resonant peak in the autospectral density using the PP approach. The method was quite efficient since the considered eigenmodes were well-detached, as is typical for simple structures such as beams. In case of eigenmodes close to each other, it is possible to filter the accelerometric data in order to improve the accuracy or to use different approaches such as FDD [7], TDD [8], or SSI [9].

In order to perform the structural identification, we developed an analytical beam model based on the Euler–Bernoulli theory with the aim of identifying the longitudinal elastic modulus. It was previously measured with a quasi-static test, so its benchmark value is known.

The Euler–Bernoulli beam theory takes into account bending stiffness and transversal inertia and assumes that plane sections remain plane and perpendicular to the beam axis after deformation, see Figure 6. In case of free vibration, the equation of motion is:

$$\frac{\partial^2 M}{\partial x^2} - \mu \frac{\partial^2 v}{\partial t^2} = 0 \tag{7}$$

where μ represents the mass per unit lenght of the beam, M is the bending moment, v is the deflection, and t is the time.

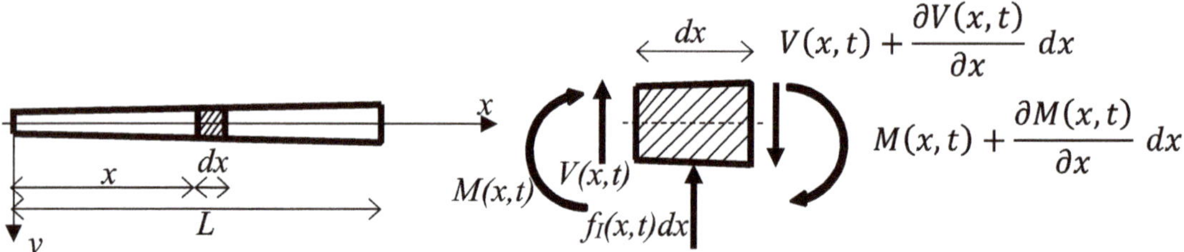

Figure 6. Internal forces acting on an infinitesimal Euler–Bernoulli beam.

Given the mass of the accelerometer (0.89 kg), of the vibrodyne (0.93 kg), and of the beam (8.82 kg), it is necessary to take into account the exact position of these masses also in the analytical model. For this reason, the Euler–Bernoulli beam Equation (7) has been integrated considering the three different fields separated by the two lumped masses of the vibrodyne and of the accelerometer, see Figure 7.

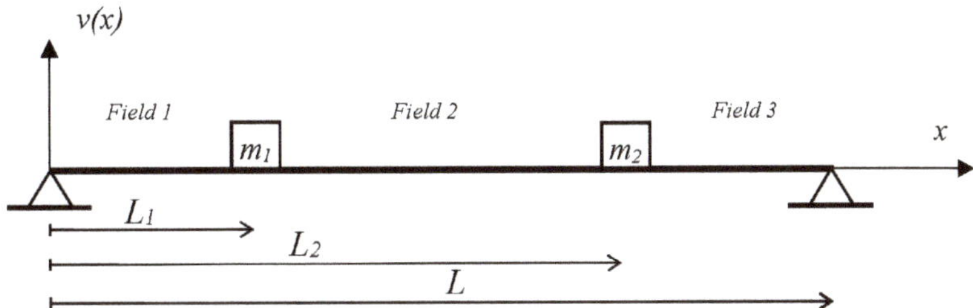

Figure 7. Integration field of the simply supported beam.

In each field, by means of the separation of variables technique, it is possible to distinguish between the time harmonic solution and the space variation of the solution:

$$v\,(x,t) = \sum_{j}^{\infty} v_j(x)\,\sin\left(\omega_j t\right) \tag{8}$$

where t is the time, x represents the position along the beam, v_j is the j-th eigenmode, and ω_j is the i-th natural circular frequency of the beam. The relationship between frequencies and circular frequencies is: $\frac{\omega_j}{2\pi} = f_i n$. Given $\lambda_j = L^4 \frac{\mu\omega_j^2}{EJ}$ and considering the three fields of integration, it is possible to prove (see [35]) that:

$$v_{1j}(x) = A_{1j}\cos\left(\frac{\lambda_j x}{L}\right) + A_{2j}\sin\left(\frac{\lambda_j x}{L}\right) + A_{3j}\cosh\left(\frac{\lambda_j x}{L}\right) + A_{4j}\sinh\left(\frac{\lambda_j x}{L}\right)$$

$$v_{2j}(x) = A_{5j}\cos\left(\frac{\lambda_j x}{L}\right) + A_{6j}\sin\left(\frac{\lambda_j x}{L}\right) + A_{7j}\cosh\left(\frac{\lambda_j x}{L}\right) + A_{8j}\sinh\left(\frac{\lambda_j x}{L}\right) \tag{9}$$

$$v_{3j}(x) = A_{9j}\cos\left(\frac{\lambda_j x}{L}\right) + A_{10j}\sin\left(\frac{\lambda_j x}{L}\right) + A_{11j}\cosh\left(\frac{\lambda_j x}{L}\right) + A_{12j}\sinh\left(\frac{\lambda_j x}{L}\right)$$

where v_{ij} denotes the spatial solution of the i-th field related to the j-th eigenmodes, and A_{ij} represents the generic integration constant that can be determined using the boundary conditions. The 12 boundary conditions expressing static and cinematic compatibility are presented in the following system of equations expressed in matrix notation:

$$D \cdot A = 0 \tag{10}$$

where D is the matrix of the system, and A is the vector containing the unknown integration constant.

In order to avoid the uniqueness of null solution, it is necessary that:

$$\det D = 0 \tag{11}$$

Equation (11) represents the frequency equation whose roots are the circular eigenfrequencies of the beam. Unfortunately, this equation can be solved only with a numerical approach; consequently, it is not possible to find a close form solution. For this reason, an iterative semianalytical algorithm was developed in Matlab^{TM} 2013 [36] that finds the eigenfrequency of the beam using Equation (11).

Thus, now it is possible to use the above mentioned experimental eigenfrequencies to set up an iterative procedure capable of finding the longitudinal elastic modulus value E_s:

1. Select a value of E_s;
2. Calculate theoretical eigenfrequencies using Equation (11);
3. Estimate the quadratic error e of Equation (3) comparing experimental eigenfrequencies and theoretical ones.

The steps are repeated till a minimum value of error e is obtained. Figure 8 presents the trend of error function e that easily points at the optimal value of $E_{s\text{-}opt}$ = 205 GPa. In this way, it was possible to have very little relative error (less than 1%) between the real $E_{s\text{-}real}$ = 203 GPa and the ones that minimize the difference in the eigenfrequencies, see Table 4, confirming the accuracy of the developed approach. Thus, when measuring the experimental eigenfrequencies, it is possible to set up an inverse problem that can allow to determine the unknown mechanical characteristics.

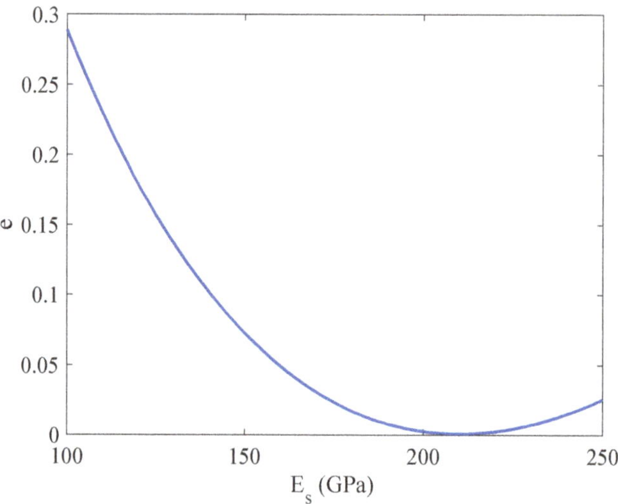

Figure 8. Quadratic error e as a function of the elastic modulus for the simply supported beams.

Table 4. Comparison between experimental eigenfrequencies of the simply supported beam and the corresponding numerical ones with the optimal value of the longitudinal elastic modulus E = 205 GPa.

Mode	$f_{i,s}$ [Hz]	$f_{i,n}$ [Hz]	Δ %
1	6.00	5.91	−1.55
2	23.23	22.48	−3.22
3	57.52	58.15	1.08
4	100.12	102.48	2.36
5	142.93	144.71	1.24

This first application represents a validation for both the modal analysis and the parametric identification. Indeed, it was developed with known eigenfrequencies and material mechanical properties.

3.3.2. Cantilever Beam

The cantilever beam geometrical characteristics are different from those of the simply supported beam. Indeed, its density is 7652.02 kg/m^3, the steel longitudinal elastic modulus E_s is still 203 GPa, while geometrical dimensions are L = 820.00 mm, b = 119.97 mm, and h = 10.08 mm. To create the full constraint, the beam was clamped for a length of 120 mm. Figure 9 presents the experimental set up for the cantilever beam, which shows the positions of the vibrodyne and of the accelerometer on the free side in order to avoid, once again, the nodes of the beam eigenmodes and maximize the vibration amplitude. The vibrodyne was still used while varying randomly the rotation speed (in order to mimic the white noise condition) and the test was repeated 6 times with a sampling frequency equal to 500 Hz.

Figure 9. Cantilever beam experimental tests scheme. The vibrodyne is on the bottom while the accelerometer is on the top side of the free end.

For this case study, the first three flexural modal frequencies were considered. Figure 10 shows the considered beam eigenmode shapes, and Table 5 shows the respective eigenfrequencies measured in the experimental tests.

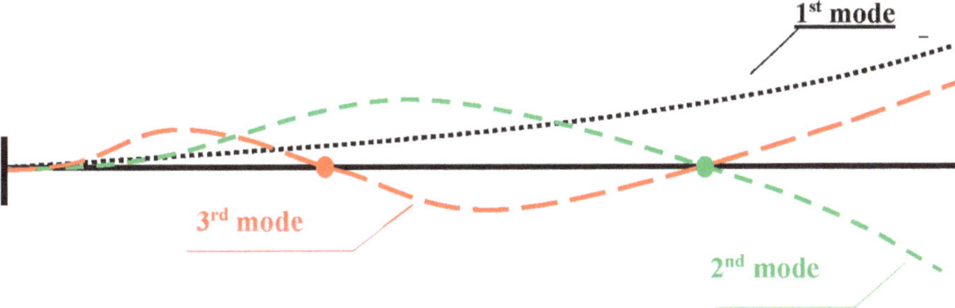

Figure 10. Considered eigenmode shapes for the cantilever beam.

Table 5. Experimental eigenfrequencies of the cantilever beam and the corresponding standard deviation (SD).

Mode	$f_{i,s}$ [Hz]	SD [Hz]
1	11.41	0.17
2	86.68	0.01
3	257.26	0.35

In order to take into account the uncertainties due to the experimental constraint, a flexural spring k_1 and a translational one k_2 were introduced in the analytical model.

Given the position of the mass of the vibrodyne and the accelerometer in this experimental case, the Euler–Bernoulli beam Equation (7) was integrated while considering only one field with a lumped mass on the free end, see Figure 11.

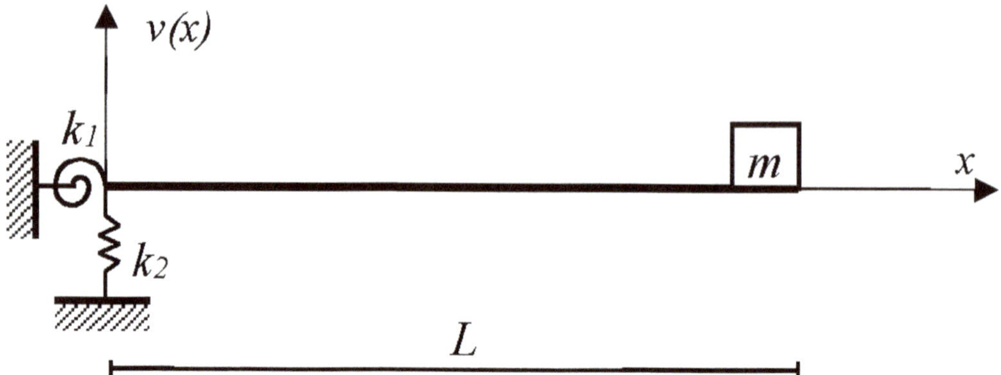

Figure 11. Cantilever beam analytical model.

Following what was performed in Section 3.3.1, the space variation of the solution is expressed by Equation (12):

$$v_j(x) = A_{1j}\cos\left(\frac{\lambda_j x}{l}\right) + A_{2j}\sin\left(\frac{\lambda_j x}{l}\right) + A_{3j}\cosh\left(\frac{\lambda_j x}{l}\right) + A_{4j}\sinh\left(\frac{\lambda_j x}{l}\right) \quad (12)$$

In this case, the four boundary conditions express the static and cinematic compatibility represented by Equation (10) already shown in Section 3.3.1.

Additionally, in this case, the frequency function was obtained by enforcing the singularity of the system matrix in Equation (10) by Equation (11). In this way, by enforcing Equation (11), it is possible to calculate the beam eigenfrequencies as a function of Young's modulus E_s and the constraint stiffnesses k_1 and k_2. Given its strongly implicit form, the roots of Equation (11) must be found with a numerical approach. In this case, again, the optimization procedure is based on minimizing the difference between the model eigenfrequencies (depending on E_s, k_1, k_2) and the experimental ones obtained by OMA. The optimization was performed by means of the simulated annealing algorithm.

For the described case of a cantilever beam, the minimum difference in eigenfrequencies (see Table 6) is reached for a value of the longitudinal elastic modulus equal to $E_s = 201$ GPa and for values of the constraint stiffnesses equal, respectively, to $k_1 = 6.89 \times 10^{15}$ Nm and $k_2 = 9.86 \times 10^{19} \frac{\text{N}}{\text{m}}$.

Table 6. Comparison between experimental eigenfrequencies of the cantilever beam and the corresponding numerical ones with the best value of the longitudinal elastic modulus $E = 201$ GPa.

Mode	$f_{i,s}$ [Hz]	$f_{i,n}$ [Hz]	Δ %
1	11.41	11.62	1.86
2	86.68	85.59	−1.25
3	257.26	255.56	−0.66

In order to check the consistency of the obtained results, different optimizations were developed varying the initial ranges of the unknown parameters.

Table 7 reports the main results showing a good consistency of E_s and k_2, while the values of k_1 presents larger variation.

Table 7. Different variation ranges and optimal values of the unknown parameters for the cantilever beam case.

E_s [GPa] Init. Range	k_1 [Nm] Init. Range	k_2 [N/m] Init. Range	E_s [GPa]	k_1 [Nm]	k_2 [N/m]
$1 \div 210$	$1 \cdot 10^{-2} \div 1 \cdot 10^{20}$	$1 \cdot 10^{-2} \div 1 \cdot 10^{20}$	201.0	6.859×10^{17}	9.720×10^{19}
$10 \div 210$	$1 \cdot 10^{-2} \div 1 \cdot 10^{20}$	$1 \cdot 10^{-2} \div 1 \cdot 10^{20}$	200.9	6.891×10^{15}	9.971×10^{19}
$1 \div 400$	$1 \cdot 10^{-2} \div 1 \cdot 10^{20}$	$1 \cdot 10^{-2} \div 1 \cdot 10^{20}$	201.0	1.213×10^{16}	9.967×10^{19}
$1 \div 300$	$1 \cdot 10^{-2} \div 1 \cdot 10^{20}$	$1 \cdot 10^{-2} \div 1 \cdot 10^{20}$	201.0	6.784×10^{15}	9.700×10^{19}
$200 \div 210$	$1 \cdot 10^{11} \div 1 \cdot 10^{20}$	$1 \cdot 10^{11} \div 1 \cdot 10^{20}$	201.0	$1.398 \cdot 10^{16}$	9.828×10^{19}
$1 \div 400$	$1 \cdot 10^{10} \div 1 \cdot 10^{20}$	$1 \cdot 10^{10} \div 1 \cdot 10^{20}$	201.0	$1.214 \cdot 10^{16}$	9.487×10^{19}

In addition, it is interesting to analyze the performance of the method when less eigenfrequencies are considered in the target function. For this reason, the optimization was performed considering just the first eigenfrequency, the first two eigenfrequencies, or all three eigenfrequencies.

Table 8 reports the results of this analysis using as initial variation ranges: $1 \div 400$ GPa for E_s, $1 \cdot 10^{10} \div 1 \cdot 10^{2}$ Nm for k_1 and $1 \cdot 10^{10} \div 1 \cdot 10^{20}$ N/m for k_2. Looking at Table 8, it is quite clear how for the specific problem it is possible to reach a good estimation of the parameters while also considering just the first eigenmode. The information added from the second and third eigenfrequencies does not significantly change the solution.

Table 8. Different variation ranges and optimal values of the unknown parameters for the cantilever beam case. Optimal solution with different number of eigenfrequencies considered in the target function.

Considered Eigenfrequencies	E_s [GPa]	k_1 [Nm]	k_2 [N/m]
1	201.0	6.893×10^{15}	9.781×10^{19}
1—2	201.0	1.662×10^{15}	9.783×10^{19}
1–2–3	201.0	1.213×10^{16}	9.967×10^{19}

3.3.3. Comparison to Other Optimization Algorithms

In order to test the efficiency of the proposed methodology based on the SA optimization algorithm, the same cantilever beam case was analyzed using two different heuristic algorithms: ant colony [19] and particle swarm [20].

The ant colony (AC) algorithm is a type of swarm intelligence algorithm inspired by the behavior of ant colonies. It is used to find the shortest path between two points in a graph. The algorithm simulates the behavior of ants as they search for food. In fact, each ant drops a pheromone trail as it traverses the graph, so that other ants are more likely to follow the paths with stronger pheromone trails. Over time, the pheromone trails will converge on the shortest path. The algorithm also includes a pheromone evaporation mechanism to prevent the trails from becoming too strong.

The particle swarm optimization (PS) algorithm is a type of optimization algorithm inspired by the behavior of bird flocks and fish schools. It is used to find the optimal solution of a problem by simulating the behavior of a group of particles, where each represents a possible solution. The particles move in the search space, guided by their current position and the best position encountered so far by any particle in the group (global best) and by the best position encountered by that particular particle (personal best). The movement of the particles is, therefore, determined by a combination of their velocity and acceleration, which are updated at each iteration based on the current global and individual best positions.

These algorithms have been implemented in Matlab^{TM} 2013 [36] and applied to the cantilever beam case in the model updating phase of the identification strategy, i.e., in the minimization of the difference between the model eigenfrequencies (depending on E_s,

k_1, k_2) and the experimental ones obtained by OMA. This difference represents the target function to the minimization of e, see Equation (3).

All the algorithms were set up to use the following variables' initial variation ranges: $1 \div 4 \cdot 10^2$ GPa for E_s, $1 \cdot 10^{-2} \div 1 \cdot 10^{-2}$ Nm for k_1 and $1 \cdot 10^{-2} \div 1 \cdot 10^{20}$ N/m for k_2.

The results of the optimization are shown in Table 9 where the error function value e, the elastic modulus E_s, the flexural spring k_1, and the translational one k_2 values are reported.

Table 9. Performance of different optimization algorithms for the cantilever beam case.

Algorithm	e	E_s [GPa]	k_1 [Nm]	k_2 [N/m]
Simulated Annealing (SA)	0.023352	201.0	6.783×10^{15}	9.867×10^{19}
Ant Colony (AC)	0.028223	204.7	5.372×10^{19}	7.082×10^{19}
Particle Swarm (PS)	0.023353	201.0	4.451×10^{19}	5.748×10^{19}

Looking at Table 9, it is clear that the minimum value of the error function was reached by the SA algorithm, while similar values of the elastic modulus E_s and of translational spring k_2 were obtained by the three algorithms. However, largest differences have been found in the values of flexural spring k_1.

Looking at these results, the last case study presented in Section 3.4 was analyzed using the SA algorithm.

3.4. Steel Frame

The third experimental case is focused on a steel loading frame (Figure 12) located in the Materials Strength Laboratory of the University of Cagliari.

The frame is composed of two columns fully constrained at the base, connected by a cross beam hinged on the above-mentioned columns at adjustable heights. The latter connection is obtained with a pin, while at the columns base, welded and bolted connections ensure a full constraint for displacements, though small rotations are allowed. Thus, the rotation constraint can be represented by a set of flexural springs. The unknown parameters that are necessary for structural identification with the proposed methodology are the constraint rotational stiffnesses at the column bases.

Figure 12. *Cont.*

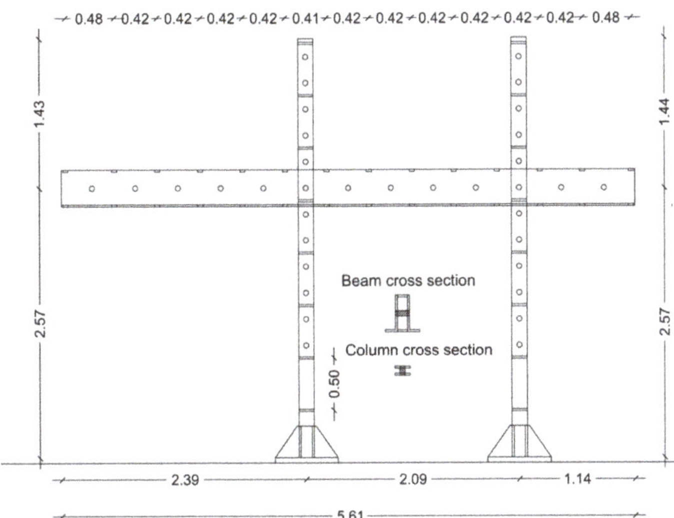

Figure 12. Steel frame picture (**top**) and geometrical dimensions (**bottom**), measures are in m.

All structural components are made of S275 steel grade, according to UNI EN 10025 [37], except for the pins, which are made of S355 steel grade. Each column is made of two 25×200 mm holed plates 40 mm apart, joined by 40 25×45 mm welded battens 250 mm apart and has a total height of 4965 mm. Fifteen holes with a diameter of 50 mm along each column allow the crossbar to be placed variably from a minimum height of 1105 mm to a maximum of 4605 mm. The crossbar consists of two L-shaped welded profiles made of two plates: 120×25 mm for the vertical leg and 325×25 mm for the horizontal one. The two angles are, thus, coupled by means of welded battens 420 mm apart. The beam has a total length of 5910 mm and can be employed at the maximum span of 5040 mm.

To determine the eigenfrequencies, five accelerometers were installed in the frame, three of which were mounted at one end of the beam and two at the top of the columns (Figure 13). The OMA yields the first 3 eigenfrequencies of the frame, once again determined by the peak picking method (see Table 10).

Figure 13. Detail of the positioning of the accelerometers at the end of the beam.

Table 10. Experimental eigenfrequencies of the steel frame.

Mode	$f_{i,s}$ [Hz]
1	1.97
2	3.25
3	4.23

In this case, the frame model has been developed with finite element analysis (FE). In fact, the presented method is suitable for the update of a finite element model, aimed at the minimization of the difference between real modal parameters (e.g., eigenfrequencies) and the numerical ones. In this application, the stiffness k_{ij} of the springs representing the constraint conditions at the base of the column was chosen as the unknown parameter of the optimization process. For this purpose, a proper Fortran code was developed to call the finite element analysis execution for each iteration of the optimization procedure, along with the execution of the simulated annealing algorithm employing the numerical results.

The 3D finite element model was developed using the commercial software Strand7 (see Figure 14). For the modeling of the columns and the crossbeam, 15,948 four-node bilinear isoparametric plate elements were used. Mutual and base connections of steel members were modeled by means of multipoint constraints employing a total of 234 link elements. In detail, the pin connections were modeled as hinged internal constraints involving translational rigid links, while the base column connections were carried out by constraining the base section to a master node where flexural base springs were located. The examination of structural details suggested that no translations were likely to occur at the column bases, thus no translational base springs were taken into account. For the same reason, no torsional springs were accounted for. In both cases, the degrees of freedom were rigidly restrained. The total number of model nodes is 17,702.

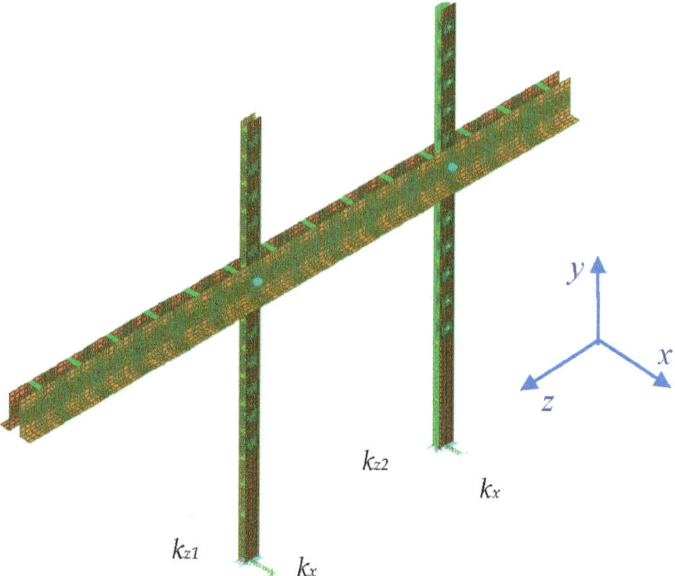

Figure 14. Finite element model of the steel frame in the configuration adopted during the tests.

This modeling strategy represents the best compromise between accuracy of results and computational efforts. The modulus of elasticity adopted for the material is $E_s = 210{,}000$ MPa and the specific weight is 7870 kg/m^3.

Therefore, the optimization process was implemented with the aim of calibrating the stiffness of the flexural base springs, only one around x-axis k_x and the two flexural stiffnesses k_{z1} and k_{z2} around the z-axis, pertaining to the base master nodes. The choice of having two different flexural springs (k_{z1} and k_{z2}) for rotations around the z-axes of each column and only one k_x stiffness for the rotations around x-axis, stemmed from evaluation of structural details and their effects on the measured eigenfrequencies.

The algorithm performed 10,018 iterations (Figures 15 and 16), identifying at the iteration number 6047 the parameters that minimize the difference between the experimental and model eigenfrequencies, see Equation (3).

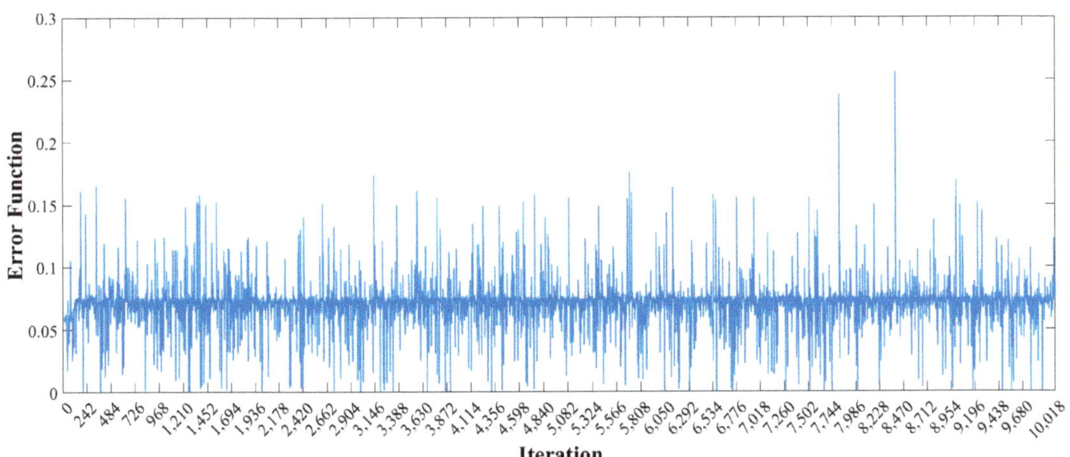

Figure 15. Trend of the error function during the 10,018 iterations performed by the simulated annealing optimization algorithm varying k_x, k_{z1}, and k_{z2} parameters.

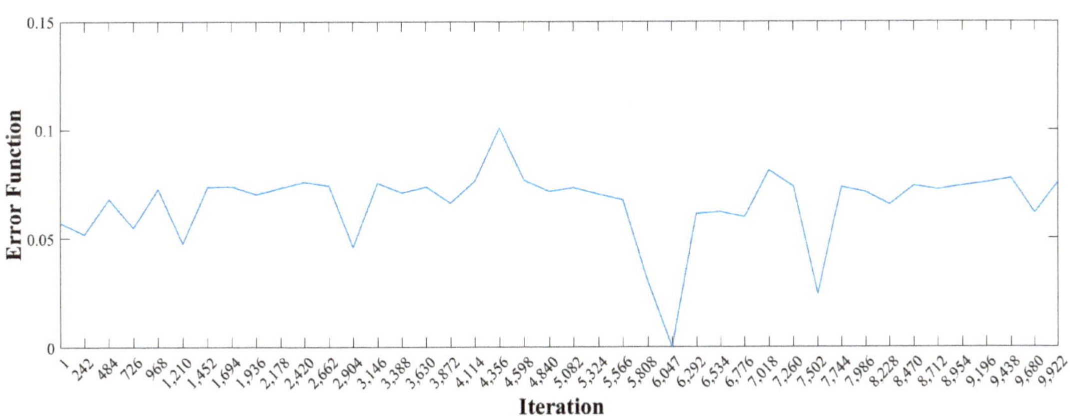

Figure 16. Detail of the trend of the error function showing that the minimum was reached at iteration number 6047.

Table 11 shows the values of k_x, k_{z1}, and k_{z2} that make the error function equal to 4.83×10^{-5}, while the comparison between the experimental frequencies and the frequencies determined with the calibrated model is shown in Table 12.

Table 11. Values of k_x, k_{z1}, and k_{z2} that minimize the error function.

Flexural Stiffness	Value [kNmm/rad]
k_x	7.93×10^7
k_{z1}	1.63×10^7
k_{z2}	1.54×10^6

Table 12. Comparison between experimental eigenfrequencies of the steel frame and the corresponding numerical ones with the best value of the k_x, k_{z1}, and k_{z2} [kNmm/rad].

Mode	$f_{i,s}$ [Hz]	$f_{i,n}$ [Hz]	Δ %
1	1.97	1.96	0.62
2	3.25	3.25	−0.05
3	4.23	4.24	−0.30

4. Discussion and Conclusions

This paper presented an innovative methodology for structural identification using OMA by the minimization of the difference between experimental and theoretical modal parameters, specifically the eigenfrequencies, using simulated annealing. Properly chosen unknown structural parameters are the variables of the minimization procedure that, by means of analytical or numerical models, allows to improve the knowledge of the structure with negligible damage to the structure itself and to its service. Indeed, OMA can be performed without any service interruption and allows to exploit environmental vibrations and service dynamic loads to evaluate experimental eigenfrequencies. The critical part of the methodology lies in the modeling and in the model optimization procedure used to tune the unknown parameters.

The method has been validated considering three new experimental cases: two small-scale steel beams and one full-scale steel frame. In the case of the simply supported beam, it was possible to assess the beam steel's elastic modulus, while in the case of the cantilever beam, some constraint characteristics were evaluated as well. In the steel frame case, the focus was on the constraint conditions of the structure determining the flexural stiffness of the springs representing the column base constraints.

Section 3.3.3 presented a comparison between three different heuristic optimization algorithms: simulated annealing, ant colony, and particle swarm. Each algorithm was used to minimize the difference between the model eigenfrequencies (depending on structural mechanical parameters) and the experimental ones obtained using OMA. For the considered case, the SA algorithm obtained the best performance but good results were also obtained with the other two.

In the beam cases, the computational cost was very small and the code could yield the optimal solution in few seconds. A completely different situation was faced for the frame case. Indeed, each run of the finite element model required about 15 s, which needed to be repeated tens of thousands of times for the complete process, resulting in almost 2 days of computational time in a common PC equipped with 8 Gb of Ram and an intel i5 processor.

Another important aspect is the experimental eigenfrequencies identification. It is of paramount relevance to acquire and measure the highest number of eigenmodes and eigenfrequencies to improve the benchmark data necessary for parameter identification. In the considered cases, the eigenmodes of the beams were clearly detached and also the first modes of the frame were quite distant. This approach can be already developed in the case of close eigenmodes using more advanced techniques to analyze the accelerometer's data.

In addition, it is not possible to find general relationships between the number of unknown parameters and the number of known eigenfrequencies considered in the target function that is always valid. Clearly, it is always better to have the largest amount of available information in the target function. The authors are persuaded that criticalities should be regarded in the general frame of ill-conditioned or rank deficient linear systems through spatial parameter estimation, see for example [38,39].

Finally, it is important to underline that the proposed case-studies are just examples for a general methodology that can be applied to larger structures where the values of elastic constants or constraint stiffness are not known or cannot be measured with simpler techniques. In particular, the estimation of constraint stiffness can be very important for existing steel structures where aging effects can modify the starting boundary conditions.

For this reason, further developments of this methodology are expected considering both damping property estimation and larger structures such as bridges or buildings that can represent interesting applications for this family of problems. The measurements of the modal parameters can be developed using also innovative techniques such as sound pressure as an excitation source and a laser doppler vibrometer as a sensor [10,11].

Author Contributions: Conceptualization, F.S. and F.M.; methodology, F.S., D.M. and F.M.; software, F.S., D.M., M.S. and A.N.; validation, F.S. and F.M.; investigation, F.S., M.S. and A.N.; resources, F.M.; data curation, M.S. and A.N.; writing—original draft preparation, F.S. and A.A.; writing—review and editing, F.S., A.A., M.S., A.N., D.M. and F.M.; funding acquisition, F.M. All authors have read and agreed to the published version of the manuscript.

Funding: This research was funded by MUR, the Italian Ministry of University and Research (Grant Number PRIN 2020; Project 2020CLBMYL, "Smart Monitoring for Safety of Existing Structures and infrastructures (S-MoSES)"); such support is gratefully acknowledged.

Data Availability Statement: Data will be available upon request.

Conflicts of Interest: The authors declare no conflict of interest.

References

1. Stochino, F.; Fadda, M.L.; Mistretta, F. Low cost condition assessment method for existing RC bridges. *Eng. Fail. Anal.* **2018**, *86*, 56–71. [CrossRef]
2. Stochino, F.; Fadda, M.L.; Mistretta, F. Assessment of RC Bridges integrity by means of low-cost investigations. *Frat. Ed Integrità Strutt.* **2018**, *46*, 216–225. [CrossRef]
3. Franchi, A.; Napoli, P.; Crespi, P.; Giordano, N.; Zucca, M. Unloading and Reloading Process for the Earthquake Damage Repair of Ancient Masonry Columns: The Case of the Basilica di Collemaggio. *Int. J. Archit. Herit.* **2022**, *16*, 1683–1698. [CrossRef]
4. Zahid, F.B.; Ong, Z.C.; Khoo, S.Y. A review of operational modal analysis techniques for in-service modal identification. *J. Braz. Soc. Mech. Sci. Eng.* **2020**, *42*, 1–18. [CrossRef]
5. Bendat, J.; Piersol, A. *Engineering Applications of Correlation and Spectral Analysis*; Wiley: New York, NY, USA, 1993.
6. Felber, A.J. *Development of a Hybrid Bridge Evaluation System*; University of British Columbia: Vancouver, BC, Canada, 1994.
7. Brincker, R.; Zhang, L.; Andersen, P. Modal identification from ambient responses using frequency domain decomposition. In Proceedings of the 18th International Modal Analysis Conference (IMAC), San Antonio, TX, USA, 7–10 February 2000.
8. Kim, B.H.; Stubbs, N.; Park, T. A new method to extract modal parameters using output-only responses. *J. Sound Vib.* **2005**, *282*, 215–230. [CrossRef]
9. De Moor, B.; Van Overschee, P.; Suykens, J. Subspace algorithms for system identification and stochastic realization. In Proceedings of the MTNS, Kobe, Japan, 17–21 June 1991.
10. Scislo, L.; Guinchard, M. Non-invasive measurements of ultra-lightweight composite materials using Laser Doppler Vibrometry system. In Proceedings of the 26th International Congress on Sound and Vibration, ICSV, Montreal, QC, Canada, 7–11 July 2019.
11. Scislo, L. Quality Assurance and Control of Steel Blade Production Using Full Non-Contact Frequency Response Analysis and 3D Laser Doppler Scanning Vibrometry System. In Proceedings of the 11th IEEE International Conference on Intelligent Data Acquisition and Advanced Computing Systems: Technology and Applications, Cracow, Poland, 22–25 September 2021.
12. Yuan, K.; Zhu, W.D. Estimation of modal parameters of a beam under random excitation using a novel 3D continuously scanning laser Doppler vibrometer system and an extended demodulation method. *Mech. Syst. Signal Process.* **2021**, *155*, 107606. [CrossRef]
13. Kirkpatrick, S.G.C.; Al, E. Optimization by Simulated Annealing. *Science* **1983**, *1*, 671–680. [CrossRef]
14. Nikolaev, A.G.; Jacobson, S.H. *Simulated Annealing*; Gendreau, M., Potvin, J.Y., Eds.; Handbook of Metaheuristics; International Series in Operations Research & Management Science: New York, NY, USA, 2010; Volume 146.
15. Bertsimas, D.; Tsitsiklis, J. Simulated Annealing. *Statist. Sci.* **1993**, *8*, 10–15. [CrossRef]
16. Jenkins, W.M. Towards structural optimization via the genetic algorithm. *Comput. Struct.* **1991**, *40*, 1321–1327. [CrossRef]
17. Mei, L.; Wang, Q. Structural optimization in civil engineering: A literature review. *Buildings* **2021**, *11*, 66. [CrossRef]
18. Greco, R.; Vanzi, I. New few parameters differential evolution algorithm with application to structural identification. *J. Traffic Transp. Eng.* **2019**, *6*, 1–14. [CrossRef]

19. Angelo, J.S.; Bernardino, H.S.; Barbosa, H.J. Ant colony approaches for multiobjective structural optimization problems with a cardinality constraint. *Adv. Eng. Softw.* **2015**, *80*, 101–115. [CrossRef]
20. Poli, R.; Kennedy, J.; Blackwell, T. Particle swarm optimization. *Swarm Intell.* **2007**, *1*, 33–57. [CrossRef]
21. Sivori, D.; Lepidi, M.; Cattari, S. Structural identification of the dynamic behavior of floor diaphragms in existing buildings. *Smart Struct. Syst.* **2021**, *27*, 173–191.
22. Gesualdo, A.; Fortunato, A.; Penta, F.; Monaco, M. Structural identification of tall buildings: A reinforced concrete structure as a case study. *Case Stud. Constr. Mater.* **2021**, *15*, e00701. [CrossRef]
23. Lacanna, G.; Betti, M.; Ripepe, M.; Bartoli, G. Dynamic identification as a tool to constrain numerical models for structural analysis of historical buildings. *Front. Built Environ.* **2020**, *6*, 40. [CrossRef]
24. Şahin, A.; Bayraktar, A. Forced-Vibration Testing and Experimental Modal Analysis of a Steel Footbridge for Structural Identification. *J. Test. Eval.* **2014**, *42*, 695–712. [CrossRef]
25. Ren, W.X.; Zhao, T.; Harik, I.E. Experimental and analytical modal analysis of steel arch bridge. *J. Struct. Eng.* **2004**, *130*, 1022–1031. [CrossRef]
26. Rinaldi, C.; Ciambella, J.; Gattulli, V. Image-based operational modal analysis and damage detection validated in an instrumented small-scale steel frame structure. *Mech. Syst. Signal Process.* **2022**, *168*, 108640. [CrossRef]
27. Mousavi, A.A.; Zhang, C.; Masri, S.F.; Gholipour, G. Structural damage localization and quantification based on a CEEMDAN Hilbert transform neural network approach: A model steel truss bridge case study. *Sensors* **2020**, *20*, 1271. [CrossRef]
28. van Laarhoven, P.J.M.; Aarts, E.H.L. *Simulated Annealing, Simulated Annealing: Theory and Applications*; Mathematics and Its Applications, Springer: Dordrecht, The Netherlands, 1987; Volume 37.
29. Metropolis, N.R.A.; Al, E. Equation of state calculations by fast computing machines. *J. Chem. Phys.* **1953**, *21*, 1087–1092. [CrossRef]
30. Stochino, F.; Lopez Gayarre, F. Reinforced Concrete Slab Optimization with Simulated Annealing. *Appl. Sci.* **2019**, *9*, 3161. [CrossRef]
31. Oldenhuis, R.P. Trajectory Optimization for a Mission to the Solar Bow Shock and Minor Planets. Master's Thesis, University of Delft, Delft, The Netherlands, 2010.
32. Welch, P. The use of fast Fourier transform for the estimation of power spectra: A method based on time averaging over short, modified periodograms. *IEEE Trans. Audio Electroacoust.* **1967**, *15*, 70–73. [CrossRef]
33. Gentile, C.; Saisi, A. Ambient vibration testing of historic masonry towers for structural identification and damage assessment. *Constr. Build. Mater.* **2007**, *21*, 1311–1321. [CrossRef]
34. Levin, R.I.; Lieven, N.A.J. Dynamic finite element model updating using simulated annealing and genetic algorithms. *Mech. Syst. Signal Process.* **1998**, *12*, 91–120. [CrossRef]
35. Frýba, L. *Dynamics of Railway Bridges*; Thomas Telford Ltd.: London, UK, 1996.
36. *Math Works*; MATLAB 2013.a Documentation: Natick, MA, USA, 2013.
37. *CEN/TC 459/SC 3*; 10025-Hot Rolled Products of Structural Steels Hot rolled Products of Structural Steels—Part 2: Technical Delivery Conditions for Non-Alloy Structural Steels. European Committee for Standardisation: Brussels, Belgium, 2019.
38. Mottershead, J.E.; Foster, C.D. On the treatment of ill-conditioning in spatial parameter estimation from measured vibration data. *Mech. Syst. Signal Process.* **1991**, *5*, 139–154. [CrossRef]
39. Mottershead, J.E.; Friswell, M.I. Model updating in structural dynamics: A survey. *J. Sound Vib.* **1993**, *167*, 347–375. [CrossRef]

Article

Assessment of Web Panel Zone in Built-up Box Columns Subjected to Bidirectional Cyclic Loads

Ramón Mata [1], Eduardo Nuñez [1,*], Frank Sanhueza [1], Nelson Maureira [1] and Ángel Roco [2,*]

[1] Department of Civil Engineering, Universidad Católica de La Santísima Concepción, Concepción 4090541, Chile
[2] Facultad de Salud y Ciencias Sociales, Universidad de las Américas, Providencia, Santiago 7500975, Chile
* Correspondence: enunez@ucsc.cl (E.N.); aroco@udla.cl (Á.R.)

Abstract: The behavior of the web panel zone has a direct effect on the cyclic performance of steel moment connections. While the mechanisms of web panel zone failure are known under cyclic load, little is known about the behavior of the web panel zone under bidirectional loads in bolted connections. Using experimental tests and calibrated numerical models, this research evaluated the web panel zone behavior under unidirectional and bidirectional cyclic loads. The results showed that bidirectional load can modify the stress and strain distribution in the web panel zone. Moreover, the increasing of the width-to-thickness ratio of the column influences the failure mechanism of the joint configuration and increases the plastic incursion in the column. These data demonstrate that bidirectional effects improve the web panel zone performance under cyclic loads.

Keywords: web panel zone; cyclic behavior; tubular columns; finite elements; bidirectional load

Citation: Mata, R.; Nuñez, E.; Sanhueza, F.; Maureira, N.; Roco, Á. Assessment of Web Panel Zone in Built-up Box Columns Subjected to Bidirectional Cyclic Loads. *Buildings* **2023**, *13*, 71. https://doi.org/10.3390/buildings13010071

Academic Editors: Chiara Bedon, Flavio Stochino, Mislav Stepinac and Binsheng (Ben) Zhang

Received: 21 October 2022
Revised: 12 December 2022
Accepted: 15 December 2022
Published: 28 December 2022

1. Introduction

The seismic provisions in [1] establish requirements for the use of moment-resisting frames in high-risk seismic zones to avoid brittle failure mechanisms. Within these requirements, the use of prequalified moment connections according to [2] is necessary to guarantee the dissipation of energy from the structure because of the inelastic bending incursion of the beam. All these connections are designed to avoid damage to the column, the welding fracture or shear yielding in the web panel zone (WPZ). If the WPZ is weak in shear, the joint will be a weak link with plastic deformations, while the beams will not develop their flexural resistance under cyclic loading, reducing the strength and stiffness of the structure significantly [3].

The investigation conducted by [4] assessed the deformation demands on the WPZ of steel moment frames through a parametric analysis using numerical models and design requirements for AISC 360 [5], FEMA-355D [6], and Eurocode 3 [7]. The results found that the research in [6] provides lower capacity and less inelastic incursion than [5,7], which present similar capacities. Similarly, the research developed in [8] proposed a numerical model for beam-to-column connections with the inelastic behavior of the WPZ in moment frames being verified through nonlinear static and dynamic analyses. The results show an increase in inelastic response and energy dissipation capacity with lower maximum structure strength without indication of a soft-story collapse mechanism in the range of deformation under 15 times the panel yield deformation. Nevertheless, this research considers a uniaxial bending effect on the panel zone and higher levels of inelastic deformation than the studies.

On the other hand, [9] shows the importance of the WPZ strength in slotted-web beam connections through a combined numerical and experimental study. The results show that high participation of the panel zone effect causes weld fracture in the beam web and decreases the connection. In this sense, several studies [10–13] have evaluated the strength and performance of the panel zone, noting that both short and long web panels have substantial post-buckling capacities. Furthermore, when the panel zone is subjected to extremely large inelastic demands, this could cause a potential material fracture and thus brittle failure in the column. Further research developed in [14] analyzed the behavior of corrugated web panel zones, obtaining a limited influence of the corrugated forms in the connection shear strength. In addition, the corrugated panels achieved a residual strength of 50% the buckling shear strength; therefore, the use of residual capacity in the design to prevent brittle failure was recommended.

To improve the behavior of the panel zone, intermediate and panel stiffeners are attached to the plate girder web panel to increase the shear strength or reduce the panel aspect ratio to limit the local buckling. The research conducted in [15] proposed a design procedure for intermediate transverse stiffeners through a set of design equations to predict the post-buckling shear strength, emphasizing the necessity of transverse stiffeners specially when the panel aspect ratio is greater than 1.0. On the other hand, [16] experimentally evaluated the effect of these stiffeners on the WPZ shear stability on cruciform beam-to-column connections. The results showed a decrease of the shear buckling in the WPZ and hysteresis behavior improved.

Based on these results, several steel moment connections have been proposed, including stiffeners in their configurations to improve the panel zone performance in investigations developed in [17–24]. Specifically, in the research conducted in [25], different joint assemblies of a new welded moment connection with horizontal stiffeners and tubular columns were performed to estimate their bidirectional cyclic behavior. An important reduction in terms of shear and bending was reached in the column. Moreover, higher deformation of the WPZ was correctly distributed by the internal diaphragm. Likewise, the study developed in [26] assessed the WPZ shear strength in cruciform columns and box columns with continuity plates using the finite element method (FEM). The results showed that cruciform sections have similar plastic capacities to box sections and more shear strength in the WPZ.

More specifically, the experimental research conducted in [27] studied steel tube panels under lateral cyclic load and focused on the evaluation of stiffness and dissipated energy on each cycle of loading. The tubes were tested as boundary elements of the shear wall had local buckling close to the base due to high compression forces and fracture near the top close to the beam. In addition, the behavior of the tubes was similar to a cantilever beam with a hysteretic behavior that showed a slight pinching without stiffness and strength degradations. Furthermore, the research conducted in [28] evaluated experimentally, numerically, and analytically the shear behavior of the panel zone in connection to hollow structural section columns (HSS) in the plane. The results reveal that for this connection, the main failure mode was a shear failure in the plane zone where shear deformation and local buckling occur. The hysteretic behavior exhibited good ductility and stability, showing that an increasing thickness of the tube increases the ultimate shear resistance.

However, this research does not evaluate the behavior under bidirectional cyclic behavior, which is a limitation in assessing the behavior of 3D joints. In this sense, the research conducted in [23] studies the seismic behavior of steel beam-column joint connections with outer stiffeners and welding connections under bidirectional cyclic loads. The prototypes were HSS columns with welded connections to annular stiffeners and beams, which presented inward buckling of the column and fracture of the ring stiffener during the test. In addition, the bidirectional cyclic load reduced the strength connection capacity compared to unidirectional loading. Finally, a stress concentration occurs along the diagonal line of HSS columns in 3D joints with a possibility of brittle failure if the outer annular stiffener is not present. Nevertheless, this research did not contemplate the use of a bolted connection to prevent brittle failure as a fracture or local buckling in the column.

On the other hand, as the number of prequalified connections is limited and most of them are used for wide-flange columns, there was a need for a new proposal for tubular connections. In addition, the implementation of tubular columns brings some limitations due to their commercial availability, which commonly does not have sufficient thickness to comply with seismic limitations for use in special and intermediate moment frames. One of these proposals is the connection developed in [22], which demonstrates experimental and numerical compliance with the requirements of [1,2] without the presence of brittle failures. However, this study did not evaluate the behavior of the proposed connection under bidirectional loads. In this sense, two numerical studies, [29,30], allow us to evaluate the behavior of the mentioned connection under bidirectional cyclic loads and axial loading for five connection configurations. The results of these investigations showed that the axial load was not critical for the connection and the damage was limited to beams in the joint configurations subjected to bidirectional cyclic loads. Nevertheless, these studies were limited in terms of configurations and the sizes of the specimens; therefore, propagation of the results to other configurations is not recommended.

In this context, an extensive parametric study was developed by [31] analyzing the effects of different column sizes, the number of beams, clear span-to-depth beam ratio (L/d), and axial load parameters on cyclic behavior using numerical models in FE. The results showed that all configurations analyzed conformed to the criteria established in [1], even for L/d ratios in the range between 7–20. In addition, the assemblies designed with low axial load are controlled by the design of the WPZ shear, while high levels are controlled by the strong column–weak beam relationship.

All these studies ensure that the connection proposed in [22] has a successful performance for every configuration even under bidirectional loads. However, these numerical studies calibrated their models from unidirectional cyclic tests and do not assess the web panel zone behavior in the column, which can control the joint design.

The aim of this research is the assess the cyclic behavior of the web panel zone in built-up box columns using moment connection subjected to bidirectional load. In this sense, the cyclic response is characterized through the full-scale experimental tests and extended to different sizes using numerical models calibrated from experimental data. Two experimental tests of the joints subjected to unidirectional plane and bidirectional cyclic loads are performed to compare the cyclic behavior. In addition, six numerical models were validated and calibrated with the results of the experimental program, allowing to extrapolate the study to other scenarios untested experimentally. Finally, conclusions are presented in reference to the web panel zone shear behavior and its influence on the hysteretic responses of the joints, which are not commonly evaluated using bidirectional moment connections.

2. Experimental Study

2.1. Test Specimens

Experimental tests on two specimens subjected to pseudo-static cyclic loading were performed. In Figure 1, the setup of specimens and joint assembly tested are shown. The assembly of specimens are based on a prequalification test according to Chapter K [1], applying the cyclic load to capture the bidirectional effect at the top of the column according to [31]. The joint studied consisted of I-beams connected to the built-up box column through the end-plate moment connection proposed in [22], using outer stiffeners and A325 bolts. To consider the bidirectional effect, interior joint configuration with two beams in the plane (2BI) and exterior joint configuration with two beams in corner position (2BC) were tested. Beams and columns satisfy the requirements established in [1] for high-ductility members. The beams and columns were designed to comply with the design requirements of [1,2], resulting in a 220 mm × 220 mm × 14 mm square built-up box column and IPE-200 beam. The end-plate thickness, bolts, and outer stiffener are shown in Figure 1. A bolt pretension was applied in bolts using a calibrated torque wrench to achieve a 70% load pretension according to [5].

To prevent displacement outside of the plane loaded, lateral restrictions are imposed at the center of the beams and the top of the column. In this study, the length of the column used is 2562 mm, while the length of the beam was estimated at 1315 mm.

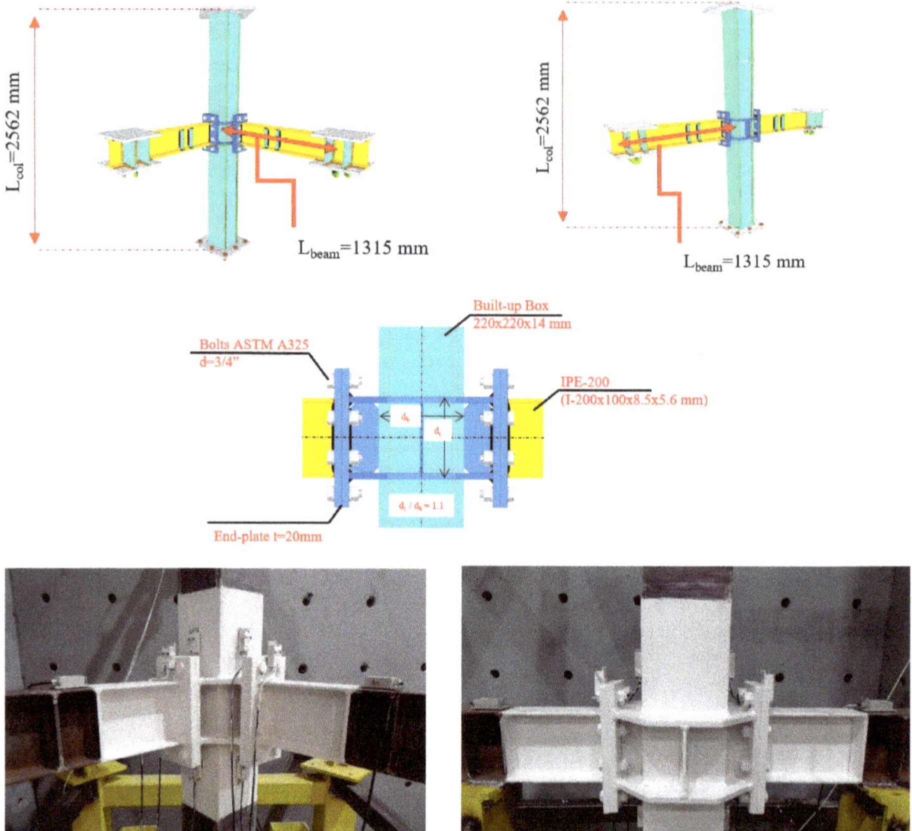

Figure 1. Test assemblies and geometries of the specimens.

2.2. Material Properties

In Table 1, the material properties obtained from the coupon tests are shown. The yield strength (Fy), tensile strength (Fu), and Young's modulus (E) of steel for beam, column, and bolts are reported. A view of the coupon steel tested is shown in Figure 2.

Table 1. Material properties of steel.

Element	Steel Type	Yield Stress (MPa)	Ultimate Stress (MPa)
Beam flange	A36	351	454
Beam flange	A36	349	432
Beam flange	A36	347	439
Beam web	A36	307	403
Beam web	A36	332	407
Beam web	A36	322	410
Column wall	A572 Gr.50	354	550
Column wall	A572 Gr.50	512	575
Column wall	A572 Gr.50	393	559
Bolt	A325	607	801
Bolt	A325	625	814
Bolt	A325	612	807

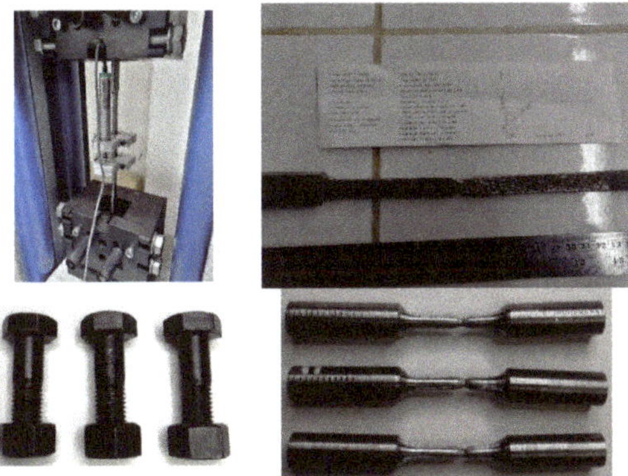

Figure 2. Test of steel plates samples.

2.3. Test Setup and Loading Procedure

The experimental setup for the cycle load is shown in Figure 3. All specimens were tested using a hydraulic testing machine with a capacity of 25 tons and a stroke of ±130 mm. The velocity of the load application used in the test reached a maximum of 20 mm/min to avoid the presence of inertial forces during the test. The horizontal displacement applied was measured by a linear variable displacement transducer (LVDT) installed between the top of the column and the actuator. The applied load was measured by a load cell in the actuator and the reaction of the beams was measured similarly using a 10-ton capacity load cell. A loading protocol was employed according to the established method in Chapter K of seismic provisions [1].

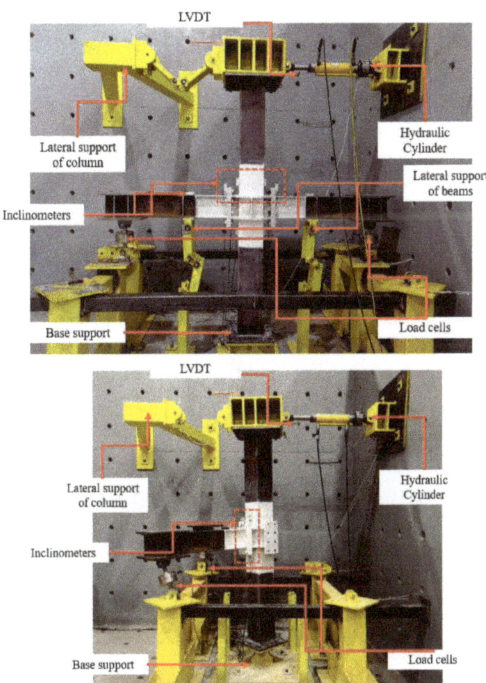

Figure 3. Experimental setup.

2.4. Experimental Results and Discussion

In this section, the results of the experimental program are presented in terms of failure mechanism, hysteretic behavior, dissipation capacity, and stiffness. The key parameters of the tests are shown in Table 2.

Table 2. Summary of the parameters obtained in the experimental tests.

Test #	Ko_West_Beam (kN/rad)	Dissipated Energy (kJ)	Mmax/Mp	θmax (rad)
1 (2BC)	2935	416	1.08	0.04
2 (2BI)	7153	722	1.55	0.04

As shown in Table 2, the dissipated energy is greater in the 2BI specimen in comparison to the 2BC specimen. Therefore, the bidirectional effect reduces the capacity of energy dissipation in steel joint configuration. In Figure 4, the failure modes are shown for the specimen tested. As is expected according to seismic design, the plastic hinges were concentrated in beams, which is required by [1]. Moreover, no failures were reported in the connection components.

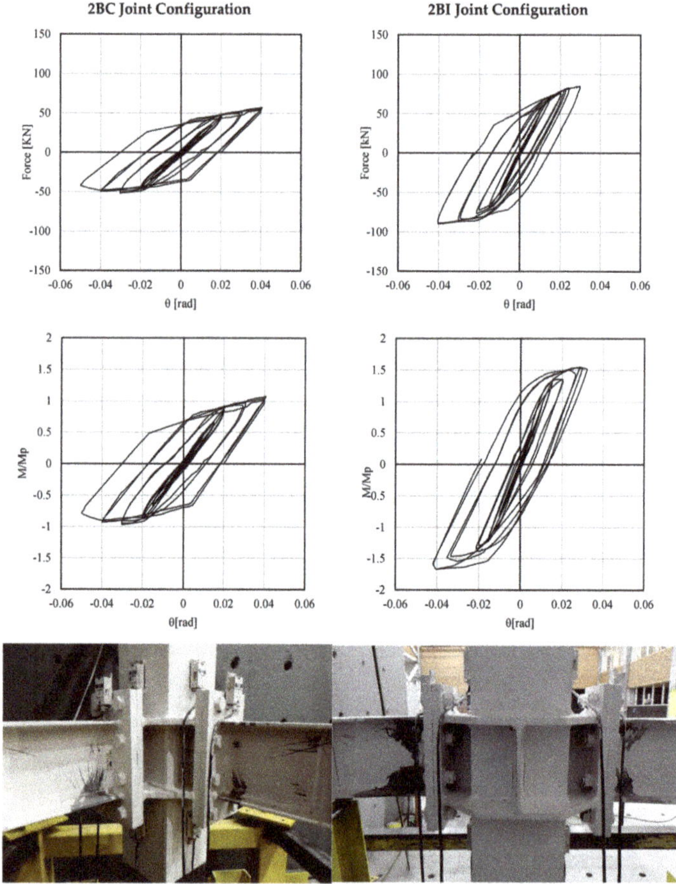

Figure 4. Force-rotation curves and normalized moment-rotation curves of beam and failure mechanisms at 0.04 rad of the specimens tested.

A ductile behavior was exhibited by all specimens tested. However, comparing both joint configurations, greater degradations in the stiffness and strength were observed in 2BC joint in comparison to 2BI joint. Finally, the developed test allows us to verify that the connection proposed in [22] has a successful performance under cyclic loads, complying with the requirements of prequalified moment connections mentioned in [2], and ensuring a ductile failure mechanism without failure in the connection components.

3. Numerical Modeling

The numerical models were developed using the finite element method through the ANSYS software [32]. The original geometry of the tests was considered with the goal to reproduce the configurations tested in the experimental program and extrapolating them to two additional scenarios. The elements and components were modeled explicitly. Loading, constitutive laws of materials, mesh size, and boundary conditions were used to resolve the nonlinear behavior of joints studied. The nonlinearities were considered by means of sub-steps in each load step using the incremental Newton–Raphson method. According to the convergence criterion [32], the convergence force and residual force obtained out of equilibrium must be below the convergence values. Furthermore, the augmented Lagrange method was used to achieve a numerical convergence, according to the investigation carried out in [33].

In the numerical models, the following assumptions were established: the length of the column is considered as the distance between the points of zero moment for each case (the points of zero moment in the columns are assumed to be at half height). The welds are excluded from the model because inelastic incursion is not expected in these elements. The diameters of the holes are estimated as standard holes according to the requirements established in [5]. Additionally, the axial load was not established as its effect was not contemplated in the experimental program. These considerations were verified and employed in [22,29,31].

Numerical models calibrated from the experimental tests were extended to evaluate the cyclic performance of columns with moderate ductility members according to [1]. Moreover, built-up box columns with slender walls according to [5] were studied to evaluate the performance of the WPZ shear. Table 3 shows the dimensions of the elements considered in the numerical model.

Table 3. Geometrical dimensions of the models.

Models	Axial Load (Py)	Configuration	Beam	Column	End-Plate Thickness (mm)	Outer Stiffener Thickness (mm)
Mod-00		2BI	IPE-200	Box 220 × 220 × 14	20	16
Mod-01	0%	2BI	IPE-200	Box 220 × 220 × 9	20	16
Mod-02		2BI	IPE-200	Box 220 × 220 × 4	20	16
Mod-03		2BC	IPE-200	Box 220 × 220 × 14	20	16
Mod-04	0%	2BC	IPE-200	Box 220 × 220 × 9	20	16
Mod-05		2BC	IPE-200	Box 220 × 220 × 4	20	16

3.1. Constitutive Laws of Materials

The constitutive law was considered by means of a bilinear kinematic law and a von Mises yield criterion was used; therefore, no variations in magnitude and location of the yield surface are established. The constitutive law of material was considered from steel coupon tests reported in Section 2. Average values were used to consider the constitutive law according to [34]. For example, in Beam elements, the Young´s modulus, E = 211,908 MPa, yielding stress σy = 334 MPa, and maximum stress σu = 424 MPa were used. The Young´s modulus, E = 196,696 MPa, yielding stress σy = 419 MPa, and maximum stress σu = 561 MPa were considered for square built-up box column elements. In addition, for the bolts, a Young´s modulus, E = 210,625 MPa, yielding stress σy = 615 MPa, and maximum stress σu = 808 MPa were used. The material for the beam is assumed for end plates and horizontal and vertical diaphragms. Posteriorly, the material properties were transformed to true stress and strain values for their use in FE models.

3.2. Boundary Conditions and Loading

To allow for rotation in the base of the column, a pinned restraint was used at the base. The ends of the beams were established as simply supported, with limited movements in Z direction. Moreover, lateral supports to the beam were applied to comply with the stability bracing members of beams according to [1,5]. The loading was applied at the top of the column for the models studied through lateral displacements. Additionally, instability by displacements out of the plane was solved by incorporating lateral support at the top of the column. In Figure 5, support conditions used in the configuration of the two beams corner (2BC) model are shown. A pretension equivalent to 70% of the bolt tension resistance was applied according to [5] to reproduce the real conditions between the plates (see Figure 6).

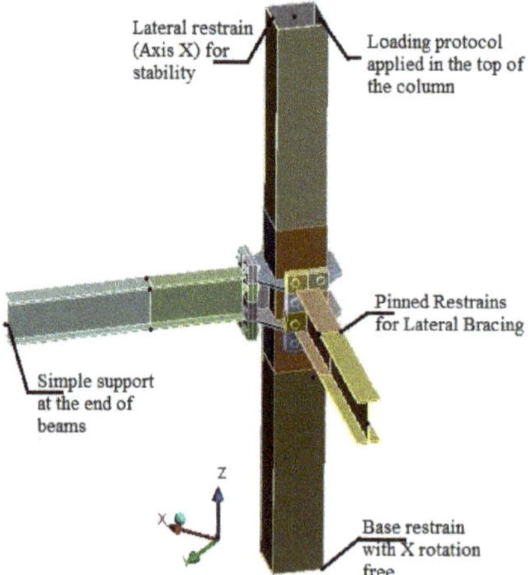

Figure 5. Boundary conditions used in the numerical model.

Figure 6. Bolt pretension used in the numerical models.

3.3. Mesh and Element Type

In this study, SOLID 186 elements were employed to simulate all elements. The element is defined by 20 nodes with 3 degrees of freedom per node given from translations in the x, y, and z directions. This element is ideal to model the plasticity, hyperelasticity, creep, stress stiffening, large deflection, and large strain capability, which accurately represent the expected stresses and deformations.

Different mesh sizes have been used to provide accurate results without great computational efforts. In this sense, the inelastic behavior in plastic hinge zones was modeled using a fine mesh, while in other zones, an expected elastic behavior was modeled as a coarser mesh. Furthermore, a sensitivity analysis was performed to obtain the accuracy of the developed model and its convergence with mesh refinement, varying the mesh element's size in the range of 5 mm to 40 mm. Finally, a maximum element size of 5 mm was selected for elements with expected nonlinear behavior, together with 25 mm for those with linear behavior, obtaining a better computational efficiency.

The FEM solves the nonlinear equations using the Newton–Raphson method. Posteriorly, the number of equations is obtained from the number of structure degrees of freedom. Different meshes, sizes, and forms have been used to determine a rational mesh to improve the computational cost. Therefore, the solution adopted uses a fine mesh in zones of interest as high stress or strain regions, specifically in the central zone of the node where the inelastic response is expected, and a coarser mesh in the other regions, as shown in Figure 7.

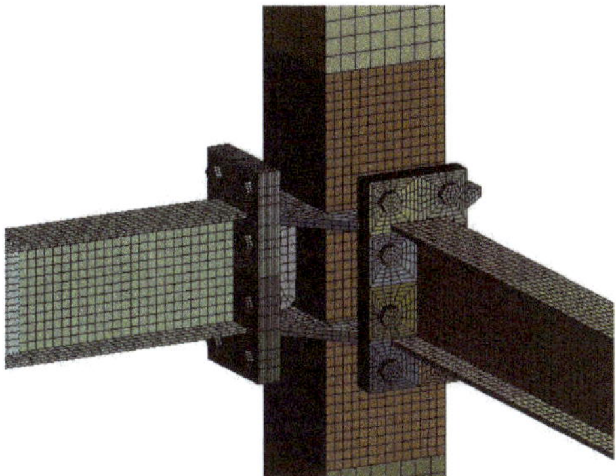

Figure 7. Meshing used in the numerical models.

Additionally, geometrical imperfections were considered according to the geometrical limits established in [35,36]. However, a limited effect of the imperfections on cyclic behavior was obtained. More investigations on these effects and their consequences in the connections can be found in [37].

3.4. Type of Contacts

In this research, a "Bonded" contact was employed to simulate welding conditions and restrained contact in all directions. The contact between the end-plates was considered through "frictional" contact to allow relative displacements between them. A friction coefficient equal to 0.3 was established according to [22]. Trial and error was performed to obtain this value [29,30]. A schematic view of contacts in the joint is shown in Figure 8. Moreover, the type of contact by region is reported in Table 4.

Table 4. Contacts simulated by using elements.

N°	Contact between Elements	Type of Contact
1	End-plates	
2	Bolt shank in contact to end-plate	Frictional ($\mu = 0.3$)
3	Bolt head in contact to end-plate	
4	Nut in contact to end-plate	

Table 4. *Cont.*

N°	Contact between Elements	Type of Contact
5	End-plate in contact to horizontal and vertical stiffener	
6	Horizontal stiffener in contact to vertical stiffener	
7	Stiffeners in contact to column	
8	End-plate in contact to beam	Bonded
9	Bolt shank in contact to bolt head	
10	Bolt shank in contact to nut	
11	Bolt shank in contact to bolt head	
12	Beam in contact to beam	

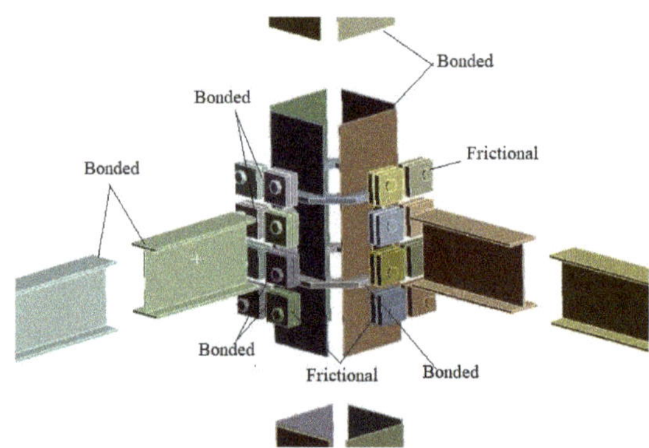

Figure 8. Schematic view of the contacts employed.

3.5. Validation of FEM

Both typologies (2BI and 2BC) of the numerical models were calibrated from the experimental program. The hysteresis curves between the experimental test and FE model were compared. In Figure 9, an acceptable match of curves was obtained; however, slight differences in terms of strength were also obtained. The differences reported may be due to higher overstrength of material in some uncharacterized components. It is important to note that material characterization is performed on limited zones of the components but not on all components.

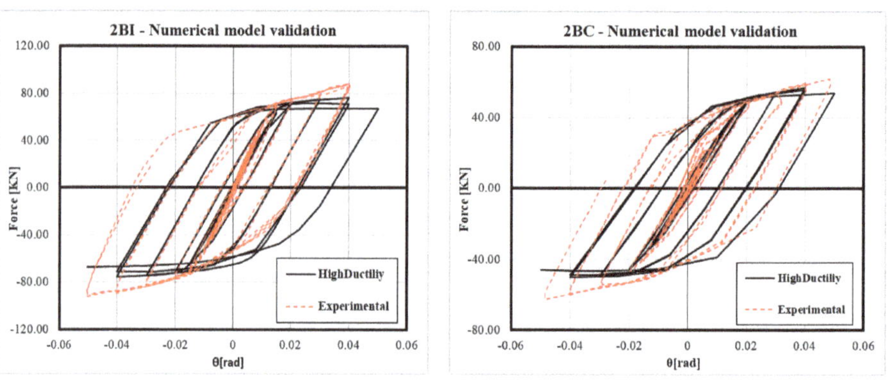

Figure 9. Comparison of the hysteresis curves of the FEM and experimental results.

4. Results

4.1. Stress Distribution

The analysis of the models described in the previous sections displays an equivalent von Mises stress distribution in Figures 10 and 11. These distributions for the different width-to-thickness column ratios show how the stresses increase as the width-to-thickness column ratio decreases.

Model	Front View	Posterior View
High-ductility column		
Moderate-ductility column		
Slender column		

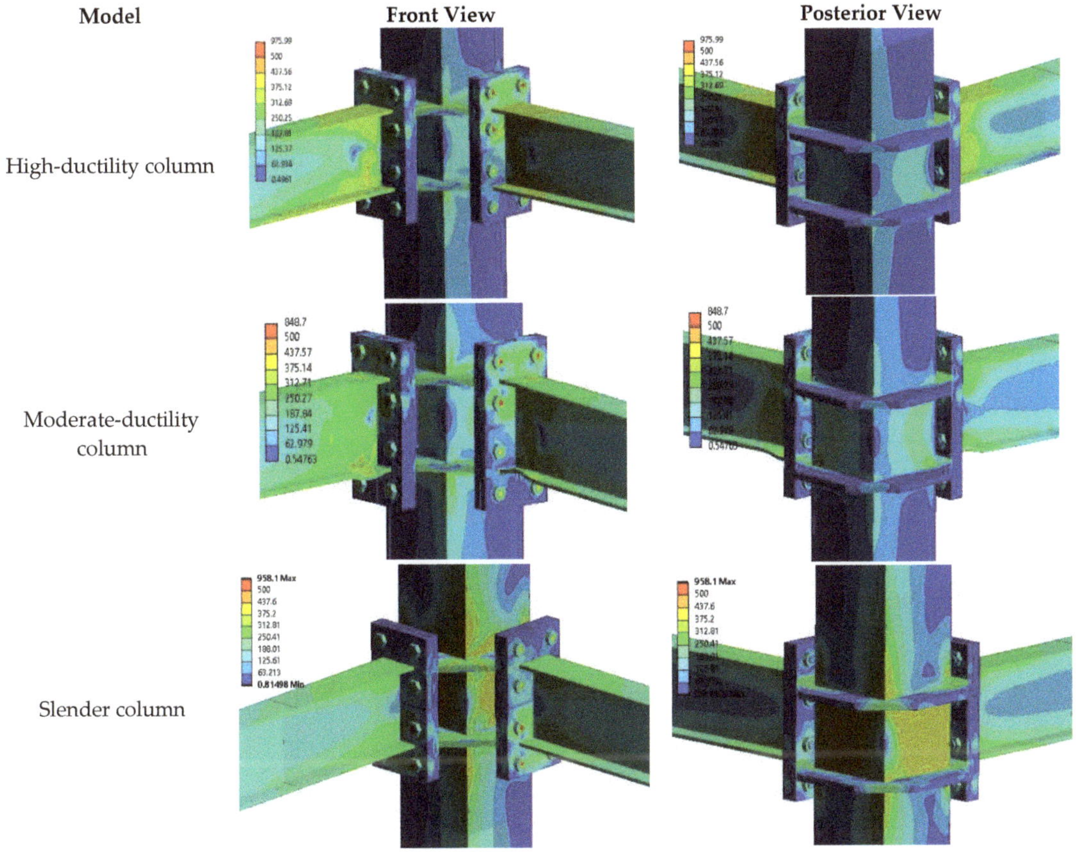

Figure 10. Stress distribution in the 2BC models (Units: MPa).

For 2BC models, the distributions of the stresses are mainly uniform in the web panel zone shear with an increase in the outer annular stiffener as the width-to-thickness column ratio decreases. In addition, the beams are less stressed when the stresses increase in the web panel zone. This demonstrates how the failure mechanism of the connection changes in the zones where the stresses exceed the yielding stress (Fy). Furthermore, the stress distribution in the panel zone is concentrated in the central area, decreasing as it approaches the edges of the column.

For the 2BI models, high-ductility models show a uniform stress distribution concentrated in the center of the web panel zone with a slight concentration in the union of the column with the outer annular stiffener as shown in Figure 11. On the other hand, the moderate-ductility model exhibited an increase in stress in all web panel zones with points of stress concentration in the column outside the web panel zone without exceeding Fy. Likewise, the slender column model shows a stress distribution similar to a tension field action in the panel zone, yielding in shear with values up to Fy. Furthermore, a stress concentration is present in the union of the column with the outer annular stiffener and column edges. Finally, in these 2BI models, the stress distribution in the web panel zone changes from a uniform normal stress distribution to a diagonal shear distribution as the width-to-thickness column ratio increases.

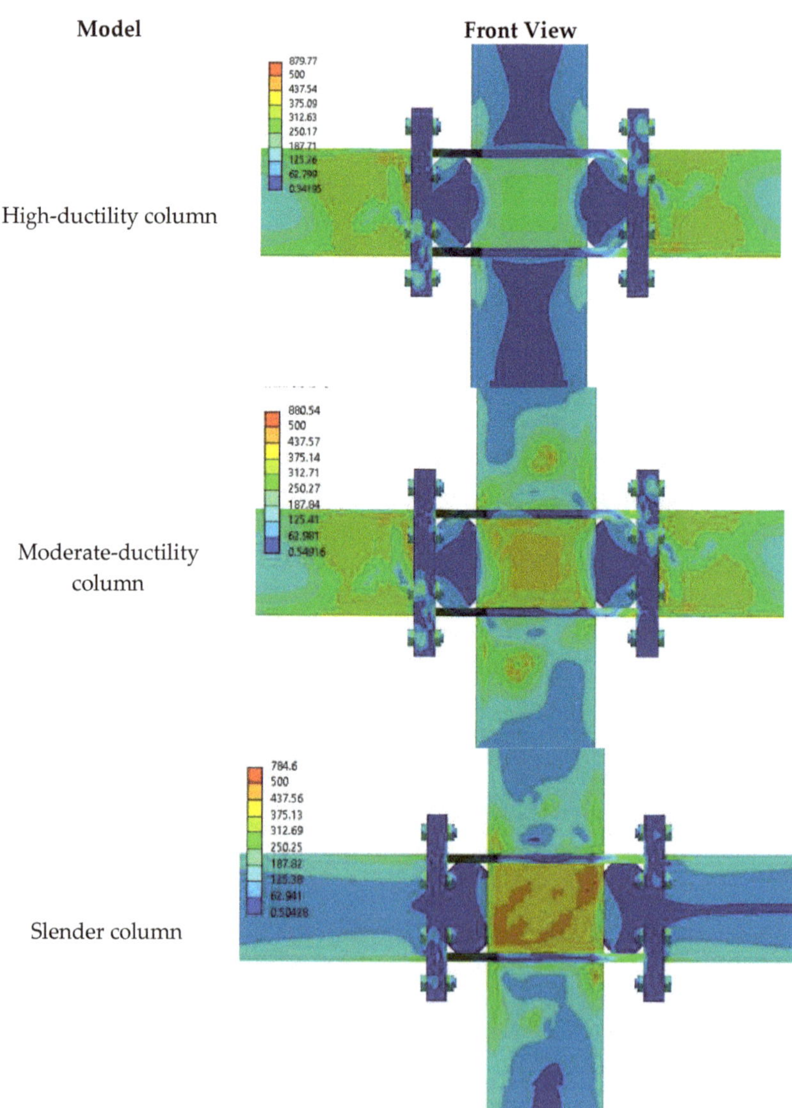

Figure 11. Stress distribution in the 2BI models (Units: MPa).

The comparison of stress distributions between the 2BI and 2BC models demonstrates that when a joint is loaded bidirectionally, the distribution of stresses becomes more uniform. This is translated to a global yielding mechanism for the entire area of the web panel zone if it occurs, contrary to the 2BI case where the shear stresses in the web panel predominate.

4.2. Plastic Strain Distributions

Similar to the previous section, equivalent plastic strains were obtained for all models analyzed and are shown in Figures 12 and 13 for 2BC and 2BI models, respectively. The plastic strain distributions for all high-ductility models show failure mechanisms as plastic hinges at the beams starting from the web yielding to the flange yielding. On the other hand, for moderate-ductility models, plastic hinges are present at the beams. However, a slightly inelastic incursion is shown in the union of the column with the outer annular stiffener and column edges for 2BC models and in the web panel zone for 2BI models.

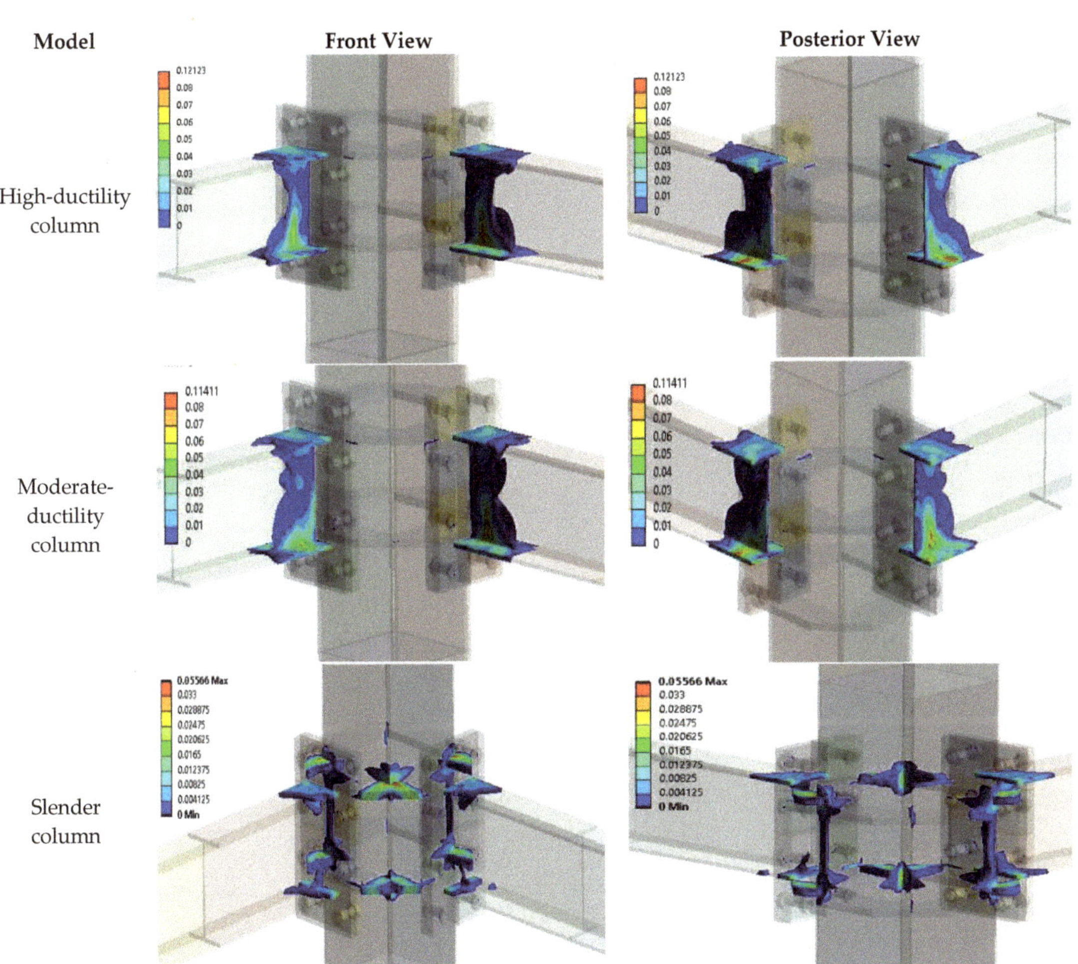

Figure 12. Plastic strain distributions in the 2BC models (Units: mm/mm).

The 2BC slender column model shows an inelastic incursion meaningfully concentrated in the outer annular stiffeners and its union with the column. However, a slight plastic strain appears in the flange of the beams instead of the failure mechanism controlled by the outer annular stiffeners. These stiffeners were designed to the maximum-concentrated beam flange force and had sufficient thickness to avoid high damage in the column and panel zones where plastic strain is not present.

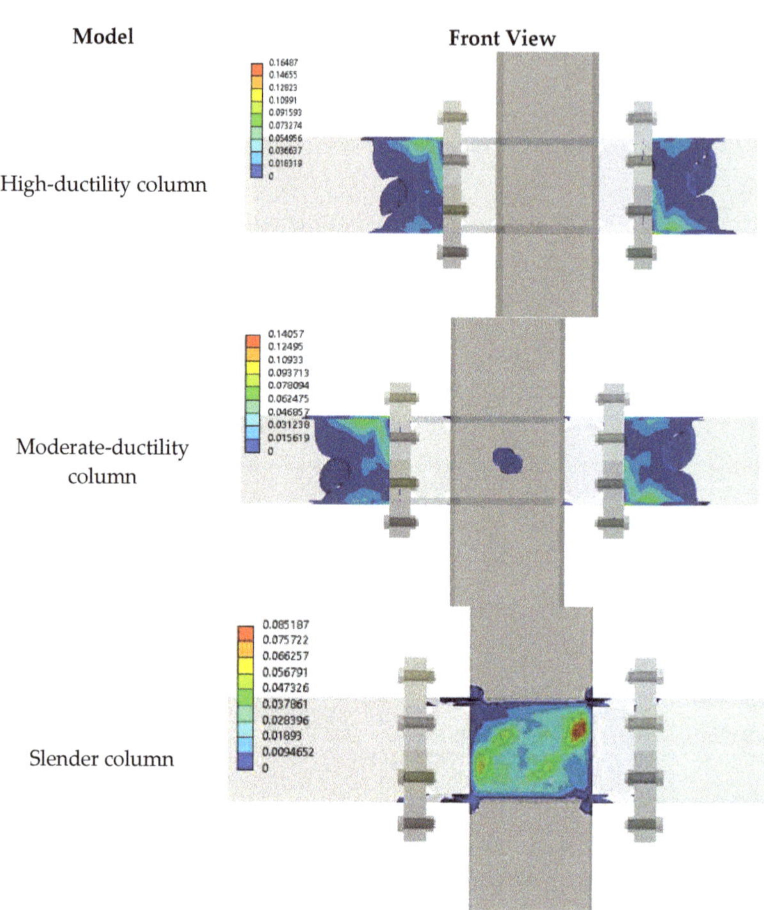

Figure 13. Plastic strain distributions in the 2BI models (Units: mm/mm).

The results of the 2BI slender model show how the plastic strains are concentrated in the web panel zone inducing the shear yielding. Likewise, slight plastic incursion is present in the outer annular stiffeners and in the zone outside the web panel zone. In addition, the strain distribution shows how inelastic behavior starts from the column edges and extends diagonally through an edge-to-edge diagonal stress field. Finally, for all high-ductility models, higher plastic strains are present in comparison to the configurations of other models. Likewise, the 2BI exhibits more plastic strains than the 2BC models, which shows how the bidirectional load limits the inelastic incursion.

4.3. Hysteretic Behavior of the Connections

The hysteretic response curves of all models analyzed are shown in Figure 14, where the vertical coordinate F is the load applied on the top of the column and the horizontal coordinate is the drift rotation of the system. In general, for all models, a ductile and stable hysteretic behavior without pinching and degradations in the strength and stiffness. However, as the width-to-thickness column ratio increases, degradations in the stiffness and strength are noted for all models. The strength degradation is greater in the 2BI models, and the stiffness degradation is greater in the 2BC models. This difference is associated with the failure model exhibited by each model, where the 2BI slender model was controlled by panel shear yielding and the 2BC by the outer annular stiffener bending yield. Additionally, the difference between the high-ductility and moderate-ductility models is small in terms of strength and stiffness; the large difference was for the high inelastic cycles of load.

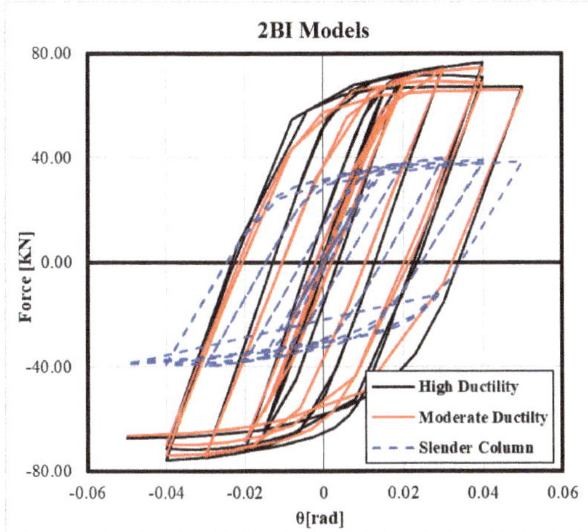
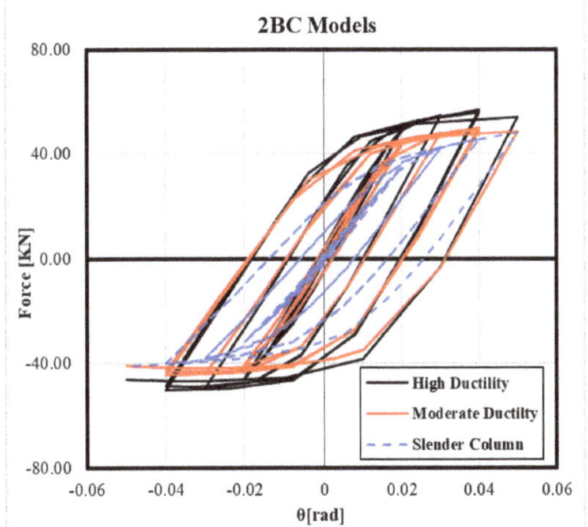

Figure 14. Hysteretic curves for all models analyzed.

A comparison between the 2BI and 2BC models was developed for each width-to-thickness column ratio and is shown in Figure 15. These results indicate that the bidirectional effect causes significant decreases in the strength and stiffness of the joints for the same number of connected beams. Nevertheless, this bidirectional effect had no greater difference between the high-ductility and moderate-ductility models. Furthermore, the slender models continue to show high stiffness degradations when the bidirectional effect is present instead of the strengths being similar. To summarize quantitatively the hysteretic behavior shown previously on hysteretic curves, Table 5 shows the key parameters of the models analyzed.

Table 5. Summary of the parameters obtained in numerical models.

Joint	Width-to-Thickness Ratio	P_{max} (kN)	θ_{max} (rad)	K_o (kN/rad)	$K_{sec\_0.04}$ (kN/rad)	Dissipated Energy (kJ)	ε_{max} (mm/mm)	σ_{max} (MPa)
	High-ductility	56.953	0.05	2718	1423	634	0	255.7
2BC	Moderate-ductility	50.088	0.05	2407	1252	575	0	326.9
	Slender	48.189	0.05	1783	1141	390	0.019	412.2
	High-ductility	76.543	0.05	4572	1913	1118	0	295.1
2BI	Moderate-ductility	74.712	0.05	3935	1867	942	0.007	398
	Slender	39.988	0.05	2580	980	649	0.085	508.3

These results show how stiffness and strength are higher in the models without a bidirectional effect (2BI), even when a slender column is present. In terms of dissipated energy, the difference between the high-ductility and moderate-ductility models is small, in the order of 9%. However, for slender models, the difference between the high-ductility and moderate-ductility models can sustain even a 40% reduction. Additionally, higher dissipated energy is observed for the 2BI slender model in comparison to the 2BC slender model; [8] suggests that shear panel failure is highly ductile and can improve the connection energy dissipation capacity. Moreover, the inelastic incursion of the outer annular stiffeners of the 2BC models is limited, which limits its energy dissipation.

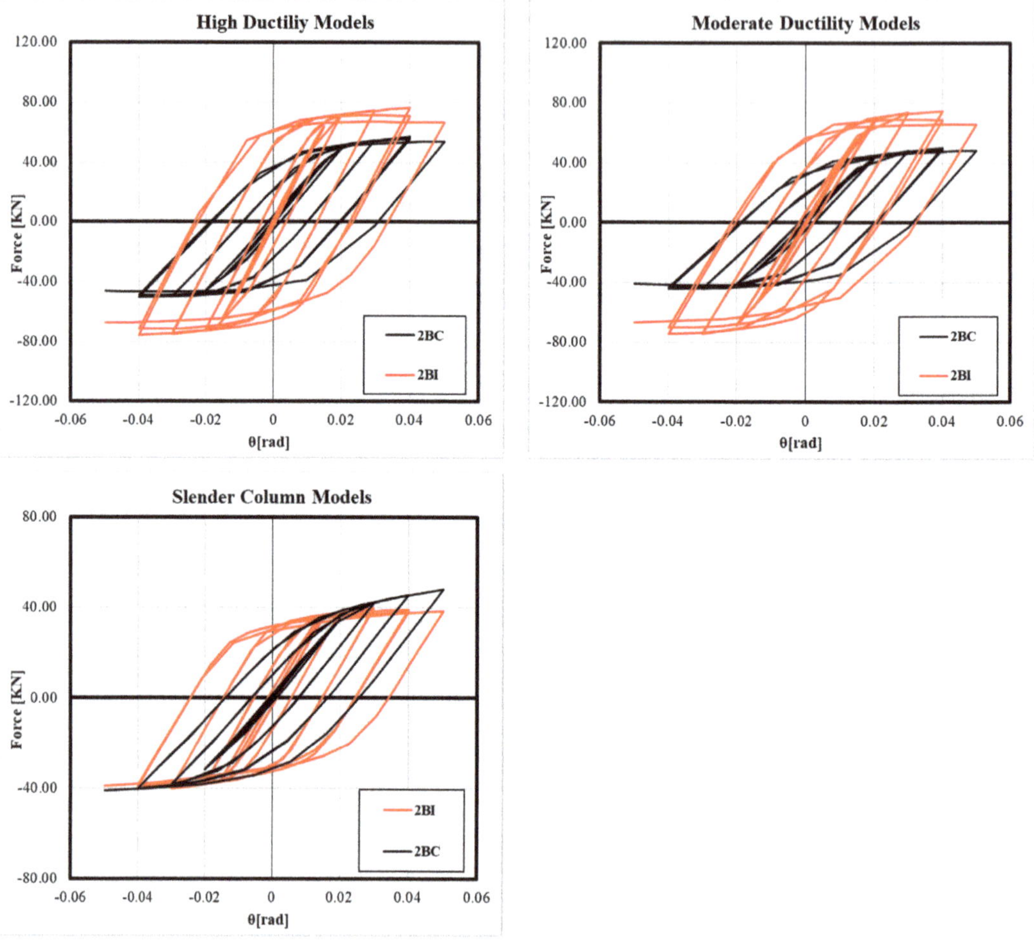

Figure 15. Comparison of the hysteretic curves between joint configurations.

Finally, a high demand in the web panel zone is denoted for interior joints (2BI) and for models with bidirectional effect (2BC) in terms of strain and stress. However, the stability of the system in terms of secant stiffness reduction is higher in 2BC models, in the order of 36%, in comparison to 2BI where the reduction is in the order of 62%.

5. Conclusions

In this research, the assessment of the cyclic behavior of the web panel zone in built-up box columns with different width-to-thickness ratios was performed. An end-plate moment connection proposed by [22] subjected to a bidirectional cyclic loading was used to connect beams to columns. The cyclic response of 3D steel joint configurations was obtained from experimental tests. Moreover, a numerical study using FEM was employed to extend the evaluation to different thicknesses in the panel zone. The effect of bidirectional loading was studied considering the failure mechanism, hysteretic response, stress distribution, strain distribution, energy dissipation capacity, stiffness, and strength of specimens and FE models. Finally, the main conclusions are described as follows:

1. The bidirectional effect reduces the strength and stiffness capacities of the connection; in the absence of the bidirectional effect, the connection can reach 0.8 Mp (plastic moment) and 0.04 rad of rotation as long as the members and components are designed according to the seismic philosophy;
2. High-ductility columns and moderate-ductility columns allow ductile failure mechanisms with plastic hinges in the beams to be reached. Nevertheless, for interior joints, slight plastic deformations could appear without system instability;
3. A combined failure mechanism with plastic hinges in the beams and plastic strains in the web panel zone is achieved for slender web columns. Therefore, the strength and stiffness of the joint configurations decreased;
4. Slender columns reached a the hysteretic behavior of the connection was still ductile until the 0.05 rad rotation. However, a reduction in the energy dissipation capacity occurred until 62% of the original capacity was reached when yielding in shear appeared in the web panel zone;
5. The presence of outer annular stiffeners and vertical stiffeners allows the local buckling on the web panel zone to be avoided, modifying the stress distribution such that is it similar to a diagonal tension field for interior joints and a uniform stress distribution for joints with bidirectional effect. However, these stiffeners should be designed to support the concentrated beam capacity load to transfer the stress properly.

Author Contributions: Conceptualization, E.N.; methodology, E.N. and R.M.; software, R.M. and E.N.; validation, E.N. and F.S.; formal analysis, E.N. and R.M.; investigation, E.N., N.M. and F.S.; resources, N.M. and Á.R.; data curation, E.N.; writing—original draft preparation, E.N. and R.M.; writing—review and editing, E.N. and Á.R.; visualization, E.N.; supervision, E.N.; project administration, E.N.; funding acquisition, Á.R. All authors have read and agreed to the published version of the manuscript.

Funding: This research was funded by Fondecyt N° 11200709 (Agencia Nacional de Investigación y Desarrollo, Chile) and DINNOVA 03/2020-II (UCSC).

Institutional Review Board Statement: Not applicable.

Informed Consent Statement: Not applicable.

Data Availability Statement: Not available.

Acknowledgments: Dirección de postgrado de la UCSC.

Conflicts of Interest: The authors declare no conflict of interest.

References

1. *AISC341-16*; Seismic Provisions for Structural Steel Buildings. American Institute of Steel Construction: Chicago, IL, USA, 2016.
2. *AISC358-16*; Prequalified Connections for Special and Intermediate Steel Moment Frames for Seismic Applications. American Institute of Steel Construction: Chicago, IL, USA, 2016.
3. Krawinkler, H.; Mohasseb, S. Effects of panel zone deformations on seismic response. *J. Constr. Steel Res.* **1987**, *8*, 233–250. [CrossRef]
4. Tuna, M.; Topkaya, C. Panel Zone deformation demands in steel moment resisting frames. *J. Constr. Steel Res.* **2015**, *110*, 65–75. [CrossRef]
5. *ANSI/AISC 360-16*; Specification for Structural Steel Buildings. American Institute of Steel Construction: Chicago, IL, USA, 2016.
6. *FEMA 355D*; State of the Art Report on Connection Performance. Federal Emergency Management Agency: Washington, DC, USA, 2000.
7. *European Standard EN 1993-1-8*; Eurocode 3: Design of Steel structures—Part 1–8: Design of Joints. European Committee for Standardization: Brussels, Belgium, 2005.
8. Skiadopoulos, A.; Lignos, D. Seismic demands of Steel moment resisting frames with inelastic beam-to-column web panel zones. *Earthq. Eng. Struct. Dyn.* **2022**, *51*, 1591–1609. [CrossRef]
9. Adlparvar, M.R.; Vetr, M.G.; Ghaffari, F. The importance of panel zone shear strength on seismic behavior of improved slotted-web beam connection. *Int. J. Steel Struct.* **2017**, *17*, 307–318. [CrossRef]
10. Hanna, M.T. Failure Loads of web panels loaded in pure shear. *J. Constr. Steel Res.* **2015**, *105*, 39–48. [CrossRef]
11. Brandonisio, G.; de Luca, A.; Mele, E. Shear inestability of panel zone in beam-to-column connections. *J. Constr. Steel Res.* **2011**, *67*, 891–903. [CrossRef]

12. Augusto, H.; da Silva, L.S.; Rebelo, C.; Castro, J. Characterization of web panel components in double-extended bolted end-plate steel joints. *J. Constr. Steel Res.* **2016**, *116*, 271–903. [CrossRef]
13. Choi, Y.S.; Kim, D.; Lee, S.C. Ultimate Shear behavior of web panels of HSB800 plate griders. *Constr. Build. Mater.* **2015**, *101*, 828–837. [CrossRef]
14. Leblouba, M.; Barakat, S.; Al-Saadon, Z. Shear Behavior of corrugated web panels and sensitivity analysis. *J. Constr. Steel Res.* **2018**, *151*, 94–107. [CrossRef]
15. Lee, S.; Lee, D.; Yoo, C. Design of intermediate transverse stiffeners for shear web panels. *Eng. Struct.* **2014**, *75*, 27–38. [CrossRef]
16. Pan, L.; Chen, Y.; Chuan, G.; Jiao, W.; Xu, T. Experimental evaluation of the effect of vertical connecting plates on panel zone shear stability. *Thin-Walled Struct.* **2016**, *99*, 119–131. [CrossRef]
17. Xu, Y.; Hao, J. Seismic performance of spatial beam-column connections in steel frame. *J. Constr. Steel Res.* **2021**, *180*, 106586. [CrossRef]
18. Fadden, M.; McCormik, J. HSS-to-HSS seismic moment connection performance and design. *J. Constr. Steel Res.* **2014**, *101*, 373–384. [CrossRef]
19. Nia, Z.S.; Ghassemieh, M.; Mazroi, A. WUF-W connection performance to box column subjected to uniaxial and biaxial loading. *J. Constr. Steel Res.* **2013**, *88*, 90–108.
20. Chen, X.; Shi, G. Experimental study of end-plate joints with box columns. *J. Constr. Steel Res.* **2018**, *143*, 307–319. [CrossRef]
21. Mohammadi, S.; Ghassemieh, M.; Mirghaderi, S.R. Cyclic behavior of steel moment connections with built-up columns in weak direction. *J. Constr. Steel Res.* **2020**, *172*, 106224. [CrossRef]
22. Nuñez, E.; Torres, R.; Herrera, R. Seismic performance of moment connections in steel moment frames with HSS columns. *Steel Comp. Struct.* **2017**, *25*, 271–286.
23. Bai, Y.; Wang, S.; Mou, B.; Wang, Y.; Skalomenos, K. Bi-directional seismic behavior of steel beam-column connections with outer annular stiffener. *Eng. Struct.* **2021**, *227*, 111443. [CrossRef]
24. Qin, Y.; Chen, Z.; Yang, Q.; Shang, K. Experimental seismic behavior of through diaphragm connections to concrete-filled rectangular steel tubular columns. *J. Constr. Steel Res.* **2014**, *93*, 32–43. [CrossRef]
25. Nia, Z.S.; Ghassemieh, M.; Mazroi, A. Panel zone evaluation of direct connection to box column subjected to bidirectional loading. *Struct. Des. Tall Spec. Build.* **2014**, *23*, 833–853.
26. Sarfarazi, S. Evaluation of panel zone shear strength in cruciform columns, box-columns and double web-columns. *Int. J. Struct. Civ. Eng.* **2016**, *5*, 52–56. [CrossRef]
27. Pazmiño, C.; Moscoso, M.; Arévalo, D.; Guaminga, E.; Gómez, C.; Herrera, M.; Vintimilla, J. Experimental study of Steel tube panel under in-plane lateral cyclic load. *Case Stud. Constr. Mater.* **2019**, *11*, e00270. [CrossRef]
28. Rong, B.; Liu, S.; Yan, J.; Zhang, R. Shear Behavior of panel zone in through-diaphragm connections to steel tubular columns. *Thin-Walled Struct.* **2018**, *122*, 286–299. [CrossRef]
29. Gallegos, M.; Nuñez, E.; Herrera, R. Numerical study on cyclic response of end-plate biaxial moment connection in box columns. *Metals* **2020**, *10*, 523. [CrossRef]
30. Nuñez, E.; Lichtemberg, R.; Herrera, R. Cyclic performance of end-plate biaxial moment connection with HSS columns. *Metals* **2020**, *10*, 1556. [CrossRef]
31. Mata, R.; Nuñez, E. Parametric study of 3D steel moment connections with built-up box column subjected to biaxial cyclic loads. *J. Constr. Steel Res.* **2022**, *197*, 107453. [CrossRef]
32. ANSYS, Inc. *Theory Reference for ANSYS and ANSYS Workbench Release 19.0*; ANSYS, Inc.: Canonsburg, PA, USA, 2019.
33. Moura, P.; Carvalho, H.; Figueiredo, L.; Aires, P.; Bártolo, R. Unitary model for the analysis of bolted connections using the finite element method. *Eng. Fail. Anal.* **2019**, *104*, 308–320.
34. Amadio, C.; Bedon, C.; Fasan, M.; Pecce, M. Refined numerical modelling for the structural assessment of steel-concrete composite beam-to-column joints under seismic loads. *Eng. Struct.* **2017**, *138*, 394–409. [CrossRef]
35. *ASTM-A6/A6M*; Standard Specification for General Requirements for Rolled Structural Steel Bars, Plates, Shapes, and Sheet Piling. ASTM: West Conshohocken, PA, USA, 2011.
36. *ASTM-A500/A500M*; Standard Specification for Cold-Formed Welded and Seamless Carbon Steel Structural Tubing in Rounds and Shapes. ASTM: West Conshohocken, PA, USA, 2021.
37. Tartaglia, R.; D'Aniello, M.; Rassati, G.A.; Swanson, J.A.; Landolfo, R. Full strength extended stiffened end-plate joints: AISC vs recent European design criteria. *Eng. Struct.* **2018**, *159*, 155–171. [CrossRef]

Perspective

Smart Textiles in Building and Living Applications: WG4 CONTEXT Insight on Elderly and Healthcare Environments

Enrico Venturini Degli Esposti [1], Chiara Bedon [2,*], Vaida Jonaitiene [3], Jan K. Kazak [4], Leonarda Francesca Liotta [5], Georgios Priniotakis [6] and Urszula Stachewicz [7]

[1] Next Technology Tecnotessile, 59100 Prato, Italy
[2] Department of Engineering and Architecture, University of Trieste, 34127 Trieste, Italy
[3] Faculty of Mechanical Engineering and Design, Kaunas University of Technology, 44249 Kaunas, Lithuania
[4] Institute of Spatial Management, Wrocław University of Environmental and Life Sciences, 50-375 Wrocław, Poland
[5] Istituto per Lo Studio Dei Materiali Nanostrutturati (ISMN)-CNR, 90146 Palermo, Italy
[6] Department of Industrial Design and Production Engineering, University of West Attica, 122 43 Athens, Greece
[7] Faculty of Metals Engineering and Industrial Computer Science, AGH University of Science and Technology, 30-059 Krakow, Poland
* Correspondence: chiara.bedon@dia.units.it; Tel.: +39-040-558-3837

Abstract: Over the past 30 years, the development of new technologies and especially of smart textiles has unavoidably led to new applications of traditional textiles in the built environment. Depending on special constructional needs (i.e., acoustic insulation, thermal insulation, shading system, etc.) or health monitoring and supporting needs (i.e., for patients with chronical disease, etc.), an increasing number of possible applications has been proposed to improve human well-being. This is especially the case for healthcare environments (like elderly or nursing homes, etc.), but also educational environments (like schools, etc.) where young or old customers can benefit from technological innovation in several ways. As an ongoing activity of WG4 members for the CA17107 "CONTEXT" European research network, this study presents a review on selected applications for building and living solutions, with special attention to healthcare environments, giving evidence of major outcomes and potentials for smart textiles-based products.

Keywords: textiles; smart textiles; new technologies; buildings; healthcare environments; health monitoring

Citation: Venturini Degli Esposti, E.; Bedon, C.; Jonaitiene, V.; Kazak, J.K.; Liotta, L.F.; Priniotakis, G.; Stachewicz, U. Smart Textiles in Building and Living Applications: WG4 CONTEXT Insight on Elderly and Healthcare Environments. *Buildings* **2022**, *12*, 2156. https://doi.org/10.3390/buildings12122156

Academic Editor: Antonio Caggiano

Received: 8 November 2022
Accepted: 5 December 2022
Published: 7 December 2022

Publisher's Note: MDPI stays neutral with regard to jurisdictional claims in published maps and institutional affiliations.

1. Introduction

Buildings comfort and pollution reducing systems is a topic taking arising importance. Researches and studies are developed to investigate the acoustic, thermal as well as the visual comfort of buildings and methods to improve comfort standards. In this context, specific needs of different target persons have to be explored: aging people mostly but even youngsters attending educational buildings. Traditional building textile materials include fibres that are mixed with concrete, fibreglass reinforcement meshes, insulators, etc. Textile architectures can typically cover permanent tensile structures based on polyester or glass fibre fabrics, with a plastic-based coatings and awnings generally supported by polyester or polyolefin fabrics, with or without coating, especially for gardening [1].

In the building and living sectors, one of the main concepts that is being explored is the zero-energy building, based on the improvement of the materials used, construction methods and architectural design. In this scope, the use of textile-based materials, such as composite structures, that can contribute to both the reduction of the raw materials used and lowering energy consumption, can be seen as a determinant factor in order to achieve this goal.

However, where are the building/architectural trends and healthcare tools going, thanks to continuous technological innovation and the availability of new solutions of "smart textiles"? What is the actual connotation and definition of smart textiles?

This review document aims at opening a discussion on recent trends, challenges, and gaps on the topic. More precisely, this presents selected review studies carried out from Working Group 4 "Building and Living" of EU-COST Action CA17107 "CONTEXT"—the European network to connect research and innovation efforts on advanced smart textiles (Horizon 2020) [2].

Principal features, applications, and challenges/gaps for the use of smart textiles, both in building components but also on customers, are discussed in following sections. In doing so, differing from earlier literature contributions, special care is given to smart textiles for healthcare environments, such as elderly homes, nursey homes, or even educational buildings, and smart textiles with specific application towards health monitoring for customers with special needs.

2. Background on Building Performance Indicators and Human, Health, Social Needs

Considering recent applications, there are no doubts that textiles can offer robust support for the development of enhanced and innovative solutions for building engineering and architectural applications (Figure 1). As far as building components and systems are taken into account, typical functions of both textiles and smart textiles can be conducted to insulation, shading, and even retrofitting of traditional constructional materials (see for example [3]).

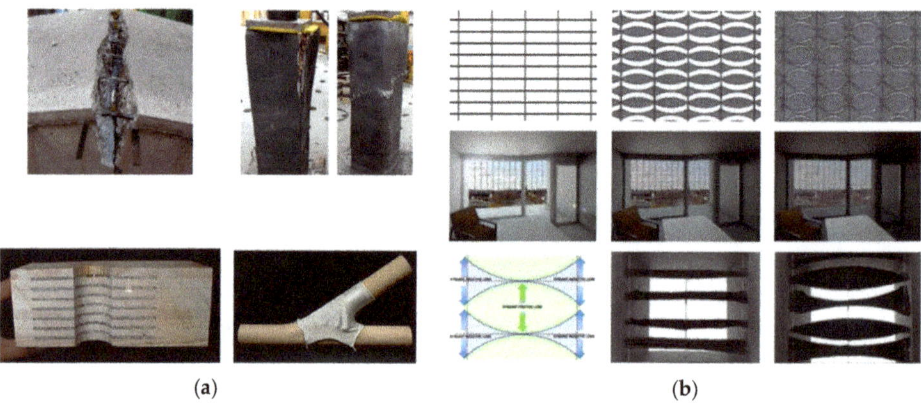

(a)   (b)

Figure 1. Examples of textiles and smart textiles: (**a**) construction retrofit; (**b**) smart shading. Figures reproduced from [3] under the terms and conditions of a Creative Commons CC-BY license agreement.

Generally speaking, worldwide researchers and producers recognize that "smart textiles" can be defined as those textiles which are able to sense, and thus react, to environmental conditions and external stimuli (i.e., mechanical, thermal, and chemical stimuli) thanks to a number of sensors incorporated in the textiles. One of the common strategies to obtain the best performance and multifunctionality of materials is using the biomimetic approach to optimize building constructions [4], architecture patterns [5], thermal insulation [6], heat dissipation systems [7], mechanical properties [8,9], and water and energy harvesting [10].

In this sense, healthcare environments—as far as they are intended as building systems—can take major advantage of a multitude of uses and smart textile solutions, which are expected to improve acoustic insulation, or thermal insulation, or shading functions, or even indoor air quality, etc.

In parallel, healthcare environments can gain a primary advantage from smart textiles as far as human needs are taken into account, and most importantly specific diseases are considered [11]. New technologies for addressing non-physical- (i.e., Alzheimer's, loneliness) and physical- (i.e., stroke, bedsores, and falls) related challenges are known to represent powerful tools. While there are no doubts about the primary role of textiles in supporting devices and components [12,13], the present review wants to open discussion on new design concepts [14], in which smart textiles are jointly used for multi-optimization purposes.

3. Smart Textiles for Buildings and Healthcare Environments

So far, textiles have been used for centuries to increase the quality of life of residents in the home environment and other living spaces. Textiles have been, for example, applied to furnishings, windows, and floors not only to control temperature, sunlight, and sound but also to offer pleasure through the colour, the texture and even the way of installing the fabric [3]. A multitude of literature examples also proved the efficiency of textiles for strengthening traditional constructional materials [15–18].

In this context, the development of smart textiles, since 1990, has unavoidably led to new applications of textiles in the built environment. Innovations in nanotechnology have attributed smart and active qualities to textiles increasing their use in architecture, especially in facades [19]. Nanomaterials enable textiles to function as a smart coating with self-healing, antimicrobial, anti-fouling, self-thermo-regulating, and other characteristics, thus contributing to increasing the sustainable qualities of the building. There is discussion on introducing textiles in architecture for signage within buildings, to provide a sustainable solution for producing ever-changing information combined with sounds and images [20]. Many of these applications of smart textiles are effective contributors of human-centred design principles that aim to improve the quality of life of end users. One of the most common applications of textiles affecting the indoor environmental quality is represented by electrospun nanofibres with sound absorbing characteristics, which can be used in architecture sound-proofing applications [21]. There is also an increasing development of smart textiles in the realm of safety, using sensors to detect vibration, sound, and movement to activate alarm systems. Additionally, in the realm of indoor pollution, smart textiles are proposed as tools to measure and filter pollutants and bacteria, thus contributing to indoor environmental quality improvement. Antioxidant and antibacterial smart textiles can be efficiently introduced into rooms in houses, with a new level of antibacterial and antioxidant systems that can help a person during sleep time [20]. There is thus an open discussion of applying smart textiles in the healthcare environment, to improve the quality of life of patients. Apart from the functional potential of smart textiles to aid architecture aspects of safety, way-finding and movement, there is the unexplored potential for contributing to the quality of life of patients by providing stimulation and engagement.

4. Applications, Potentials, Challenges for Buildings and Healthcare Environments

4.1. Acoustic Insulation by Natural Fibres and Textiles

The World Health Organization says that noise disrupts our rest and sleep, prevents customers from concentrating while working or studying, and this causes people in a noisy environment to become more upset, leading to deteriorating mental health, hypertension, and ischemic heart disease. Rooms in which people spend most of their lives must be well insulated both from the noise emanating from the environment and from its sources in the buildings. In the recent years, the European Commission has set a green and sustainable strategic direction, turning to natural raw materials found in nature, their processing, and use for the production of building materials [22,23]. Researchers of materials acoustics are also interested in various natural fibres and textile waste, as well as such acoustic materials that are gradually finding their niche in the market. Composite materials are developed using natural fibres from hemp, coconut, and reed, as well as sludge boards (Figure 2).

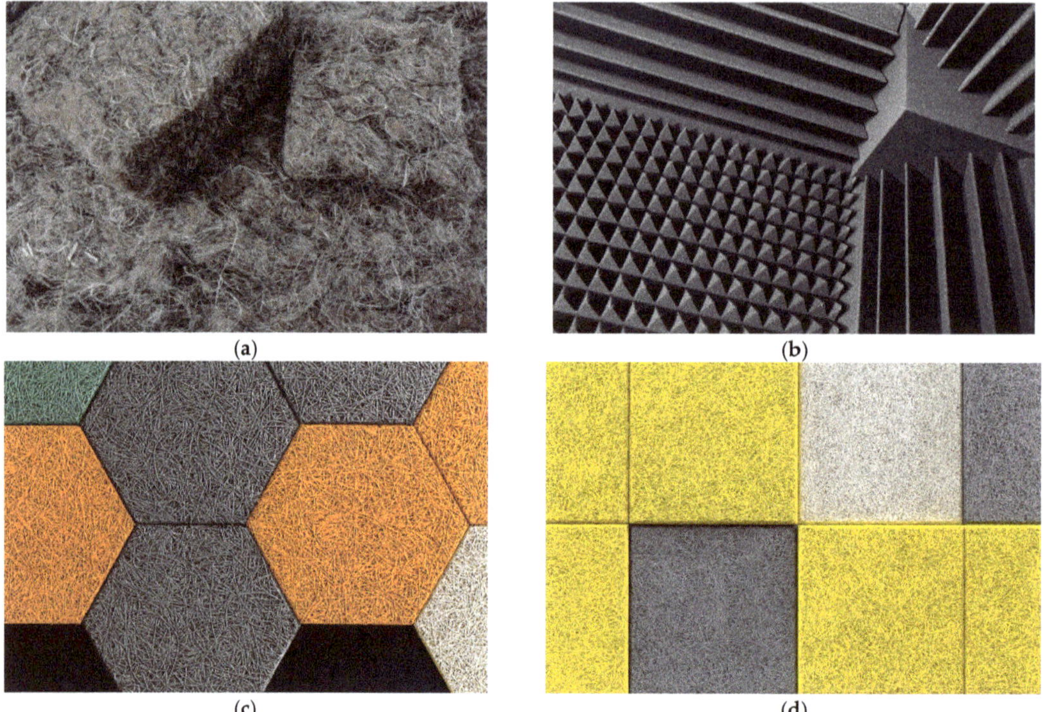

Figure 2. Examples of possible solutions for acoustic insulation: (**a**) hemp isolation material; (**b**) textile acoustic panels; (**c**) wool acoustic panels; (**d**) synthetic textile waste panels.

Acoustic panels combine in general micelles and recycled textiles. Acoustic wall panels, partitions, and ceilings are made from recycled plastic bottles, or new acoustic materials are created by recycling cotton and cellulose [22]. From a technical point of view, the development of panels from textile fibres and waste textile materials poses an open problem, because the desired ratio of sound absorption to panel strength must be achieved [22]. To get the board to absorb sound well, the porosity of the board must be properly increased. However, increasing the porosity reduces the strength of the plate and makes it particularly brittle.

Recently, based on increasing sustainability needs for buildings and components, a considerable number of research efforts have been focused on the use of recycled textiles for acoustic (and even thermal) insulation systems [24,25]. The potential of industrial waste nonwoven polyester textile has been also addressed in [26] in terms of sound absorbing capacity, thermal conductivity, and also reaction to fire.

Residents of apartment buildings often suffer from the noise from other floors and the neighbours. The vibrations, bumping, and shouting of neighbours can cause severe discomfort and even health problems. There are a number of companies who have developed a unique acoustic technology—acoustic panels made of fungal micelles and recycled textile waste. This is one of the first projects of its kind to be commercialized. Acoustic panels are made of soft, foam-like mushroom materials. Acoustic panels in natural light tones stand out with their elegant velvet finish and 3D relief for better sound absorption [27].

In order to successfully carry out the acoustic redevelopment of old, first-generation apartments, it is necessary to evaluate the existing indoor noise environment, identify problem areas and eliminate the main identified sources of noise, such as shock and plumbing noise [28]. After listening to residents living in old apartments, the following problems became clear: the noise of a high-weight floor impact and plumbing.

Various fibres are used to strengthen concrete, improving the properties of cement. Textile or mineral wool inserts with acoustic properties provide an aesthetically attractive and functional finish, improve the acoustics of the space, and absorb noise in the room. A critical analysis is reported in [29,30]. In terms of natural fibres, bamboo coir and jute are analysed. Also, the effect of alkali present in cement mixture on the degradation of natural fibres is detailed. Critical observations such as changes in crack patterns, the effect of the nature of fibres, and the environment in which they are reinforced is then discussed. For example, the use of different sealing materials for hydrophobic fibres has direct consequences on the ultimate property of reinforced concrete. In terms of synthetic fibres, predominantly used solutions in such reinforcements—such as polypropylene (PP), polyethylene (PE), and nylon (PA6)—have also specific features. The fibre–matrix interface studies is discussed and further research areas are suggested.

Modern technologies for acoustics and noise reduction are typically represented by wood wool acoustic panels/multifunctional solutions (Figure 2). Wood and wool panels are suitable for use where environmental preference is a priority. They are made of simple, natural materials, such as wool, wood or cement, are of high quality, inexpensive and, most importantly, absorb sound well. Such panels not only protect well from noise, but also embellish the interior. It is for example claimed that sheep's wool can absorb up to 30% of its weight in moisture without losing any insulating properties. Wool also has another advantage—with higher humidity and condensate content, due to its unique properties, its protein fibre does not create conditions for mould to breed [30]. They also absorb echoes and do not release heat from the premises [31]. Acoustic panels are most suitable for large rooms with high ceilings or industrial spaces adapted for offices. They effectively absorb sound, form cosy spaces and complement interior design solutions. Since this material absorbs sound from both sides, it protects against noise twice as well. The system of acoustic panels can consist of wall panels and their great advantage is a wide range of different colours, sizes, and shapes (see Figure 2 and [30]).

Acoustic partitions dividing the space on the floor also help to absorb the sound. They are light, flexible, made of softer material and can be easily transferred from one place to another. In addition, they are easy to assemble or disassemble, according to special needs [23].

Currently, hemp concrete, which is characterized by a high-volume ratio of added hemp fibres, is used in buildings and it is gaining popularity (Figure 3 and [31,32]). Unlike other substances, hemp concrete does not wear with age, but only strengthens. It contains some CO_2 locked inside, in the form of hemp fibres, and because organic compounds stay locked into its matrix, the release of CO_2 from hemp decaying cannot occur.

Another very important aspect is that hemp concrete has antiseptic properties. Lime, which is an integral part of the said concrete, has the same characteristics. It is also important to mention the fact that rodents do not come into a hemp house, and insects do not breed in it. Such a house also breathes; it does not need to be covered with steam-insulating films.

Whoever has asthma and other respiratory problems, would be wise to choose to build a hemp concrete house. It is worth noting that hemp concrete is not combustible. Since hemp fibre shives are soaked in lime, this substance practically does not burn [32].

4.2. Acoustic Insulation by Nanofibrous Layers and Textiles

In 2017, the Next Technology Tecnotessile (NTT) research centre supported an Italian technical textile company in the implementation of a project aimed to improve the acoustics in classrooms and healthcare environments [34]. Many schools in Italy are hosted in historic buildings, and for this reason are affected by poor sound comfort. The concept of the innovative solution for the improved acoustic performance lay in pre-defined absorbing panels supported by novel nanofibrous layers, which, thanks to a specific surface, porosity, and tortuosity, allow the reduction of the energy of sound waves. It was designed as a complete kit for the acoustic correction of the classrooms, which could be installed on site by the members of the school environment (teachers, pupils), avoiding wasted time and costs for the involvement of professionals. The choice of the right kit and the expected effectiveness in terms of performance acoustics are based on a device made within the project that, given the sound source and the size of the classroom, or in any case of the environment under consideration, proposes the most appropriate solution in terms of the number of panels and related positioning.

(a)

(b)

Figure 3. Example of (**a**) hemp blocs and (**b**) positioning of hemp concrete (reproduced from [33]).

The activity focused on the development of innovative solutions to increase the acoustic performance of sound-absorbing panels and further nanotechnological functionalization, necessary both for the final application sector and for any new markets, by producing nanofibrous layers. Today, these nanomaterials find application in industries such as textiles, electronics, catalysis, and filtration. However, some recent studies highlight their possible benefits in terms of sound absorption. The parameters that describe the effectiveness in terms of sound absorption are typically density, porosity, and geometry. Such a set of aspects are intrinsic to the morphology that non-woven nanofibres assume with the electrospinning process. Features like the completely random distribution, the high volume/surface ratio, and the considerable porosity, are in fact optimal factors for the increased acoustic performance of panels. Among the solutions which are known to the literature as "electrifiable", the attention goes to polyamide (PA6), polyfluoride-polyvinylidene (PVDF), and other polyesters.

An experimental investigation has been carried out for the optimization of recycled polyester electrospinning. This activity required particular attention to the optimization of process parameters, as the material undergoes a phase of degradation that alters and worsens intrinsic properties, such as the collapse of the average molecular weight of polymer chains. An extended study followed for the functional characterization of the panels made by coupling conventional materials with PVDF nanofibrous layers. The instrument used for the characterization and comparison of the materials is the "Kundt tube" (Figure 4 and [34]).

The Kundt tube realized for the NTT project was used to measure, according to ASTM E2611, the so-called "transmission loss", i.e., the share of sound absorption in transmission. Thanks to the use of the fabric outside the acoustic panel, it was possible to protect the nanostructured layer from any areas of imperfect adhesion by inserting the nanofibres inside the panel/fabric sandwich. The new systems were made from non-woven textile structures in polyester of different thicknesses and densities depending on the specific use. Textile structures made using non-woven production technologies were superficially modified by smoothing and printing processes. The experiments involved the development of surface treatments of non-woven panels by thermal and physical-mechanical processes and subsequent printing processes to achieve optimal aesthetic quality, to be used at sight. It was possible to validate the new material consisting of non-woven wadding—nanofibres—that responded well to the requirements for an application in the technical field of acoustics. In particular, maximum acoustic performance was achieved by using low-thickness panels while greatly reducing the volume of nanofibre-free components. With the introduction of a textile matrix that remains on the outer side of the prototype it was also possible to work on a series of chemical finishing treatments for the introduction of additional technical properties.

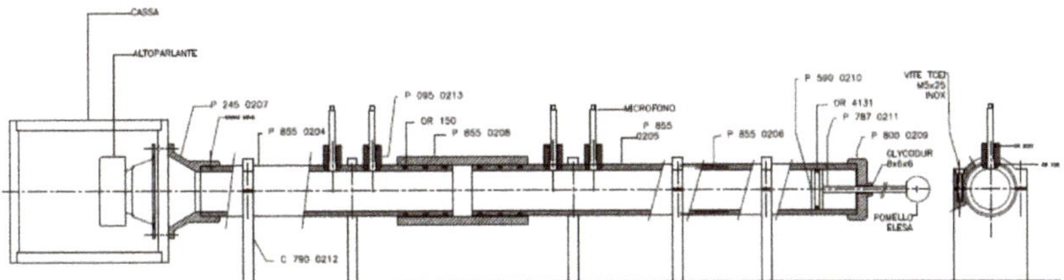

Figure 4. Layout of "Kundt tube" developed in NTT funded research project.

As mentioned above, it was chosen to couple the sound-absorbing materials to a thin, non-woven textile. For self-cleaning and sanitization of the interior spaces, a formula containing inorganic photoactive nanofillers was developed, and the application was made. This finishing is capable, when activated by the UV component of light, to generate hydroxyl radicals capable of degrading stains and any harmful compounds as bacteria.

In general, a homogeneous distribution of materials on the surfaces of the room is preferable in sound absorbent classrooms, where panels have to be placed at an adequate height from the ground. Sound reverberation correction was achieved in the pilot classroom through the application of sound-absorbing and acoustically reflective panels, gaining an optimal reverberation time.

4.3. Acoustic and Thermal Insulation by Phase Change Materials (PCM) and Textiles

The phase change materials (PCM) are known for the ability to absorb or release energy required for heating or cooling. PCM can be encapsulated in electrospun fibres [35]. Most commonly, PCMs are various paraffins, oils, fatty acids, or ionic liquids [36,37]. If the electrospun fibres are well designed they have great thermal management ability [35] especially for energy storage applications [38], thermal insulation [13], and thermal comfort [39]. Thermo-physiologically comfortable clothing supports the thermoregulation of the body and helps the wearer to keep a comfortable temperature. The thermal properties can also be achieved by biomimicry of natural materials [40] as the high surface area and porosity are desired in thermal and sound insulation [41,42].

The acoustic insulation in the current regulations and European Union policies requires novel material. We need more environmentally friendly, lighter, and thinner absorbers that meet the best acoustic performance specifications. The typical materials are composites based on fibreglass or mineral wool, but still have durability issues. The design and fabrication of composite fibrous materials with multifunctional interior core structures are increasingly attractive because of the incorporation of different inside materials. One of them is electrospun fibres, having the great advantage of designing composite fibres with multifunctionality by constructing, for example, core–shell fibres. The incorporation of non-Newtonian or shear thickening fluids exhibits increased energy dispensation on impact or the ability to damp sounds. The core–shell fibres have a greater sound-absorbing ability than the single-phase fibres. One of the typical core materials used in co-axical electrospinning is polyethylene glycol (PEG), especially for low frequency sounds [43]. Low frequency sounds include car or aircraft engines, but also many electronic and medical devices often used in homes for the elderly [44]. The electrospun fibres showed great resistivity for acoustic wave flow [45] and enhancement of sound transmission loss [46].

4.4. Smart Textiles for Indoor Air Quality Improvement

Indoor air quality, as known, is strongly affected by indoor materials, such as carpet and cloth, which may act as sources or sinks of gas-phase air pollutants. In this regard, smart textiles can efficiently contrast the typical phenomena of degradation and aging of traditional materials, with direct effects on air quality levels and improvements.

The review study in [47] reports that the associations between ozone concentrations measured outdoors and both morbidity and mortality may be partially due to indoor exposures to ozone and ozone-initiated oxidation products. It is observed that indoor exposures to ozone and its oxidation products can be efficiently reduced by filtering ozone from ventilation air and limiting the indoor use of products and materials whose emissions react with ozone. It is thus clear that such steps might be especially valuable in schools, hospitals, and childcare centres in regions that routinely experience elevated outdoor ozone concentrations.

Lakey et al. [48] emphasized the importance of quantifying the impact of clothing on ozone–human interactions and indoor air quality. As such, the authors developed a kinetic multilayer model of surface and bulk chemistry of the skin, which includes mass transport through the skin and chemical reactions of skin lipids and ozone in the skin and secondary chemistry in the gas phase (Figure 5).

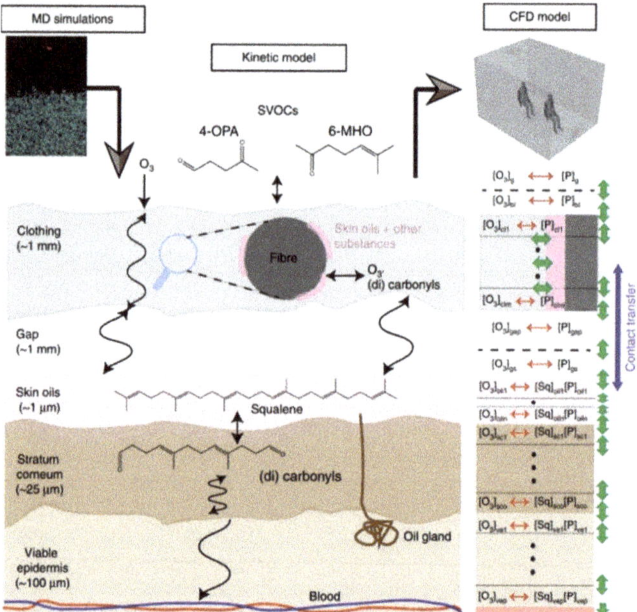

Figure 5. Schematic representation of the kinetic multilayer model developed in [48] for skin–cloth–ozone interactions. Figure reproduced from [48] under the terms and conditions of a Creative Commons license agreement.

Textiles and fibres in carpets have a key role in terms of indoor air quality in the same order of clothes. The study presented in [49], in this regard, investigates the behaviour and potential of different solutions for carpets against ozone. The comparative study in [49] indicates that carpets are good sinks for ozone with potential to lower harmful ozone levels indoors. On the other hand, carpets can emit significant levels of volatile organic compounds, and these emissions can be amplified in the presence of ozone. Special attention is hence required for dedicated healthcare housing or educational applications.

In [50], a new technological textile has been proposed and investigated, based on TiO_2 nanoparticles and optical fibres, for the degradation of pollutants potentially existing in the indoor air of hospitals. As shown, the use of such technology for the treatment of indoor air, not only in hospitals but also in various other sectors, looks very promising and is expected to be deepened in the near future. In addition, the study confirms that using optical fibres can be extremely efficient in the inactivation of microorganisms present in air, and thus in the degradation of different types of air pollutants.

The study presented by Zhu et al. [51] investigated an effective and environmentally friendly technology, which has been introduced to reduce the malodours from textiles. Results suggest that the proposed solution can play a significant role in the removal of various smells. Thus, it has the potential to clean the indoor environment from odour pollution, and this is particularly of interest for healthcare and educational applications. Given such a high potential for deodorization, the same technology may also be used in a wide range of fields including wastewater odour treatment in the chemical or the livestock breeding industry.

4.5. Smart Shading Devices

According to several literature studies, there is evidence that immobility and a lack of physical activity are widespread issues among seniors over the age of 65. Long durations of not leaving the house may increase mortality risk, social isolation, depression, cognitive impairment, and other health issues. Going out of the house, even just to spend some rest time outside, can indeed improve mental health, increase levels of vitamin D, and provide opportunities for participation in a variety of activities such as psychosocial, emotional, cultural, therapeutic, leisure, or even physical activity, since resting is an important part of physical activity for some seniors [52].

According to specific findings from Taiwan, for example, old adults who live near greenways with high levels of neighbourhood social capital and high-quality routes, natural components, and seats, engage a lot in outdoor activities [53].

Studies conducted in Madrid showed that old adults represent 26.35% of the users of outdoor public spaces (like parks, squares, streets, etc.). However, their decision to stay in those places depends mostly on environmental variables, such as mean radiant temperature and air temperature [54]. One of the solutions supporting the presence of old adults outdoors is creating shaded spaces. Results from Turkey confirm that outdoor shaded places provide a better microclimate during hot and dry summers, as well as a social space boosted by pleasant conditions [55]. The impact of a shade is significant, which can be confirmed by thermovision measurements. In Polish conditions during the summer period, the difference in the temperatures of two benches in an urban park exceeded 17 °C [56], which influenced the comfort of users of the public space.

The shading function, increasing the quality of life of old people, could be served by smart textiles. First of all, they can offer shade for the space underneath (as in shading sails, see Figure 6), and at the same time have additional functions, such as, for instance, energy production from solar radiation.

For example, perovskite solar cells are a promising photovoltaic technology solution. There has been limited research into 3D (wire-shaped) flexible perovskite solar cells. Through composite integration, more work is needed to actualize technologies such as self-powering woven textiles and multifunctional materials [57]. However, some studies have proved that on the lab scale, perovskite solar cells present high power conversion efficiency, simple device fabrication, all solid-state structure, and the possibility to integrate traditional devices into a fibre format [58]. All experiments with perovskite solar cells showed that by now this solution is recognized as a strong contender for next generation photovoltaic technology [59]. The other alternative could be to use photo-thermoelectric textiles, which were already analysed as solar energy source in the case of clothes, [60,61] or combined solutions integrating different technological approaches [62].

Regardless of the technology adopted, in the case of shading sails made of smart textiles, there is no need to struggle with the typical technical difficulties which can be found when introducing smart textiles into clothes (i.e., with comfort and wearability challenges), [63]. This makes production easier on an industrial scale. An open question is how the produced energy could be used: outside or inside a nursing home. Considering that some studies focus directly on the preferences of old adults on the outdoor environment in nursing homes [64], future studies could analyse how this specific community cooperates with smart shading sails, including comfort or a sense of security.

Figure 6. Example of shading sails (Elche, Spain). Photo © Jan Kazak.

5. Applications, Potentials, Challenges for Life Quality Improvement

5.1. Smart Textiles for Health Monitoring

Among many elderly people, seeking comfort and an independent lifestyle are key reasons for using health monitoring or smart home systems. In particular, the shortage of hospitals and nursing facilities opens new development opportunities for smart textile technologies and intelligent sensors. One of them is pressure sensitive mats used in bedding to signify any physiological conditions, indicating movements and respiratory rates or alternating pressure on different parts of the body [65]. The COVID-19 pandemic also resulted in enormous development in protective clothing [39,66,67], especially in protective face masks [68,69].

Electrospun polymer membranes are excellent materials owing to their versatility, tremendous range of possible physic-chemical properties, and their tunability. Electrospun membranes are very useful for any biosensor applications, especially concerning health monitoring. The biosensors consist apart of the bifunctional membrane of transducers for biological substances detection, where the sensing comes through substances recognition process, and transducer converts it into output signals. Here, the electronic devices can be incorporated to textiles and membranes and act as skin-like sensors. The electrospun membranes give high permeability and flexibility [70]. One of the expanding technologies based on textiles are triboelectric nanogenerators (TENG), which can capture the energy from the charges generated at the fibres' surfaces by motion [71].

There are also some nanogenerators (NG) that are able to harvest energy from the environment via mechanical, thermal or other processes [72]. An example of incorporating nanotechnology based on electrospun fibre yarns in a real-life, smart textile applications, showcasing both the energy harvesting and motion sensing potential of the triboelectric yarn, is presented in Figure 7. The smart textile application is demonstrated for wearable and implantable devices based on the ore–shell structure of carbon nanotubes (CNT) yarn and electrospun PVDF fibres [73].

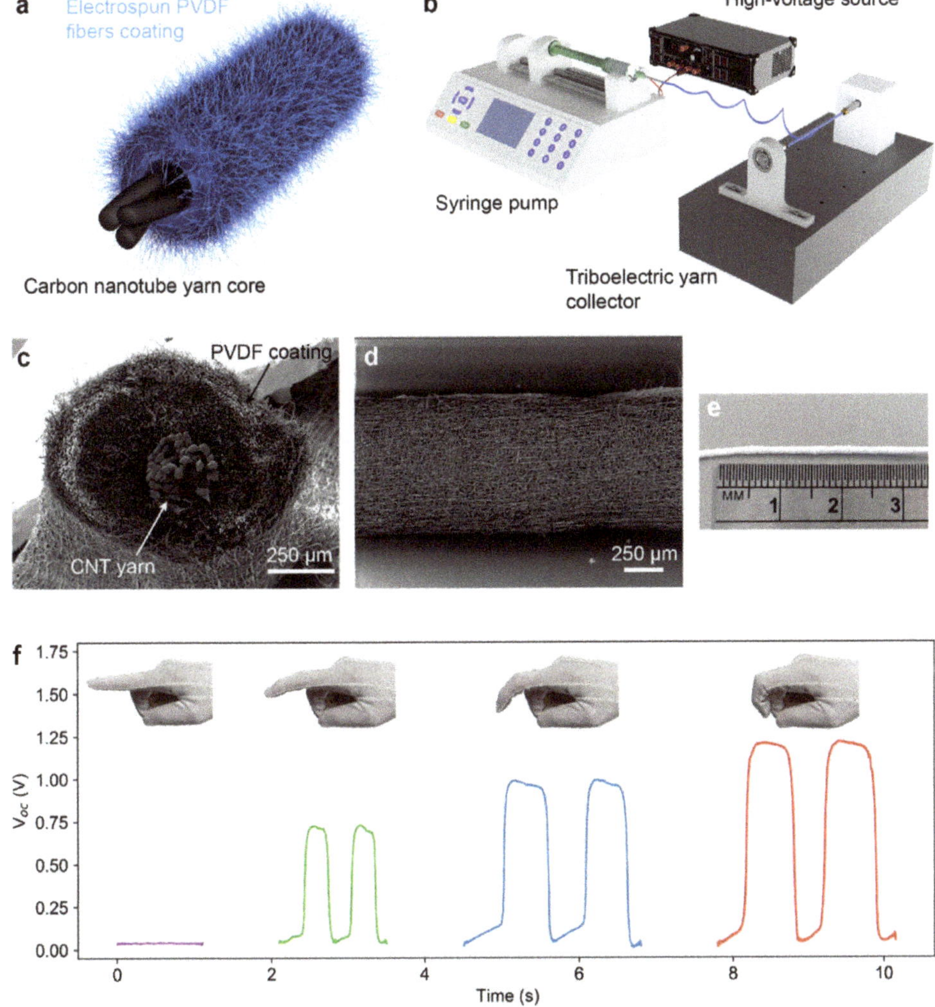

Figure 7. Triboelectric core–shell structure of yarns: (**a**) scheme of the yarn; (**b**) fabrication setup of yarn, including electrospinning setup and rotating collector; (**c**,**d**) SEM micrographs of the cross section and along the yarn; (**e**) image of triboelectric yarn above the ruler; (**f**) demonstration of the triboelectric yarn in a glove with the output measured as the haptic potential. The yarn (5 cm in length) was attached to the index finger of the glove. All figures are reproduced from [73] under the terms and conditions of an open access CC-BY license agreement.

While electrospinning is basically a fibre formation technique, which uses electrostatic forces to draw ultrafine fibres from a wide variety of polymers to create smart and hybrid fibres featuring piezoelectric, triboelectric, or multi-responsive mats. In electrospinning, high voltage is applied to the polymer solution that is typically pushed through a stainless-steel nozzle. Both type of voltage polarities can be applied to the nozzle and counter electrode, usually a collector in various configurations [74]. Due to the electrostatic forces, the polymer solution forms the cone jet [75] and charges distribute at the liquid interface, causing the spinning and stretching of the polymer jet [76]. During this process solvents evaporate and the solid fibres are produced [77], typically in random arrangements as a nonwoven membrane with porosity reaching above 90% [78]. All the parameters affecting the morphology and properties of electrospun fibres can be found in [79–81].

The collectors used for the deposition of fibres [82] allows the creation of 3D structures [83], not only for biomedical applications [84–86] but also for water [10,87–90] and energy harvesting [73,91,92]. However, modifying the physic-chemical properties of polymers may go along with an undesired change in their mechanical properties [93,94].

5.2. Smart Textiles for Life Quality Improvement in Dementia Disease

According to the literature, it has been proved that wearable textile sensors can benefit the home monitoring of several chronic diseases, including dementia [95,96] (Figure 8). There are today also new possibilities to incorporate smart textiles as large area displays with sensors and activators, that could be involved in the built environment (to serve as safety detectors), to enable way finding, and serve as functional communication tools thus improving the health care environment [97].

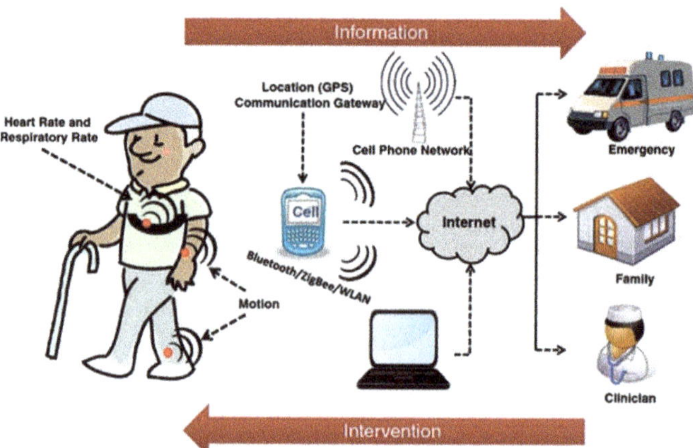

Figure 8. Illustration of a remote health monitoring system based on wearable sensors. Health-related information is gathered via body-worn wireless sensors and transmitted to the caregiver via an information gateway such as a mobile phone. Caregivers can use this information to implement interventions as needed. Figure reproduced from [96] under the terms and conditions of a Creative Commons CC-BY license agreement.

The purpose of the design is to develop through active textile designs and products that stimulate the senses, promote positive emotions, and offer comfort in both individual and social contexts. The design community is facing the challenge of developing custom products aimed at people with dementia, as global population aging is observed. It is important to demonstrate the benefits of creative design through research and development for old people. It is a challenge for social care providers to focus on prosperity and well-being, and not just on medical care. Through research, it seems that design solutions are urgently needed that can help the elderly and especially those suffering from dementia, to live well and enjoy life.

One current open challenge is, among others, the assessment of the possibilities of smart textiles in the care home environment for people living with dementia. Many old people, after a prolonged period of institutionalisation, feel lonely, socially isolated, and bored, and they are exposed to the risk of developing depression. The potential of a nursing home as a therapeutic resource is increasingly recognised through research on the impact of design and other environmental features and practices that support independence, enhance personal identity, and enhance quality of life. Other non-clinical practices that have been shown to be beneficial to the well-being of people with Alzheimer's and other related diseases include multi-sensory stimulation environments.

There is a growing body of research in the use of textiles including multi-sensory and e-textiles for care home residents, mostly as aprons, cushions, and small size objects, but this research is mostly focused on producing and evaluating objects for handling [98,99], see Figure 9.

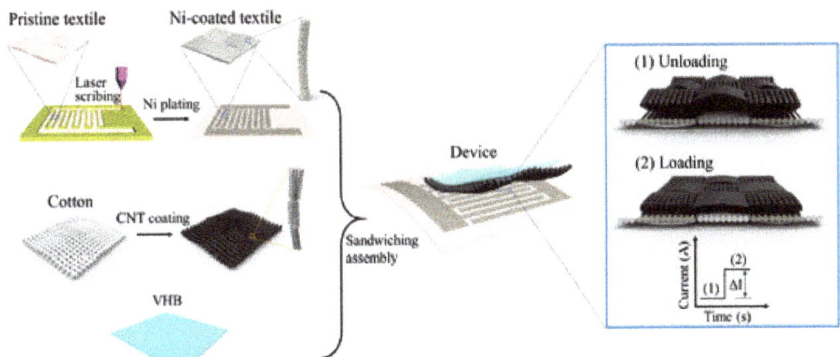

Figure 9. Detail example for the fabrication of textile pressure sensor based upon nickel (Ni) and carbon-nanotube-coated textiles. Figure reproduced from [99] under the terms and conditions of a Creative Commons CC-BY license agreement.

The use of smart textiles in interior architecture as multi-sensory stimulators, in this regard, looks like a promising solution that could engage people with dementia, thus increasing their quality of life. Smart textiles can appeal to the senses that are remaining active even in the last stages of dementia by providing auditory, visual, haptic, and other sensory stimulation as, for example, through vibration and light electro stimulation. Using large displays of smart textiles as part of the interior design or incorporated within furniture and soft furnishing, but also in especially designed art-sculptures, could provide a source of interactive engagement, thus increasing communication, pleasure, and consequently the quality of life of people living with dementia.

6. Conclusions

Over the decades, the development of new technologies, and especially smart textiles, has unavoidably led to new applications of traditional textiles in the built environment. Depending on special constructional requirements (i.e., acoustic insulation, thermal insulation, shading systems, indoor air quality improvement, etc.) or health monitoring and supporting needs (i.e., for patients with chronical disease, etc.), an increasing number of possible applications has been proposed. For instance, the use of smart textiles in interior architecture as multi-sensory stimulators is a promising solution that could engage people with dementia, thus increasing their quality of life. Using large displays of smart textiles incorporated in especially designed art-sculptures could provide a source of interactive engagement, thus increasing communication, pleasure, and consequently the quality of life of people living with dementia.

Smart textiles can also be efficiently used for improving the air quality in indoor atmospheres and contrast the typical phenomena of degradation and aging of traditional materials, with direct effects on air quality levels, especially in hospitals, nurseries, schools, and elderly and healthcare environments. Textiles and fibres in carpets as well as in clothes have a key role in terms of indoor air quality for special application in healthcare buildings, improving human well-being.

Another important point concerning the use of textiles in building is the shading function of smart materials increasing the quality of life of old people.

In conclusion, in this paper, as an ongoing activity of WG4 members for the CA17107 "CONTEXT" European research network (Horizon 2020), a review on selected applications for building and living solutions was presented and discussed with literature support, with special attention to healthcare environments, giving evidence of major outcomes and potentials for smart textiles-based products.

Author Contributions: This review article results from a joint collaboration of all the involved authors. All authors have read and agreed to the published version of the manuscript.

Funding: This research received no external funding.

Institutional Review Board Statement: Not applicable.

Informed Consent Statement: Not applicable.

Data Availability Statement: Not applicable.

Acknowledgments: The EU-COST Action CA17107 "CONTEXT"—European network to connect research and innovation efforts on advanced smart textiles (Horizon 2020), is gratefully acknowledged for facilitating the networking and scientific collaboration of the involved authors (WG4 members) and for providing financial support to their mobility for research interaction.

Conflicts of Interest: The authors declare no conflict of interest.

References

1. Echeverria, C.A.; Handoko, W.; Pahlevani, F.; Sahajwalla, V. Cascading use of textile waste for the advancement of fibre reinforced composites for building applications. *J. Clean. Prod.* **2019**, *208*, 1524–1536. [CrossRef]
2. CONTEXT. CA17107—European Network to Connect Research and Innovation Efforts on Advanced Smart Textiles. 2017. Available online: https://www.context-cost.eu/ (accessed on 6 November 2022).
3. Bedon, C.; Rajčić, V. Textiles and Fabrics for Enhanced Structural Glass Facades: Potentials and Challenges. *Buildings* **2019**, *9*, 156. [CrossRef]
4. Prabhakaran, R.T.D.; Spear, M.J.; Curling, S.; Wootton-Beard, P.; Jones, P.; Donnison, I.; Ormondroyd, G.A. Plants and architecture: The role of biology and biomimetics in materials development for buildings. *Intell. Build. Int.* **2019**, *11*, 178–211. [CrossRef]
5. Gruber, P.; Imhof, B. Patterns of Growth—Biomimetics and Architectural Design. *Buildings* **2017**, *7*, 32. [CrossRef]
6. Zhao, Y.; Fang, F. A Biomimetic Textile with Self-Assembled Hierarchical Porous Fibers for Thermal Insulation. *ACS Appl. Mater. Interfaces* **2022**, *14*, 25851–25860. [CrossRef] [PubMed]
7. Zuazua-Ros, A.; Martín-Gómez, C.; Ramos, J.C.; Gómez-Acebo, T. Bio-inspired Heat Dissipation System Integrated in Buildings: Development and Applications. *Energy Procedia* **2017**, *111*, 51–60. [CrossRef]
8. Fratzl, P. Biomimetic materials research: What can we really learn from nature's structural materials? *J. R. Soc. Interface* **2007**, *4*, 637–642. [CrossRef]
9. Stachewicz, U. Microstructure study of fractured polar bear hair for toughening, strengthening, stiffening designs *via* energy dissipation and crack deflection mechanisms in materials. *Mol. Syst. Des. Eng.* **2021**, *6*, 997–1002. [CrossRef]
10. Knapczyk-Korczak, J.; Szewczyk, P.K.; Ura, D.P.; Bailey, R.J.; Bilotti, E.; Stachewicz, U. Improving water harvesting efficiency of fog collectors with electrospun random and aligned Polyvinylidene fluoride (PVDF) fibers. *Sustain. Mater. Technol.* **2020**, *25*, e00191. [CrossRef]
11. Oatley, G.; Choudhury, T.; Buckman, P. Smart Textiles for Improved Quality of Life and Cognitive Assessment. *Sensors* **2021**, *21*, 8008. [CrossRef]
12. Blaylock, A.; Constantin, F.; Ligabue, L.; Bocaletti, L.; Siroka, B.; Siroky, J.; Wright, T.; Bechtold, T. Caregiver's vision of bedding textiles for elderly. *Fash. Text.* **2015**, *2*, 6. [CrossRef]
13. Gerhardt, L.-C.; Lenz, A.; Spencer, N.D.; Munzer, T.; Derler, S. Skin-textile friction and skin elasticity in young and aged persons. *Ski. Res. Technol.* **2009**, *15*, 288–298. [CrossRef] [PubMed]
14. Yao, M.; Li, L. Design Thinking Applied to Home Textiles Innovation: A Case Study in an Elderly Centre in Hong Kong. *Designs* **2022**, *6*, 49. [CrossRef]

15. Bournas, D.A.; Triantafillou, T.; Zygouris, K.; Stavropoulos, F. Textile-Reinforced Mortar versus FRP Jacketing in Seismic Retrofitting of RC Columns with Continuous or Lap-Spliced Deformed Bars. *J. Compos. Constr.* **2009**, *13*, 360–371. [CrossRef]

16. POLYMAST Project, 2011. Polyfunctional technical textiles for the protection and monitoring of masonry structures against earthquakes. Final report. Transnational access within the SERIES project: SEVENTH FRAMEWORK PROGRAM capacities specific programme research infrustructures (project No.: 227887).

17. Bournas, D.A. Concurrent seismic and energy retrofitting of RC and masonry building envelopes using inorganic tex-tile-based composites combined with insulation materials: A new concept. *Compos. Part B Eng.* **2018**, *148*, 166–179. [CrossRef]

18. Furtado, A.; Rodrigues, H.; Arêde, A.; Varum, A. Cost-effective analysis of textile-reinforced mortar solutions used to re-duce masonry infill walls collapse probability under seismic loads. *Structures* **2020**, *28*, 141–157. [CrossRef]

19. Jalil, W.D.A. Smart textiles for the architectural façade. In *IOP Conference Series: Materials Science and Engineering*; IOP Publishing: Bristol, UK, 2020; Volume 737. [CrossRef]

20. Oliveira, A. Smart Textile for Architecture: Living with Technology, Advances in Intelligent Systems and Computing Human Interaction, Emerging Technologies and Future Applications II. In Proceedings of the 2nd International Conference on Human Interaction and Emerging Technologies: Future Applications (IHIET—AI 2020), Lausanne, Switzerland, 23–25 April 2020.

21. Priniotakis, G.; Stachewicz, U.; van Hoof, J. Smart textiles and the indoor environment of buildings. *Indoor Built Environ.* **2022**, *31*, 1443–1446. [CrossRef]

22. Gounni, A.; Mabrouk, M.T.; El Wazna, M.; Kheiri, A.; El Alami, M.; El Bouari, A.; Cherkaoui, O. Thermal and economic evaluation of new insulation materials for building envelope based on textile waste. *Appl. Therm. Eng.* **2019**, *149*, 475–483. [CrossRef]

23. Gravagnuolo, A.; Angrisano, M.; Girard, L.F. Circular Economy Strategies in Eight Historic Port Cities: Criteria and Indicators Towards a Circular City Assessment Framework. *Sustainability* **2019**, *11*, 3512. [CrossRef]

24. Briga-Sá, A.; Nascimento, D.; Teixeira, N.; Pinto, J.; Caldeira, F.; Varum, H.; Paiva, A. Textile Waste as an Alternative Thermal Insulation Building Material Solution. *Constr. Build. Mater.* **2013**, *38*, 155–160. [CrossRef]

25. Binici, H.; Eken, M.; Dolaz, M.; Aksogan, O.; Kara, M. An environmentally-friendly thermal insulation material from sunflower stalk, textile waste and stubble fibres. *Constr. Build. Mater.* **2014**, *51*, 24–33. [CrossRef]

26. Antolinc, D.; Filipič, K.E. Recycling of Nonwoven Polyethylene Terephthalate Textile into Thermal and Acoustic Insulation for More Sustainable Buildings. *Polymers* **2021**, *13*, 3090. [CrossRef] [PubMed]

27. Iaşnicu, I.; Vasile, O.; Iatan, R. Thickness influence on absorbing properties of stratified composite materials. *J. Eng. Stud. Res.* **2015**, *21*, 28–34. [CrossRef]

28. Oh, Y.K. An assessment model for the indoor noise environment of aged apartment houses. *J. Asian Archit. Build. Eng.* **2014**, *13*, 445–451. [CrossRef]

29. Mukhopadhyay, S.; Khatana, S. A review on the use of fibers in reinforced cementitious concrete. *J. Ind. Text.* **2015**, *45*, 239–264. [CrossRef]

30. Danihelová, A.; Němec, M.; Gergeľ, T.; Gejdoš, M.; Gordanová, J.; Sčensný, P. Usage of Recycled Technical Textiles as Thermal Insulation and an Acoustic Absorber. *Sustainability* **2019**, *11*, 2968. [CrossRef]

31. Enache, F.; Vasile, V.; Gruin, A.; Bolborea, B. Use of wool in the composition of building materials-sustainable solution for improving thermal, acoustic performance and indoor air quality. *Constructii* **2020**, *21*, 43–46.

32. Jami, T.; Karade, S.R.; Singh, L.P. A review of the properties of hemp concrete for green building applications. *J. Clean. Prod.* **2019**, *239*, 117852. [CrossRef]

33. Pixabay GmbH. Available online: https://pixabay.com/ (accessed on 6 November 2022).

34. Next Technology Tecnotessile (NTT). Available online: https://www.tecnotex.it (accessed on 6 November 2022).

35. Kizildag, N. Smart composite nanofiber mats with thermal management functionality. *Sci. Rep.* **2021**, *11*, 4256. [CrossRef]

36. Chalco-Sandoval, W.; Fabra, M.J.; López-Rubio, A.; Lagaron, J.M. Use of phase change materials to develop electrospun coatings of interest in food packaging applications. *J. Food Eng.* **2017**, *192*, 122–128. [CrossRef]

37. Paroutoglou, E.; Fojan, P.; Gurevich, L.; Afshari, A. Thermal Properties of Novel Phase-Change Materials Based on Tamanu and Coconut Oil Encapsulated in Electrospun Fiber Matrices. *Sustainability* **2022**, *14*, 7432. [CrossRef]

38. Mao, X.; Hatton, T.A.; Rutledge, G.C. A Review of Electrospun Carbon Fibers as Electrode Materials for Energy Storage. *Curr. Org. Chem.* **2013**, *17*, 1390–1401. [CrossRef]

39. Ivanoska-Dacikj, A.; Stachewicz, U. Smart textiles and wearable technologies—Opportunities offered in the fight against pandemics in relation to current COVID-19 state. *Rev. Adv. Mater. Sci.* **2020**, *59*, 487–505. [CrossRef]

40. Metwally, S.; Comesaña, S.M.; Zarzyka, M.; Szewczyk, P.K.; Karbowniczek, J.E.; Stachewicz, U. Thermal insulation design bioinspired by microstructure study of penguin feather and polar bear hair. *Acta Biomater.* **2019**, *91*, 270–283. [CrossRef]

41. Altay, P.; Uçar, N. Comparative analysis of sound and thermal insulation properties of porous and non-porous polystyrene submicron fiber membranes. *J. Text. Inst.* **2022**, *113*, 2177–2184. [CrossRef]

42. Si, Y.; Yu, Y.; Tang, X.; Ge, J.; Ding, B. Ultralight nanofibre-assembled cellular aerogels with superelasticity and multifunctionality. *Nat. Commun.* **2014**, *5*, 5802. [CrossRef]

43. Bertocchi, M.J.; Vang, P.; Balow, R.B.; Wynne, J.H.; Lundin, J.G. Enhanced Mechanical Damping in Electrospun Polymer Fibers with Liquid Cores: Applications to Sound Damping. *ACS Appl. Polym. Mater.* **2019**, *1*, 2068–2076. [CrossRef]

44. Alves, J.; Paiva, F.; Silva, L.; Remoaldo, P. Low-Frequency Noise and Its Main Effects on Human Health—A Review of the Literature between 2016 and 2019. *Appl. Sci.* **2020**, *10*, 5205. [CrossRef]

45. Hurrell, A.; Horoshenkov, K.; King, S.; Stolojon, V. On the relationship of the observed acoustical and related non-acoustical behaviours of nanofibers membranes using Biot- and Darcy-type models. *Appl. Acoust.* **2021**, *179*, 108075. [CrossRef]

46. Rabbi, A.; Bahrambeygi, H.; Shoushtari, A.M.; Nasouri, K. Incorporation of Nanofiber Layers in Nonwoven Materials for Improving Their Acoustic Properties. *J. Eng. Fibers Fabr.* **2013**, *8*, 155892501300800412. [CrossRef]

47. Weschler, C.J. Ozone's Impact on Public Health: Contributions from Indoor Exposures to Ozone and Products of Ozone-Initiated Chemistry. *Environ. Health Perspect.* **2006**, *114*, 1489–1496. [CrossRef] [PubMed]

48. Lakey, P.S.J.; Morrison, G.C.; Won, Y.; Parry, K.M.; Von Domaros, M.; Tobias, D.J.; Rim, D.; Shiraiwa, M. The impact of clothing on ozone and squalene ozonolysis products in indoor environments. *Commun. Chem.* **2019**, *2*, 56. [CrossRef]

49. Abbass, O.A.; Sailor, D.J.; Gall, E.T. Effect of fiber material on ozone removal and carbonyl production from carpets. *Atmos. Environ.* **2017**, *148*, 42–48. [CrossRef]

50. Abidi, M.; Hajjaji, A.; Bouzaza, A.; Lamaa, L.; Peruchon, L.; Brochier, C.; Rtimi, S.; Wolbert, D.; Bessais, B.; Assadi, A.A. Mod-eling of indoor air treatment using an innovative photocatalytic luminous textile: Reactor compactness and mass transfer enhancement. *Chem. Eng. J.* **2022**, *430*, 132636. [CrossRef]

51. Zhu, L.; Liu, Y.; Ding, X.; Wu, X.; Sand, W.; Zhou, H. A novel method for textile odor removal using engineered water nanostructures. *RSC Adv.* **2019**, *9*, 17726–17736. [CrossRef]

52. Akinci, Z.S.; Marquet, O.; Delclòs-Alió, X.; Miralles-Guasch, C. Urban vitality and seniors' outdoor rest time in Barcelona. *J. Transp. Geogr.* **2022**, *98*, 103241. [CrossRef]

53. Chang, P.-J. Effects of the built and social features of urban greenways on the outdoor activity of older adults. *Landsc. Urban Plan.* **2020**, *204*, 103929. [CrossRef]

54. Larriva, M.T.B.; Higueras, E. Health risk for older adults in Madrid, by outdoor thermal and acoustic comfort. *Urban Clim.* **2020**, *34*, 100724. [CrossRef]

55. Yıldırım, M. Shading in the outdoor environments of climate-friendly hot and dry historical streets: The passageways of Sanliurfa, Turkey. *Environ. Impact Assess. Rev.* **2019**, *80*, 106318. [CrossRef]

56. Szopińska, E.; Kazak, J.; Kempa, O.; Rubaszek, J. Spatial Form of Greenery in Strategic Environmental Management in the Context of Urban Adaptation to Climate Change. *Pol. J. Environ. Stud.* **2019**, *28*, 2845–2856. [CrossRef]

57. Adams, G.R.; Okoli, O.I. A review of perovskite solar cells with a focus on wire-shaped devices. *Renew. Energy Focus* **2018**, *25*, 17–23. [CrossRef]

58. Hussain, I.; Chowdhury, A.R.; Jaksik, J.; Grissom, G.; Touhami, A.; Ibrahim, E.E.; Schauer, M.; Okoli, O.; Uddin, M.J. Conductive glass free carbon nanotube micro yarn based perovskite solar cells. *Appl. Surf. Sci.* **2019**, *478*, 327–333. [CrossRef]

59. Yang, Y.; Hoang, M.T.; Bhardwaj, A.; Wilhelm, M.; Mathur, S.; Wang, H. Perovskite solar cells based self-charging power packs: Fundamentals, applications and challenges. *Nano Energy* **2022**, *94*, 106910. [CrossRef]

60. Zhang, X.; Shiu, B.; Li, T.-T.; Liu, X.; Ren, H.-T.; Wang, Y.; Lou, C.-W.; Lin, J.-H. Photo-thermoelectric nanofiber film based on the synergy of conjugated polymer and light traps for the solar-energy harvesting of textile solar panel. *Sol. Energy Mater. Sol. Cells* **2021**, *232*, 111353. [CrossRef]

61. Zhang, X.; Shiu, B.; Li, T.-T.; Liu, X.; Ren, H.-T.; Wang, Y.; Lou, C.-W.; Lin, J.-H. Synergistic work of photo-thermoelectric and hydroelectric effects of hierarchical structure photo-thermoelectric textile for solar energy harvesting and solar steam generation simultaneously. *Chem. Eng. J.* **2021**, *426*, 131923. [CrossRef]

62. Zhou, Y.; Chen, Y.; Zhang, Q.; Zhou, Y.; Tai, M.; Koumoto, K.; Lin, H. A highly-efficient concentrated perovskite solar cell-thermoelectric generator tandem system. *J. Energy Chem.* **2021**, *59*, 730–735. [CrossRef]

63. Du, K.; Lin, R.; Yin, L.; Ho, J.S.; Wang, J.; Lim, C.T. Electronic textiles for energy, sensing, and communication. *iScience* **2022**, *25*, 104174. [CrossRef]

64. Liu, J.; Wei, Y.; Lu, S.; Wang, R.; Chen, L.; Xu, F. The elderly's preference for the outdoor environment in Fragrant Hills Nursing Home, Beijing: Interpreting the visual-behavioural relationship. *Urban For. Urban Green.* **2021**, *64*, 127242. [CrossRef]

65. Arcelus, A.; Jones, M.H.; Goubran, R.; Knoefel, F. Integration of Smart Home Technologies in a Health Monitoring System for the Elderly. In Proceedings of the 21st International Conference on Advanced Information Networking and Applications Workshops (AINAW'07), Niagara Falls, ON, Canada, 21–23 May 2007.

66. Karagoz, S.; Kiremitler, N.B.; Sarp, G.; Pekdemir, S.; Salem, S.; Goksu, A.G.; Onses, M.S.; Sozdutmaz, I.; Sahmetlioglu, E.; Ozkara, E.S.; et al. Antibacterial, Antiviral, and Self-Cleaning Mats with Sensing Capabilities Based on Electrospun Nanofibers Decorated with ZnO Nanorods and Ag Nanoparticles for Protective Clothing Applications. *ACS Appl. Mater. Interfaces* **2021**, *13*, 5678–5690. [CrossRef]

67. Goel, S.; Hawi, S.; Goel, G.; Thakur, V.K.; Agrawal, A.; Hoskins, C.; Pearce, O.; Hussain, T.; Upadhyaya, H.M.; Cross, G.; et al. Resilient and agile engineering solutions to address societal challenges such as coronavirus pandemic. *Mater. Today Chem.* **2020**, *17*, 100300. [CrossRef]

68. Naragund, V.S.; Panda, P.K. Electrospun nanofiber-based respiratory face masks—A review. *Emergent Mater.* **2022**, *5*, 261–278. [CrossRef] [PubMed]

69. Zakrzewska, A.; Bayan, M.A.H.; Nakielski, P.; Petronella, F.; De Sio, L.; Pierini, F. Nanotechnology Transition Roadmap toward Multifunctional Stimuli-Responsive Face Masks. *ACS Appl. Mater. Interfaces* **2022**, *14*, 46123–46144. [CrossRef] [PubMed]

70. Wang, Y.; Tan, P.; Wu, Y.; Luo, D.; Li, Z. Artificial intelligence-enhanced skin-like sensors based on flexible nanogenerators. *View* **2022**, *3*, 20220026. [CrossRef]

71. Babu, A.; Aazem, I.; Walden, R.; Bairagi, S.; Mulvihill, D.M.; Pillai, S.C. Electrospun nanofiber based TENGs for wearable electronics and self-powered sensing. *Chem. Eng. J.* **2022**, *452*, 139060. [CrossRef]

72. Pang, Y.; Cao, Y.; Derakhshani, M.; Fang, Y.; Wang, Z.L.; Cao, C. Hybrid Energy-Harvesting Systems Based on Triboelectric Nanogenerators. *Matter* **2021**, *4*, 116–143. [CrossRef]

73. Busolo, T.; Szewczyk, P.K.; Nair, M.; Stachewicz, U.; Kar-Narayan, S. Triboelectric Yarns with Electrospun Functional Polymer Coatings for Highly Durable and Washable Smart Textile Applications. *ACS Appl. Mater. Interfaces* **2021**, *13*, 16876–16886. [CrossRef] [PubMed]

74. Ura, D.P.; Stachewicz, U. The Significance of Electrical Polarity in Electrospinning: A Nanoscale Approach for the Enhancement of the Polymer Fibers' Properties. *Macromol. Mater. Eng.* **2022**, *307*, 2100843. [CrossRef]

75. Stachewicz, U.; Dijksman, J.F.; Yurteri, C.; Marijnissen, J.C.M. Experiments on single event electrospraying. *Appl. Phys. Lett.* **2007**, *91*, 254109. [CrossRef]

76. Stachewicz, U.; Stone, C.A.; Willis, C.R.; Barber, A.H. Charge assisted tailoring of chemical functionality at electrospun nanofiber surfaces. *J. Mater. Chem.* **2012**, *22*, 22935–22941. [CrossRef]

77. Szewczyk, P.K.; Stachewicz, U. The impact of relative humidity on electrospun polymer fibers: From structural changes to fiber morphology. *Adv. Colloid Interface Sci.* **2020**, *286*, 102315. [CrossRef]

78. Stachewicz, U.; Modaresifar, F.; Bailey, R.J.; Peijs, T.; Barber, A.H. Manufacture of Void-Free Electrospun Polymer Nanofiber Composites with Optimized Mechanical Properties. *ACS Appl. Mater. Interfaces* **2012**, *4*, 2577–2582. [CrossRef] [PubMed]

79. Ura, D.P.; Rosell-Llompart, J.; Zaszczyńska, A.; Vasilyev, G.; Gradys, A.; Szewczyk, P.K.; Knapczyk-Korczak, J.; Avrahami, R.; Šišková, A.O.; Arinstein, A.; et al. The Role of Electrical Polarity in Electrospinning and on the Mechanical and Structural Properties of As-Spun Fibers. *Materials* **2020**, *13*, 4169. [CrossRef] [PubMed]

80. Mailley, D.; Hébraud, A.; Schlatter, G. A Review on the Impact of Humidity during Electrospinning: From the Nanofiber Structure Engineering to the Applications. *Macromol. Mater. Eng.* **2021**, *306*, 2100115. [CrossRef]

81. Xue, J.; Wu, T.; Dai, Y.; Xia, Y. Electrospinning and Electrospun Nanofibers: Methods, Materials, and Applications. *Chem. Rev.* **2019**, *119*, 5298–5415. [CrossRef]

82. Lavielle, N.; Hébraud, A.; Schlatter, G.; Thöny-Meyer, L.; Rossi, R.M.; Popa, A.M. Simultaneous Electrospinning and Electrospraying: A Straightforward Approach for Fabricating Hierarchically Structured Composite Membranes. *ACS Appl. Mater. Interfaces* **2013**, *5*, 10090–10097. [CrossRef]

83. Stachewicz, U.; Qiao, T.; Rawlinson, S.C.; Almeida, F.; Li, W.-Q.; Cattell, M.; Barber, A.H. 3D imaging of cell interactions with electrospun PLGA nanofiber membranes for bone regeneration. *Acta Biomater.* **2015**, *27*, 88–100. [CrossRef]

84. Krysiak, Z.J.; Stachewicz, U. Electrospun fibers as carriers for topical drug delivery and release in skin bandages and patches for atopic dermatitis treatment. *WIREs Nanomed. Nanobiotechnol.* **2022**, e1829. [CrossRef]

85. Jun, I.; Han, H.-S.; Edwards, J.R.; Jeon, H. Electrospun Fibrous Scaffolds for Tissue Engineering: Viewpoints on Architecture and Fabrication. *Int. J. Mol. Sci.* **2018**, *19*, 745. [CrossRef]

86. Martins, A.; Reis, R.L.; Neves, N.M. Electrospinning: Processing technique for tissue engineering scaffolding. *Int. Mater. Rev.* **2008**, *53*, 257–274. [CrossRef]

87. Knapczyk-Korczak, J.; Szewczyk, P.K.; Stachewicz, U. The importance of nanofiber hydrophobicity for effective fog water collection. *RSC Adv.* **2021**, *11*, 10866–10873. [CrossRef]

88. Knapczyk-Korczak, J.; Stachewicz, U. Biomimicking spider webs for effective fog water harvesting with electrospun polymer fibers. *Nanoscale* **2021**, *13*, 16034–16051. [CrossRef] [PubMed]

89. Ura, D.P.; Knapczyk-Korczak, J.; Szewczyk, P.K.; Sroczyk, E.A.; Busolo, T.; Marzec, M.M.; Bernasik, A.; Kar-Narayan, S.; Stachewicz, U. Surface Potential Driven Water Harvesting from Fog. *ACS Nano* **2021**, *15*, 8848–8859. [CrossRef] [PubMed]

90. Knapczyk-Korczak, J.; Szewczyk, P.K.; Ura, D.P.; Berent, K.; Stachewicz, U. Hydrophilic nanofibers in fog collectors for increased water harvesting efficiency. *RSC Adv.* **2020**, *10*, 22335–22342. [CrossRef] [PubMed]

91. Szewczyk, P.K.; Gradys, A.; Kim, S.K.; Persano, L.; Marzec, M.; Kryshtal, A.; Busolo, T.; Toncelli, A.; Pisignano, D.; Bernasik, A.; et al. Enhanced Piezoelectricity of Electrospun Polyvinylidene Fluoride Fibers for Energy Harvesting. *ACS Appl. Mater. Interfaces* **2020**, *12*, 13575–13583. [CrossRef]

92. Busolo, T.; Ura, D.P.; Kim, S.K.; Marzec, M.M.; Bernasik, A.; Stachewicz, U.; Kar-Narayan, S. Surface potential tailoring of PMMA fibers by electrospinning for enhanced triboelectric performance. *Nano Energy* **2018**, *57*, 500–506. [CrossRef]

93. Karbowniczek, J.E.; Ura, D.P.; Stachewicz, U. Nanoparticles distribution and agglomeration analysis in electrospun fiber based composites for desired mechanical performance of poly(3-hydroxybuty-rate-co-3-hydroxyvalerate (PHBV) scaffolds with hydroxyapatite (HA) and titanium dioxide (TiO_2) towards medical applications. *Compos. Part B Eng.* **2022**, *241*, 110011. [CrossRef]

94. Kaniuk, Ł.; Podborska, A.; Stachewicz, U. Enhanced mechanical performance and wettability of PHBV fiber blends with evening primrose oil for skin patches improving hydration and comfort. *J. Mater. Chem. B* **2022**, *10*, 1763–1774. [CrossRef]

95. Hong, K. SMART Textiles: The Use of Embedded Technology on Tactile Textiles as Therapy for the Elderly. In *Proceedings of the International Colloquium in Textile Engineering, Fashion, Apparel and Design 2014 (ICTEFAD 2014)*; Ahmad, M., Yahya, M., Eds.; Springer: Singapore, 2014; pp. 43–47. [CrossRef]

96. Patel, S.; Park, H.; Bonato, P.; Chan, L.; Rodgers, M. A review of wearable sensors and systems with application in rehabilitation. *J. Neuroeng. Rehabil.* **2012**, *9*, 21. [CrossRef]

97. Jakob, A.; Collier, L. Sensory enrichment for people living with dementia: Increasing the benefits of multisensory environments in dementia care through design. *Des. Health* **2017**, *1*, 115–133. [CrossRef]
98. Treadaway, C.; Fennell, J.; Kenning, G.; Prytherch, D.; Walters, A. Designing for wellbeing in late stage dementia, Well-Being 2016: Co-Creating Pathways to Well-Being. In Proceedings of the Third International Conference Exploring the Multi-Dimensions of Well-Being, Birmingham, UK, 5–6 September 2016; pp. 126–129.
99. Yang, K.; Isaia, B.; Brown, L.J.; Beeby, S. E-Textiles for Healthy Ageing. *Sensors* **2019**, *19*, 4463. [CrossRef]

Article

Effect of Interlayer and Inclined Screw Arrangements on the Load-Bearing Capacity of Timber-Concrete Composite Connections

Yuri De Santis [1,*], Martina Sciomenta [1], Luca Spera [1], Vincenzo Rinaldi [1], Massimo Fragiacomo [1] and Chiara Bedon [2]

[1] Department of Civil, Architecture and Building and Environmental Engineering, University of L'Aquila, Via Giovanni Gronchi 18, 67100 L'Aquila, Italy
[2] Department of Engineering and Architecture, University of Trieste, Via Alfonso Valerio 6/1, 34127 Trieste, Italy
* Correspondence: yuri.desantis@graduate.univaq.it

Abstract: The solution of timber-to-concrete composite (TCC) floors represents a well-established construction technique, which is consistently used for both the retrofitting of existing timber floors and the realization of new diaphragms. The success of TCC floors relies on the intrinsic effectiveness in increasing both the in-plane (for lateral loads) and the out-of-plane (for gravity loads) performance of existing timber floors. As a widespread retrofit intervention, it is common to use existing floorboards as a permanent formwork for the concrete pouring. Rather few research studies of literature, in this regard, highlighted an overall reduction of load capacity and slip modulus due to the presence of such an interposed interlayer. In this regard, the present paper focuses on the use of screws as efficient mechanical connectors and analyses different configurations and inclination angles for their arrangement. This main goal is achieved by performing parametric Finite Element (FE) numerical analyses, validated on previous experimental tests, in order to specifically investigate the influence of the in-between interlayer, as well as the role of friction phenomena and the influence of the test setup and experimental protocol to achieve the basic mechanical performance indicators.

Keywords: timber-to-concrete composite (TCC) connections; self-tapping screws; inclined screws; interlayers; Finite Element (FE) models; experiments; push-out setup; friction; slip modulus

Citation: De Santis, Y.; Sciomenta, M.; Spera, L.; Rinaldi, V.; Fragiacomo, M.; Bedon, C. Effect of Interlayer and Inclined Screw Arrangements on the Load-Bearing Capacity of Timber-Concrete Composite Connections. *Buildings* **2022**, *12*, 2076. https://doi.org/10.3390/buildings12122076

Academic Editor: Humberto Varum

Received: 22 October 2022
Accepted: 20 November 2022
Published: 26 November 2022

Publisher's Note: MDPI stays neutral with regard to jurisdictional claims in published maps and institutional affiliations.

1. Introduction

Timber-to-concrete composite (TCC) systems with screws have become since the 1980s a widespread solution for floor-building and retrofitting, representing an option to the typical "timber plank" additional layer [1].

The framework of a generic TCC floor consists of a thin concrete layer overlapping timber beams mutually joined with steel fasteners (i.e., dowels, plates and screws). These can be fully or partially threaded and can be placed perpendicularly to the floor or inclined at an angle. The high efficiency of this solution compared to traditional floors lies in: (i) increased strength and stiffness, (ii) excellent acoustic/thermal insulation, (iii) improved fire resistance [2–4], (iv) capacity to reduce the sensitivity to vibrations, (v) horizontal diaphragms for structures and (vi) redistribution of more than 50% of a concentrated load to the nearest joists [5]. The overall behaviour of a composite TTC floor is mostly influenced by the interaction between timber, concrete and screws. Timber is installed in the tensile zone, the thin concrete layer in the compression zone, while the screws, which transfer the internal actions between the two members, are mainly subjected to shear stresses. A combination of shear and tensile forces develops into the screws if they are inclined. To guarantee the applicability of the section analysis, no slip should occur at the interface of load-bearing elements, and the shear connectors should be sufficiently stiff. However, a certain relative slip between concrete and timber is often present, and this phenomenon should be properly taken into account in the design process [6]. Most importantly, for accurate structural design calculations, the partial interaction between members should be accounted for when calculating the equivalent bending stiffness, the maximum stresses in each component and the maximum forces in the fasteners [7]. In most cases, however, the design process is regulated by short-term and long-term deformability checks only. A hybrid multiscale-based material model–standardized structural analysis method could be thus used for the analysis of TCC floor–creep behaviour [8].

The connection system is a fundamental part of TCC structures, and thus different types of connectors have been proposed and studied in recent years. Ceccotti [9] proposed a stiffness classification index by selecting the most common ones. In that study, he emphasized in particular that dowel-type fasteners (i.e., nails, screws and dowels) are less rigid than surface connectors, while the notched elements with connectors are less stiff than those with glued interfaces. Compared to other fasteners, screws have the advantage of being easily available and also easy to install [10].

Long threaded screws have been used for fastening and reinforcement interventions in recent years. Traditionally, screw fasteners have been typically placed perpendicular to the sliding interface plane. In such a configuration, the relative sliding between the timber beam and the concrete slab is mostly contrasted by the bending stiffness of the screw fastener and thus results in a relatively low slip modulus for the TCC system. The configurations with inclined screws, on the other hand, make possible the load transfer to the timber element, even along the axial direction, to take the most advantage of the generally high tensile capacity of screws. However, by reducing the angle between the screw and the shear plane, installation difficulties arise especially for extreme inclination angles, and an inclination of 45° represents a balanced solution between mechanical performance/capacity and ease of application in practice [11]. Overall, under a given geometric and loading condition, TCC floor configurations with inclined screws require a smaller number of connectors compared to those with perpendicular screws.

Several experimental tests and analytical formulations have been thus carried out to assess the load-carrying behaviour of screws for TCC solutions in different configurations. Among others, an interesting comparison has been proposed in terms of single-inclination screws and inclined cross-screws subjected to shear stress forces. Steinberg et al. [12] performed push-out tests with lightweight concrete slabs and different screw configurations, proving that the shear stiffness of inclined cross screws is significantly higher than a single-inclined screw. Kavaliauskas et al. [13] evaluated the possible use of Johansen's yield theory to estimate the ultimate load for TCC floors using self-tapping threaded screws at a given angle to timber-concrete interface. Bedon and Fragiacomo [14] proposed an extended FE numerical model to reconstruct the general mechanical behaviour of notched connections for TCC beams and floor components. The main purpose was to reproduce the failure mechanisms of push-out specimens made of steel, concrete and timber, giving evidence of screw arrangements and their effects on local and global mechanical performances, including damage. Other FE numerical analyses were developed by Sciomenta et al. [15], in order to understand how different concrete types (and especially Rubbercrete instead of ordinary concrete) and inclination angles could modify the connector load-bearing capacity.

In the last years, rather few authors have investigated the influence of the interlayer which is commonly placed between the timber element and the concrete slab, and more precisely its effect on the overall load-bearing capacity. This layout represents the typical structural situation in which timber reinforcing elements are positioned on existing flooring, as for example happens in common practice for most retrofit interventions. Van Der Linden [16], in this regard, investigated the effect of an OSB (Oriented Strand Board) interlayer with screws and observed a decrease of 50% on the slip modulus for the TCC system, compared to the configuration without an interposed OSB interlayer. Jorge et al. [17] carried out a set of experimental tests accounting for the presence of a timber interlayer in a conventional push-out setup, for specimens made of normal-weight concrete (NWC) as well as of lightweight concrete (LWAC). In the set of specimens made of NWC, for example, it was observed that the interlayer inclusion leads to an average reduction of 30% and 50% for the load-bearing capacity and slip modulus, respectively. In the set of specimens made by LWAC, in contrast, it was found that such a reduction falls to 10% and 30%, respectively.

The effect of a plywood interlayer coupled with self-compacting concrete (SCC) and SFS screws was investigated by Moshiri et al. [18]. Through that experimental campaign, the undesirable effect of interlayer on strength and stiffness was confirmed overall. A significant reduction of both the serviceability and ultimate slip moduli (ranging 20–40%) was also experimentally observed. The influence of an interlayer on mechanical properties of TCC structures using threaded rebars as shear connectors was finally studied in [19], and also in this case the experimental evidences highlighted a modification in shear strength and stiffness of TCC connections. In terms of modelling, a mathematical model capable of estimating the slip modulus of TCC systems was developed in [20], so as to take into account also its dependency upon the screw inclination and the interlayer flooring thickness. In [21], the same mathematical model was further elaborated and extended, confirming the limitation of inclined screws in terms of slip modulus, due to the weakness of the timber interlayer. Recently, two new analytical models have been proposed in [10,22] for the slip modulus prediction of connections with inclined screws in TCC systems. Their main advantage is the capacity of accounting for sound insulation [22] and timber board interlayer [10], respectively. In both cases, it is shown that the concrete thickness has no influence on the connection stiffness, and the interlayer has a negative impact.

In this paper, a set of FE parametric analyses is carried out on TCC connections with inclined screws and an OSB interlayer. A novel approach is proposed to efficiently calibrate the mechanical properties of connection material components, with specific evidence of the cohesive contact between the screw and the surrounding wooden element. Under these working assumptions, a parametric investigation focused on the interlayer influence on TCC load-bearing capacity is thus presented. To this aim, preliminary FE analyses are first validated based on former push-out experimental results [20]. Successively, different screw inclinations, as well as thicknesses and mechanical properties for the interlayer, are taken into account. The influence of interaction type and features on the sliding plane, as well as the test setup (i.e., direct shear, push-out or inclined shear), are also examined and quantified in terms of mechanical performance indicators for the examined TCC connections.

2. Background Experiments

An experimental campaign was carried out to assess and quantify the influence of the in-between interlayer, as well as the role of the screw length, on the basic mechanical performance indicators of inclined screw connections commonly used in TCC floors.

A total of twenty push-out tests were investigated in [20], with specimens divided into two main groups, namely "clc8160" and "clc8240" series, in which the samples differed in geometrical features, as shown in Figure 1. Both the groups were characterized by a symmetric configuration, consisting of an inner glue-laminated timber (glulam) element and two outer concrete slabs. The latter, made of concrete of strength class C25/30 according to Eurocode 2 [23], were connected to the central glulam element, with strength class GL24h according to EN 14080:2013 [24], by two fully threaded screws per side, at an inclination of 45°. Between the timber elements and the concrete ones, an OSB panel was thus interposed, acting as an existing floor panel or a disposable formwork in existing or new buildings, respectively. Regarding the fasteners and interlayers in use in the first group of specimens (clc8160), screws with a diameter d of 8 mm and a length l_p of 160 mm, and OSB panels with a thickness t_i of 22 mm, were taken into account (Figure 1a). In the second group, clc8240, screws with the same diameter (8 mm) and a length of 240 mm, with 44 mm-thick OSB panels, were considered (Figure 1b). The cross-section of concrete slabs of both groups were 50 mm by 500 mm, with a length of 450 mm. The glulam elements were

450 mm in length and 160 mm by 120 mm in cross-section for clc8160 samples, while the cross-section increased to 200 mm by 120 mm for clc8240 specimens.

Figure 1. Reference push-out specimens for the experimental tests: (**a**) clc8160 and (**b**) clc8240 series.

All the tests were conducted in accordance with UNI EN 26891:1991 [25], which describes the procedural stages of push-out protocol. After estimating F_{est}, which represents the maximum load expected based on preliminary evaluations, the load was increased to 0.4 F_{est} and maintained for 30 s, then was decreased to 0.1 F_{est} and maintained for the same time period. Subsequently the load was increased further, until the failure of the specimen or a maximum deformation of 15 mm (whichever came first).

The relative slip between concrete and timber was measured with displacement transducers placed on both the interfaces. In addition, two metal rings were adopted in order to prevent the possible separation of the elements due to the load-reaction eccentricity. Finally, all the tests were conducted by load control, and the displacements were recorded versus the total applied load, allowing the typical load-displacement curves to be obtained, as displayed in Figure 2. Relevant specimen data and results are given in Tables 1–3.

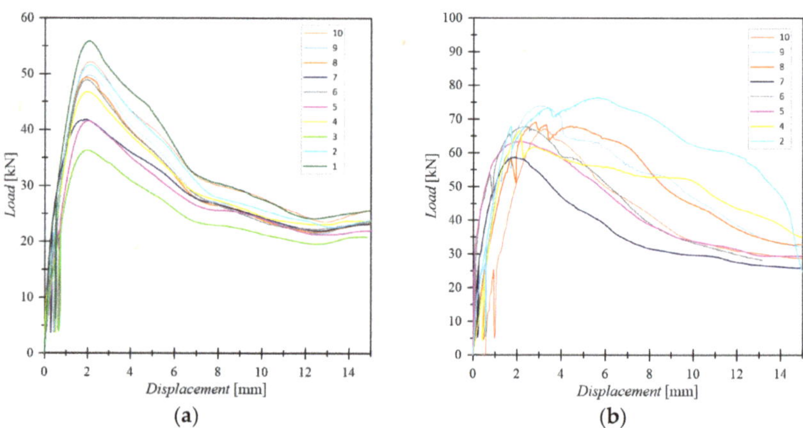

Figure 2. Experimental load-displacement curves for (**a**) clc8160 and (**b**) clc8240 specimens.

Table 1. Specimen geometric data, FE numerical results per screw and corresponding percentage scatters.

| Arrangement | | | | | | Mechanical Performance | | | | |
Model	d [mm]	l_p [mm]	t_i [mm]	θ [°]	$F_{u,max}$ [kN]	$S_{FEM\text{-}exp}$ [%]	K_s [kN/mm]	$S_{FEM\text{-}exp}$ [%]	K_u [kN/mm]	$S_{FEM\text{-}exp}$ [%]	Failure Mode
E1 (clc8160)	8	128	44	45	11.4	−4	11.3	3	10.0	−9	wt
E2 (clc8240)	8	78	22	45	17.5	3	12.2	−8	10.8	−20	wt

3. FE Numerical Investigation

3.1. Modelling Assumptions

The experimental tests were reproduced by FE numerical models implemented in the ABAQUS software package [26]. With the aim of investigating the influence of previously evidenced selected parameters, such as the interlayer mechanical properties and the interaction of the interlayer with the main load-bearing members, a modelling strategy based on three-dimensional solid brick elements was taken into account. Specifically, the effects of these parameters were investigated in terms of ultimate strength F_u, stiffness K_{ser} and ultimate stiffness K_u for a given TCC connection.

The reference simulation consisted in a static incremental, geometrically linear, displacement-controlled analysis. Two planes of symmetry were taken into account for the experimental setup in Figure 1, and this allowed modelling 1/4th of the nominal geometry for the reference specimen (see Figure 3). More precisely, the glulam member, the concrete member, the interlayer and the screws were described with C3D8R elements, which represent a general-purpose linear brick element type, with reduced integration, and is available in the ABAQUS library.

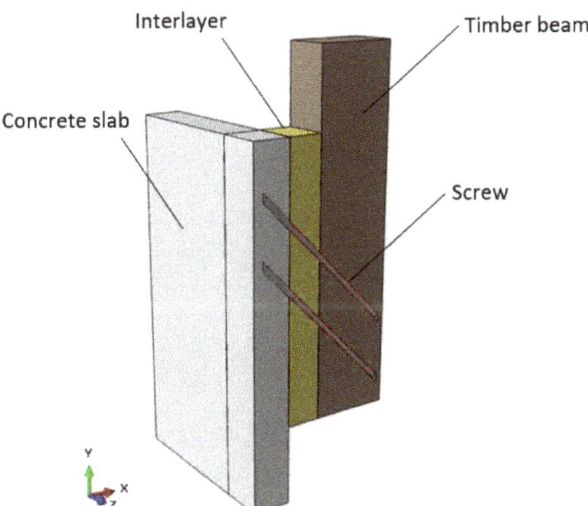

Figure 3. Example of a typical FE assembly of TCC connection (1/4th of the experimental geometry, with hidden mesh).

At the preliminary stage, the complex geometry of the examined TCC systems required numerous mesh sensitivity studies. In order to obtain an accurate and computationally efficient FE model, the spatial domain of each member was divided into several subdomains, in order to allow the definition of transition zones between regions with extremely refined mesh patterns (i.e., close to screws) or with coarser mesh schemes. Overall, sensitivity studies showed that a mesh size of approximately ≈1 mm in proximity of screws and ≈5 mm in the terminal areas of specimens should be taken into account for similar systems. The resulting FE model consisted of a total of 200,000 solid brick elements with an aspect ratio less than 3.

To avoid the separation of TCC members during the loading stage, additional restraints, as in Figure 1, were also taken into account in FE modelling. The effect of these retainers was modelled in the form of an additional translation constraint, which was applied to the concrete member base surface in the x direction (Figure 3). The displacements in the y direction were also constrained on the two base-concrete contact surfaces. Overall, the modelling of 1/4th of the nominal geometry also required the application of constraints for $u_z = u_{r,x} = u_{r,y} = 0$ on the xy symmetry plane and $u_x = u_{r,y} = u_{r,z} = 0$ on the yz symmetry plane, respectively.

3.2. Material Properties and Constitutive Models

The glulam GL24h member was mechanically described as a homogeneous orthotropic elastic-plastic material and its input properties were determined on the basis of product standards, technical specifications of the product and literature data [25,28]. In more detail, the moduli of elasticity parallel and perpendicular to grain were assumed as $E_{0,m} = 11,500$ N/mm^2 and $E_{90,m} = 300$ N/mm^2, respectively. The shear modulus and the rolling shear modulus were also quantified in $G_v = 650$ N/mm^2 and $G_r = 65$ N/mm^2, respectively. Poisson's ratios for the local directions of interest were set to $v = 0.4$, according to the extensive literature research in [27] for softwoods.

To account for damage, an anisotropic yield was defined through Hill's yield criterion. The strength values parallel and perpendicular to grain were assumed to be equal to $f_{0,m} = 37.5$ N/mm^2 and $f_{90,m} = 3.5$ N/mm^2. The shear strength and the rolling shear strength values were defined in $f_v = 3.5$ N/mm^2 and $f_r = 1.2$ N/mm^2, respectively.

The reference OSB panel that was used as interlayer can be described as an OSB/3 class, with a thickness between 18 and 25 mm and ρ = 600 kg/m^3 [28,29]. The corresponding mechanical properties were thus used for the definition of an orthotropic material. According to [28], the elastic modulus and the strength in the main and secondary directions were assumed as $E_{0,m} = 3500$ N/mm^2 and $E_{90,m} = 1400$ N/mm^2 and $f_{0,m} = 18$ N/mm^2 and $f_{90,m} = 9$ N/mm^2. According to [29], finally, the shear moduli and shear strength values were defined in $G_v = 1080$ N/mm^2 and $G_r = 50$ N/mm^2, $f_v = 7.0$ N/mm^2 and $f_r = 1.0$ N/mm^2.

The experimental investigation summarized in Section 2, for example, highlighted no significant crack on the concrete side for most specimens. Therefore, the concrete member was modelled as a homogeneous and isotropic elastic material with an elastic modulus defined in accordance with the strength class C25/30 of Eurocode 2 [23], that is $E_{c,m} = 31,476$ N/mm^2 and:

$$f_{cm} = f_{ck} + 8 \tag{1}$$

$$E_{cm} = 22000 \left(\frac{f_{cm}}{10} \right)^{0.3} \tag{2}$$

The carbon steel screw was modelled with an elastic-plastic material law, with elastic modulus $E_s = 210,000$ N/mm^2 and yielding stress such as to guarantee the same yield moment declared in the ETA-19/0244 [30], that is $f_y = 1195$ N/mm^2.

Finally, a static friction coefficient of $\mu_{T-O} = 0.5$ and $\mu_{C-O} = 0.62$, respectively, was adopted on the glulam-OSB and concrete-OSB shear planes, in accordance with previous outcomes from Air et al. [31].

3.3. Screw-Members Interaction

Due to the complex geometry of the screw represented by the presence of the thread, as well as due to the state of combined shear and axial force that occurs when the screws are inclined with respect to the shear plane for TCC systems, it is necessary to pay particular attention to the modelling of the screw-member transition area.

An efficient modelling technique herein adopted for the mechanical description of the screw-glulam interaction mechanism was originally developed by Avez et al. [32] and also applied in [33] for the numerical analysis of similar timber-timber connections, as well as further adapted for localized analysis of bonded-in-rod (BiR) adhesive connections for timber [34]. The technique consists in the use of a fictitious layer of deformable material (herein called "soft-layer") in conjunction with a cohesive surface interaction with damage initiation and propagation criteria. In this way it is possible to take into account:

- the high initial withdrawal stiffness guaranteed by the direct timber-screw interaction through the screw thread;
- the progressive degradation of this interaction due to damage to the interface at the attainment of a limit stress;
- the possibility of separation on the screw-timber interface;
- the actual axial and flexural stiffness of the screw.

According to this recalled approach, the screw is described in the present study by a cylinder with a diameter equal to the inner thread diameter of the screw. In this way, the axial stiffness EA and flexural stiffness EI of the simplified geometry can be considered to be similar to those of a real fastener.

The gap between the cylinder representative of the screw shank and the surface external to the thread on which withdrawal failure occurs is then modelled by a hollow cylinder of extremely deformable material ("soft-layer"). Therefore, the soft-layer schematizes that area in which the thread and the damaged timber fibres coexist, and consequently this complex medium can only be approximated in a fictitious way by an orthotropic elastic material.

For the present investigation, the mechanical properties of the medium were defined with reference to the main directions of the cylinder (axial, radial and tangential directions in Figure 4) and were assumed to coincide with those of glulam in the direction perpendicular to the fibres ($E_r = E_t = 300 \, \text{N/mm}^2$ and $G_{a-r} = G_{a-t} = G_{r-t} = 650 \, \text{N/mm}^2$). The chosen value of the elastic modulus in the axial direction of cylinder was calibrated in such a manner as not to affect the axial stiffness of the fastener ($E_a = 50 \, \text{N/mm}^2$).

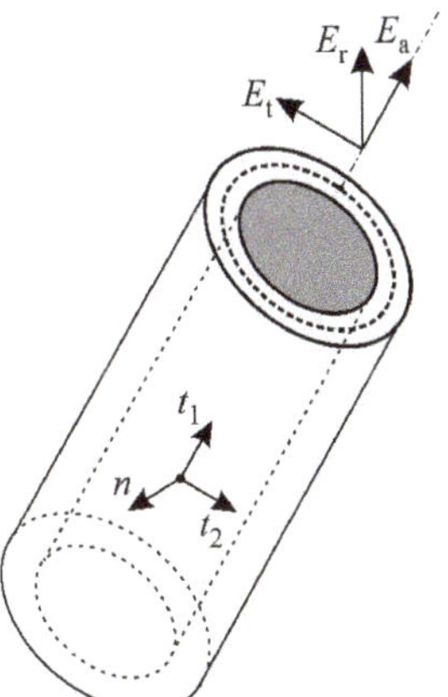

Figure 4. Soft-layer and cohesive contact geometries and reference systems.

A contact interaction with "normal-hard" mechanical behaviour was defined on the outer surface of the soft-layer, which avoids the interpenetration of the nodes of the glulam member and the nodes of the soft-layer without the possibility of transferring tensile stresses in the direction normal to the surface. In defining the cohesive contact, the stiffness and resistance in the normal direction ($K_n = 0 \, \text{N/mm}^3$, $f_n = 0 \, \text{N/mm}^2$) were set to zero, in order to allow free separation of the two surfaces. With these positions and considering a decoupled behaviour, the stiffness matrix of the interaction was hence reduced to:

$$\begin{Bmatrix} \tau_1 \\ \tau_2 \end{Bmatrix} = \begin{bmatrix} K_{s1} & 0 \\ 0 & K_{s2} \end{bmatrix} \begin{Bmatrix} \delta_{s1} \\ \delta_{s2} \end{Bmatrix} \tag{3}$$

with τ_1, τ_2 and δ_{s1}, δ_{s2} being shear stress and displacements in the first (axial) and secondary tangential directions.

The MAXS damage initiation criterion was also reduced to:

$$\max\left\{\frac{\tau_1}{f_{s1}}, \frac{\tau_2}{f_{s2}}\right\} = 1 \tag{4}$$

with f_{s1} and f_{s2} being the shear resistance values for the cohesive contact in the two main directions.

Finally, a linear law for the evolution of damage was used with an ultimate displacement equal to 4 mm [32]. A friction type interaction was added to the tangential interaction of the cohesive type, so that it could be progressively activated with the evolution of cohesive degradation and reproduce the timber-timber sliding after reaching the local breakage by withdrawal of the screw ($\mu_{T-S} = 0.5$). A normal-hard and tangential-penalty type interaction was also assigned to the sliding surface between screw and OSB with ($\mu_{OSB-S} = 0.5$), implicitly assuming a negligible withdrawal resistance in the interlayer and therefore considering the withdrawal prevented by friction between the damaged timber fibres only.

Considering the negligible frequency of withdrawal at concrete side failure in the experimental campaign of Section 2, and with the intention of creating a FE model capable of capturing the average response of those experimental tests, the screw was considered perfectly bound to the concrete slab via a rigid "tie" constraint.

3.4. Model Updating

The use of the soft-layer and cohesive contact makes it possible to consider the screw-timber interaction in a simplified way, but introduces some parameters in the model that are not directly related to the physical characteristics of the modelled object. For the configurations reproduced, a direct dependence of the numerically determined stiffness and strength ($K_{ser,FEM}$ and $F_{u,FEM}$) on the stiffness and resistance of the cohesive contact in the longitudinal direction (K_{s1} and f_{s1}), respectively, was observed. Therefore, considering K_{s1} and f_{s1} as the parameters of the model characterized by greater uncertainty and having fixed all the other parameters as described in the previous paragraph, two independent optimizations were carried out.

With a gradient-based algorithm, an optimization of the average FE-experimental percentage deviation of the stiffness K_{ser} of the two tested configurations (E1 = clc8160 and E2 = clc8240) was carried out. The function $\overline{S_K}$ to minimize as K_{s1} varies can be defined by the following equation:

$$\overline{S_K} = \frac{S_{K,E1} + S_{K,E2}}{2} \tag{5}$$

where

$$S_{K,Ei} = \frac{K_{ser,FEM,Ei} - K_{ser,exp,Ei}}{K_{ser,exp,Ei}} \tag{6}$$

The stiffness of the cohesive contact for the iteration $n + 1$ was determined using the following equation:

$$K_{s1,n+1} = K_{s1,n} - \frac{K_{s1,n} - K_{s1,n-1}}{\overline{S_{K,n}} - \overline{S_{K,n-1}}} \, \gamma \, \overline{S_{K,n}} \tag{7}$$

With the same procedure, an optimization of the average FE-experimental percentage deviation of the F_u strength for the two tested TCC configurations was thus performed. The function $\overline{S_F}$ to minimize with f_{s1} variations can be defined by the following equations:

$$\overline{S_F} = \frac{S_{F,E1} + S_{F,E2}}{2} \tag{8}$$

$$S_{F,ci} = \frac{F_{u,FEM,Ei} - F_{u,exp,Ei}}{F_{u,exp,Ei}} \tag{9}$$

$$f_{s1,n+1} = f_{s1,n} - \frac{f_{s1,n} - f_{s1,n-1}}{S_{F,n} - S_{F,n-1}} \, \gamma \, \overline{S_{F,n}} \tag{10}$$

Using as optimization stop criteria such as $\overline{S_K} \leq 0.05$ and $\overline{S_F} \leq 0.05$, and assuming as learning rate $\gamma = 0.5$, the two values able to make the FE models respond like the real systems were thus quantified in $K_{s1} = 22 \text{ N/mm}^3$ and $f_{s1} = 7.5 \text{ N/mm}^2$.

The so calibrated and updated FE models were found to be able to describe well the average experimental response in terms of force displacement curve (see Figure 5) and also in terms of synthetic parameters (Table 1). The failure mode predicted by both the FE models resulted in the screw withdrawal (Figure 6), which is also in line with the most recurrent failure mode of the experimental campaign. Given that the optimization of parameters was performed based on the pre-failure behaviour only, it can be noted that the post-failure force-displacement curves of FE numerical models slightly differ from the experimental measurements. However, it is also important to note that the present analyses were primarily focused on the until-failure behaviour of examined connections, and such an intrinsic limit of FE models was hence considered satisfactory for the purpose of current investigations.

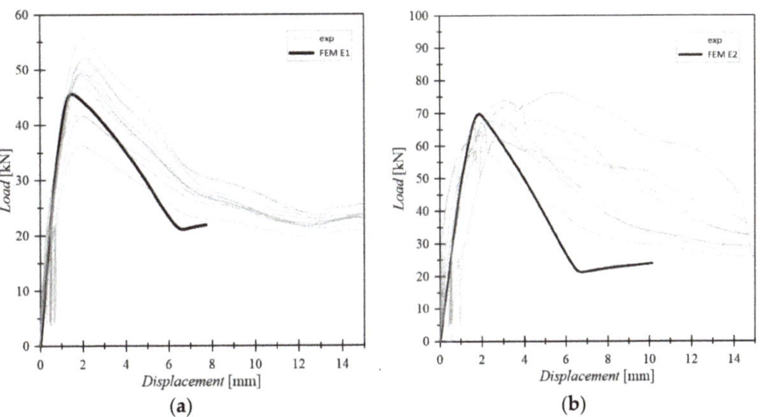

Figure 5. Experimental and FE numerical load-displacement curves for (**a**) clc8160 and (**b**) clc8240 specimens.

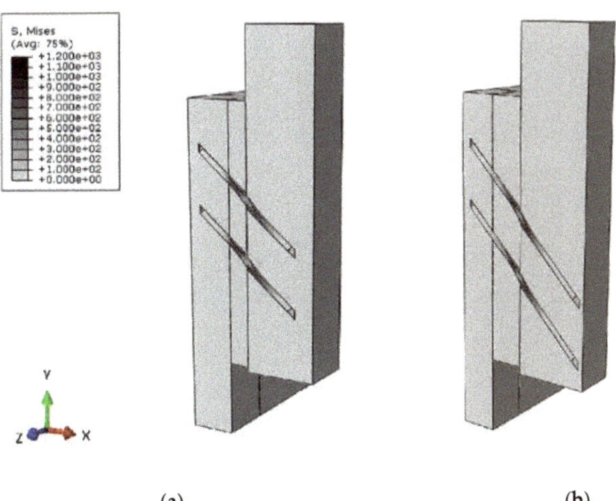

(**a**)   (**b**)

Figure 6. Typical deformed shape and Von Mises stresses for (**a**) E1 (clc8160) and (**b**) E2 (clc8240) tmodels.

4. Elaboration of Push-Out Experimental Results

4.1. Mechanical Performance Indicators

As known, the recorded experimental curves efficiently describe the behaviour of the examined TCC connections and can be used to estimate several parameters, such as K_s and K_u, which are the slip moduli used for design at serviceability limit states and ultimate limit states. In present study, both these parameters were evaluated according to UNI EN 26891:1991 [25] as:

$$K_s = \frac{0.4 \cdot F_{est}}{v_{i,mod}} \tag{11}$$

$$K_u = \frac{0.6 \cdot F_{est}}{v_{0.6} - v_{24} + v_{i,mod}} \tag{12}$$

$$v_{i,mod} = \frac{4}{3}\left(v_{0.4} - v_{0.1}\right) \tag{13}$$

where $v_{0.1}$, $v_{0.4}$ and $v_{0.6}$ represent the displacements recorded under loads of 0.1 F_{est}, 0.4 F_{est} and 0.6 F_{est}, respectively, whereas v_{24} is the displacement measured under a load of 0.4 F_{est} in the second loading branch.

4.2. Load-Bearing Capacity and Failure Mode

The values of the maximum load and of the slip moduli K_s and K_u with regard to the whole specimens are presented in Tables 2 and 3. The characteristic value $F_{u,k}$ of the maximum load was evaluated by means of the logarithmically normal distribution provided by EN 14358:2016 [35]. With reference to the failure mechanisms, the prevailing failure modes were determined by timber embedment combined with withdrawal of screws from the timber element as indicated in Tables 2 and 3, respectively. Only two specimens of the clc8240 group evinced a different failure mode, characterized by withdrawal of screws from the concrete slabs, and three specimens of the same group showed a partial splitting of the glulam element near the midline of the specimen in addition to the recurring mechanism previously described, as shown in Figure 7.

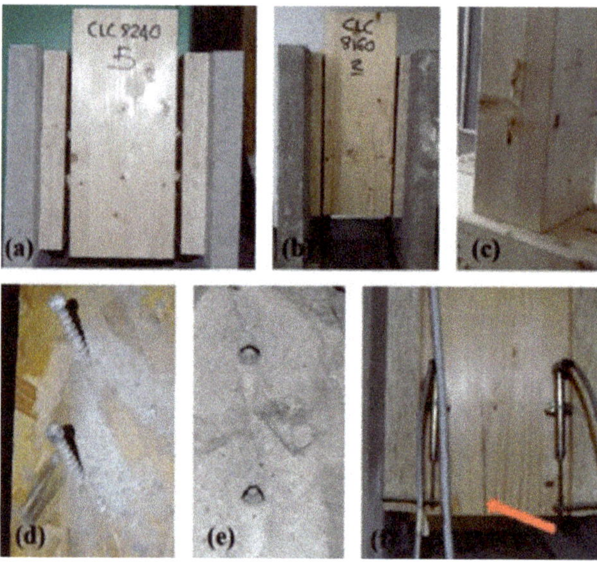

Figure 7. Failure mechanisms for the examined TCC connections in push-out experimental tests: examples of (**a–c**) recurring mechanisms or (**d–f**) infrequent failure modes. Red arrow: failure.

A significant dispersion of experimental results was found for the tested samples (see Tables 2 and 3). This effect can be caused by the variability of mechanical properties of constituent materials (especially concrete and OSB), but can also be ascribed to additional intrinsic uncertainty in the coupling process of experimental samples.

Table 2. Failure load and stiffness parameters for experimental series clc8160.

Specimen	$F_{u,max}$ [kN]	K_s [kN/mm]	K_u [kN/mm]	Failure Mode
1	55.9	69.6	77.0	c + wt
2	51.7	42.7	43.2	c + wt
3	36.3	28.3	27.2	c + wt
4	46.8	37.2	37.5	c + wt
5	41.6	32.1	32.5	c + wt
6	48.9	38.1	38,6	c + wt
7	41.8	64.8	51.2	c + wt
8	49.4	41.6	41.8	c + wt
9	49.8	43.1	42.0	c + wt
10	52.2	44.9	46.1	c + wt
max	55.9	69.6	77.0	
min	36.3	28.3	27.2	
mean value	47.4	44.2	43.7	
σ_y	0.13 *	13.2	13.5	
$F_{u,k}$	35.9	-	-	

Key: wc = withdrawal at concrete side, wt = withdrawal at timber side, c = timber embedment, s = glulam splitting; (*) = according to EN 14358:2016 [35].

Table 3. Failure load and stiffness parameters for experimental series clc8240.

Specimen	$F_{u,max}$ [kN]	K_s [kN/mm]	K_u [kN/mm]	Failure Mode
1	-	-	-	wc
2	76.3	62.6	73.8	wc
3	-	-	-	c + wt
4	66.7	45.8	48.0	c + wt
5	63.5	-	-	c + wt
6	67.8	-	-	c + wt + s
7	58.7	-	-	c + wt
8	69.1	45.6	45.3	c + wt + s
9	74.0	53.6	54.9	c + wt + s
10	67.2	57.3	46.9	c + wt
max	76.3	62.6	73.8	
min	58.7	45.6	45.3	
mean value	67.9	53.0	53.8	
σ_y	0.08 *	7.4	11.7	
$F_{u,k}$	67.7	-	-	

Key: wc = withdrawal at concrete side, wt = withdrawal at timber side, c = timber embedment, s = glulam splitting; (*) = according to EN 14358:2016 [35].

5. Results of FE Numerical Parametric Study

Parametric studies were carried out using the FE models calibrated towards the experimental configurations, with the aim of identifying the parameters of the system that mostly influence its response in terms of stiffness K_{ser} and K_u and strength F_u. To this aim, typical modifications were taken into account in terms of fastener arrangement and detailing as in Figure 8.

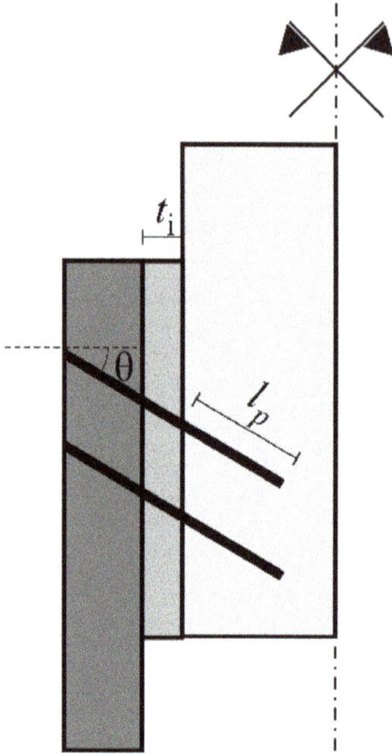

Figure 8. Schematic representation of FE model–influencing parameters (cross-section).

5.1. Screw Inclination and Interlayer Thickness

The first series of parametric studies aimed at investigating the dependence on the inclination of the screw and the thickness of the interlayer while maintaining the length of penetration of the screw into the main member l_p constant (Figure 8). Eight models have been created assuming the experimental configuration E2 as reference model and by varying the thickness of the interlayer (0, 22 mm and 44 mm) and the angle of inclination of screw (30, 45 and 60°), see Table 4.

Table 4. Parametric analysis on screw inclination and interlayer thickness, with evidence of corresponding failure modes.

Model	t_i [mm]	θ [°]	Failure Mode
G1	0	30	phs
G2	22	30	wt
G3	44	30	wt
G4	0	45	wt
G5	22	45	wt
E2 *	44	45	wt
G7	0	60	wt
G8	22	60	c + wt
G9	44	60	c + wt

Key: phs = plastic hinge on shear plane, wt = withdrawal at timber side, c = timber embedment; (*) = experimental configuration as in clc8240.

A modest dependence of the resistance on the investigated parameters was generally observed (see Figure 9a). In this regard, it is worth noting that the G1 configuration was the only one able to manifest a failure mode with the formation of a plastic hinge on the sliding surface.

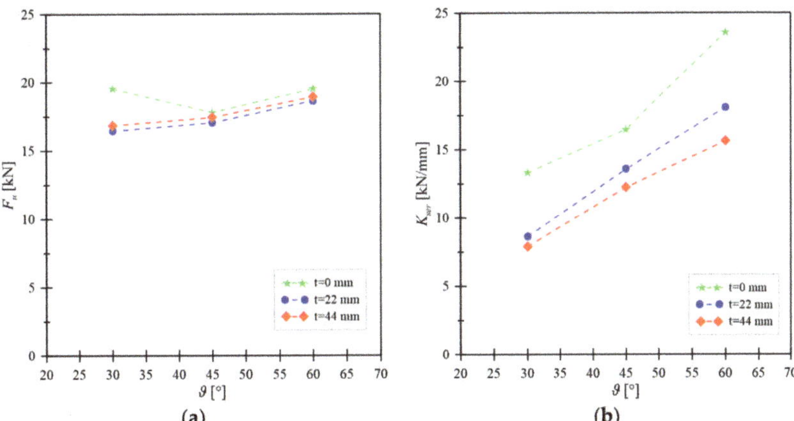

Figure 9. FE numerically predicted trends for (**a**) F_u and (**b**) K_{ser} for different values of inclination angles and interlayer thicknesses.

In this configuration, due to the low inclination angle of screw and the absence of the interlayer, the sharing of forces transferred between the members through the flexural capacity of the screw is generally greater than in the other configurations. Overall, this effect leads to the overcoming of the flexural capacity before withdrawal capacity could be exceeded.

The results in terms of stiffness confirm the results of the analytical model developed by Di Nino et al. [20], see Figure 9b. As shown, by increasing the inclination angle of screw, a significant increase in stiffness can be obtained. The insertion of an interposed interlayer, even of limited thickness, produces a significant reduction in stiffness. By increasing the thickness of the interlayer, finally, the stiffness slightly decreases.

5.2. Interlayer-Timber Member Interaction and Interlayer Mechanical Properties

At a second stage, based on previous outcomes, different types of interaction between the OSB and the glulam member were numerically considered. The two limit cases of frictionless contact and glued contact were taken into account. In addition to the case with the previously recalled friction coefficient from the literature (i.e., model E2), the case with a halved coefficient was also considered, with the aim of evaluating the sensitivity of the overall load-bearing response of TCC system to this parameter. In this regard, the analysed FE model was found to be highly sensitive to the interaction between the members, both in terms of strength and stiffness.

Figure 10, in particular, shows the strength and stiffness modifications for varying the input mechanical properties of the interlayer. In addition to the reference *E2* configuration reported in Figure 10 (i.e., with OSB/3 interlayer panel), more precisely, additional TCC configurations were defined as:

- "particleboard": interlayer made of particleboard (i.e., structural panels for use in wet areas, as defined in EN 12369-1:2001);
- "glulam": interlayer made of the same material as the main timber member;
- and "gap", that is with an interlayer consisting of a physical gap, but being still capable of keeping the timber and concrete members separated (with appropriate kinematic constraints) in the direction perpendicular to the sliding plane.

In this case as well, the investigated FE models were found to be sensitive to the mechanical properties of the interlayer, especially in terms of TCC connection stiffness. A significant reduction in strength and stiffness was observed for the FE model with "gap", but such an outcome was mostly justified by the absence of friction phenomena (i.e., $\mu = 0$) between the concrete and the glulam members and the interlayer. It can in fact be noted in

Figure 10a,b that the "gap" and the "$\mu = 0$" FE models approximately exhibit the same strength and stiffness capacities.

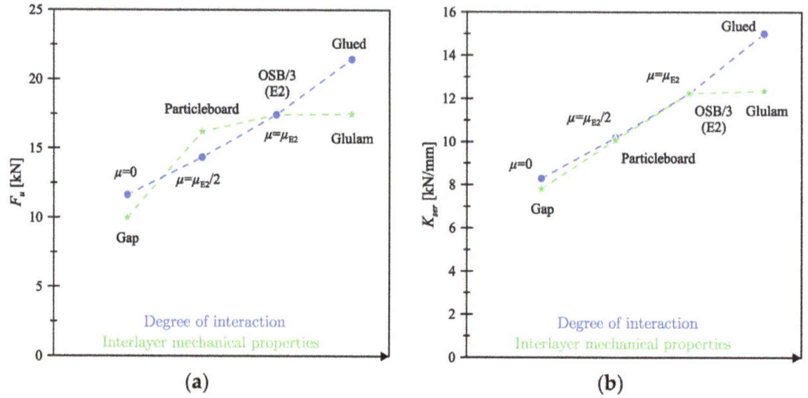

Figure 10. FE numerically predicted trends for (**a**) F_u and (**b**) K_{ser} for different interlayer-to-members interaction and interlayer mechanical properties.

5.3. Friction Contribution and Test Setup

The friction sensitivity study further highlighted the important role of this parameter. For the configurations of Table 4, in more detail, the graph in Figure 11 shows the percentages of F_u which are transferred on the sliding plane through the screw and by friction. It can be noted that:

- the share of force term which is transferred by friction is lower on the interlayer-glulam sliding plane (SL2) with respect to the concrete-interlayer sliding plane (SL1), due to the lower static friction coefficient;
- for the configuration E1, which is similar to the G5 configuration in terms of inclination angle and interlayer thickness but is characterized by a different penetration length $l_{p,\,E1}/l_{p,\,G5} = 0.61$, the FE numerical analysis shows a negligible increase in the share of force transferred by friction on the sliding plane SL1. Conversely, on the sliding surface SL2, the share of force transferred by friction slightly decreases;
- as the thickness of the interlayer increases, with constant inclination and length of penetration for the screw into the glulam member, the share of force transferred by friction clearly increases;
- finally, as the angle of inclination of the screw increases with respect to the normal to the sliding plane, for constant thickness of interlayer and length of the screw penetration into the glulam member, the share of force transferred by friction decreases.

The friction force on the sliding plane is proportional, by means of the friction coefficient, to the force perpendicular to it. The force perpendicular to the sliding plane is given by two contributions: the component of the reaction forces to the internal stress of the screw, and the external force necessary to ensure the equilibrium of the entire connection assembly [36].

Figure 12a, in this regard, shows some of the most common experimental test configurations for the mechanical characterization of connection systems. A schematic layout is proposed for direct shear (DS), push-out (PO) and inclined shear (IS), respectively, with evidence of corresponding reaction forces and effects on sliding plane behaviour.

In the direct shear configuration (DS), more precisely, external forces are applied on the sliding plane, thus eliminating any external reaction forces perpendicular to the sliding plane itself. It is clear that such a condition has direct effects on the mechanical performance and load-bearing capacity of main members for TCC systems. In the case of push-out (PO) and inclined shear (IS) test setup conditions, there is evidence of a perpendicular component $F_\perp$ induced by the geometric configuration of the setup itself. In the case of a push-out test, in more detail, the perpendicular component depends on the lever arm of $F_\parallel$ and $F_\perp$, which is unknown. In the inclined shear case, $F_\perp$ is indeed represented by the component of the external applied force and is therefore a priori known.

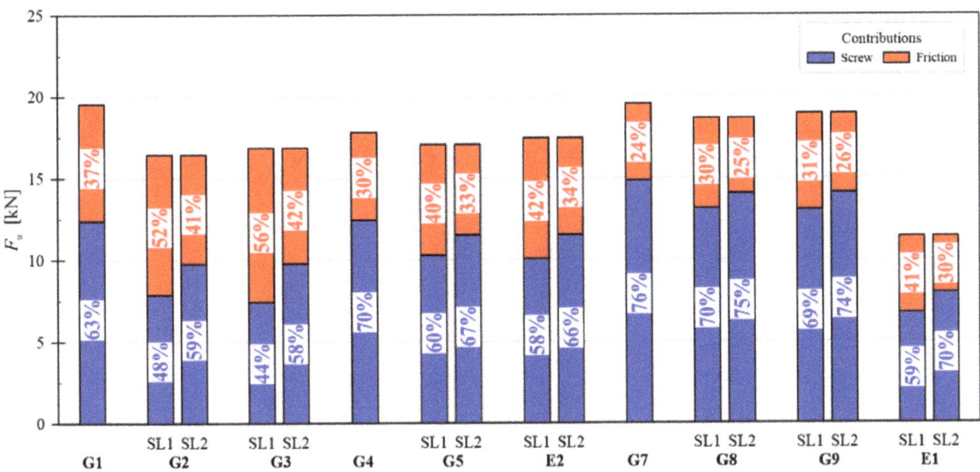

Figure 11. Analysis of F_u percentage transferred on the sliding planes through the screw and by friction. Key: SL1 = sliding plane between the concrete member and the interlayer; SL2 = sliding plane between the glulam member and the interlayer.

To further explore the effect of these setup details on TCC performance, the FE model of the G4 system was used for a sensitivity analysis of synthetic results and TCC performance indicators. The graphical results in Figure 12b, in this regard, show that through the different experimental setup conditions the force transferred between the members by the screw itself does not vary. The frictional force component undergoes an increase of 26% and 20% from the DS configuration to the PO and IS configurations, respectively, causing an overall ultimate force increase of 6% and 5%, respectively. The presence of the $F_\perp$ force term, finally, leads to increases in stiffness between 11% and 13%, respectively. In the selected configuration, a greater force $F_\perp$ was found especially in the case of inclined shear (IS), compared to push-out (with an average ratio equal to $F_{\perp,\,PO}/F_{\perp,\,IS} = 0.7$).

5.4. Perfectly Glued Fastener

Finally, the attention of parametric numerical studies was focused on the ideal condition of a perfectly bonded screw. Preventing the screw withdrawal produces an increase of 40%, both in terms of strength and stiffness, and this can be achieved by considering the screw perfectly glued on the timber side. In this case, the failure mode passes from screw withdrawal, to screw and timber plasticization, which consequently leads to a more ductile behaviour of the TCC connection.

Such a limit case was numerically created and investigated starting from the E2 model by applying a rigid "tie" constraint to the interface between the screws and the timber components. In this way, the timber nodes and the screw nodes were rigidly and perfectly bound, thus simulating a typical case in which the glue-side failure mode is prevented.

For all the parametric studies herein discussed, a K_u value between $0.8\,K_{ser}$ and $0.9\,K_{ser}$ was found in the presence of perfectly bonded fasteners. Sensitivity studies, in this regard, showed that the transition to a C45 concrete class could involve relatively small maximum variations in strength and stiffness, in the order of < 1%.

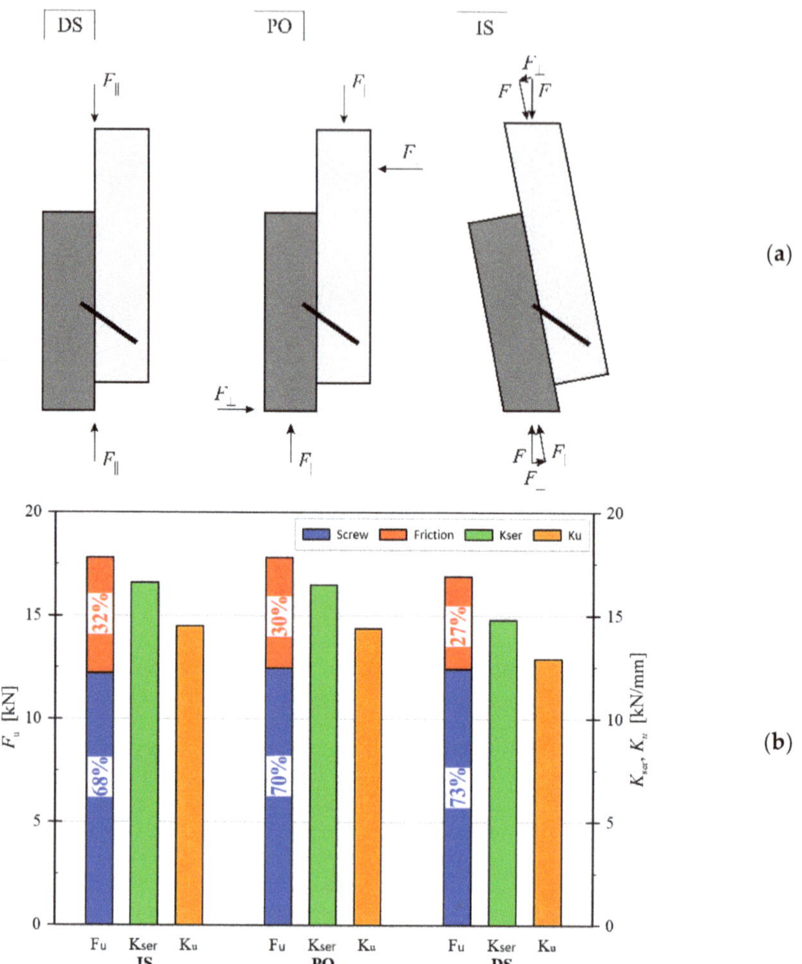

Figure 12. Evidence of (**a**) common test configurations and (**b**) FE numerical estimation of F_u, K_{ser} and K_u performance indicators. Key: DS = Direct shear; PO = Push-out; IS = inclined shear.

6. Discussion

The significant impact that the interlayer has on the load-bearing performance and mechanical parameters of timber-concrete composite connections is generally recognized by literature evidence. However, no distinction is commonly made between the different types of interlayers of typical use in constructions. In this study, the performed FE numerical simulations proved that connections with OSB and glulam interlayers have similar mechanical performance. However, in the case of highly deformable interlayers or particleboards, the strength and stiffness parameters for the examined connections can suffer from significant modifications, compared to previous cases.

Another important aspect is represented by the influence of friction, which is often neglected or minimally considered in a rough way, by simplified analytical models, for the stiffness and strength prediction. In some of the studied configurations herein presented, it was shown that friction can contribute by more than half to the strength of a given connection. Accordingly, further extended studies are needed to define empirical formulations capable of adequately taking into account such a contribution. From a practical point of view, benefits of friction phenomena could in fact lead to a considerable reduction in the number of fasteners which are commonly used in composite structures.

Finally, the role of the test setup for experimental investigations was properly emphasized by the present outcomes. Significant qualitative and quantitative differences were found between the direct shear (DS) test, the inclined shear (IS) test and the push-out (PO) test configurations. In this context, it is worth noting that all these differences are not minimally considered in current structural design regulations, and it is suggested to extend investigations and future elaborations.

7. Conclusions

In this paper, a Finite Element (FE) model capable of taking into account the main variables of a typical timber-concrete composite (TCC) system with inclined screws and interposed interlayer was developed and discussed in terms of parametric analysis. To this aim, the reference FE model was preliminarily validated based on the push-out experimental results of two different TCC configurations.

Based on discussion of FE parametric numerical results, it was highlighted that:

- the inclination of screw and the thickness of the interlayer have a modest influence on the resistance of a given TCC connection but strongly affect the expected failure mode;
- the insertion of an interlayer, even of limited thickness, produces a significant reduction in stiffness, which slightly increases with its thickness;
- the interlayer type and mechanical capacity is also an important parameter, since it further affects the TCC connection stiffness;
- the type of friction/contact interaction of main load-bearing components on the sliding plane was found to be the most significant parameter. The transition from frictionless ("$\mu = 0$") to bonded ("glued") numerical configurations could lead to increases of 84% and 81% in terms of strength and stiffness, respectively, for TCC systems. A reduction factor of strength and stiffness of 0.66 can be adopted when friction between timber and concrete is not guaranteed, whilst whenever the interlayer is glued a coefficient of 1.25 can be taken into account;
- friction alone contributes between 24% and the 56% to reference mechanical performance parameters, depending on the considered sliding plane and the specific geometric configuration;
- the push-out and inclined test configurations, finally, introduce additional forces perpendicular to the sliding plane, and this phenomenon affects the contribution of friction to the overall ultimate force and stiffness of TCC systems. In this regard, strength and stiffness correction factors equalling 0.95 and 0.90, respectively, are suggested to normalize inclined shear and push-out test results to the direct shear results.

In conclusion, the FE numerical studies herein carried out and discussed highlighted the significant impact of the interlayer and its interaction on the main design parameters for a given TCC connection. In this regard, further experimental and numerical studies are necessary for the development of suitable calculation methods that take these parameters into account.

Author Contributions: Conceptualization, Y.D.S. and M.S.; methodology, Y.D.S.; software, Y.D.S.; validation, Y.D.S.; formal analysis, Y.D.S. and L.S.; investigation, Y.D.S., M.S., L.S., V.R., L.S., M.F. and C.B.; resources, Y.D.S., M.S., L.S., V.R., L.S., M.F. and C.B.; data curation Y.D.S. and L.S; writing—original draft preparation, Y.D.S., M.S., L.S., V.R., L.S. and C.B.; visualization, Y.D.S., M.S. and L.S.; supervision, M.F. and C.B.; project administration, M.F.; funding acquisition M.F. All authors have read and agreed to the published version of the manuscript.

Funding: This research was funded by the Italian Ministry of the University (PRIN 2015, Prot. 2015YW8JWA) and Rothoblaas Srl.

Institutional Review Board Statement: Not applicable.

Informed Consent Statement: Not applicable.

Data Availability Statement: Data supporting this research article will be shared upon request.

Acknowledgments: Special thanks are due to Rothoblaas Srl for the financial and technical support provided, without which the present research would not have been possible. The Italian Ministry of the University is also gratefully acknowledged for partially funding the research presented in this paper as a part of the Research Projects of National Interest PRIN 2015, Prot. 2015YW8JWA "The short supply chain in the biomass-timber sector: procurement, traceability, certification and Carbon Dioxide sequestration".

Conflicts of Interest: The authors declare no conflict of interest.

References

1. Bedon, C.; Sciomenta, M.; Fragiacomo, M. Correlation approach for the Push-Out and full-size bending short-term performances of timber-to-timber slabs with Self-Tapping Screws. *Eng. Struct.* **2021**, *238*, 112232. [CrossRef]
2. Herzog, T.; Natterer, J.; Schweitzer, R.; Volz, M.; Winter, W. *Timber Construction Manual*, 4th ed.; Birkhauser-Publishers for Architecture: Basel, Switzerland, 2004.
3. Fontana, M.; Frangi, A. Fire behaviour of timber-concrete composite slabs using beech. *Bautechnik* **2015**, *92*, 323–329. [CrossRef]
4. Frangi, A.; Fontana, M. A design model for the fire resistance of timber–concrete composite slabs. In Proceedings of the IABSE, International Conference on Innovative Wooden Structures and Bridges, Lahti, Finland, 29–31 August 2001.
5. Monteiro, S.; Dias, A.; Lopes, S. Distribution of Concentrated Loads in Timber-Concrete Composite Floors: Simplified Approach. *Buildings* **2020**, *10*, 32. [CrossRef]
6. Frangi, A.; Knobloch, M.; Fontana, M. Fire Design of Timber-Concrete Composite Slabs with Screwed Connections. *J. Struct. Eng.* **2010**, *136*, 219–228. [CrossRef]
7. Mirdad, M.A.H.; Khan, R.; Chui, Y.H. Analytical Procedure for Timber− Concrete Composite (TCC) System with Mechanical Connectors. *Buildings* **2022**, *12*, 885. [CrossRef]
8. Binder, E.; Derkowski, W.; Bader, T.K. Development of Creep Deformations during Service Life: A Comparison of CLT and TCC Floor Constructions. *Buildings* **2022**, *12*, 239. [CrossRef]
9. Ceccotti, A. Composite concrete-timber structures. *Prog. Struct. Eng. Mater.* **2002**, *4*, 264–275. [CrossRef]
10. Du, H.; Hu, X.; Xie, Z.; Wang, H. Study on shear behavior of inclined cross lag screws for glulam-concrete composite beams. *Constr. Build. Mater.* **2019**, *224*, 132–143. [CrossRef]
11. Marchi, L.; Scotta, R.; Pozza, L. Experimental and theoretical evaluation of TCC connections with inclined self-tapping screws. *Mater. Struct. Constr.* **2017**, *50*, 180. [CrossRef]
12. Steinberg, E.; Selle, R.; Faust, T. Connectors for Timber–Lightweight Concrete Composite Structures. *J. Struct. Eng.* **2003**, *129*, 1538–1545. [CrossRef]
13. Kavaliauskas, S.; Kvedaras, A.K.; Valiūnas, B. Mechanical behaviour of timber-to-concrete connections with inclined screws. *J. Civ. Eng. Manag.* **2007**, *13*, 193–199. [CrossRef]
14. Bedon, C.; Fragiacomo, M. Three-dimensional modelling of notched connections for timber-concrete composite beams. *Struct. Eng. Int.* **2017**, *27*, 184–196. [CrossRef]
15. Sciomenta, M.; de Santis, Y.; Castoro, C.; Spera, L.; Rinaldi, V.; Bedon, C.; Fragiacomo, M.; Gregori, A. Finite elements analyses of timber-concrete and timber rubberized concrete specimens with inclined screws. In Proceedings of the World Conference Timber Engineering (WCTE), Santiago, Chile, 9–12 August 2021.
16. Van Der Linden, M.L.R. Timber-Concrete Composite Floor Systems. Ph.D. Thesis, Technische Universiteit Delft, Delft, The Netherlands, 1999.
17. Jorge, L.F.C.; Lopes, S.M.R.; Cruz, H.M.P. Interlayer Influence on Timber-LWAC Composite Structures with Screw Connections. *J. Struct. Eng.* **2011**, *137*, 618–624. [CrossRef]
18. Moshiri, F.; Shrestha, R.; Crews, K. The Effect of Interlayer on the Structural Behavior of Timber Concrete Composite Utilizing Self-Compacting and Conventional Concretes. *Int. J. Eng. Technol.* **2015**, *7*, 103–109. [CrossRef]
19. Djoubissié Denouwé, D.; Messan, A.; Fournely, E.; Bouchair, A. Influence of Interlayer in Timber-Concrete Composite Structures with Threaded Rebar as Shear Connector-Experimental Study. *Am. J. Civ. Eng. Archit.* **2018**, *6*, 38–45. [CrossRef]
20. Di Nino, S.; Gregori, A.; Fragiacomo, M. Experimental and numerical investigations on timber-concrete connections with inclined screws. *Eng. Struct.* **2020**, *209*, 109993. [CrossRef]
21. De Santis, Y.; Fragiacomo, M. Timber-to-timber and steel-to-timber screw connections: Derivation of the slip modulus via beam on elastic foundation model. *Eng. Struct.* **2021**, *244*, 112798. [CrossRef]
22. Mirdad, M.A.H.; Chui, Y.H. Stiffness Prediction of Mass Timber Panel-Concrete (MTPC) Composite Connection with Inclined Screws and a Gap. *Eng. Struct.* **2020**, *207*, 110215. [CrossRef]
23. *Eurocode 2: Design of Concrete Structures-Part 1-1: General Rules and Rules for Buildings*; EN 1992-1-1; European Committee for Standardization (CEN): Brussels, Belgium, 2004.
24. *Timber Structures-Glued Laminated Timber and Glued Solid Timber–Requirements*; EN 14080; European Committee for Standardization (CEN): Brussels, Belgium, 2013.

25. *Timber Structures. Joints Made with Mechanical Fasteners. General Principles for the Determination of Strength and Deformation Characteristics*; EN 26891; European Committee for Standardization (CEN): Brussels, Belgium, 1991.
26. Abaqus. *Computer Software*; Dassault Systèmes: Providence, RI, USA, 2021.
27. Bartolucci, B.; De Rosa, A.; Bertolin, C.; Berto, F.; Penta, F.; Siani, A.M. Mechanical properties of the most common European woods: A literature review. *Fract. Struct. Integr.* **2020**, *14*, 249–274. [CrossRef]
28. *Oriented Strand Boards (OSB)-Definitions, Classification and Specifications*; EN 300; European Committee for Standardization (CEN): Brussels, Belgium, 2006.
29. *Wood-Based Panels-Characteristic Values for Structural Design-Part 1: OSB, Particleboards and Fibreboards*; EN 12369-1; European Committee for Standardization (CEN): Brussels, Belgium, 2001.
30. European Technical Assessment 19/0244. 2019.
31. Air, J.R.; Arriaga, F.; Iniguez-Gonzalez, G.; Crespo, J. Static and kinetic friction coefficients of scots pine (*Pinus sylvestris* L.), parallel and perpendicular to grain direction. *Mater. Constr.* **2014**, *64*, 2979955. [CrossRef]
32. Avez, C.; Descamps, T.; Serrano, E.; Léoskool, L. Finite element modelling of inclined screwed timber to timber connections with a large gap between the elements. *Eur. J. Wood Wood Prod.* **2016**, *74*, 467–471. [CrossRef]
33. Bedon, C.; Fragiacomo, M. Numerical analysis of timber-to-timber joints and composite beams with inclined self-tapping screws. *Compos. Struct.* **2019**, *207*, 13–28. [CrossRef]
34. Bedon, C.; Rajcic, V.; Barbalic, J.; Perkovic, N. CZM-based FE numerical study on pull-out performance of adhesive bonded-in-rod (BiR) joints for timber structures. *Struct.* **2022**, *46*, 471–491. [CrossRef]
35. *Timber Structures–Calculation and Verification of Characteristic Values*; EN 14358; European Committee for Standardization (CEN): Brussels, Belgium, 2016.
36. Blaß, H.J.; Steige, Y. *Steifigkeit Axial Beanspruchter Vollgewindeschrauben*; KIT Scientific Publishing: Karlsruhe, Germany, 2018. [CrossRef]

 *buildings*

Article

Simplified Models to Capture the Effects of Restraints in Glass Balustrades under Quasi-Static Lateral Load or Soft-Body Impact

Emanuele Rizzi, Chiara Bedon * and Claudio Amadio

Department of Engineering and Architecture, University of Trieste, 34127 Trieste, Italy
* Correspondence: chiara.bedon@dia.units.it; Tel.: +39-040-558-3937

Abstract: Structural glass balustrades are usually composed of simple glass panels which are designed under various restraint solutions to minimize large out-of-plane deflections and prematurely high tensile/compressive stress peaks under lateral loads due to crowd. Linear supports, point-fixing systems, and others can be used to create geometrical schemes based on the repetition of simple modular units. Among others, linear restraints that are introduced at the base of glass panels are mechanically described in the form of ideal linear clamps for glass, in which the actual geometrical and mechanical details of real fixing components are reduced to rigid nodal boundaries. This means that, from a modelling point of view, strong simplifications are introduced for design. In real systems, however, these multiple components are used to ensure appropriate local flexibility and adequately minimize the risk of premature stress peaks in glass. The present study draws attention to one of these linear restraint solutions working as a clamp at the base of glass panels in bending. The accuracy and potential of simplified mechanical models in characterizing the effective translational and rotational stiffness contributions of its components are addressed, with the support of efficient and accurate Finite Element (FE) numerical models and experimental data from the literature for balustrades under double twin-tyre impact. Intrinsic limits are also emphasized based on parametric calculations in quasi-static and dynamic regimes.

Keywords: glass balustrades; laminated glass (LG); linear restraints; mechanical models; Finite Element (FE) numerical models

Citation: Rizzi, E.; Bedon, C.; Amadio, C. Simplified Models to Capture the Effects of Restraints in Glass Balustrades under Quasi-Static Lateral Load or Soft-Body Impact. *Buildings* **2022**, *12*, 1664. https://doi.org/10.3390/buildings12101664

Academic Editor: Nikolai Ivanovich Vatin

Received: 4 September 2022
Accepted: 8 October 2022
Published: 12 October 2022

Publisher's Note: MDPI stays neutral with regard to jurisdictional claims in published maps and institutional affiliations.

1. Introduction

The use of structural glass in buildings for load-bearing components is rather common [1]. Especially for transparent barriers and balustrades, glass panels can be variably assembled and arranged with a multitude of restraint types and boundary conditions (Figure 1). While commonly associated with regular and simple (i.e., squared or rectangular) flat modular shapes, glass panels for balustrades can be characterized by the presence of holes for point-fixings and should, in any case, be verified against equivalent lateral loads [2] in terms of tensile stress peaks at the Ultimate Limit State (ULS) and deflections at the Serviceability Limit State (SLS). Careful attention is also required for the Collapse Limit State (CLS) so as to include possible partial fracture mechanisms at the design stage. Besides the intrinsic geometrical simplicity and repeatability, rather complex mechanical phenomena should be taken into account for structural design. Major issues can derive from characterization in terms of material properties (including damage constitutive models, strain rate effects, etc.), load description, or even mechanical description of the effects due to restraints and connections in use [3], especially in the framework of Finite Element (FE) numerical models. For structural performance assessment, current design standards prescribe that specific impact pendulum test configurations should be taken into account to experimentally verify the load-bearing capacity of a given balustrade (EN 12600 [4], DIN 18008-4 [5], etc.).

Figure 1. Examples of glass balustrades characterized by various restraint conditions: (**a**) linear restraint at the base; (**b**) point-fixing at the base; (**c**) top-bottom point-fixings; (**d**) lateral point-fixing. Solutions (**a**,**b**) do not require glass drilling; (**c**,**d**) are characterized by the presence of glass holes.

Several research studies of the literature, in this regard, explored various mechanical and load-bearing aspects of selected glass panel solutions of practical interest under the effect of impactors, offering support to design detailing and even FE numerical modelling.

Structurally speaking, human-induced impact loads should be properly addressed for safe design purposes and—when possible—supported by accurate but computationally efficient numerical models [6]. In this regard, traditional in-service glass windows have been experimentally and numerically explored under the effects of ball drop/hard-body impact setup in [7]. The vulnerability of historic glass facades subjected to soft-body/bird-strike impact has been numerically assessed in [8], with the support of in-field experimental characterizations and Coupled Eulerian Lagrangian (CEL) numerical models. Small-scale and full-scale glass columns have been explored in [9,10] under soft-body and hard-body impactor conditions, including the analysis of preliminary damage in glass elements. New reduced models for the analysis of glass panels under soft-body impact have been presented in [11,12] with the goal of determining the maximum principal stresses of a given glass system when subjected to the dynamic impact of double twin-tyre, based on computationally efficient numerical analysis. A simplified modelling strategy has been proposed in [13] for glass panels under both spheroconical bag impact and double twin-tyre impact. Further numerical efforts have been proposed in [14] for glass balustrades under soft-body impact.

Glass balustrades have been experimentally and numerically explored in [15,16] under both static lateral loads and impact. The experimental and numerical study reported in [17] further explored the response of glass balustrades under soft-body impact by taking into account the effects of partial glass damage. The use of non-destructive tools and methods for glass systems under dynamic loads has also been experimentally assessed, such as the use of digital image correlation and video tracking in [18] for simple monolithic glass elements under random impact or optical measurements for laminated glass (LG) balustrades presented in [19].

The present study draws attention to the mechanical characterization of glass balustrades under lateral loads and impact and to the analytical derivation of simplified mechanical models (SM1 to SM4, in the following) to support their efficient and safe structural design and verification. Attention is given to glass balustrade panels with linear base restraints. As such, detailed analytical models are used for the definition of a set of empirical equations that are presented to support the calibration of computationally efficient FE numerical models, where the contribution of various restraint components could be efficiently schematized in the form of equivalent springs. In doing so, a semi-analytical calculation approach is followed, with the major result being that the finally assembled FE models can be strongly simplified in terms of geometrical components, mechanical interactions, and, thus, computational costs. These simplified components are typically used in real balustrade systems to prevent premature stress peaks in glass, accommodate the deformations due to imposed design loads, and mitigate the glass panels regarding premature damage or even collapse.

For detailed calculation purposes, the attention of the present study is given to the geometrical and mechanical features of experimental samples discussed in [17]. As shown, the reported idealized models and simplified semi-analytical calculations often suffer from approximate mechanical descriptions of basic restraint components, which have a fundamental role in the prediction of the overall bending response of glass balustrades. Comparative results are hence discussed in Section 5 to assess the accuracy of the SM1-to-SM4 models, giving evidence of their potentials and intrinsic limits.

2. Research Study

2.1. Methodology and Goal

In design practice, linearly restrained glass balustrades are often schematized according Figure 2. A given glass panel can in fact be assumed to be fixed at the base by geometrically complex metal profiles, as in Figure 2a, and subjected to horizontal load P at the unrestrained edge of its bending span L.

While such a continuous restraint is expected to offer high rigidity against lateral loads, it is also asked to preserve a local deformation capacity for the glass panel so as to prevent premature stress peaks in glass.

In terms of glass and its resistance verification check, the typical analytical analysis follows the mechanical model representative of (i) a glass cantilever under nodal load P at the unrestrained end, with corresponding maximum tensile stress peaks at the base end of glass with the ideally rigid clamp restraint (see Figure 2b). Such a model represents the simplest approach for glass stress and deflection analysis and roughly describes the effect of real fixing systems, as well as their mechanical interaction with the glass panel in out-of-plane bending.

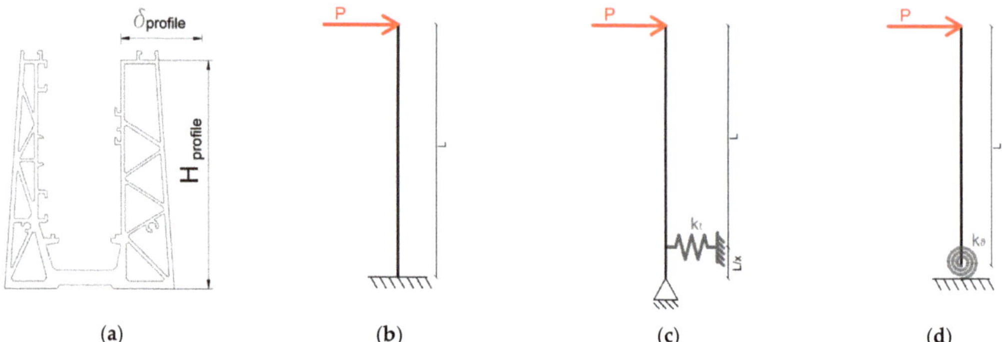

Figure 2. Schematic representation of possible mechanical models for a glass balustrade with a linear restraint at its base: (**a**) example of a metal restraint (cross-section); (**b**) cantilever mechanical model with an ideally rigid clamp restraint; or (**c,d**) mechanical models inclusive of equivalent springs to reproduce the real boundaries.

Alternatively, the structural analysis can be schematized based on still simplified but more articulated mechanical models, as, for example, in the approaches (ii) or (iii) in Figure 2c,d. There, as shown, equivalent springs are uniformly distributed (in the width of the balustrade) to capture—with specific attention to the restraints in use—the expected maximum stress peaks and out-of-plane lateral deflections for the glass panel in bending. To the extent that the fixing system of Figure 2a is schematized in the form of equivalent translational springs—as in Figure 2c, for example—it is assumed that, under the effects of a given lateral load P, the fixing system reacts to the glass panel in bending with a transversal reaction force F_k, which should be taken into account in terms of equivalent spring, as follows:

$$k_t = \frac{F_k}{\delta_{profile}} = \frac{Px}{\delta_{profile}} \tag{1}$$

with x being the distance defined in Figure 2c.

Similarly, the fixing system of Figure 2a can be mechanically characterized in the form of an equivalent rotational stiffness term, as in Figure 2d, where:

$$k_\theta = \frac{M}{\theta} = \frac{PL}{sin^{-1}\left(\delta_{profile}/H_{profile}\right)} \tag{2}$$

and the contribution of this kind of spring is still associated with reproducing the possible out-of-plane bending response of an assembled glass balustrade with a linear base restraint. However, the simplified analytical descriptions and ideal restraints above are known to introduce approximations in real boundaries and system components. For LG components, moreover, attention should be given to the use of "composite" sections with equivalent monolithic glass thicknesses, as in the so-called EET approach [20].

As an alternative to simple analytical procedures, computationally expensive FE numerical models could be developed to account for all the primary and secondary components in this kind of assembled system. In Figure 3, an example of a full three-dimensional (3D) solid brick model for the system of Figure 2 is shown, with evidence of global assembled components and restraint details. While such a modelling approach can efficiently support the analysis of a wide set of possible configurations of technical interest, it is generally characterized by long modelling and simulation times and may also involve additional uncertainties in terms of materials characterization, mechanical interaction calibrations, and kinematic constraint definitions.

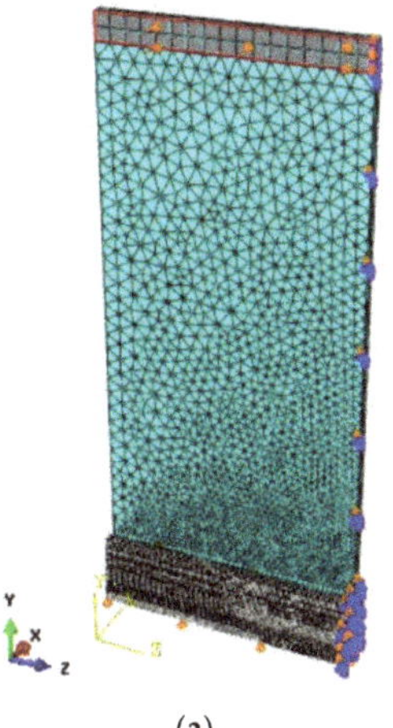 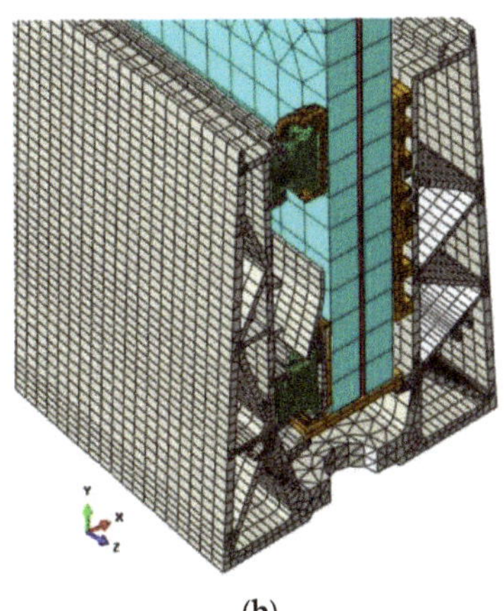 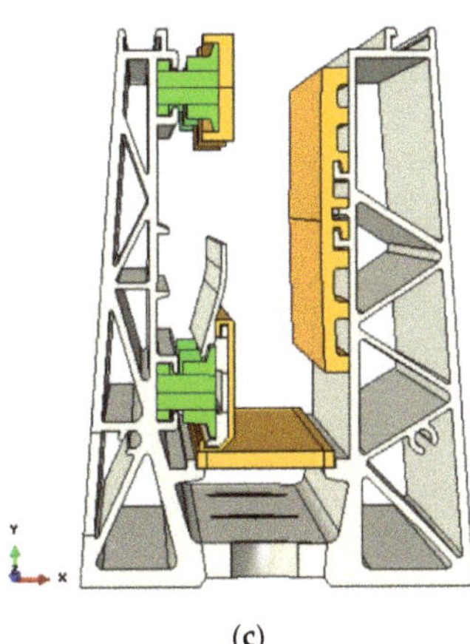

(a)   (b)   (c)

Figure 3. Example of full 3D FE numerical model of a laminated glass balustrade a with linear base metal restraint (ABAQUS): (**a**) axonometric view of the half balustrade assembly and (**b**) detailed view of the base restraint region, with (**c**) a cross-section of the metal fixing system (glass panel and mesh pattern hidden from section view).

In this paper, the attention is thus focused on the development, calibration, and assessment of different simplified mechanical models for the structural analysis of typical glass modular units in use for balustrades. Parametric numerical calculations are carried out in ABAQUS [21]. The proposed investigation shows how the effect of details and their mechanical description in FE codes can affect the overall estimation of stress and deformation trends and distributions for rather simple mechanical systems and thus possibly affect the subsequent design assumptions.

To this aim, a past experimental study, recalled in Section 2.2, is taken into account for the validation of various mechanical models. In doing so, a linear elastic constitutive behaviour is taken into account for glass components, while the attention is primarily focused on the characterization of base restraint components and on the analysis of their effects on glass balustrade performance, including stress distributions and deflection trends.

Based on literature feedback, a "Refined" FE numerical model is first developed to support accurate comparisons with simplified approaches (Section 3). A set of four different simplified mechanical models are then separately developed and investigated under quasi-static lateral loads and soft-body impact conditions. First, the models SM1 and SM2 are derived and calibrated in Section 4 based on the definition of discrete/lumped equivalent springs which are introduced in the same location of real restraint components. Successively, based on the simplified SM2 concept, two additional simplified models, agreeing, respectively, with Figure 2c,d, are developed and assessed under quasi-static and dynamic loading. In this latter case, the major feature is represented by the presence of linearly distributed springs for translation (SM3) or rotation (SM4) restraints, which are introduced at the bottom of the glass panel for the analysis of the glass modular unit under the effects of conventional lateral loads. A summary of presently developed mechanical models is reported in Table 1.

Table 1. Summary of presently developed numerical models for the analysis of the glass balustrade system described in Figure 4.

FE Numerical Model	LG Cross-Section (mm)	FE Model Features	
		LG Panel	Base Restraint
Refined		Full 3D solid brick elements (layered section)	Full 3D solid brick elements
SM1		Same as that of Refined	Lumped equivalent springs in the region of restraints
SM2	10 + 1.52 PVB + 10	2D shell elements (equivalent monolithic glass section)	Lumped equivalent springs in the region of restraints
SM3		Same as that of SM2	Linearly distributed equivalent springs (translational) at the bottom edge of the glass
SM4		Same as that of SM2	Linearly distributed equivalent springs (rotational) at the bottom edge of the glass

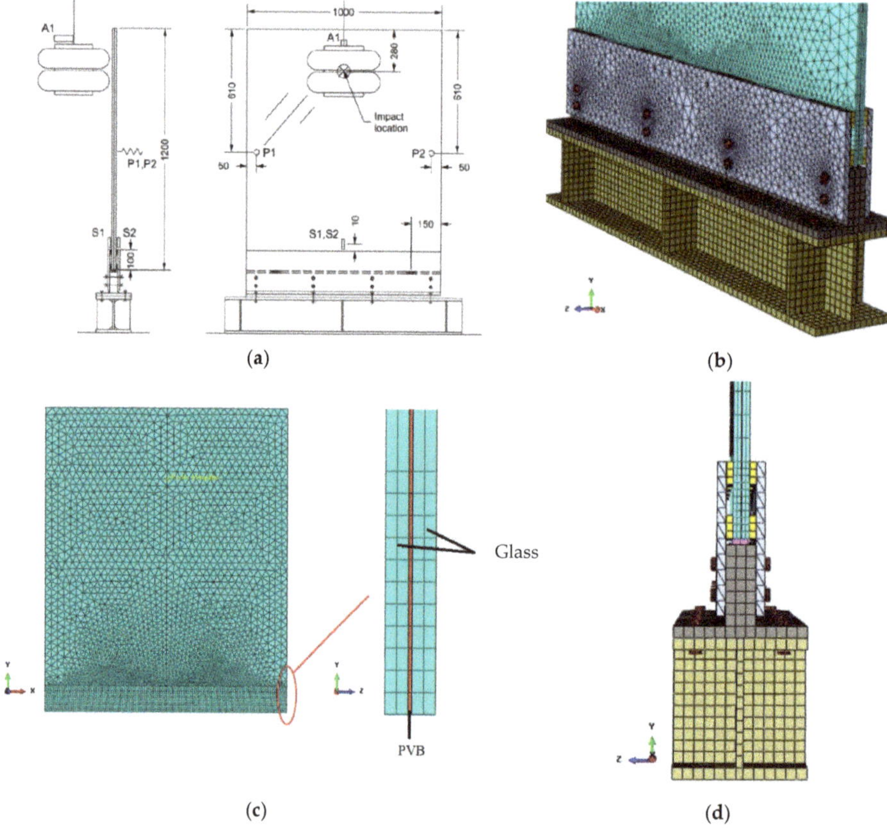

Figure 4. (**a**) Experimental setup (based on [17]), with dimensions in mm, and (**b–d**) detailed views of the presently developed "Refined" FE numerical model (ABAQUS).

2.2. Reference Glass Balustrade

The balustrade experimentally and numerically discussed in [17] is taken into account for the present study. The specimen in Figure 4a is characterized by the total dimension $B = 1000 \times L = 1200$ mm^2 and a double LG section (10/10.4 in thickness) composed of tempered glass panes (10 mm in thickness) and bonding Polyvinyl butyral (PVB®, 1.52 mm in thickness). The bottom linear connection consists of two 10 mm-thick steel plates, which are rigidly fixed to a base support via M10 class 8.8 bolts (length l_b, area A_b), distributed as schematized in Figure 4. Additional setting blocks ($A_{SB} = h_{SB} = 30 \times b_{SB} = 120$ mm^2 in dimensions, with t_{SB} being their thickness) are used at the glass–steel interface to provide soft support to the glass panel in out-of-plane bending and to avoid premature stress peaks in the region of restraints. An additional supporting system consisting of two polyurethane blocks (50 mm wide, 8 mm thick) is introduced at the bottom edge of the glass panel and placed at a distance of 150 mm from the lateral edges, with the goal of preventing further stress peaks and premature glass breakage at the base edges.

The original experiments reported in [17] were carried out in accordance with EN 12600 provisions, including various impact configurations. In doing so, a conventional double twin-tyre impactor was used, while changing its drop height. As in Figure 4a, for all the impact configurations, the analysis included the measure of lateral displacements (P1, P2), glass stresses (S1, S2), and impactor acceleration (A1). A detailed FE model was also presented in [17] to explore and support the experimental findings.

The presently developed "Refined" FE model of Figure 4b–d was thus preliminary implemented in ABAQUS [21] for validation towards the experimental data from [17], as well as for further support for simplified mechanical models. Once the accuracy of the Refined model was assessed, this latter model was in fact also used to address the detailed calibration and validation of the proposed SM1-to-SM4 mechanical models summarized in Table 1.

3. Full 3D Refined Numerical Model

3.1. Model Description

As specified in Figure 4, the Refined numerical model was used in the present study to quantify and compare some key performance indicators for the structural analysis of the examined balustrade in terms of displacements and stress distributions in glass but also local deformations, rotations, and reaction forces in base restraint components.

In particular, the nominal geometry from Figure 4 was described in the form of a set of 8-node, 10-node, or 6-node 3D solid brick elements (C3D8R, C3D6R, or C3D10 elements from the ABAQUS library). A set of surface-to-surface contact interactions was introduced at the interface of the glass panel and fixing system for all the model regions where any kind of mechanical contact could take place during bending. Mesh refinement was privileged, especially in the region of base restraints (Figure 4b). The final balustrade assembly consisted of $\approx$120,000 elements and $\approx$505,000 Degrees of Freedom (DOFs).

Linear elastic constitutive models were used for the characterization of all the balustrade components, as also reported in [17]. Most importantly, an equivalent secant modulus was used for the PVB layer under impact loading so as to take into account its viscoelastic behaviour. The input features in use for the present simulations are summarized in Table 2.

Table 2. Summary of the mechanical properties for materials in use in the Refined numerical model.

	Material Properties			
Material	**Constitutive Model**	**Modulus of Elasticity [N/mm^2]**	**Poisson Ratio**	**Density [kg/m^3]**
Steel	Linear elastic	210,000	0.3	7850
POM	Linear elastic	2413	0.45	1250
Glass	Linear elastic	70,000	0.23	2500
PVB	Linear elastic	180	0.485	1250

The attention of the structural assessment was devoted to the analysis of the linearly restrained LG balustrade under the effects of:

- L1: a quasi-static, monotonically increasing lateral load at the top edge of the glass (until a maximum value P = 4.5 kN/m), and
- L2: a twin-tyre impact loading configuration which was numerically reproduced and calibrated according to the experimental setup summarized in Section 2 (with 300 mm being the drop height).

For the L1 configuration, a quasi-static lateral load P was distributed along the width of the glass balustrade, towards the top (unrestrained) edge of the LG panel (at an average height of 1.1 m). In case of the L2 dynamic configuration, a double twin-tyre impactor was numerically described in addition to the glass balustrade components so as to reproduce the desired impact configurations, as in [17], under an imposed translational velocity.

The Refined numerical model was in fact used to capture global and local phenomena in all the model components so as to facilitate the validation of simplified models. In this regard, the analysis of results was also focused on the stress and deformation trends in the components of the base restraint. Typical examples are reported in Figure 5.

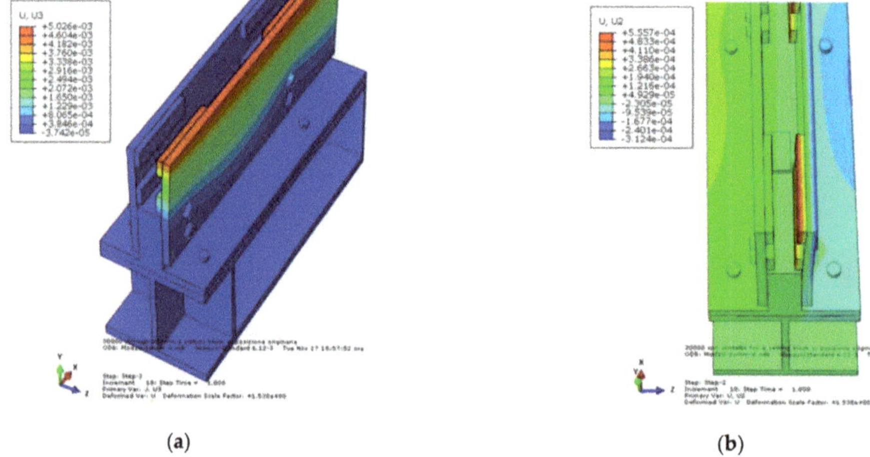

(a)   (b)

Figure 5. Example of deformation and parameters for the base restraints under lateral loads, as obtained from the Refined FE numerical model (ABAQUS): (**a**) out-of-plane and (**b**) vertical deformations (glass panel and mesh hidden from view, legend values in m).

3.2. Results

The typical behaviour of the Refined numerical model was first addressed regarding the double twin-tyre impact configurations that were experimentally investigated and numerically analyzed in [17]. Selected performance indicators can be seen in Figure 6 in the form of the lateral displacement time history for the LG panel, the principal stress measured in glass, and the acceleration time history. It can be noted, over the time of impact and contact for the double twin-tyre, a rather good correlation between the FE numerical predictions and the corresponding experimental measurements. Additionally, the presently developed Refined model proved to offer a close correlation with the numerical estimates reported in [17] (see Table 3).

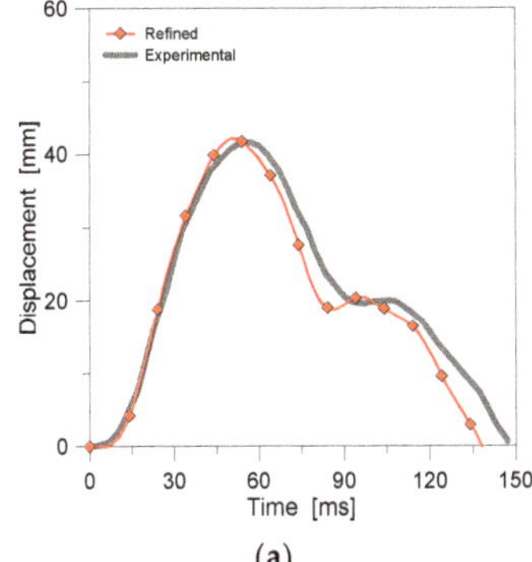
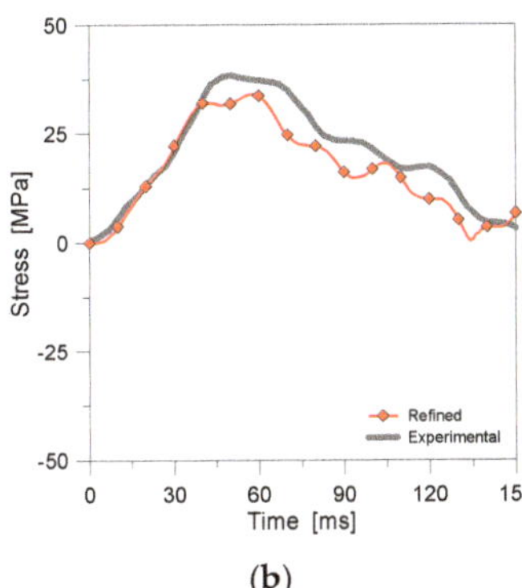
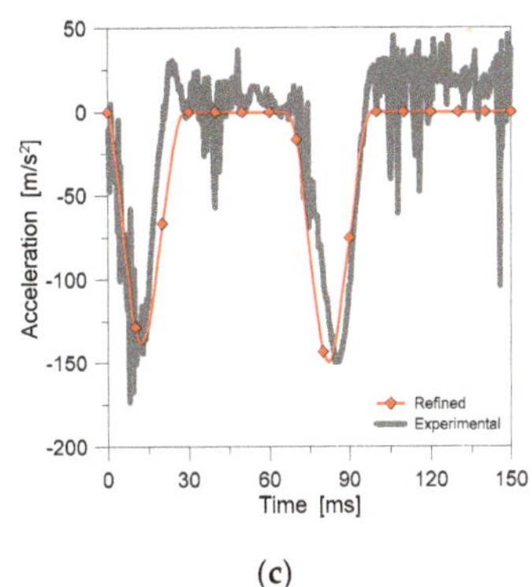

(a)   (b)   (c)

Figure 6. Numerical analysis of the Refined model (ABAQUS) under double twin-tyre impact (300 mm being the drop height), and comparison with the experimental results: (**a**) lateral displacement; (**b**) principal stress, and (**c**) acceleration time histories.

Table 3. Summary of comparative results in terms of impactor acceleration for the present Refined numerical model and past experiments with a double twin-tyre.

			Refined Model		
Drop Height [mm]	$a_{max,test}$ [17] $[\mathrm{m/s^2}]$	$a_{max,model}$ [17] $[\mathrm{m/s^2}]$	$a_{max,Refined}$ $[\mathrm{m/s^2}]$	Δ_1 [%]	Δ_2 [%]
300	223	216	216.38	−2.97	0.18
400	282	272	269.17	−4.55	−1.04
500	332	334	345.36	4.02	3.40

For the 300 mm drop height configuration in Figure 6a, for example, maximum principal stresses were measured in the order of 33 MPa, which denotes a linear elastic behaviour of glass. The percentage scatter of the Refined numerical estimates compared to the experimental outcomes was quantified in the order of +2.4%. Similar trends were observed for the lateral displacement of the glass balustrade, with approximately 42 mm of maximum deformation in the control point and a scatter of +2.3% of the numerical and experimental findings. Further, the numerical analysis was carried out by varying the drop height of the impactor. In this case, the Refined numerical model generally proved to capture the past experimental outcomes, as well as the FE numerical estimates reported in [17]. In Table 3, a summary of comparisons is, for example, proposed in terms of the maximum measured acceleration for the double twin-tyre under impact conditions, as calculated from past experiments ($a_{max,test\,[17]}$), from the FE numerical model presented in [17] ($a_{max,model\,[17]}$), and from the presently developed FE numerical model. Moreover, the percentage scatter of the present Refined model is calculated towards past experiments (Δ_1) or towards past numerical simulations (Δ_2), respectively.

4. Derivation and Calibration of Simplified Numerical Models

4.1. Simplified Characterization of the Base Restraint—SM1 Model

The first simplified procedure (SM1) assumes that all the steel components of the base restraint are removed from the FE model in Figure 4. In other words, only the layered LG panel and the lateral setting blocks, still reproduced by 3D solid brick elements, are kept in position (see Figure 7). From a practical point of view, such a kind of assumption means that the steel base restraint is replaced with equivalent translational springs.

Figure 7. SM1 simplified model: (**a**) model concept and 3D assembly (axonometric view from ABAQUS), with (**b**) the reference mechanical system (cross-section) and evidence of lumped equivalent springs.

These springs are properly calibrated to possibly reproduce a rather complex mechanism. According to the original layout of assembled steel components, the springs are, in fact, expected to reproduce the mechanical effects of the lateral steel plates (1) and (2) for the LG panel subjected to out-of-plane bending (i.e., Figure 8a, sub-scheme SS1), as well as the possible additional contribution due to the local deformation of the base steel flange (i.e., Figure 8b, sub-scheme SS2).

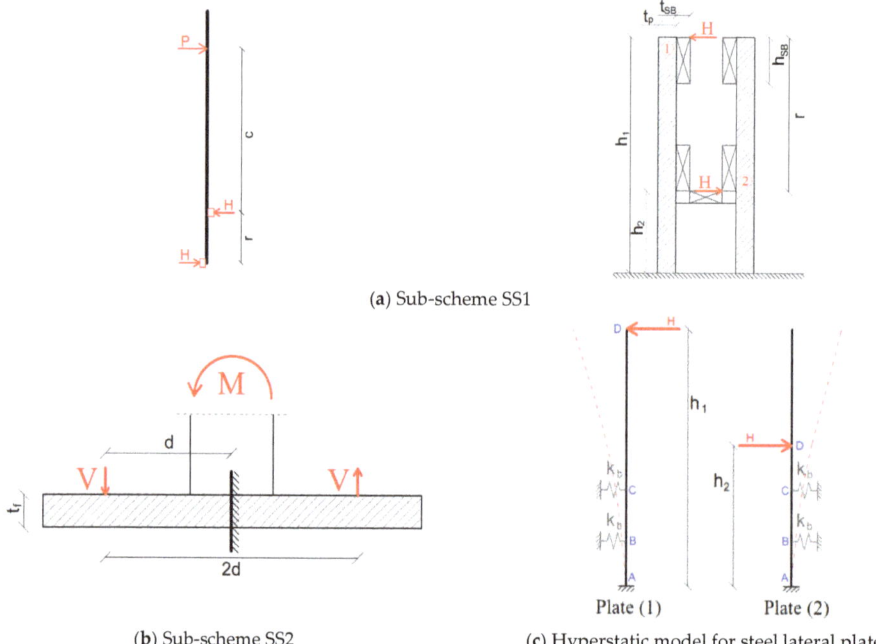

(**a**) Sub-scheme SS1

(**b**) Sub-scheme SS2

(**c**) Hyperstatic model for steel lateral plates

Figure 8. SM1 simplified numerical model: details of sub-schemes SS1 and SS2, and simplified local mechanical models to calculate the stiffness of steel lateral plates and the steel base flange for the LG balustrade in bending.

For the sub-scheme SS1 in Figure 8a, the top lateral displacement of steel plates (1) and (2) could be calculated as for two cantilevers under lateral force H, that is:

$$\delta_p^{(i)} = \frac{H\,h_i^3}{3E_s J_p}, \quad \text{with } i = 1,\,2 \text{ plates} \tag{3}$$

where:

$$H = \frac{P(r+c)}{r} \tag{4}$$

is the horizontal reaction force for sub-scheme SS1, with $r = (h_1 - h_2)$, from Figure 8a, and:

$$J_p = \frac{B\,t_p^3}{12} \tag{5}$$

is the second moment of the area for each t_p-thick steel plate, with $B = 1$ m being the extension of plates in the width of the balustrade. Such a kind of calculation, with the reference input parameters, would result in a relatively small top lateral displacement for both steel plates (1) and (2), namely, $\delta_p^{(1)} = 1.95$ mm and $\delta_p^{(2)} = 0.08$ mm, respectively, for the present study.

However, the so-derived displacement amplitudes are based on a roughly simplified schematization of the real system, reciprocal mechanical interactions, and, thus, a strong overestimation of bending stiffness terms compared to the Refined model parameters and the corresponding estimates. In this sense, the need for a more detailed calculation for the sub-scheme SS1 in Figure 8a could follow the hyperstatic schematic model presented in Figure 8c, in which:

$$k_b = \frac{A_b\,E_s}{l_b} \tag{6}$$

is the axial stiffness of base bolts, while the control points D represent the location of contact forces H at the interface of the LG panel in bending with the lateral supporting plates. The extended analytical solution of the hyperstatic scheme in Figure 8c (herein omitted) gives an improved—but still approximate—estimation for the lateral displacement in the control point D of plates (1) and (2). To further assess the accuracy of such a calculation approach, in this regard, a dedicated FE numerical analysis was developed in ABAQUS on 1D wire models, whose typical results are reported in Figure 9. For the LG balustrade under a given quasi-static top lateral load P, the solution of the hyperstatic scheme would result in lateral displacements of steel restraining plates equal to:

$$\delta_{p,D}^{(1)} = 3.17 \text{ mm} \qquad \delta_{p,D,FE}^{(1)} = |3.56| \text{ mm} \qquad \Delta = -10.95\%$$

$$\delta_{p,D}^{(2)} = 0.32 \text{ mm} \qquad \delta_{p,D,FE}^{(2)} = |0.45| \text{ mm} \qquad \Delta = -28.8\%$$

that is, in a scatter up to $\approx -11\%$ for steel plate (1) and $\approx -29\%$ for steel plate (2), respectively, compared to the 1D numerical model.

Regarding the sub-scheme SS2 of Figure 8b, it is also assumed for the base steel flange that the total rotation under top lateral loads for the LG balustrade can be considered by:

$$\theta_2 = \frac{\delta_{tot}}{d} = \frac{\delta_f + \delta_b}{d} \tag{7}$$

where:

$$\delta_f = \frac{V\,d^3}{3\,E_s\,J_f} \quad \text{is the deformation of the base flange,} \tag{8}$$

with J_f being its second moment of the area:

$$J_f = \frac{B t_f^3}{12} \tag{9}$$

and:

$$\delta_b = \frac{V \, l_b}{E_s \, A_b} \quad \text{being the contribution in deformation for bolts,} \tag{10}$$

with:

$$M = H \, (h_1 - h_2) \tag{11}$$

and

$$V = \frac{M}{2 \, d} \tag{12}$$

The so-defined θ_2 rotation term of the base flange under moment M manifests on the clamped lateral steel plates (1) and (2) in the form of an additional lateral displacement, which—even being relatively small—can be calculated as:

$$\delta_\theta^{(i)} = \theta_2 \, (d + h_i), \quad \text{with } i = 1, \, 2 \tag{13}$$

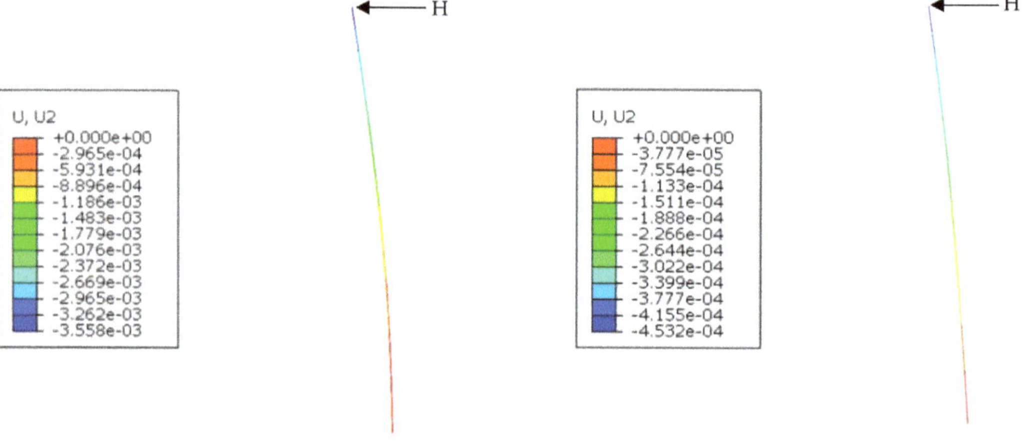

(**a**) Steel plate (1) (**b**) Steel plate (2)

Figure 9. Simplified numerical model SM1, with evidence of the local analysis of lateral deflections measured in steel plates (1) and (2). Comparative calculations for the assessment of simplified empirical formulations in use for sub-scheme SS1 (ABAQUS). Legend values in m (out-of-scale deformed shapes).

For the present application, the contribution of Equation (13) can be quantified in $\delta_\theta^{(1)} = 0.45$ mm and $\delta_\theta^{(2)} = 0.24$ mm, respectively, which should be added to the previously calculated SS1 displacement estimates.

In the above conditions, it is in fact possible to uniformly distribute a set of equivalent translational springs on a glass surface of total extension A_{SB}, which corresponds to the location and size of setting blocks, namely:

$$k_{t,sup,Winkler} = \frac{2 \, k_{t,sup}}{8 \, A_{SB}} \tag{14}$$

and

$$k_{t,inf,Winkler} = \frac{2 \, k_{t,inf}}{8 \, A_{SB}} \tag{15}$$

where the translational stiffness terms

$$k_{t,sup} = \frac{H}{\delta_p^{(1)} + \delta_\theta^{(1)}}$$ (16)

and

$$k_{t,inf} = \frac{H}{\delta_p^{(2)} + \delta_\theta^{(2)}}$$ (17)

find correspondence in the original mechanical model of Figure 7.

4.2. Simplified Characterization of the Base Restraint and LG Panel—SM2 Model

Differing from the SM1 system, the SM2 model introduces additional simplifications on the side of the layered glass panel. More precisely, the 3D solid brick sandwich section for the LG panel is replaced by an equivalent thickness monolithic glass panel composed of 2D shell elements, in which the total thickness is calculated based on nominal LG features, with the support of the EET formulation [2]. More specifically, assuming the input geometrical and mechanical parameters summarized in Section 2, the EET analytical calculation results in a total thickness of glass equal to 21.3 mm.

Figure 10 shows the corresponding FE model under the double twin-tyre setup, with evidence of 8-node, S8R-type monolithic shell elements used for the description of the glass panel. The material characterization, in the framework of an EET-based analysis of LG panes, is still based on material properties according to Table 2.

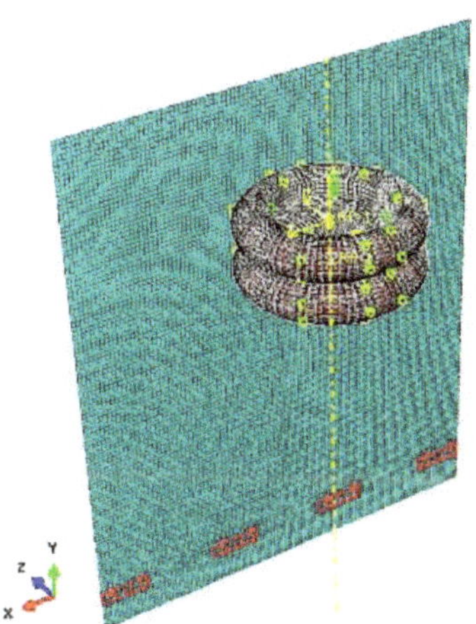

Figure 10. Axonometric view of the SM2 simplified model under the double twin-tyre setup (ABAQUS).

Overall, the FE numerical model of Figure 10 can be schematized as in Figure 11a. Another relevant feature of the SM2 simplified model is represented by the introduction of equivalent translational stiffness terms, which are calculated for a set of springs and introduced (uniformly distributed on lumped regions, as in Figure 10) at the base of the 2D shell-based monolithic glass panel (i.e., in the region of setting blocks) so as to capture a more realistic distribution of stresses for glass in bending.

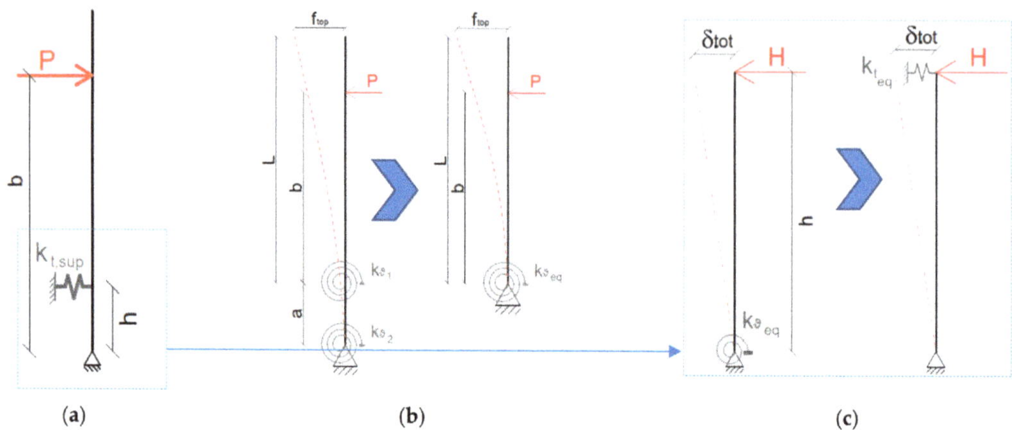

Figure 11. Derivation of equivalent stiffness parameters, as for the SM2 simplified model: (**a**) required translational stiffness, with (**b**) equivalent rotational and (**c**) translational terms.

More precisely, the translational stiffness contribution required for this kind of approach is defined in Figure 11a. This input value can be rationally obtained based on some considerations of the real mechanical system and, in particular, based on two additional rotational contributions, which are first used to represent the sub-schemes SS1 and SS2 in Figure 8 (see Figure 11b) and cumulated in an equivalent rotational spring, which is introduced at the bottom edge of the glass. Successively, once this equivalent rotational stiffness term is placed, as in Figure 11c, additional mechanical considerations based on rigid-body rotation assumptions for the h-span cantilever and an equivalent translational stiffness term to reproduce the effect of original restraints is estimated.

In Figure 11, the schematic model of Figure 11c represents a detailed view of the base restraint region only, where the distance h is the same as that in Figure 11a. Most importantly, the advantage of the overall analytical procedure schematized in Figure 11 is that the resulting equivalent translational spring stiffness, $k_{t,eq} = k_{t,sup}$, can be derived from multiple considerations in the rotational stiffness form (Figure 11c), that is:

$$k_{t,eq} = k_{t,sup} = \frac{H}{\delta_{tot}} = \frac{k_{\vartheta,eq}}{h^2} \tag{18}$$

and where $k_{\vartheta,eq}$ is implicitly representative of the rotational contributions of sub-schemes SS1 and SS2. From Equation (18), the translational springs can thus be distributed as in the FE numerical model of Figure 10, given that:

$$k_{t,sup,Winkler} = \frac{k_{t,sup}}{4\,A_{SB}} = \frac{k_{t,eq}}{4\,A_{SB}} \tag{19}$$

The SM2 procedure assumes that, for the detailed scheme in Figure 11c:

$$k_{\vartheta,eq} = \frac{H\,h}{\theta} \text{ and } H = \frac{k_{\vartheta,eq}\,\theta}{h} \tag{20}$$

where:

$$\theta = \frac{\delta_{tot}}{h} \text{ and } \delta_{tot} = \theta\,h \tag{21}$$

To this aim, basic assumptions similar to those of SM1 are taken into account for sub-schemes SS1 and SS2. It is observed that, as far as a rigid link is used to connect the two rotational springs in Figure 11b, these two contributions can be expressed as:

$$k_{\theta 1} = \frac{Hh}{\theta_1} \tag{22}$$

and

$$k_{\theta 2} = \frac{Vd}{\theta_2} \tag{23}$$

For the rotational spring "1" of Equation (22), in particular, the deformation terms to take into account to express the expected rotation amplitude still derive from a quantification of the lateral displacements of steel plates (1) and (2) under lateral load P at the top edge of the balustrade, but they also derive from the possible crushing of lateral setting blocks which are compressed due to the progressively increasing deformation of the glass panel in out-of-plane bending. This means that, for Equation (22), it is:

$$\theta_1 = \frac{\delta_{p,1}^{tot}}{h} = \frac{\delta_p^{(1)} + \delta_p^{(2)} + \delta_{SB}^{(1)} + \delta_{SB}^{(2)}}{h} \tag{24}$$

where:

$$\delta_{SB}^{(1)} = \delta_{SB}^{(2)} = \frac{Ht_{sb}}{E_{sb}\,A_{sb}/3} \tag{25}$$

while θ_2 for Equation (23) is given by Equation (7).

Overall, the final value for the equivalent rotational spring to be introduced at the base of the glass panel (Figure 11b) comes from:

$$\frac{1}{k_{\theta,eq}} = \frac{1}{k_{\theta 1}} + \frac{(a+b)(a+L)}{b\,L}\frac{1}{k_{\theta 2}} \tag{26}$$

where:

$$\delta_{\theta,eq} = \delta_{\theta 1} + \delta_{\theta 2} = \frac{PbL}{k_{\theta,eq}} \tag{27}$$

$$\delta_{\theta,1} = \frac{PbL}{k_{\theta 1}} \tag{28}$$

$$\delta_{\theta,2} = \frac{P(a+b)\,(a+L)}{k_{\theta 2}} \tag{29}$$

and:

$$\delta_{EJ} = \frac{P(a+b)^2}{6\,EJ}\left[3\,(a+L) - (a+b)\right] \tag{30}$$

δ_{EJ} in Equation (30) represents the displacement of the glass panel under top-edge lateral load P and depends on the bending stiffness of the equivalent monolithic glass panel only (i.e., cantilever response analysis for the composite LG panel); thus, the top lateral deflection at the unrestrained edge of glass can be quantified in:

$$f_{top} = \delta_{EJ} + \delta_{\theta 1} + \delta_{\theta 2} \tag{31}$$

4.3. Linearly Distributed Base Springs—SM3 and SM4 Models

In conclusion, two additional simplified mechanical models are taken into account, namely, the SM3 and SM4 models. Differing from the SM2 model, the equivalent springs are moved from the actual region of the setting blocks and are equally distributed. For SM3, this spring distribution is performed in the same height as that of the real restraints (see Figure 12a) so as to possibly capture the same translational effect of the Refined system. For the SM4 model with rotational equivalent springs, these are uniformly distributed along the bottom edge of the glass panel (see Figure 12b). In both cases, the LG panel is still described as for SM2, that is, in the form of equivalent monolithic shell elements made of glass.

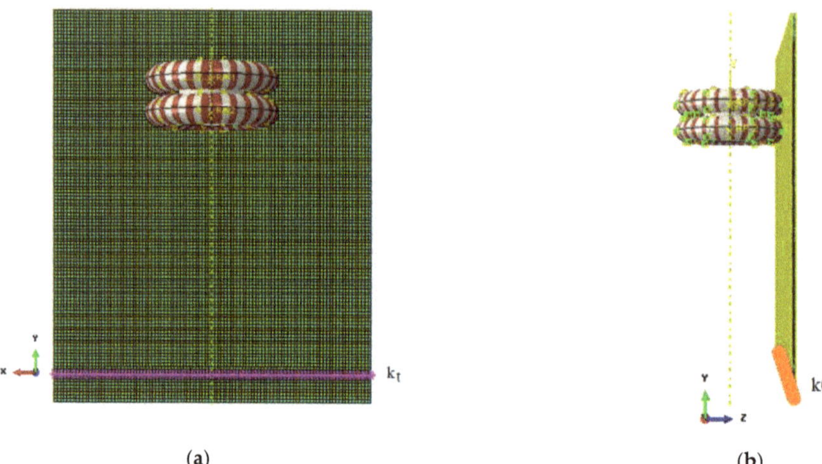

(a)   (b)

Figure 12. Front and axonometric views of (**a**) SM3 and (**b**) SM4 models under double twin-tyre impact (ABAQUS).

The comparative results are discussed in Section 5 for the balustrade under quasi-static lateral load or double twin-tyre impactor, while Table 4 gives evidence of the computational cost of explored simplified numerical models compared to that of the Refined one.

Table 4. Summary of the computational cost of the Refined and SM1-to-SM4 simplified numerical models (ABAQUS) in terms of the total number of elements and DOFs required to reproduce the nominal geometry of the examined balustrade.

	FE Model Features	
FE Numerical Model	**Number of Elements**	**Number of DOFs**
Refined	≈120,000	≈505,000
SM1	≈76,000	≈214,000
SM2	≈12,000	≈73,000
SM3 and SM4	≈12,000	≈73,000

In this regard, it is worth noting that the SM1 simplified model mostly reduces to half the number of elements and DOFs, with substantial benefits in terms of structural analysis. For the SM2 model, the number of elements is in the order of the 1/10th part compared to the Refined model. The use of monolithic shell elements for the LG panel in place of 3D solid brick elements can be quantified in computational benefits from the comparison of SM2 features with SM1 features.

On the other hand, it must be recalled that the LG panels described by the monolithic shell section made of glass are able to provide accurate results for limited loading and boundary conditions only. Finally, the SM3 and SM4 models are characterized by the same order of magnitude of elements and DOFs as in the case of the SM2 model, where minimum variations are quantified by the removal of setting blocks.

5. Discussion of Numerical Results

5.1. Simplified Models SM1 and SM2

A first comparison is presented in terms of top-lateral displacement for the examined glass balustrade under quasi-static lateral load ($P = 4.5$ kN). Numerical results are proposed in Figure 13 for the simplified models SM1 and SM2 in comparison to the Refined FE model, with evidence of top lateral displacement (Figure 13a) and a calculated percentage scatter of SM1 or SM2 models compared to the Refined one (Figure 13b), as a function of imposed lateral load P, where:

$$\Delta_f = 100 \cdot \frac{\left(f_{SM(i)} - f_{Refined}\right)}{f_{Refined}} \quad \text{with } i = 1, 2 \tag{32}$$

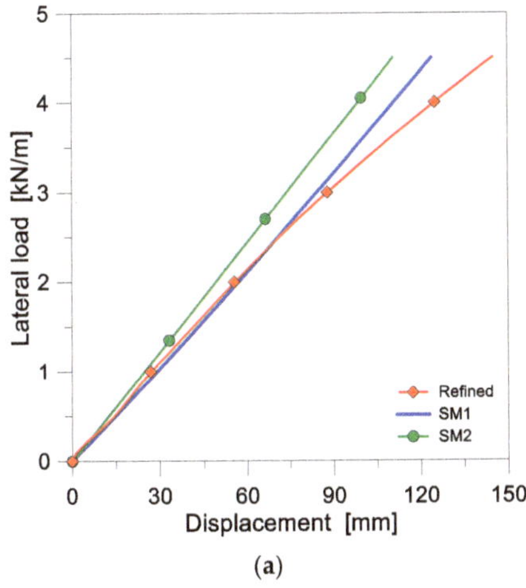

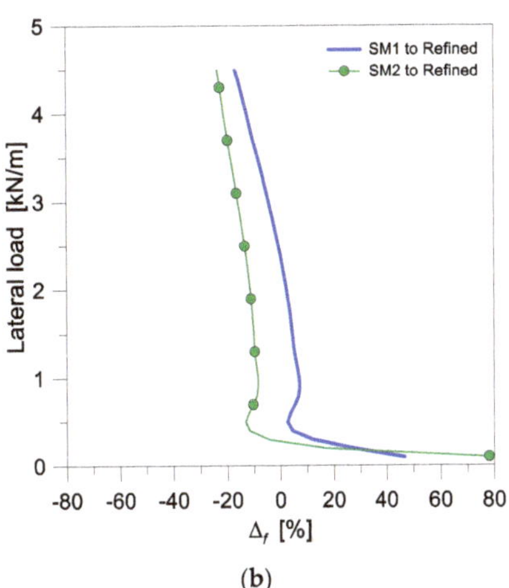

Figure 13. Numerical analysis of Refined, SM1, and SM2 simplified models under quasi-static lateral load (P = 4.5 kN/m) at the top edge of glass (ABAQUS): (**a**) top lateral displacement and (**b**) calculated percentage scatter of SM to Refined models (Equation (32)).

Regarding SM1, it can be noted that its linear deflection trend in Figure 13a corresponds to a very large percentage scatter in Equation (32) compared to the Refined model, in which local behaviours of primary and secondary fixing components can be efficiently taken into account. Especially in the first loading stage, it is evident that the SM1 model overestimates up to ≈50% the initial deformation of the LG balustrade, while the calculated percentage scatter progressively decreases in Figure 13b as the imposed load P increases. In this regard, the major sensitivity of the calculated scatter in small amplitudes of the loading stage must be attributed to the linear mechanical characterization taken into account in the SM1 approach for the fixing system, which disregards any possible non-linear effect included by the Refined assembly. In the early loading stage, small adjustments of the LG panel within the fixing restraining system can manifest, for simplified mechanical models, in a strong overestimation of top lateral deflections, whilst the glass panel is still not subjected to relevant bending action.

Most importantly, the trend in Figure 13b depends on a combination of local and global effects, and for higher imposed lateral loads, it can also include the out-of-plane bending deformation of glass in addition to the phenomena in the base connection. In this way, the scatter values in Figure 13b for the SM1-to-Refined model comparisons have both positive (i.e., conservative, compared to Refined model estimates) and negative (i.e., unconservative) values. Compared to the computational costs summarized in Table 4, this means that reducing the number of FE elements and DOFs down to approximately −36% and −57% for the SM1 model induces a large scatter (but on the safe side) for small lateral load amplitudes (i.e., $P < 0.3$ kN/m in Figure 13b). For high lateral load amplitudes (i.e., $P > 1.5$ kN/m in Figure 13b), however, the expected top lateral deflection for the LG panel could be sensitively underestimated, with a possible risk for design applications.

When the SM2 model is taken into account (see Figure 13b), an even larger percentage scatter is achieved compared to that of the Refined model, as a major effect of the mechanical characterization of equivalent springs.

Further parametric numerical results are presented in Figure 14 for the LG balustrade under double twin-tyre impact in terms of (a) P1-P2 displacement and the corresponding (b) stress peaks in glass or (c) impactor acceleration in time.

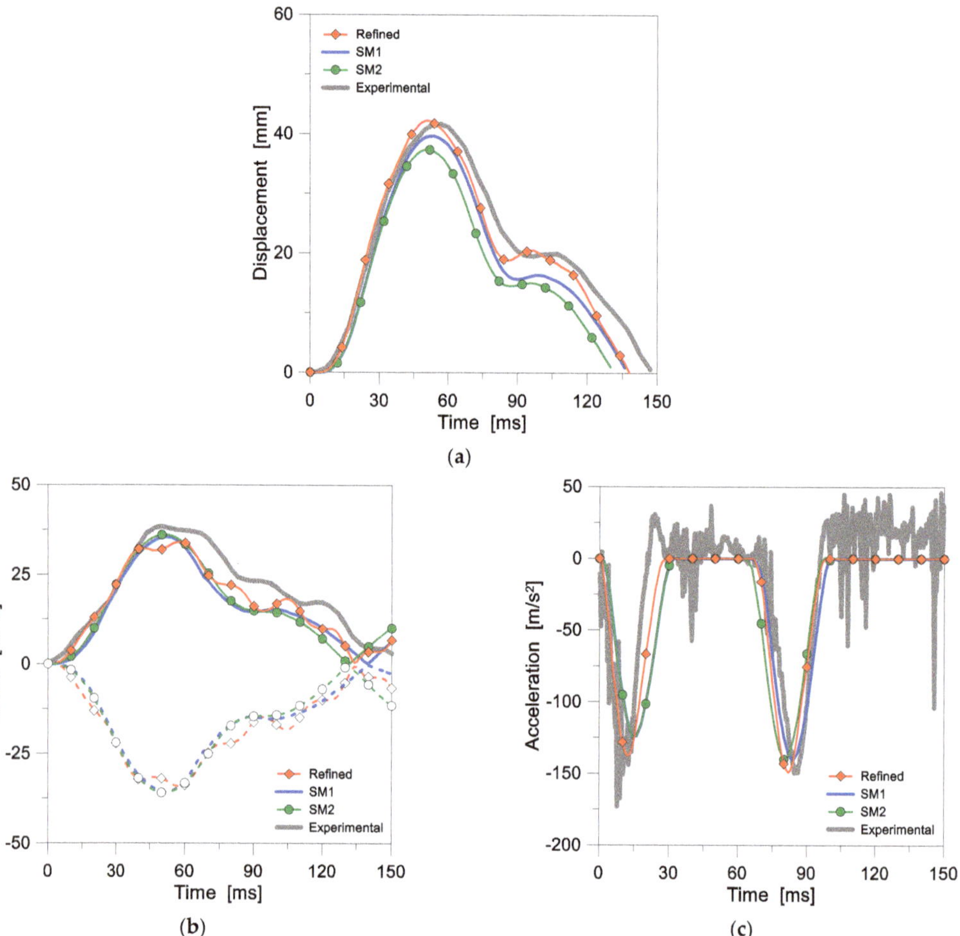

Figure 14. Numerical analysis of Refined, SM1, and SM2 simplified models (ABAQUS) under double twin-tyre impact (300 mm being the drop height): (**a**) lateral displacement; (**b**) maximum principal stress in glass (dashed lines for compression side), and (**c**) acceleration time histories.

In terms of simplified modelling under dynamic loads, it is possible to see that both the SM1 and SM2 approaches are overall characterized by a relatively higher stiffness compared to the Refined model. The percentage scatter on the maximum lateral displacements of the LG panel, moreover, was predicted as -6% for the SM1 model compared to the Refined one. An even larger scatter is observed in Figure 14a for the SM2 model (-13%), which reduces the number of FE elements and DOFs down to -90% and -85%, respectively. Overall, the SM2 model underestimates the SM1 results in the order of -5% in terms of deflections.

In terms of stress analysis in glass (double twin-tyre impact setup) and, most importantly, stress peak analysis compared to the experimental feedback in [17], a rather close correlation can be observed for the contour plot distribution trends, as it is for the selected examples reported in Figure 15. It is thus easy to note that the presented contour plots are qualitatively similar, and stress peaks can be detected in the region of setting blocks/equivalent springs. Such a finding enforces the need for specific attention for even local assembly details that could have a primary role in verification procedures for structural safety. In terms of stress values and quantitative analysis, however, it can be seen that both the SM1 and SM2 simplified models are rather approximate, given that they tend to overestimate or underestimate, respectively, the expected stress peaks in glass (+29.8% for SM1 and −7.1% for SM2, based on the maximum envelope).

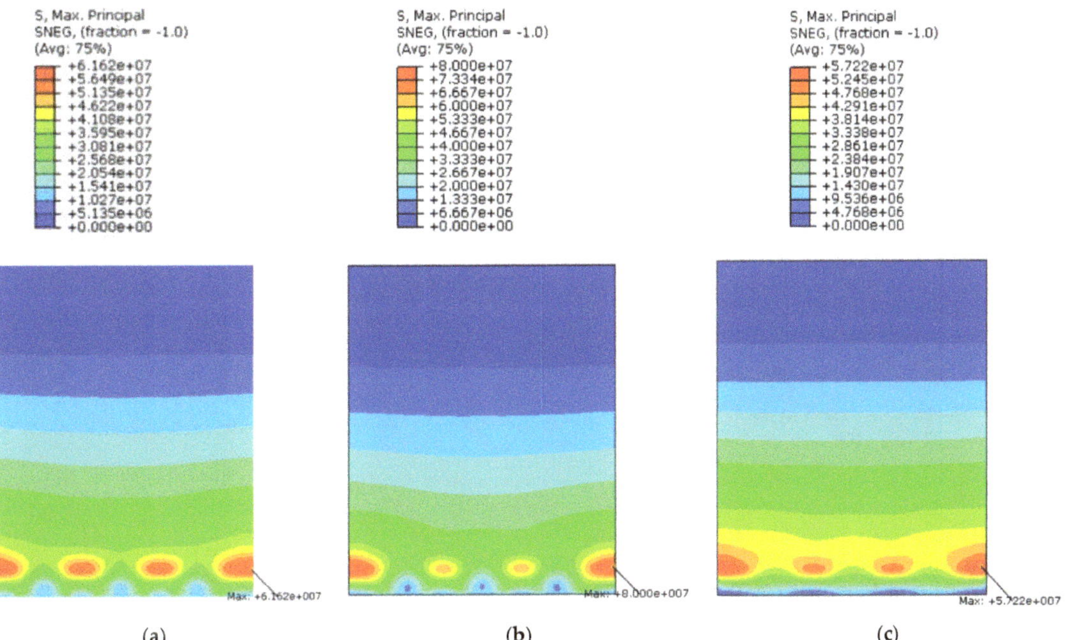

Figure 15. Numerical analysis of principal stress distribution and peaks in glass for the (**a**) Refined model, (**b**) SM1 simplified model, and (**c**) SM2 simplified model (ABAQUS) under double twin-tyre impact (300 mm being the drop height), with legend values in Pa.

Overall, the best quantitative correlation in terms of local stress analysis is found at control points S1, S2 of Figure 4a, with an average scatter of +9% for the SM1 and SM2 models (Figure 15c). Such an outcome strictly depends on the calibration of discrete equivalent springs and, thus, on the analysis of local mechanisms and phenomena which are expected from the real assembled system. Another important limitation is represented by boundary conditions for the simplified models in use, given that these equivalent springs are rigidly connected to the ground and are thus unable to follow and accommodate the relative deformations of steel plates, while setting blocks are still able to adapt to possible local deformations of the composite assembly (as for the full 3D Refined model).

5.2. Effect of Linearly Distributed Equivalent Springs—SM3 and SM4

A conclusive assessment attempt is carried out by the derivation of linearly distributed equivalent springs in terms of translational (SM3) or rotational (SM4) terms, respectively. Such a procedure requires, from an analytical and numerical point of view, rather quick calibration steps and a rather fast computational time of analysis. From a practical point of view, this means that the SM2 simplified model (where the EET-based monolithic glass section is still taken into account for 2D shell elements) is further roughly simplified, in terms of boundaries, towards a calculation approach which agrees with Figure 2c,d. Actually, the equivalent springs of the SM2 approach are distributed in the width of the balustrade to create a bed of equivalent springs for the LG panel. Structurally speaking, the LG panel is thus analyzed in out-of-plane bending with a relatively flexible/partially rigid base restraint, but the SM3 and SM4 simplified procedures are weak in terms of precision for the localization and size of real fixing system components.

A direct effect of such a kind of modelling approach relies on the reliable analysis of intrinsic limits for these procedures, as it is clear that further major approximations are introduced to the original assembled system. Most importantly, this assumption suggests that the reference performance indicators for structural design (and, especially, the stress analysis in glass) should be examined at both the local and global levels.

In Figure 16, for example, a set of selected response parameters for the SM3 or SM4 balustrades under quasi-static load or soft-body impact is compared to the Refined model. It can be seen from Figure 16a—quasi-static load $P = 4.5$ kN—that both the SM3 (translational springs) and SM4 (rotational springs) models are globally characterized by high stiffness compared to the Refined model, even more than the previously investigated SM1 and SM2 models. Such a stiffening effect can be also noted in Figure 16b in terms of the percentage scatter trend calculated from Equation (32) for increasingly lateral load P amplitudes.

This observed numerical outcome suggests that the presence of discrete components and soft gaskets for typical restraints (as for most practical applications with glass) is clearly associated with local flexibility contributions and displacement accommodation capacities that hardly match with the linearized equivalent mechanical restraints. From Figure 16a,b, it is also possible to notice that the use of distributed rotational springs (SM4) is indeed less rigid than SM3 (translational springs). The two idealized SM3 and SM4 balustrade descriptions are thus not fully mechanically equivalent in terms of glass panel behaviour in out-of-plane bending.

As far as the dynamic response is analyzed, the SM3 and SM4 models with linearly distributed springs were again subjected to a double twin-tyre impact with an imposed drop height of 300 mm. Typical comparative results from the numerical analysis can be seen in Figure 16c, in terms of lateral displacements of glass, and in Figure 16d, in terms of impactor accelerations in time.

In this case, it is worth noting that a local analysis and comparison of the stress evolution in glass as a function of time (i.e., as for control points S1, S2) is not meaningful due to the strongly different loading condition for the examined systems. Such comparative evidence is clearly representative of a major intrinsic limit of both the SM3 and SM4 procedures, given that the estimated distribution of stresses in glass strongly differs from the original one. In other words, basic considerations in similar conditions could only be drawn in terms of displacement analysis, with rather weak numerical results for the stress verification of glass components.

Both the local and global stress analyses for these systems should be carefully explored, as they are representative of a key parameter for the overall design process. Evidence of these major limits of the SM3 and SM4 procedures can be noticed in Figure 17, where it is clear that the typical stress distribution is mostly different compared to that in Figure 15. Most importantly, the SM3 and SM4 stress peaks are observed to be misplaced. Additionally, they largely underestimate the stress estimates of the reference Refined model (-21.7% for SM3 and -13.6% for SM4). Further, no mechanical equivalence can be noted for the SM3 and SM4 approaches, and all these intrinsic limits may result in unsafe design choices.

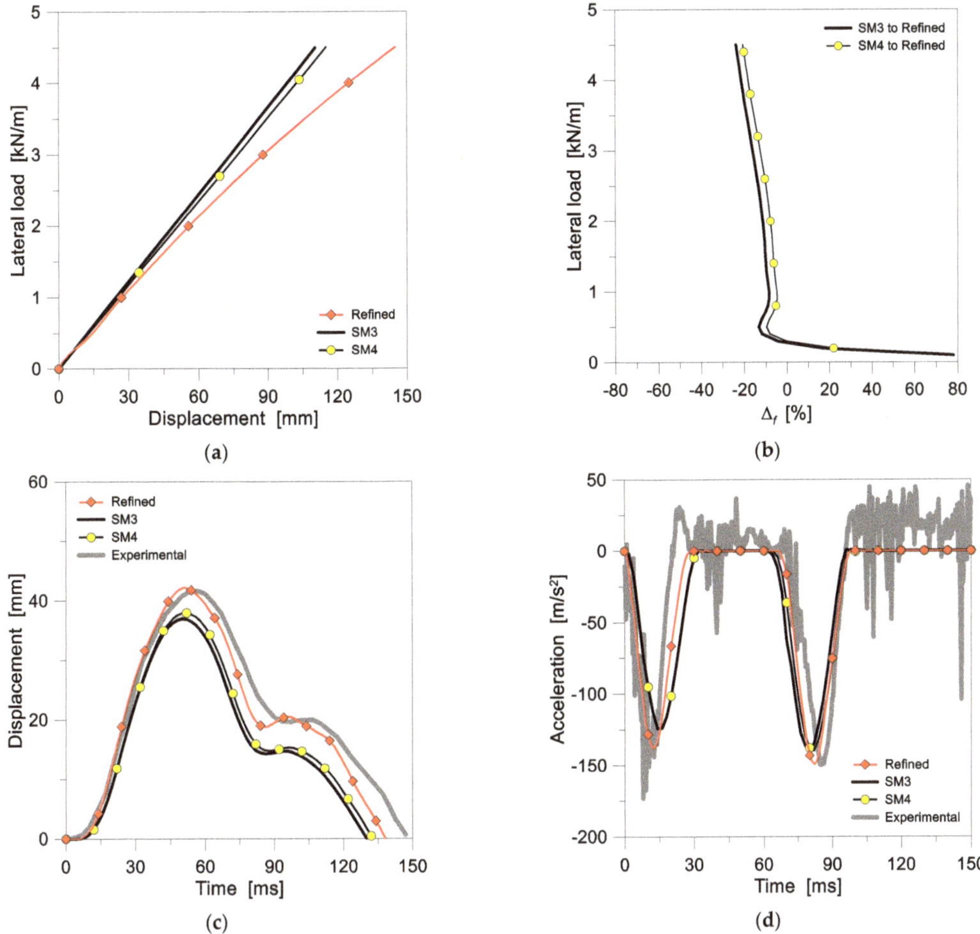

Figure 16. Numerical analysis of the Refined model, SM3 (translational springs) simplified model, and SM4 (rotational springs) simplified model (ABAQUS): (**a**) top lateral displacement under quasi-static load (P = 4.5 kN/m) at the top edge of glass and (**b**) calculated percentage scatter of SM to Refined models (Equation (32)), with (**c**,**d**) response analysis under double twin tyre impact (300 mm being the drop height).

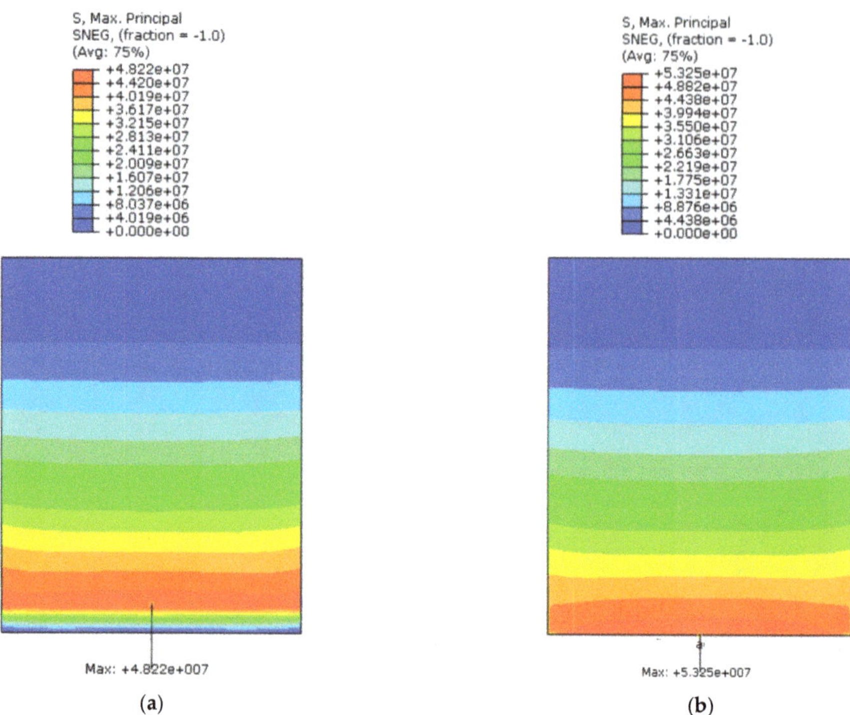

(a)   (b)

Figure 17. Numerical analysis of the principal stress distribution and peaks in glass for the (**a**) SM3 (translational springs) and (**b**) SM4 (rotational springs) simplified models (ABAQUS) under double twin-tyre impact (300 mm being the drop height), with legend values in Pa.

6. Conclusions

The structural design of glass balustrades, as is known, requires basic knowledge on material and mechanical aspects. From a mechanical point of view, rather simple analytical considerations can be drawn, based, for example, on cantilever assumptions for rough stress analysis estimates. Most of the applications and calculations for linearly restrained glass balustrades are in fact carried out under the assumption of idealized mechanical model base-clamped glass plates subjected to the effects of a quasi-static top lateral load representative of crowd. However, compared to real restraints, the use of idealized or simplified mechanical models can result in misleading interpretations and predictions for glass verification purposes. Additionally, design protocols may also require the verification of structural capacity for these assembled systems under soft-body dynamic impact events.

In this paper, the attention was focused on a case-study system of the literature, consisting of a base-restrained laminated glass (LG) balustrade subjected to quasi-static lateral loads or double twin-tyre impact. As a reference, a full 3D solid brick "Refined" model has been developed to support the calibration and validation stages.

Four different simplified mechanical models (SM1 to SM4) have thus been developed and addressed based on the validation of past experimental data of the literature so as to assess the potential and limits in the use of equivalent discrete or distributed translational and/or rotational springs to restrain LG panels for similar load-bearing applications. The comparative numerical analysis reported in this paper, as expected, confirmed that both primary and secondary components and soft members of typical use in glass applications to create ad hoc restraints have a key role in preserving premature stress peaks and allow for the accommodation of possible local deformations under external design loads.

The use of equivalent springs to describe real restraints confirmed the high computational efficiency of simplified mechanical models compared to expensive full 3D solid brick assemblies or, even more so, compared to full-size experimental tests. In some cases, the local analysis of performance indicators gave a rather good correlation compared to a more refined numerical description of constituent components (as, for example, in the case of the SM1 and SM2 local stress peak estimates in glass and displacement predictions). On the other side, as was noted for the SM3 and SM4 simplified procedures, the use of mechanically equivalent, linearly distributed equivalent springs is not able to reproduce the effective mechanical features and behaviour of real assemblies and can thus result in misleading numerical performances and even unsafe stress estimates for glass verification, with potential risks for safety check purposes.

Author Contributions: Conceptualization, E.R., C.B. and C.A.; methodology, C.B. and C.A.; software, E.R. and C.B.; validation, E.R. and C.B.; formal analysis, E.R.; investigation, E.R. and C.B.; data curation, E.R.; writing—original draft preparation, E.R. and C.B.; supervision, C.B. and C.A.; project administration, C.B. All authors have read and agreed to the published version of the manuscript.

Funding: This research received no external funding.

Data Availability Statement: Data will be shared upon request.

Acknowledgments: M. Kozłowski (Silesian University of Technology, Poland) is acknowledged for sharing experimental data.

Conflicts of Interest: The authors declare no conflict of interest.

References

1. Feldmann, M.; Kasper, R.; Abeln, B.; Cruz, P.; Belis, J.; Beyer, J. *Guidance for European Structural Design of Glass Components—Support to the Implementation, Harmonization and Further Development of the Eurocodes*; Report EUR 26439; Dimova, P., Feldmann, D., Eds.; Joint Research Centre–Institute for the Protection and Security of the Citizen: Ispra, Italy, 2014. [CrossRef]
2. *CNR-DT 210/2013*; Istruzioni Per la Progettazione, L'esecuzione ed il Controllo di Costruzioni con Elementi Strutturali di Vetro. National Research Council of Italy (CNR): Roma, Italy, 2013; (In Italian–English Version Freely Available).
3. Bedon, C.; Santarsiero, M. Transparency in Structural Glass Systems Via Mechanical, Adhesive, and Laminated Connections. *Adv. Eng. Mater.* **2018**, *20*, 1700815. [CrossRef]
4. *EN 12600:2002*; Glass in Building. Pendulum Test. Impact Test Method and Classification for Flat Glass Method and Classification for Flat Glass. European Committee for Standardization: Brussels, Belgium, 2002.
5. *DIN 18008-4*; Glass in Building—Design and Construction Rules—Part 4: Additional Requirements for Barrier Glazing. DIN: Berlin, Germany, 2013.
6. Bedon, C.; Zhang, X.; Santos, F.; Honfi, D.; Kozłowski, M.; Arrigoni, M.; Figuli, L.; Lange, D. Performance of structural glass facades under extreme loads—Design methods, existing research, current issues and trends. *Constr. Build. Mater.* **2018**, *163*, 921–937. [CrossRef]
7. Figuli, L.; Papan, D.; Papanova, Z.; Bedon, C. Experimental mechanical analysis of traditional in-service glass windows subjected to dynamic tests and hard body impact. *Smart Struct. Syst.* **2021**, *27*, 365.
8. Bedon, C.; Santi, M.V. Vulnerability and Structural Capacity Assessment of Historic Glass Facades under Bird-Strike. *Math. Probl. Eng.* **2022**, *2022*, 6059466. [CrossRef]
9. Kalamar, R.; Bedon, C.; Eliasova, M. Experimental investigation for the structural performance assessment of square hollow glass columns. *Eng. Struct.* **2016**, *113*, 1–15. [CrossRef]
10. Bedon, C.; Kalamar, R.; Eliasova, M. Low velocity impact performance investigation on square hollow glass columns via full-scale experiments and Finite Element analyses. *Compos. Struct.* **2017**, *182*, 311–325. [CrossRef]
11. Fröling, M.; Persson, K.; Austrell, P.-E. A reduced model for the design of glass structures subjected to dynamic impulse load. *Eng. Struct.* **2014**, *80*, 53–60. [CrossRef]
12. Andersson, L.; Kozłowski, M.; Persson, P.; Austrell, P.-E.; Persson, K. Reduced order modeling of soft-body impact on glass panels. *Eng. Struct.* **2022**, *256*, 113988. [CrossRef]
13. Bez, A.; Bedon, C.; Manara, G.; Amadio, C.; Lori, G. Calibrated Numerical Approach for the Dynamic Analysis of Glass Curtain Walls under Spheroconical Bag Impact. *Buildings* **2021**, *11*, 154. [CrossRef]
14. Schneider, J.; Schula, S. Simulating soft body impact on glass structures. *Proc. Inst. Civ. Eng. Struct. Build.* **2016**, *169*, 416–431. [CrossRef]
15. Baidjoe, Y.; Van Lancker, B.; Belis, J. Calculation methods of glass parapets in aluminum clamping profiles. *Glass Struct. Eng.* **2018**, *3*, 321–334. [CrossRef]

16. Biolzi, L.; Bonati, A.; Cattaneo, S. Laminated Glass Cantilevered Plates under Static and Impact Loading. *Adv. Civil. Eng.* **2018**, *2018*, 7874618. [CrossRef]
17. Kozłowski, M. Experimental and numerical assessment of structural behaviour of glass balustrade subjected to soft body impact. *Compos. Struct.* **2019**, *229*, 111380. [CrossRef]
18. Bedon, C.; Fasan, M.; Amadio, C. Vibration Analysis and Dynamic Characterization of Structural Glass Elements with Different Restraints Based on Operational Modal Analysis. *Buildings* **2019**, *9*, 13. [CrossRef]
19. Portal, N.W.; Flansbjer, M.; Honfi, D.; Kozłowski, M. The dynamic structural response of a laminated glass balustrade analysed with optical measurements. Ce/papers, Special Issue: Engineered Transparency 2021: Glass in Architecture and Structural Engineering. *Glass Struct. Des.* **2021**, *4*, 251–261. [CrossRef]
20. Galuppi, L.; Royer-Carfagni, G. Effective thickness of laminated glass beams: New expression via a variational approach. *Eng. Struct.* **2012**, *38*, 53–67. [CrossRef]
21. Simulia. *ABAQUS Computer Software*, Dassault Systémes: Providence, RI, USA, 2021.

Article

Body CoM Acceleration for Rapid Analysis of Gait Variability and Pedestrian Effects on Structures

Chiara Bedon

Department of Engineering and Architecture, University of Trieste, 34127 Trieste, Italy; chiara.bedon@dia.units.it

Abstract: Knowledge of body motion features and walk-induced effects is of primary importance for the vibration analysis of structures, especially low-frequency slabs and lightweight and/or slender systems, as well as for clinical applications. Structurally speaking, consolidated literature procedures are available for a wide set of constructional solutions and typologies. A basic assumption consists in the description of walking humans' effects on structures through equivalent deterministic loads, in which the ground vertical reaction force due to pedestrians depends on their mass and motion frequency. However, a multitude of additional parameters should be taken into account and properly confirmed by dedicated laboratory studies. In this paper, the focus is on the assessment of a rapid analysis protocol in which attention is given to pedestrian input, based on a minimized sensor setup. The study of gait variability and related effects for structural purposes is based on the elaboration of single Wi-Fi sensor, body centre of mass (CoM) accelerations. A total of 50 walking configurations was experimentally investigated in laboratory or in field conditions (for more than 500 recorded gaits), with the support of an adult volunteer. Parametric gait analysis is presented considering different substructure conditions and motion configurations. Body CoM acceleration records are then used for the analysis of a concrete slab, where the attention is focused on the effects of (i) rough experimental body CoM input, or (ii) experimentally derived synthetized gait input. The effects on the structural side of rough experimental walk time histories or synthetized experimental stride signals are discussed.

Keywords: vibrations; body center of mass (CoM); vertical acceleration; laboratory experiments; in-field experiments; micro electro-mechanical systems (MEMS) sensor

Citation: Bedon, C. Body CoM Acceleration for Rapid Analysis of Gait Variability and Pedestrian Effects on Structures. *Buildings* **2022**, *12*, 251. https://doi.org/10.3390/buildings12020251

Academic Editor: Elena Ferretti

Received: 13 January 2022
Accepted: 18 February 2022
Published: 21 February 2022

Publisher's Note: MDPI stays neutral with regard to jurisdictional claims in published maps and institutional affiliations.

1. Introduction

As is known, special attention is required for the analysis under random walks of possible vibration issues in low-frequency slabs (i.e., with a fundamental vibration frequency lower than 8 Hz), or slender and lightweight slabs characterized by limited bending stiffness or even reduced mass, compared to occupants. In this regard, several consolidated calculation approaches can be found in the literature in support of design [1–3].

From a structural point of view, special attention may be required by pedestrian systems which can be more sensitive to walk-induced effects compared to other structural typologies. Traditional vibration serviceability assessment is based on consolidated structural health monitoring procedures and techniques [4–9]. Recently, various studies have started to assess the dynamic behaviour of laminated glass slabs under pedestrians, showing that structural dynamic parameters and performance indicators are rather different from other constructional solutions, as a major effect of flexibility, slenderness and (often) limited mass, compared to occupants [10–13].

When attention is given to the characterization of the effect of walking occupants, however, even more complex calculation models are needed to account for realistic parameters [14–16], and the use of wearable or Wi-Fi sensors can result in efficient support for the analysis. For example, the assumption of a basic inverted pendulum model to describe non-rigid-leg pedestrians (Figure 1a) offers strong insight regarding walking mechanics, but also has intrinsic limitations [17]. Many factors are known to typically affect walking features, including age, medical issues, etc. [18], and psychological discomfort [19,20], in addition to dynamic mechanical parameters of the substructure.

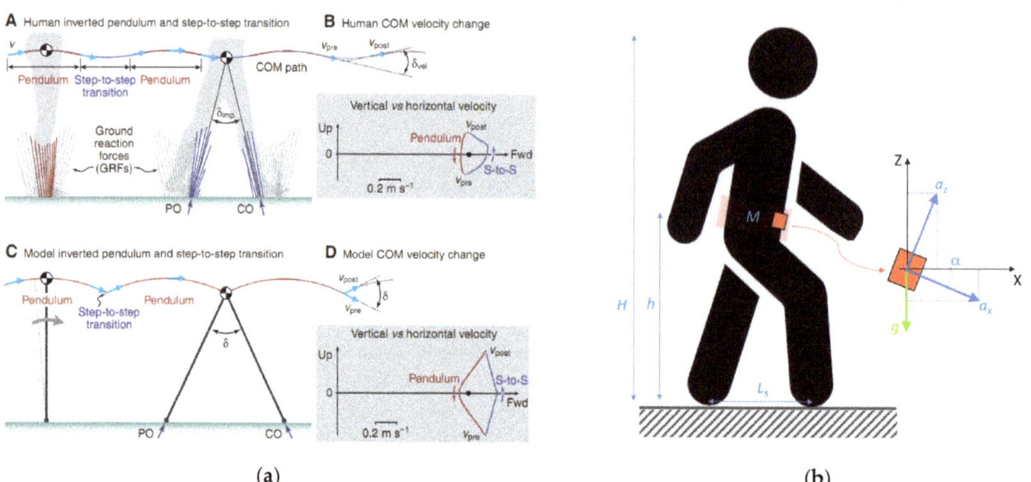

Figure 1. Experimental method: (**a**) reference literature pendulum model (figure reproduced from [17] with permission from The Company of Biologists®, order number 1175262, January 2022) and (**b**) herein adopted measurement system for body CoM acceleration records, with detail of lumbar belt to keep in position the used Wi-Fi sensor.

Among others, the use of various Wi-Fi sensors and instruments, especially accelerometers, has been addressed by several studies to explore the motion variability and features of pedestrians [21–24]. Compared to classical deterministic models to describe loads and ground reaction forces due to pedestrians on structures, there are multiple aspects to take into account. In [25], the use of body centre of mass (CoM) measures was assessed with experimental measures of a volunteer. The optimal use of multiple body sensors to capture motion records for the estimation of the vertical ground reaction force of pedestrians for structural vibration purposes has been assessed in [26]. Bocian et al. [27] demonstrated with experimental studies that a single point inertial measurement can accurately capture the pedestrian vertical induced force, and thus efficiently support the structural vibration analysis process. Different accelerometer locations can offer consistent gait parameters [28]. Smartphone inertial measurement units (IMUs) can also efficiently capture the gait characteristics of pedestrians [29].

In this paper, attention is focused on the analysis of body motion parameters for vibration serviceability assessment of structures, in order to use body measurements for rapid structural evaluation. This is carried out based on rough experimental signals, and by the derivation of synthetized stride signals. More precisely, experimental records are proposed for body CoM of an invited adult volunteer (Figure 1b), with the support of a single high-precision micro electro-mechanical systems (MEMS) sensor, and statistical analysis of acceleration time histories (50 walking configurations, for more than $n_s = 500$ recorded gaits) is presented. The effects of different substrate conditions (i.e., rigid or flexible slab), different types of shoes, and various motion speed intervals are studied. Correlation analysis is carried out for several walking features, using the linear regression method, to quantify the variability of motion parameters. The experimental body CoM input is then applied to a case-study, low-frequency concrete slab, to quantify the structural vibration effects and address them with respect to conventional loading protocols.

2. Background and Goals

Locomotion mechanics have been investigated by a multitude of research studies, for different purposes. In general terms, the ground reaction force (GRF) and its vertical component (vGRF) are the most meaningful parameters to describe walk and pedestrian-induced effects on structures [6]. Over the years, for structural applications, knowledge of GRF features and trends has facilitated the derivation and calibration of deterministic load models for the reliable description of human-induced effects on pedestrian systems [14]. A major advantage is that the conventional gait effect can be analytically described, as in Figure 2a, and the primary influencing parameter—for a given pedestrian with mass M—is represented by the average walking frequency f_s. For structural analysis, the reference stride module, as in Figure 2a, is repeatedly applied on the structure to cover n_s gaits [30]. On the other hand, gait variability and possible irregularities, such as asymmetry, which are typical of normal walks [31], and other aspects, are disregarded. Average vibration effects are, thus, predicted for the structure, disregarding any interaction with pedestrians. Recently, the availability of WiFi or wearable sensors has suggested their use for gait analysis and locomotion studies [27–29]. Three different positions for inertial sensors were considered in [32]. Among others, laboratory studies presented in [26,27] demonstrated that a single inertial sensor in body CoM can capture motion features. The investigation reported in [33] also showed that a single IMU sensor can be efficiently applied to in-field and clinical situations, without the need of dedicated laboratory instruments (i.e., force plates, instrumented treadmills, etc.). Compared to more accurate instrument acquisitions, or multiple sensors, such a possibility represents a major advantage in support of monitoring procedures and diagnostic analyses.

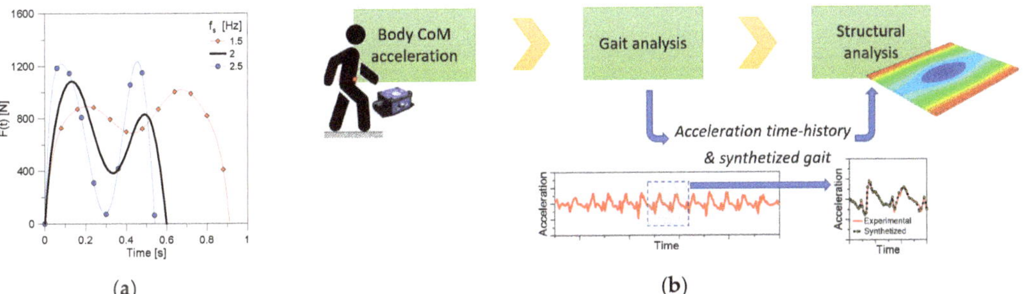

Figure 2. Gait variability and walk-induced effects on structures based on (**a**) deterministic loads (examples of gait-induced loads for different walking frequencies f_s) or (**b**) present procedure.

According to Figure 1 and the literature, the present investigation takes advantage of the fact that the vertical force transferred by a pedestrian on a substructure can be estimated from Newton's second law of motion, in which the body CoM-induced dynamic term can be expressed as:

$$F(t) = Ma_Z(t) \tag{1}$$

When a single Wi-Fi sensor is used as in Figure 1b, experimental records can be disconnected from any kind of laboratory setup, and can thus facilitate the definition of a rapid protocol for diagnostic and structural assessment purposes.

The proposed analysis follows the concept layout in Figure 2b, where body CoM accelerations are explored to find correlations in locomotion indicators, and successively elaborated to derive basic input for structural analysis. Attention is given to (i) extended walking paths (i.e., with variable motion speed) and (ii) synthetized modular gaits. The accuracy of such a procedure is quantified based on numerical comparisons of dynamic performance indicators for a case-study concrete slab [30], calculated as from (i) or (ii) input, or even (iii) a deterministic loading protocol, as in Figure 2a.

3. Experimental Investigation

3.1. Instruments and Setup

The present experimental study involved a single volunteer (38-year female, $M = 80$ kg, $H = 1.85$ m, $h = 1.10$ m) asked to walk while using a high-precision, Wi-Fi triaxial sensor to track acceleration and inclination time histories (BeanDevice® WiLow® type [34] based on micro electro-mechanical systems (MEMS) technology for structural health monitoring). The sampling rate was set at 200 Hz and a lumbar belt was used to keep the sensor in position during body motion (Figure 1b).

In total, 50 walk patterns were measured under various movement frequencies f_s (W1 to W5 in Table 1). The volunteer was asked to walk naturally and to keep a rather constant speed for each test repetition. For each walking configuration, a minimum of $n_{s,min} = 10$ steps was taken into account as a reference (average steps can be seen in Table 1). Such an assumption resulted in a total of more than $n_s = 500$ recorded gaits to post-process for structural vibration analysis. Moreover, this choice allowed recording to be started from a resting position, with progressive increase in the motion speed and successively reaching a final position at rest (Figure 3).

Table 1. Set of walking configurations for experimental acquisitions. SLAB#1 = rigid (laboratory), SLAB#2 = flexible (in-field).

	Samples	Walk	Steps (avg) n_s	Substructure	Shoes/Features
W0	2	Straight	14	SLAB#1 & #2	S1/GRIP outsoles + heel air cushions (low hiking)
W1	10	Straight	15	SLAB#1	S2/Croslite^{TM} resin
W2	10	Straight	15	SLAB#1	S2/Croslite^{TM} resin
W3	10	Straight	12	SLAB#1	S3/Kalensole^{TM} foam (running)
W4	10	Straight	13	SLAB#1	S4/Sneakers
W5	10	In-place	12	SLAB#1	S2/Croslite^{TM} resin

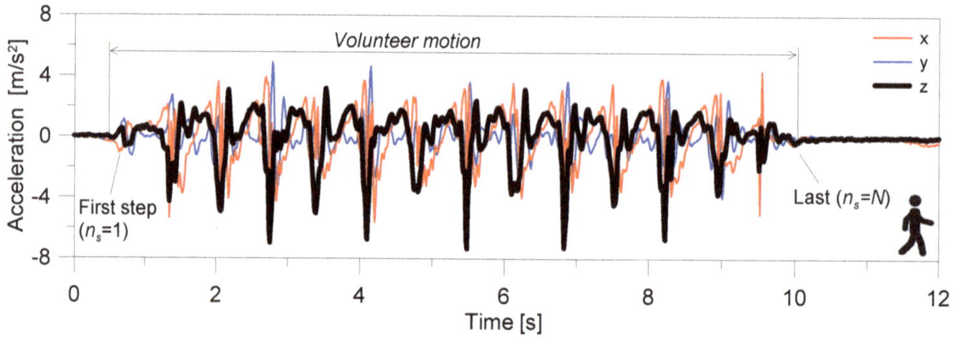

Figure 3. Example of experimental records for body CoM accelerations in time.

In doing so, under the limited condition of a single volunteer for the whole investigation, the experimental program evaluated some potential influencing parameters on walk features. Different types of shoes were used throughout the study, and randomly distributed during the repetition of measurements (S1 to S4 types in Table 1). Most importantly, W0 signals of Table 1 were collected for the invited volunteer who was asked to walk with constant speed above two different substructures, corresponding to a rigid concrete foundation (SLAB#1) or a flexible slab system (SLAB#2). This was obtained in laboratory conditions for SLAB#1 (Figure 4a) or based on in-field records for SLAB#2 (Figure 4b). Finally, a last set of records (W5 in Table 1) was collected during in-place/stationary walks of the volunteer.

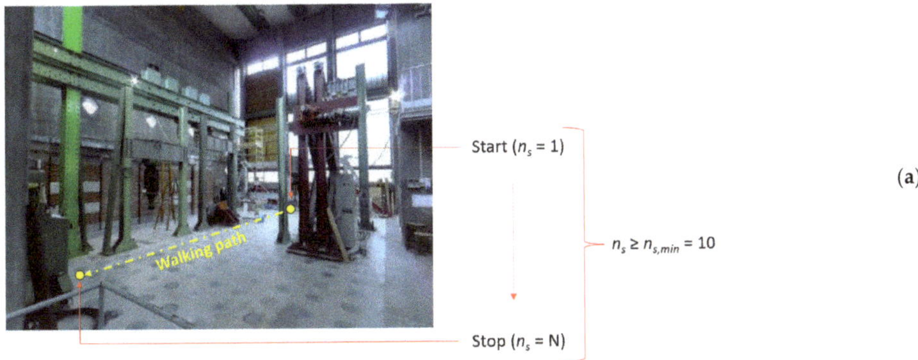

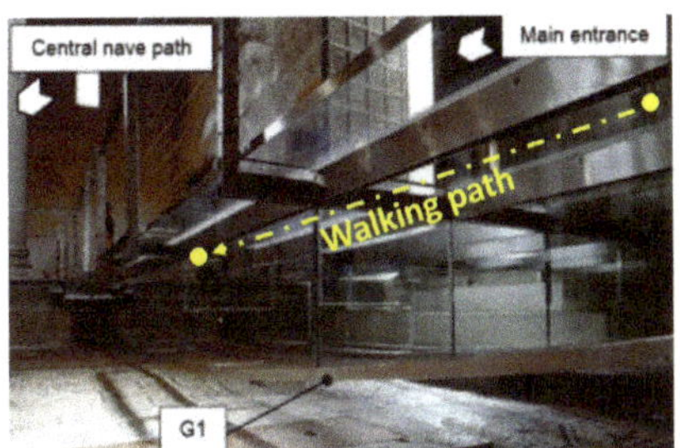

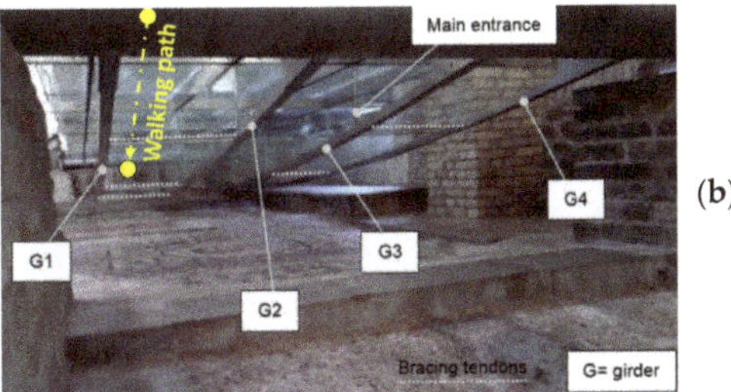

Figure 4. Reference (**a**) rigid concrete floor (SLAB#1) and (**b**) flexible glass SLAB#2 system for W0 body CoM measures (figure (**b**) adapted from [10] with permission from Elsevier®, license agreement n. 5223670279902, January 2022).

Assuming the MEMS sensor oriented as in Figure 1b, the vertical acceleration component $a_Z(t)$ during walks was first calculated taking into account the sensor inclination due to body movements, thus:

$$a_Z(t) = a_z(t)\cos\alpha(t) - a_x(t)\sin\alpha(t) \tag{2}$$

Following Equation (1), the calculated output of Equation (2) was elaborated as primary input for gait analysis and structural performance assessment. A typical body CoM output is proposed in Figure 3, with evidence of acceleration components.

3.2. Derivation of Motion Parameters

All the experimental records collected as in Section 3.1 were elaborated to calculate the corresponding walking frequency f_s and other relevant motion parameters. The gait length L_s was calculated on the basis of covered distance (L), divided by the number of steps n_s, for each test repetition:

$$L_S = \frac{L}{n_s} \tag{3}$$

The walking speed v_s was measured based on average values as:

$$v_S = L_s\, f_s \tag{4}$$

while the stride interval s_i (in seconds) was defined as in Figure 5a.

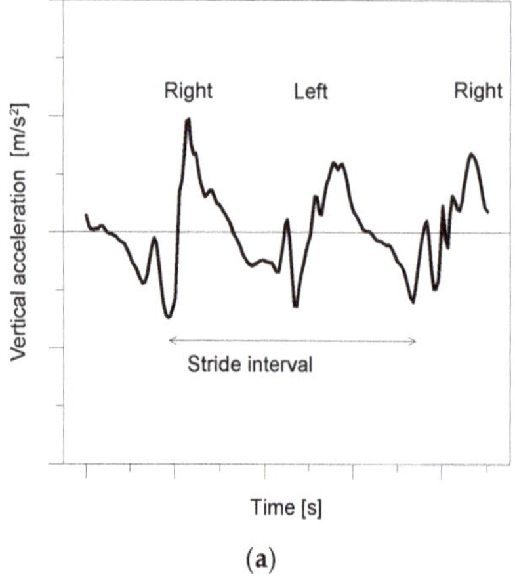

(a)

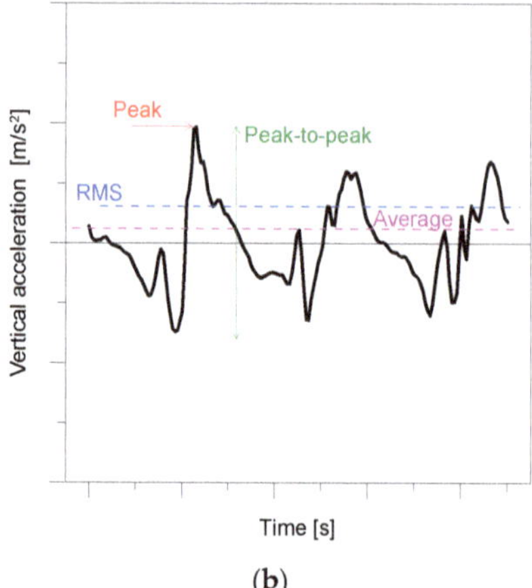

(b)

Figure 5. Analysis of body CoM records: example of (**a**) stride interval definition and (**b**) reference acceleration parameters.

According to Figure 5b, additional gait parameters included—with special attention to the vertical direction—the derivation for each record of the absolute acceleration value ($a_{Z,max}$), the peak-to-peak value ($a_{Zp\text{-}p}$), the average value ($a_{Z,avg}$), and the RMS value:

$$a_{Z,RMS} = \sqrt{\frac{1}{t_n - t_{n-1}} \int_{t_{n-1}}^{t_n} a_Z^2(t)\, dt} \tag{5}$$

4. Analysis of Experimental Signals

4.1. Substructure (W0)

Preliminary measurement was carried out when the volunteer was asked to walk on the rigid floor (SLAB#1) and on the flexible slab system (SLAB#2), as reported in Figure 4. These measurements were carried out for the volunteer equipped with the same shoes (S1 type), and asked to walk naturally, with a rather uniform gait length, frequency and in a straight path.

The rigid SLAB#1 setup was designed to coincide with the laboratory environment in Figure 4a, which consisted of a massive 80 cm thick reinforced concrete contrast floor. For the analysis of body CoM accelerations on SLAB#2, the suspension laminated glass walkway already investigated in [10], and reproduced in Figure 4b, was taken into account for in-field measurements. The structure consisted of a composite slab in which the laminated glass section layout included three 12 mm thick glass panels and interposed PVB® foils (0.76 mm thick). An additional glass layer, 6 mm in thickness, was used to protect the laminated section. The glass panels were linearly supported along the edges by a metal grid composed of C-shaped steel members. Such a solution was used to cover a total surface of 14.5 m × 2.8 m. The overall slab system was then sustained by four longitudinal steel-glass girders, spanning over the full bending length of 14.5 m. Most importantly, the flexible SLAB#2 system was characterized by a total mass for glass panels of the order of $M_{glass} \approx 4020$ kg and a vibration frequency $f_{1,e} = 7.28$ Hz (experimental measure for the empty structure [10]).

Typical experimental acquisitions from W0 walking configurations on SLAB#1 or SLAB#2 can be seen in Figure 6, divided by the acceleration component as a function of time. The average walking frequency was measured as $f_s = 1.46$ Hz, which corresponds to conventional slow motion. It is easy to see the progressive increase in acceleration peaks from rest, as well as the typical trend of acceleration modules corresponding to each gait.

Each W0 walk consisted of $n_s = 15$ gaits and was calculated at a walking speed $v \approx 1$ m/s (0.983 m/s), with $L_s = 0.67$ m the average gait length.

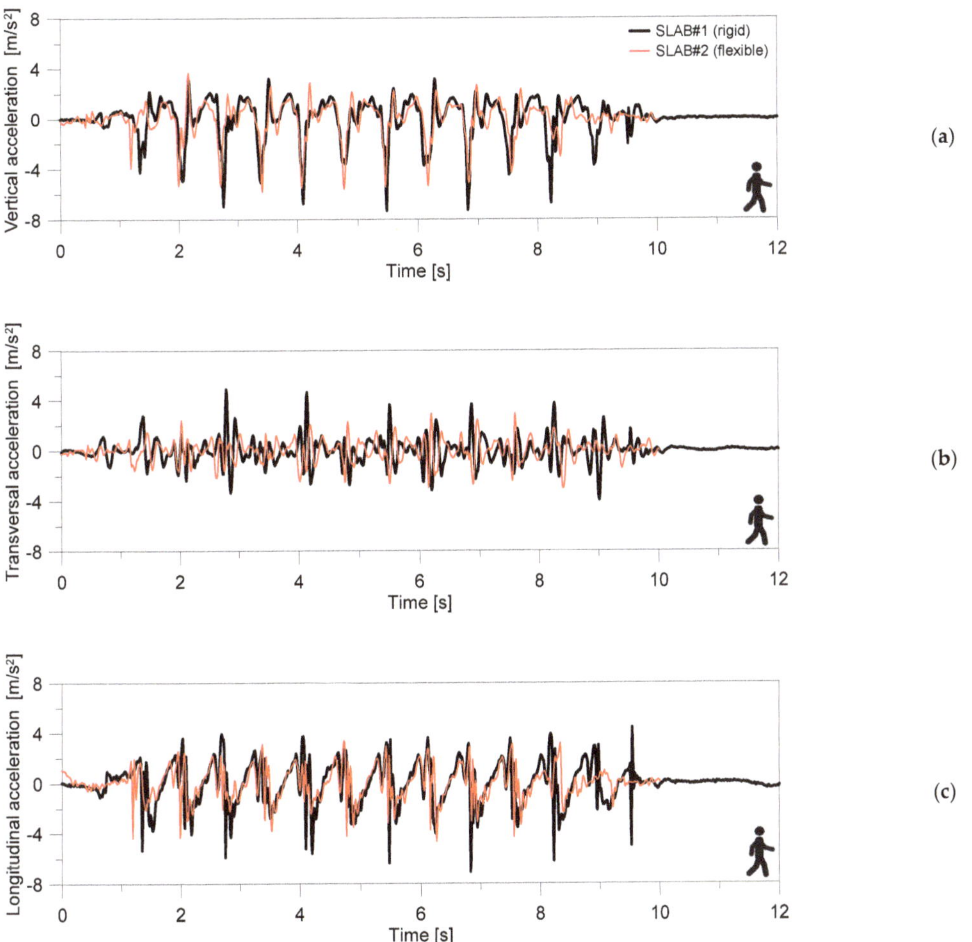

Figure 6. Example of experimental body CoM accelerations in time (W0 setup) for the volunteer moving on a rigid slab (SLAB#1) or a flexible substructure (SLAB#2).

Especially for the vertical acceleration component, which is of primary interest for the present study, limited modifications of acceleration records can be noticed for the volunteer walking on SLAB#1 or #2. The RMS values were calculated as 0.441 m/s^2 and 0.444 m/s^2, respectively. For SLAB#1, the vertical acceleration peak (absolute value) was measured as $a_{Z,max} = 7.33$ m/s^2 ($a_{Z,avg} = 0.00086$ m/s^2), with a maximum peak-to-peak value $a_{Z,p-p} = 10.52$ m/s^2. For SLAB#2, the walking records resulted in $a_{Z,max} = 5.76$ m/s^2, $a_{Z,avg} = 0.0041$ m/s^2 and $a_{Z,p-p} = 9.43$ m/s^2. Body CoM acceleration trends were thus found to be rather uniform in time but with slightly more pronounced peaks in the presence of the rigid substructure (SLAB#1). The peak-to-peak value was calculated for SLAB#1 up to +11.5% compared to SLAB#2 (+27% for the absolute maximum value).

High sensitivity can be noticed in Figure 6b for the transversal component of CoM acceleration, which was more severely affected by body motion. Maximum acceleration peaks were also higher for SLAB#1 in terms of transversal and longitudinal acceleration, due to different interaction effects of the volunteer with the substrate.

The post-processing fast transform analysis of signals in Figure 7 further confirms the rather good agreement of trends but with limited peaks for walks on flexible SLAB#2.

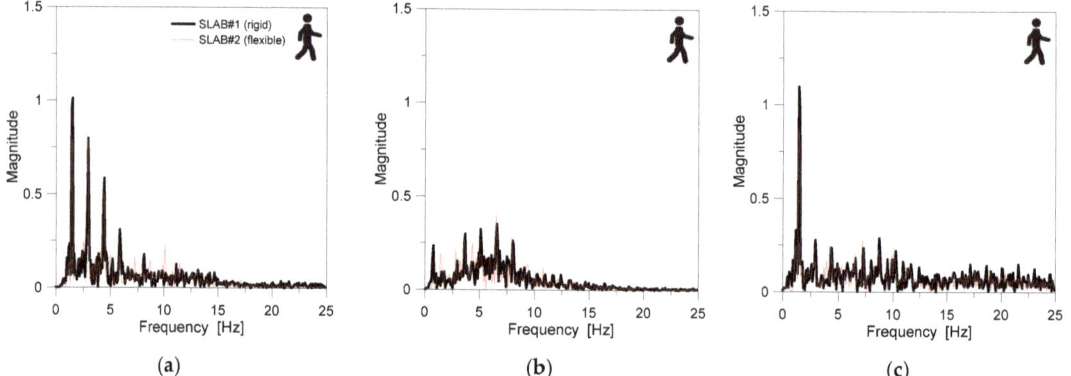

Figure 7. Analysis of experimental body CoM acceleration records (W0 signals) in terms of fast transform magnitude, with evidence of (**a**) vertical, (**b**) transversal, and (**c**) longitudinal components.

4.2. Straight Walks (W1 to W4)

Acceleration measures from linear walk patterns, as for the W1 to W4 schemes in Table 1, were successively analysed. Typical records were found to present several modifications in signal content and trend (see for example Figure 8, W2_2 signal).

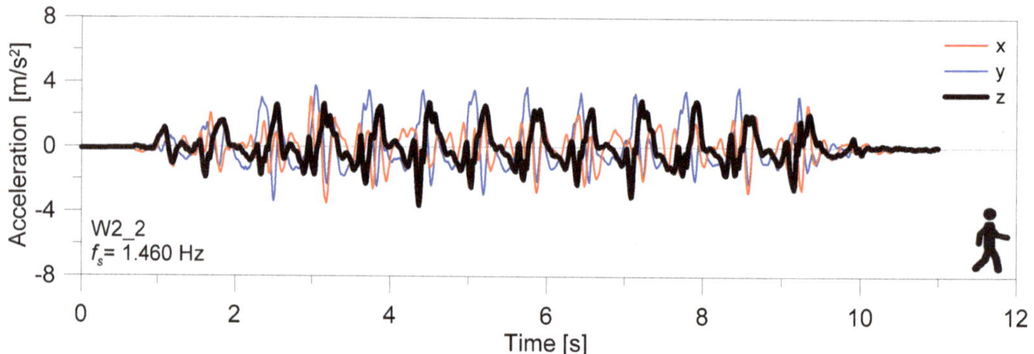

Figure 8. Example of experimental body CoM accelerations in time (W2 selection) for the volunteer moving on SLAB#1.

As reported in Figure 9, for the selected W2_2 signal (with f_s = 1.460 Hz) and W3_1 signal (with f_s = 1.433 Hz), while being characterized by similar walking frequency for the invited volunteer, multiple experimental signals were found to be associated with a certain scatter in magnitude and trend. This suggests that walking frequency f_s alone does not allow the obtaining of an unequivocal description of corresponding motion for a given pedestrian with mass M.

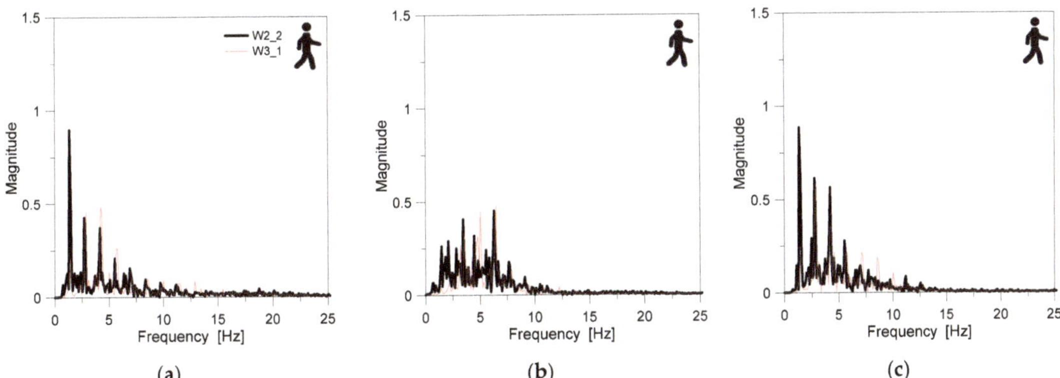

Figure 9. Analysis of experimental body CoM acceleration records (W2–W3 selection) in terms of fast transform magnitude, with evidence of (**a**) vertical, (**b**) transversal, and (**c**) longitudinal components.

In this regard, a summary of experimental outcomes for W1-to-W4 configurations (40 records) is shown in Figure 10. The average walking frequency was measured as 1.51 Hz ($\pm$0.16 Hz, with p-value = 0.012), see Figure 10a.

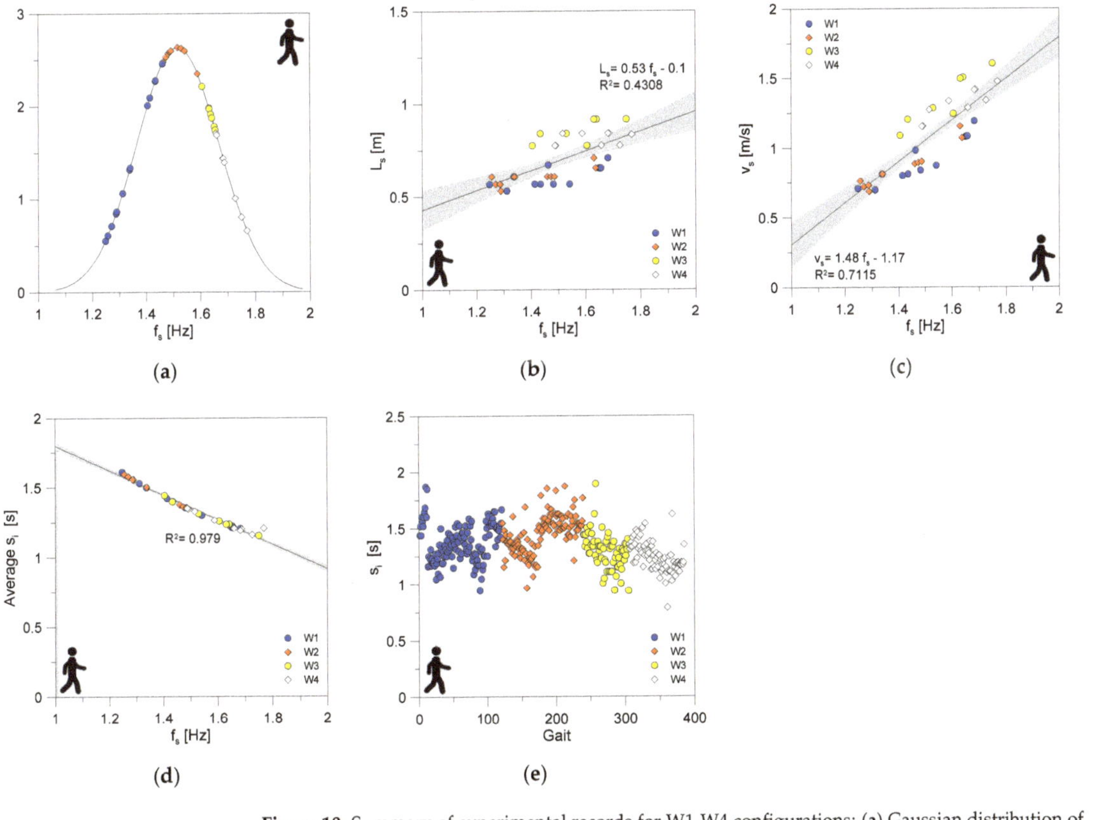

Figure 10. Summary of experimental records for W1-W4 configurations: (**a**) Gaussian distribution of samples (40 in total), (**b**) stride length, (**c**) walking speed, (**d**) average stride interval, (**e**) stride interval variation. The 95% confidence interval is shown in grey colour.

Further walking characteristics are proposed in Figure 10b–d, as a function of f_s. A linear regression model is also presented, with evidence of R^2 values, to analyse motion features. The grey fill shows the corresponding 95% confidence interval. The stride interval, calculated as in Figure 4a, is shown in Figure 10e for all the collected records. The mean value was estimated in $s_{i,avg}$ = 1.354 s. Regarding the acceleration trend and peaks for the same experimental records, comparative results are summarized in Figure 11 for the vertical component. Additional comparative analysis of collected body CoM data is also presented in Tables 2 and 3.

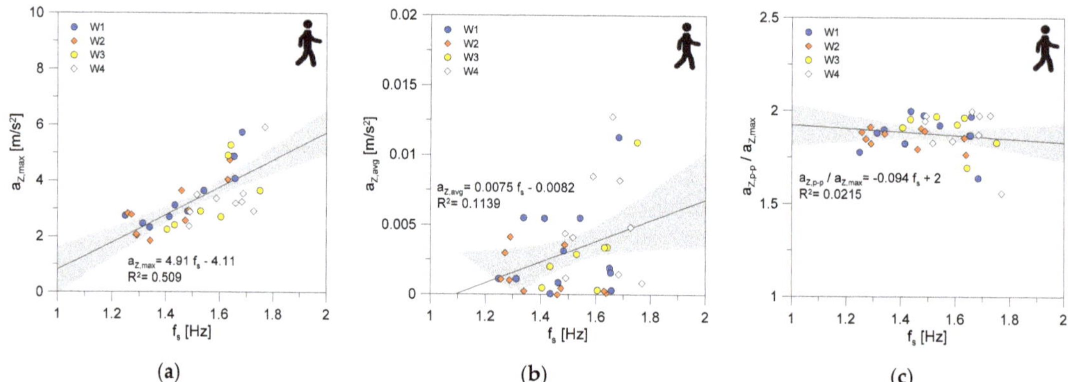

(a)　　　　　　　(b)　　　　　　　(c)

Figure 11. Summary of experimental records for W1–W4 configurations, as a function of calculated walking frequency: (**a**) peak of vertical acceleration, (**b**) average vertical acceleration, (**c**) peak-to-peak versus maximum acceleration. The 95% confidence interval is shown in grey colour.

Table 2. Analysis of W1-to-W4 features (SLAB#1), with standard deviation in brackets and correlation coefficient R.

	Samples	Frequency		Length		Velocity	
		f_s [Hz]	Range [Hz]	L_s [m]	R	V [m/s]	R
W1	10	1.492 (±0.156)	1.248–1.683	0.604 (±0.055)	0.80	0.907 (±0.171)	0.95
W2	10	1.414 (±0.145)	1.257–1.637	0.607 (±0.049)	0.83	0.862 (±0.154)	0.96
W3	10	1.571 (±0.122)	1.405–1.750	0.854 (±0.064)	0.69	1.346 (±0.189)	0.92
W4	10	1.623 (±0.106)	1.488–1.769	0.810 (±0.033)	0.26	1.316 (±0.138)	0.88
Total	40	1.517 (±0.153)	1.248–1.769	0.702 (±0.124)	0.66	1.078 (±0.270)	0.84

Table 3. Analysis of W1-to-W4 vertical acceleration data (SLAB#1), with standard deviation in brackets and correlation coefficient R.

	Samples	Peak			Average		Peak-to-Peak	
		$a_{Z,max}$ [m/s²]	Range [m/s²]	R	$a_{Z,avg}$ [m/s²]	R	$a_{Z,p\text{-}p}$ [m/s²]	R
W1	10	3.586 (±1.147)	2.322–5.733	0.90	0.0033 (±0.003)	0.24	6.674 (±1.941)	0.95
W2	10	2.947 (±0.935)	1.840–4.743	0.81	0.0013 (±0.002)	−0.42	5.438 (±1.613)	0.82
W3	10	3.439 (±1.213)	2.249–5.268	0.66	0.0033 (±0.004)	0.74	6.447 (±2.081)	0.65
W4	10	3.431 (±0.999)	2.366–5.913	0.60	0.0051 (±0.004)	0.06	6.353 (±1.253)	0.67
Total	40	3.348 (±1.057)	1.840–5.913	0.71	0.0032 (±0.003)	0.34	6.219 (±1.736)	0.75

From comparative data in Figure 11 and Tables 2 and 3, it is possible to see a large variation in walking parameters from the involved volunteer. The overall trend of vertical acceleration peak from body CoM motion had a weak correlation with walking frequency (R^2 = 0.509, Figure 11a), while the average vertical acceleration was measured at less than 0.015 m/s^2. (Figure 11b).

Finally, the peak-to-peak versus maximum ratio was calculated to be of the order of ≈1.8, but with high sensitivity to walking configurations and with a rather scattered distribution (Figure 11c). The best correlation of body CoM records with walking frequency was generally found for the W1 and W2 sets (with lower average f_s) for stride length and speed in Table 2, but also for vertical acceleration trends in Table 3. The exception is represented by average acceleration values in Table 3, which were calculated from the CoM position at rest, and thus do not provide useful feedback for analysis. The peak-to-peak value, for all the tested scenarios, was generally found to be of the order of less than twice the absolute peak of vertical acceleration.

Regarding the transversal acceleration of body CoM during walks, the experimental records showed that maximum peaks can be relevant and higher than in the vertical direction, and thus more complex biomechanical models should be considered, in addition to the vertical component only. For the present study, transversal acceleration peaks were found to exceed the vertical component (absolute terms) in ≈56% of available records. Typical trends are shown in Figure 12 as a function of (a) walking frequency, (b) average speed, or (c) vertical acceleration peak, respectively.

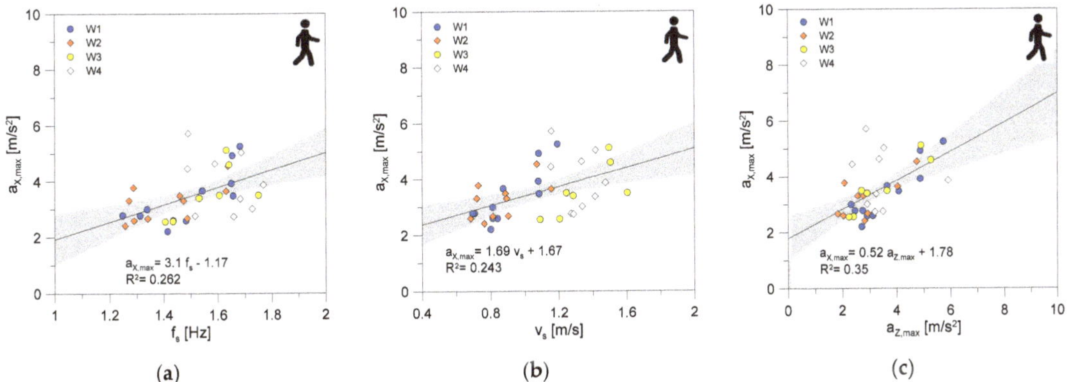

<table>
<tr><td>(a)</td><td>(b)</td><td>(c)</td></tr>
</table>

Figure 12. Summary of experimental records for W1–W4 configurations, in terms of transversal acceleration peak as a function of (a) walking frequency, (b) walking speed, and (c) vertical acceleration peak. The 95% confidence interval is shown in grey colour.

4.3. In-Place Walks (W5)

The analysis of experimental records was successively extended to W5 configurations characterized by in-place motion of the volunteer. Typical acceleration records were found to have, as expected, large modifications from W1–W4 conditions. An example can be seen in Figures 13 and 14 for selected data, while Figure 15 and Table 4 summarize the comparative analysis of acquired time histories. A less pronounced correlation of motion features was generally observed with walking frequency f_s.

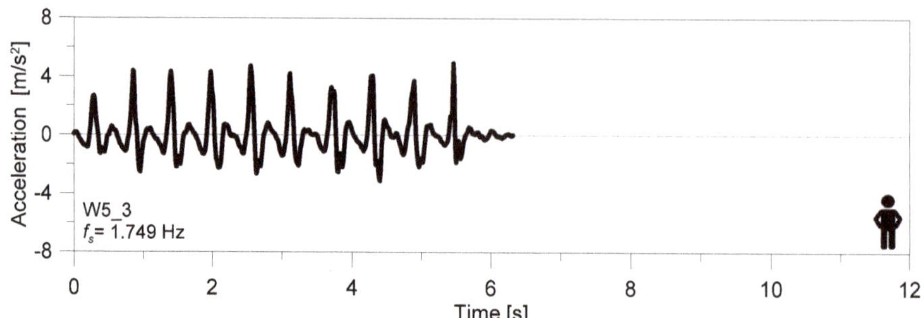

Figure 13. Example of experimental body CoM accelerations in time (W5 selection) for the volunteer on SLAB#1.

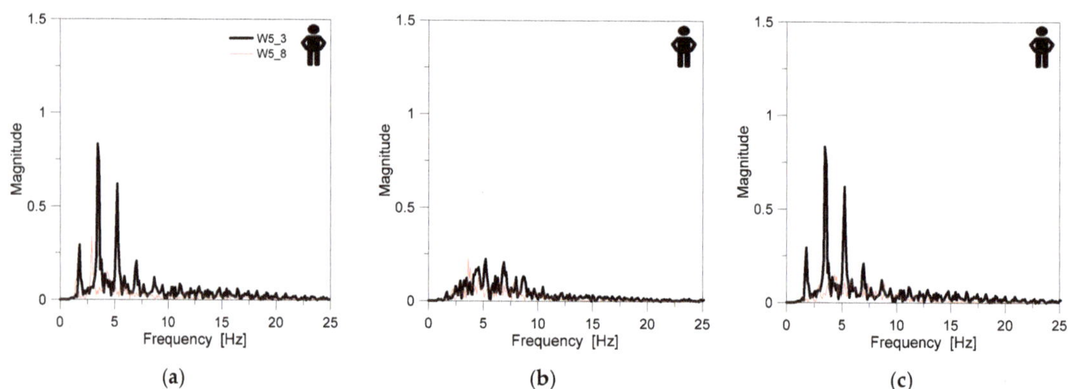

Figure 14. Analysis of experimental body CoM acceleration records (W5 selection) in terms of fast transform magnitude, with evidence of (**a**) vertical, (**b**) transversal, and (**c**) longitudinal components.

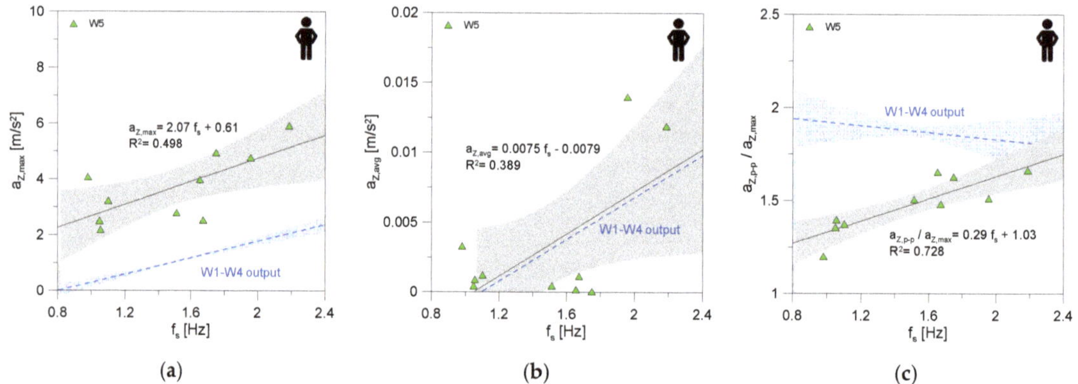

Figure 15. Summary of experimental records for W5 configurations (10 in total, SLAB#1), as a function of calculated walking frequency: (**a**) peak of vertical acceleration, (**b**) average vertical acceleration, (**c**) peak-to-peak versus maximum acceleration. The 95% confidence interval is shown in grey colour.

Table 4. Analysis of W5 vertical acceleration data (SLAB#1, 10 samples), with standard deviation in brackets and correlation coefficient R.

Frequency		Peak			Average		Peak-to-Peak	
f_s [Hz]	Range [Hz]	$a_{Z,max}$ [m/s²]	Range [m/s²]	R	$a_{Z,avg}$ [m/s²]	R	$a_{Z,p\text{-}p}$ [m/s²]	R
1.491 (±0.425)	0.980–2.187	3.694 (±1.246)	2.192–5.927	0.71	0.0033 (±0.005)	0.62	5.546 (±2.266)	0.81

As in the case of W1–W4 configurations, W5 output showed that the vertical acceleration peaks had weak linear correlation with motion frequency (Figure 15a), the average CoM acceleration was not meaningful (Figure 15b) and the peak-to-peak versus maximum acceleration ratio was rather well fitted by the linear regression model (Figure 15c). For qualitative comparison to the experimental trends of W1 to W4 configurations, previous data are also reported in Figure 15. It is possible to notice a substantial increase in the vertical component for W5 (Figure 15a), the rather close correlation of average acceleration values (Figure 15b), and the marked variation in the peak-to-peak versus maximum value ratio (Figure 15c).

In Figure 16, final comparisons are presented in terms of stride interval, with evidence of individual W5 outcomes and W1–W4 trends. As far as the average stride interval is concerned, Figure 16a shows a rather good correlation of average values for in-place walks or straight walks. Individual measurements in Figure 16b, however, show higher peaks (maximum and minimum values) for W5 configurations, compared to W1–W4 outcomes.

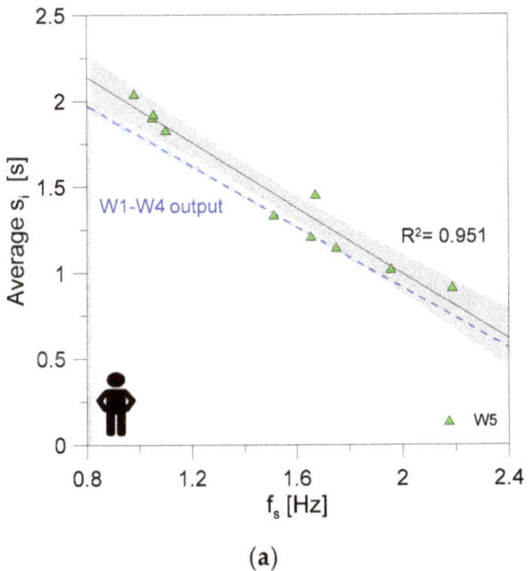

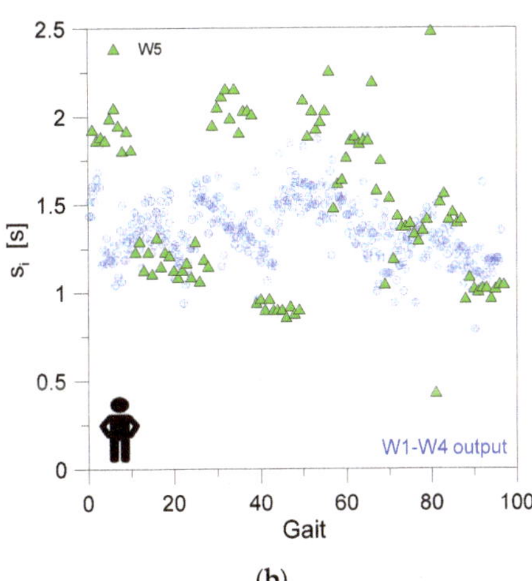

Figure 16. Summary of experimental records for W5 configurations: (**a**) average stride interval, and (**b**) stride interval variation. The 95% confidence interval is shown in grey.

5. Structural Use of Body CoM Acceleration Signals

5.1. Slected Slab

A series of finite element numerical analyses was carried out in ABAQUS [35], to quantify the effects of experimentally derived CoM acceleration time histories on the dynamic response of structural slabs. A set of geometrical nonlinear dynamic analyses was performed to predict the accelerations of pedestrian systems under random walking excitations. To this end, a concrete slab from [30], with dimensions 5 × 6 m and 150 mm in thickness, with linear simple supports along the two maximum edges (and short edges unrestrained), was used.

The use of the selected slab was inspired by the study presented in [30], where extended discussion of the numerical application of deterministic walking loads, as in Figure 2a, was presented. In this sense, the study in [30] was used for validation and quantitative comparison of the present body CoM input. Table 5 summarizes some major features, such as section properties, size, structural mass-to-occupant ratio R_M (with $M = 80$ kg), and the vertical fundamental vibration frequency of the empty ($f_{1,e}$) or occupied slab ($f_{1,o}$, with $M = 80$ kg). A conventional Rayleigh approach was applied to define the mass-proportional and stiffness-proportional damping terms through parametric numerical analysis [36], with x = 3% as the damping term [30].

Table 5. Features of selected slab for the present investigation.

Material	Size [m^2]	Thickness [m]	Mass [kg]	R_M	$f_{1,e}$ [Hz]	$f_{1,o}$ [Hz]
Concrete	5×6	0.15	11,250	≈140	7.23	7.1

5.2. Loading

Each FE numerical analysis consisted of two sub-steps, the first to apply quasi-static permanent loads of structural members (gravity), and the second for dynamic analysis under the single pedestrian/volunteer.

Careful consideration was paid to quantification of the walk-induced effects of a volunteer pedestrian, based on different input signals.

As a reference:

(i) The vertical load in time due to a pedestrian was defined from experimental acceleration histories discussed previously (vertical component), with $M = 80$ kg, by taking into account all the available W0 to W5 records. Acceleration time histories were applied as in Figure 17a. A total of 52 dynamic analyses were carried out.

(ii) Successively, the effect of gait variability for rough experimental body CoM input as in (i) was also numerically assessed. According to Figure 17b, an average stride record (vertical component) was derived from W1 to W4 experimental signals and fitted with different sine curves to obtain an average synthetized gait module, based on experimental measures. The so-derived synthetized stride signal was applied to the concrete slab, as in Figure 17a, and repeated to cover n_s gaits through the total time of dynamic analyses.

(iii) For comparison, the deterministic loading approach, according to Figure 2a, and already taken into account in [30] for the same concrete slab was also considered (with $M = 80$ kg).

In this manner, the numerical outcomes based on present body CoM experimental accelerations were applied to a traditional solving approach for vibration serviceability analysis of pedestrian structures. Given that the conventional frequency range $f_s = 1.5$–2.5 Hz of normal walks was taken into account for the deterministic procedure (with 0.1 Hz the increment), a set of 10 nonlinear simulations was carried out in the present study for concrete slabs loaded as in (iii).

5.3. Results

5.3.1. Substructure Effect (W0 Setup)

Figure 18a shows the typical distribution of vertical accelerations under the imposed W0 walking paths, while Figure 18b shows the evolution of vertical acceleration at the center of the slab.

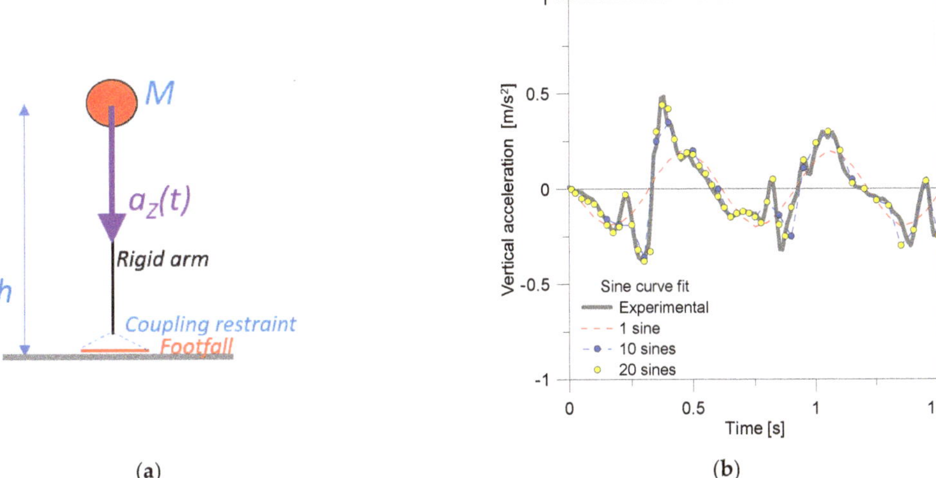

Figure 17. Loading protocol: (**a**) input body CoM acceleration and (**b**) sine curve fit of an experimental stride record (signal #7, W1) to derive a synthetized signal.

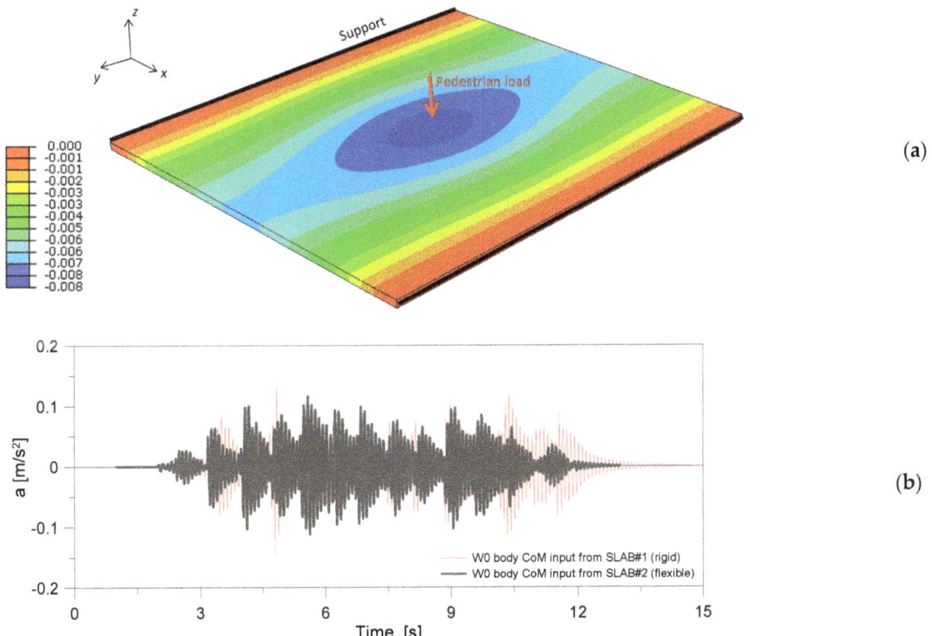

Figure 18. Summary of numerical performance indicators of the examined concrete slab under W0 experimental body CoM accelerations or deterministic pedestrian loads (ABAQUS): (**a**) typical acceleration distribution (values in m/s²), with (**b**) acceleration at the center of slab.

The overall performance of the concrete slab under body CoM input based on rigid substructure (SLAB#1) could be quantified as a_{max} = 0.14 m/s², $a_{p\text{-}p}$ = 0.28 m/s², a_{RMS} = 0.0086 m/s² and CREST = 16.92. It was shown in Section 4.1 that the available W0 signals were characterized by a mostly identical RMS value for the vertical component.

The effect of pedestrian movement on the W0 setup with flexible support (SLAB#2) could be quantified, on the structural side, in terms of marked modifications of performance indicators, with $a_{max} = 0.11$ m/s^2 (-20% the response of the same concrete slab under W0 input on SLAB#1), $a_{p\text{-}p} = 0.23$ m/s^2 (-17%), $a_{RMS} = 0.0111$ m/s^2 ($+30\%$) and CREST $= 10.38$ (-38%).

5.3.2. Straight (W1–W4) or In-Place Walks (W5)

The acceleration peaks in the vertical direction (a_{max}) or the RMS value (a_{RMS}) were taken into account over the time of each simulation for the slab under (i) experimental time histories or corresponding (ii) deterministic pedestrian loads. The trend of parametric results can be seen in Figure 19a,b, as a function of the input walking frequency f_s. A summary of performance indicators is also presented in Table 6. It can be seen that quite good correlations were obtained, especially for structural parameters and performance indicators based on W1 and W2 body CoM input. W5 body CoM input tended to underestimate the structural response for walking frequencies higher than 1.3 Hz (normal walk).

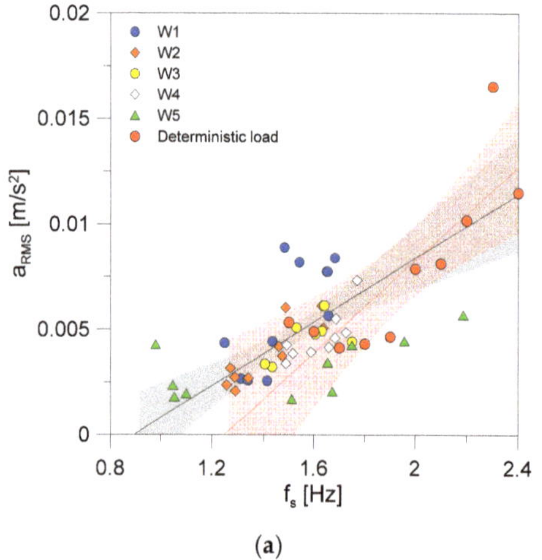

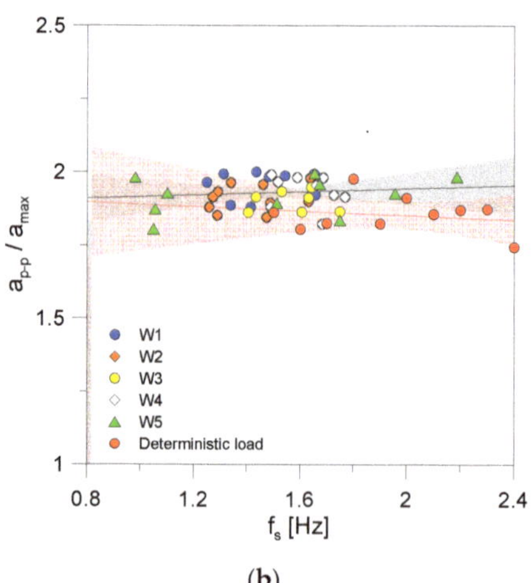

Figure 19. Summary of numerical performance indicators of the examined concrete slab under W1–W4 and W5 experimental body CoM accelerations or deterministic pedestrian loads (ABAQUS): (**a**) RMS value, and (**b**) peak-to-peak versus maximum acceleration. The 95% confidence interval is shown in grey and red colour, respectively.

Table 6. Analysis of the examined concrete slab under W1–W5 experimental body CoM accelerations, with average values of maximum, RMS and peak-to-peak vertical accelerations of slab (±standard deviation) and R-square correlation coefficient.

		Peak		RMS		Peak-to-Peak	
	Simulations	a_{max} [m/s^2]	R^2	a_{RMS} [m/s^2]	R^2	$a_{p\text{-}p}$ [m/s^2]	R^2
W1	10	0.098 (±0.045)	0.78	0.0057 (±0.003)	0.55	0.193 (±0.091)	0.77
W2	10	0.072 (±0.034)	0.87	0.0038 (±0.001)	0.77	0.138 (±0.065)	0.88
W3	10	0.091 (±0.025)	0.60	0.0045 (±0.001)	0.41	0.173 (±0.049)	0.55
W4	10	0.093 (±0.038)	0.67	0.0046 (±0.001)	0.61	0.180 (±0.072)	0.66
Total W1–W4	40	0.088 (±0.037)	0.63	0.0048 (±0.002)	0.40	0.171 (±0.073)	0.61
W5	10	0.068 (±0.029)	0.41	0.0032 (±0.001)	0.37	0.130 (±0.058)	0.42

5.3.3. Synthetized Stride Signals Based on Body CoM Measures

Finally, for the slab under synthetized stride signal, further comparative results are collected in Table 7 and Figure 20. The fit residual is also shown in Table 7, as the higher number of sine curves resulted in a smaller fitting error compared to the rough experimental curve. More precisely, Figure 20a shows the typical trend of vertical acceleration at the center of slab under synthetized strides from experimental records, while normalized performance indicators are summarized in Figure 20b, where each FE result is compared to the dynamic response of the slab under rough experimental stride input.

Table 7. Analysis of the examined concrete slab under synthetized body CoM stride acceleration (example for experimental stride from signal #7-W1), with maximum acceleration, RMS value, peak-to-peak acceleration, and CREST factor.

Body CoM Input	Fit Residual (max) [m/s^2]	a_{max} [m/s^2]	a_{RMS} [m/s^2]	a_{p-p} [m/s^2]	CREST
Experimental stride	-	0.01190	0.00081	0.02304	14.67
Sine curve fit 20	0.0722	0.00758	0.00040	0.01473	18.74
Sine curve fit 10	0.1269	0.00709	0.00042	0.01373	16.87
Sine curve fit 6	0.1765	0.00466	0.00030	0.00911	15.51
Sine curve fit 3	0.2909	0.00352	0.00016	0.00704	21.61
Sine curve fit 1	0.3946	0.00093	0.00006	0.00182	16.01

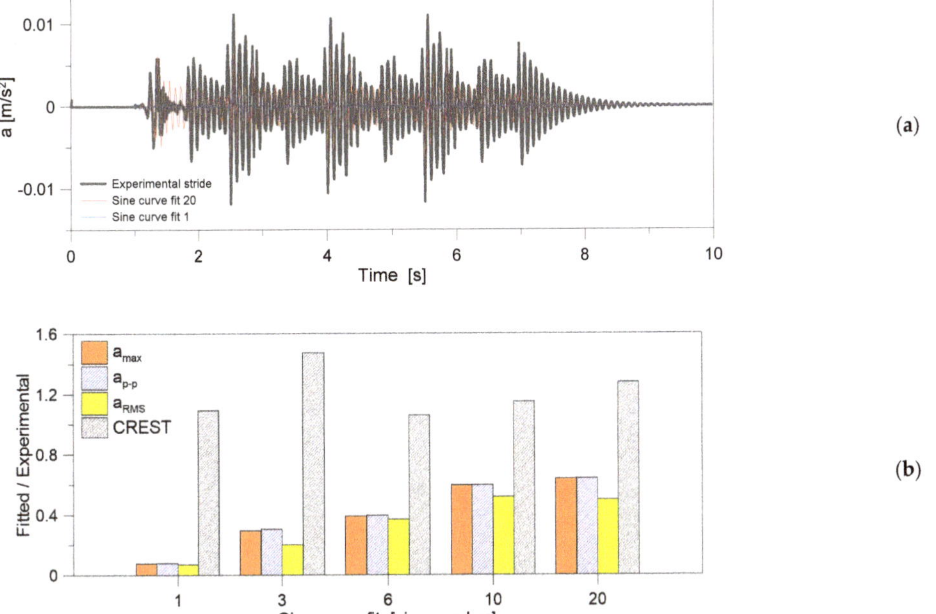

Figure 20. Numerical response of the examined concrete slab under experimental body CoM accelerations—synthetized stride signal #7-W1 (ABAQUS): (**a**) acceleration at the center of slab, and (**b**) normalized performance indicators based on the fitted stride signal.

A rather high sensitivity of structural performance indicators can be seen in Figure 20b, compared to the rough experimental stride signal, even in the presence of relatively small residual fitting. This is also in line with numerical outcomes in Figure 19. The exception in Figure 20b is represented by CREST factor estimates, which are more conservative or unconservative compared to the rough experimental stride signal

6. Summary and Conclusions

The vibration analysis of the pedestrian system, especially for slabs with low vibration frequency, which are highly sensitive to walk-induced effects, represents an open challenge for designers. In this sense, the use of reliable pedestrian models can support efficient decision stages in design, monitoring procedures, etc., in conjunction with in-field experiments or expensive laboratory protocols. Knowledge of walk-induced effects, including a multitude of parameters, should be taken into account.

In this paper, attention was focused on the analysis of body centre of mass (CoM) motion acceleration parameters under several walking conditions, for structural analysis purposes. A total of 50 configurations was experimentally measured (for more than 500 recorded gaits) with the support of a triaxial Wi-Fi accelerometer and an adult volunteer. The sensitivity of motion parameters was explored in terms of body CoM measures, as well as in terms of the quantification of structural effects on a case-study concrete slab subjected to experimentally derived, uncoupled body CoM input.

In terms of body CoM measures, it was shown that:

- A substructure effect (W0), which results in adaptation of body movements, and thus variation of measured acceleration peaks and trends during walks, can be predicted for a given pedestrian;
- For straight walks on a rigid substructure (W1–W4), high variability in motion features and performance indicators was generally captured for a given walking frequency. This suggests that walking frequency alone is sometimes a weak parameter for generalization of synthetized stride models;
- The transversal acceleration component was measured in the same order or higher than the vertical component, for the majority of walking patterns;
- For in-place/stationary walks (W5), a substantial modification of body CoM acceleration parameters was observed, even under similar motion frequency and stride intervals;
- The vertical acceleration component of W5 walks was significantly higher than W1–W4 configurations.

On the structural side, the use of uncoupled body CoM acceleration input for the vibration analysis of a concrete slab showed that:

- Similar body CoM input (i.e., walking frequency and stride interval) but based on different substructures (i.e., rigid or flexible, as for the W0 setup) can result in strong modifications of structural performance indicators, for a given pedestrian system. In this sense, further extended scenarios should be investigated to quantify this kind of sensitivity;
- Body CoM input calculated from straight walks (W1–W4 setup) was generally associated with a tendency to overestimate the structural performance indicators of a given slab, compared to structural estimates for the same slab based on deterministic pedestrian loads. However, a rather close match was observed in the present study in terms of average structural effects for the examined slab (i.e., linear regression method applied to structural acceleration peaks, etc.);
- Most of the W5 body-CoM input (stationary walks) was found to underestimate a large number of examined performance indicators for the investigated concrete slab. Such an outcome was observed both for normal walking frequency, or higher frequencies;
- Finally, the derivation of synthetized stride acceleration input based on uncoupled body CoM acceleration measures was found to be severely affected by fitting accuracy, resulting in possible unsafe underestimation of structural performance indicators compared to rough experimental time histories.

Funding: This research received no external funding.

Institutional Review Board Statement: Not applicable.

Informed Consent Statement: Informed consent was obtained from all subjects involved in the study.

Data Availability Statement: Supporting data will be shared upon request.

Conflicts of Interest: The author declares no conflict of interest.

References

1. Bachmann, H.; Ammann, W. *Vibrations in Structures Induced by Man and Machines*; IABSE—International Association for Bridge and Structural Engineering: Zurich, Switzerland, 1987; ISBN 3-85748-052-X.
2. Bachmann, H.; Ammann, W. Vibrations in structures induced by man and machines. *Can. J. Civ. Eng.* **1987**, *15*, 1086–1087.
3. Sedlacek, G.; Heinemeyer, C.; Butz, C.; Veiling, B.; Waarts, P.; Duin, F.; Hicks, S.; Devine, P.; Demarco, T. *Generalisation of Criteria for Floor Vibrations for Industrial, Office, Residential and Public Building and Gymnasium Halls*; European Commission: Luxembourg, 2006.
4. Busca, G.; Cappellini, A.; Manzoni, S.; Tarabini, M.; Vanali, M. Quantification of changes in modal parameters due to the presence of passive people on a slender structure. *J. Sound Vib.* **2014**, *333*, 5641–5652. [CrossRef]
5. Setareh, M.; Gan, S. Vibration testing, analysis and human-structure interaction studies of a slender footbridge. *J. Perform. Constr. Facil.* **2018**, *32*, 040018068. [CrossRef]
6. Muhammad, Z.; Reynolds, P.; Avci, O.; Hussein, M. Review of Pedestrian Load Models for Vibration Serviceability Assessment of Floor Structures. *Vibration* **2018**, *2*, 1–24. [CrossRef]
7. Gheitasi, A.; Ozbulut, O.E.; Usmani, S.; Alipour, M.; Harris, D.K. Experimental and analytical vibration serviceability assessment of an in-service footbridge. *Case Stud. Nondestruct. Test. Eval.* **2016**, *6*, 79–88. [CrossRef]
8. Salgado, R.; Branco, J.M.; Cruz, P.J.; Ayala, G. Serviceability assessment of the Góis footbridge using vibration monitoring. *Case Stud. Nondestruct. Test. Eval.* **2014**, *2*, 71–76. [CrossRef]
9. Lee, J.H.; Park, M.J.; Yoon, S.W. Floor Vibration Experiment and Serviceability Test of iFLASH System. *Materials* **2020**, *13*, 5760. [CrossRef]
10. Bedon, C. Experimental investigation on vibration sensitivity of an indoor glass footbridge to walking conditions. *J. Build. Eng.* **2020**, *29*, 101195. [CrossRef]
11. Bedon, C. Diagnostic analysis and dynamic identification of a glass suspension footbridge via on-site vibration experiments and FE numerical modelling. *Compos. Struct.* **2019**, *216*, 366–378. [CrossRef]
12. Bedon, C.; Noè, S. Post-Breakage Vibration Frequency Analysis of In-Service Pedestrian Laminated Glass Modular Units. *Vibration* **2021**, *4*, 836–852. [CrossRef]
13. Gong, M.; Li, Y.; Shen, R.; Wei, X. Glass Suspension Footbridge: Human-Induced Vibration, Serviceability Evaluation, and Vibration Mitigation. *J. Bridg. Eng.* **2021**, *26*, 05021014. [CrossRef]
14. Shahabpoor, E.; Pavic, A.; Racic, V. Interaction between Walking Humans and Structures in Vertical Direction: A Literature Review. *Shock Vib.* **2016**, *2016*, 1–22. [CrossRef]
15. Živanović, S.; Pavic, A. Probabilistic Modeling of Walking Excitation for Building Floors. *J. Perform. Constr. Facil.* **2009**, *23*, 132–143. [CrossRef]
16. Barela, A.M.; Duarte, M. Biomechanical characteristics of elderly individuals walking on land and in water. *J. Electromyogr. Kinesiol.* **2008**, *18*, 446–454. [CrossRef] [PubMed]
17. Adamczyk, P.G.; Kuo, A.D. Redirection of center-of-mass velocity during the step-to-step transition of human walking. *J. Exp. Biol.* **2009**, *212*, 2668–2678. [CrossRef]
18. Coughlin, P.; Kent, P.; Turton, E.; Byrne, P.; Berridge, D.; Scott, D.; Kester, R. A New Device for the Measurement of Disease Severity in Patients with Intermittent Claudication. *Eur. J. Vasc. Endovasc. Surg.* **2001**, *22*, 516–522. [CrossRef] [PubMed]
19. Bedon, C.; Fasan, M. Reliability of field experiments, analytical methods and pedestrian's perception scales for the vibration serviceability assessment of an in-service glass walkway. *Appl. Sci.* **2019**, *9*, 1936. [CrossRef]
20. Bedon, C.; Mattei, S. Facial Expression-Based Experimental Analysis of Human Reactions and Psychological Comfort on Glass Structures in Buildings. *Buildings* **2021**, *11*, 204. [CrossRef]
21. Miyazaki, S. Long-term unrestrained measurement of stride length and walking velocity utilizing a piezoelectric gyroscope. *IEEE Trans. Biomed. Eng.* **1997**, *44*, 753–759. [CrossRef]
22. Veltink, P.H.; Bussmann, H.B.J.; de Vries, W.; Martens, W.L.J.; Van Lummel, R.C. Detection of Static and Dynamic Activities Using Uniaxial Accelerometers. *IEEE Trans. Rehabil. Eng.* **1996**, *4*, 375–385. [CrossRef]
23. Sabatini, A.M.; Martelloni, C.; Scapellato, S.; Cavallo, F. Assessment of Walking Features from Foot Inertial Sensing. *IEEE Trans. Biomed. Eng.* **2005**, *52*, 486–494. [CrossRef] [PubMed]
24. Godfrey, A.; Del Din, S.; Barry, G.; Mathers, J.; Rochester, L. Instrumenting gait with an accelerometer: A system and algorithm examination. *Med Eng. Phys.* **2015**, *37*, 400–407. [CrossRef] [PubMed]
25. Simonetti, E.; Bergamini, E.; Vannozzi, G.; Bascou, J.; Pillet, H. Estimation of 3D Body Center of Mass Acceleration and Instantaneous Velocity from a Wearable Inertial Sensor Network in Transfemoral Amputee Gait: A Case Study. *Sensors* **2021**, *21*, 3129. [CrossRef] [PubMed]

26. Van Nimmen, K.; Zhao, G.; Seyfarth, A.; Van den Broeck, P. A Robust Methodology for the Reconstruction of the Vertical Pedestrian-Induced Load from the Registered Body Motion. *Vibration* **2018**, *1*, 250–268. [CrossRef]
27. Bocian, M.; Brownjohn, J.; Racic, V.; Hester, D.; Quattrone, A.; Monnickendam, R. A framework for experimental determination of localised vertical pedestrian forces on full-scale structures using wireless attitude and heading reference systems. *J. Sound Vib.* **2016**, *376*, 217–243. [CrossRef]
28. Rispens, S.M.; Pijnappels, M.; van Schooten, K.S.; Beek, P.J.; Daffertshofer, A.; van Dieen, J.H. Consistency of gait characteristics as determined from acceleration data collected at different trunk locations. *Gait Posture* **2014**, *40*, 187–192. [CrossRef]
29. Shahar, R.T.; Agmon, M. Gait Analysis Using Accelerometry Data from a Single Smartphone: Agreement and Consistency between a Smartphone Application and Gold-Standard Gait Analysis System. *Sensors* **2021**, *21*, 7497. [CrossRef]
30. Cai, Y.; Gong, G.; Xia, J.; He, J.; Hao, J. Simulations of human-induced floor vibrations considering walking overlap. *SN Appl. Sci.* **2019**, *2*, 19. [CrossRef]
31. Herzog, W.; Nigg, B.M.; Read, L.J.; Olsson, E. Asymmetries in ground reaction force patterns in normal human gait. *Med. Sci. Sports Exerc.* **1989**, *21*, 110–114. [CrossRef]
32. Wouda, F.J.; Giuberti, M.; Bellusci, G.; Maartens, E.; Reenalda, J.; Van Beijnum, B.-J.F.; Veltink, P.H. Estimation of Vertical Ground Reaction Forces and Sagittal Knee Kinematics During Running Using Three Inertial Sensors. *Front. Physiol.* **2018**, *9*, 218. [CrossRef]
33. Jiang, X.; Napier, C.; Hannigan, B.; Eng, J.J.; Menon, C. Estimating Vertical Ground Reaction Force during Walking Using a Single Inertial Sensor. *Sensors* **2020**, *20*, 4345. [CrossRef] [PubMed]
34. Beanair GmbH. *BeanDevice®Wilow®User Manual—Wilow®Wireless Sensor—Version 2.3.2*; Beanair GmbH: Berlin, Germany, 2019.
35. *ABAQUS*, Computer Software; Simulia: Dassault, RI, USA, 2021.
36. Clough, R.W.; Penzien, J. *Dynamics of Structures*; McGrawHill: New York, NY, USA, 1975.

 *buildings*

Article

Lightweight Composite Floor System—Cold-Formed Steel and Concrete—LWT-FLOOR Project

Ivan Lukačević *, Ivan Ćurković, Andrea Rajić and Marko Bartolac

Faculty of Civil Engineering, University of Zagreb, 10000 Zagreb, Croatia; ivan.curkovic@grad.unizg.hr (I.Ć.); andrea.rajic@grad.unizg.hr (A.R.); marko.bartolac@grad.unizg.hr (M.B.)
* Correspondence: ivan.lukacevic@grad.unizg.hr; Tel.: +385-146-391-55

Abstract: In the last few decades, the application of lightweight cold-formed composite steel–concrete structural systems has constantly been increasing within the field of structural engineering. This can be explained by efficient material usage, particularly noticeable when using cold-formed built-up sections and the innovative types of shear connections. This paper summarises an overview of the development of the cold-formed composite steel–concrete floor systems. Additionally, it provides the background, planned activities, and preliminary results of the current LWT-FLOOR project, which is ongoing at the University of Zagreb, Faculty of Civil Engineering, Croatia. The proposed structural system is formed of built-up cold-formed steel beams and cast-in-place concrete slabs that are interconnected using an innovative type of shear connection. Preliminary analytical and numerical results on the system bending capacity are presented. Obtained results are mutually comparable. The resistance of the fixed beam solution is governed by the resistance of the steel beam, while pinned beam solution is governed by the degree of shear connection without the influence of the increased number of spot welds in the steel beam.

Keywords: cold-formed built-up steel; spot welding; steel–concrete composite system; floor system; finite element (FE) modelling

Citation: Lukačević, I.; Ćurković, I.; Rajić, A.; Bartolac, M. Lightweight Composite Floor System—Cold-Formed Steel and Concrete—LWT-FLOOR Project. *Buildings* **2022**, *12*, 209. https://doi.org/10.3390/buildings12020209

Academic Editors: Chiara Bedon, Flavio Stochino and Mislav Stepinac

Received: 31 December 2021
Accepted: 10 February 2022
Published: 12 February 2022

Publisher's Note: MDPI stays neutral with regard to jurisdictional claims in published maps and institutional affiliations.

1. Introduction

One of the key strategies to reduce human impact on Earth is to completely rethink our present lifestyle, especially the one led in industrialised countries. One of the critical aspects of this new lifestyle is sustainability, including reducing raw material and energy consumption. This can be achieved through innovations that will maximise the values of the structural components and building materials during their lifecycle. Although a widespread, systematic approach is still lacking, individual scientific projects are well aware of this problem and provide relevant solutions. An excellent example of this is the application of the composite cold-formed steel–concrete solutions that significantly reduce material consumption and contribute to the aforementioned change of the present unsustainable lifestyle.

The known fact is that composite steel–concrete systems are generally one of the most cost-effective structural systems applied in multi-storey buildings. The main reason behind this is that composite steel–concrete solutions integrate structural efficiency and the speed of construction. The structural efficiency results from the effective usage of structural materials, namely steel and concrete, thus omitting their inherent disadvantages. On the other hand, construction speed is enhanced since propping and formwork installation can be significantly reduced or even completely avoided.

However, despite all the advantages and benefits of composite structural systems, so far, they have not had the chance to be applied to any greater extent, i.e., certainly not to the extent that they deserve [1]. According to Ahmed et al. [2], the main forces driving the research within the field of composite steel–concrete structural systems are related to the development of innovative construction methods and new structural products, the best applications of new as well as underdeveloped materials, and considerations of socioeconomic and environmental consequences towards sustainability and resilience. The proposed structural flooring systems analysed in this work that use cold-formed steel (CFS) profiles combined with concrete cover all the aforementioned research aspects.

The purpose of this paper is to provide an overview of the development of the cold-formed composite steel–concrete structural systems. Furthermore, the paper gives insight into the background, planned activities, and preliminary results of the current LWT-FLOOR project, which is ongoing at the University of Zagreb, Faculty of Civil Engineering, Croatia. The proposed structural system is formed of built-up cold-formed steel beams and cast-in-place concrete slabs that are interconnected using an innovative type of shear connection. Preliminary analytical and numerical results on the system bending capacity are presented.

2. Materials and Methods

2.1. Overview of the Cold-Formed Steel Application in Composite Steel–Concrete Structural Systems

Generally, the interaction between steel and concrete has been widely used in structural engineering due to multiple benefits that result from combining desirable mechanical properties of each material. This explains why concrete is reinforced with steel rebars, or composite steel–concrete structural systems are used when higher resistance values or larger spanning capabilities are required. The advantages of structural steel are high tensile strength and ductility, while concrete ones are high stiffness and high compressive strength [3]. Therefore, using composite beams as structural flooring systems has a couple of benefits over the non-composite ones [4]:

- Material savings between 30% and 50% of structural steel can be achieved.
- Increased stiffness can reduce beam height for the same span, which can lead to benefits either as lower storey heights, decrease in cladding costs, more space for mechanical services, and increased usable space for the same building height.

Today, the most common version of composite beam comprises a hot-rolled steel beam and a floor slab, either concrete or composite steel–concrete, that interact through the application of shear connectors. In this form, the floor slab enhances the resistance of the steel beam by increasing its local and global stability. This observation can be further implemented to optimise the structural steel material so that cold-formed steel sections could be utilised instead of hot-rolled sections.

Application of cold-formed profiled steel sheeting has, for some time now, extensively been used in the construction of structural floor systems, i.e., the composite steel–concrete slabs [2]. Similarly, cold-formed steel profiles have also been investigated as soffits within the composite steel–concrete beams and columns [5–7]. However, such structural systems have not been applied to a greater extent. Nevertheless, the profiled steel sheeting used in the proposed solutions becomes an integral part of a structural system where it performs many different roles in the construction and the exploitation phases, such as [4]:

- Provides a working platform and protects the workers below;
- Supports the loads during construction and may eliminate the need for propping;
- Acts as permanent formwork for the concrete slab;
- If either mechanical or frictional interlock is being realised, the sheet can be considered to contribute to the required area of the primary tension reinforcement calculated for the slab;
- When through-deck welded stud shear connectors are used, the composite slab may be considered to restrain the steel beams.

The stated facts show enormous potential for further development of composite structural solutions using cold-formed steel elements and innovative shear connections. This potential is the result of the engineering point-of-view and the socioeconomic and environmental point-of-view to develop a sustainable and resilient ecosystem related to the built environment.

An extensive overview of the current development in composite steel–concrete beams and flooring systems can be found in [2]. Cold-formed steel has seen extensive research and application in composite slabs. However, it has not been extensively used in composite beam solutions where the downstand beam is made of cold-formed built-up steel sections and works together with the concrete flange, although such an idea has been around for a couple of decades [8]. Some advantages of composite beams using cold-formed built-up steel sections and concrete slabs are flexibility in architectural and beam cross-section design, the possibility of shallow slab depths, easy adaptation to irregular geometry, enabling reduction of self-weight, etc. However, the behaviour of such structural solutions has not been, up until a recent couple of years, investigated to a greater extent.

Further optimisation of composite systems can also be achieved by reducing self-weight with compact concrete with lightweight aggregates [9]. The authors of the paper [9] conducted a parametric study and summarised the results of the numerical analyses on the composite steel and concrete beams with standard and lightweight concrete slabs. In the cases of reducing composite beam self-weight, systems tend to vibrate more easily under human activities, and vibrational aspects of such systems are very important [10].

Hanaour [8] was among the first to examine the behaviour of composite beams with a cold-formed section using various shear connector types. Tests were conducted on composite beams using a steel cross-section consisting of two cold-formed channel sections. These channel sections were connected back-to-back by self-drilling screws with welded or screwed channel shear connectors over which the concrete slab is cast. Additionally, tests were conducted on composite beam with two cold-formed Z sections connected on the bottom with a plate and on top with a concrete plank generating a box section using 10 mm post-installed bolts as shear connectors. The obtained results show high ductility and capacity of beams when proper design and execution is applied.

Lakkavalli and Liu [11] conducted tests on twelve large-scale composite slab joists consisting of cold-formed steel C-sections and concrete to investigate behaviour and strength and to assess the effectiveness of the shear transfer mechanism. C-sections were partially embedded into the concrete slab for all specimens, so that top flanges were set into the concrete by the distance, which corresponded to half of the slab thickness. In addition to variation of the shear connection type and its spacing, the C-section thickness was also varied within the specimens. Results showed that the shear transfer mechanism using bent-up tabs performed the best when ultimate capacity is compared, followed by drilled holes and self-drilling screws. The results also showed that reduction of shear transfer spacing did not result in capacity increase which may be a consequence of overlapping of stress fields. Finally, the obtained experimental capacities of beam specimens were approximately 19% higher than the theoretically calculated capacities based on push-out test results.

Hsu et al. [12] proposed a new composite beam system consisting of a reinforced concrete slab on a corrugated metal deck, back-to-back cold-formed steel joists, and a continuous cold-formed furring shear connector. A shear connector is screwed to the joist top flange through the metal deck. Conducted tests on composite beams showed that the composite sections could reach ultimate strength without local shear or buckling failure when the proposed furring shear connector is used. In addition, the paper proposed analysis and design methods that can predict load-deflection behaviour and flexural strength of the beams and shear strength of the fasteners.

In [13,14], authors studied the structural behaviour of composite beams that integrate cold-formed steel with Ferro-cement slabs through a bolted-type shear connection. Parametric studies on several wire mesh variations in a Ferro-cement slab, the thickness of the steel section, and bolt diameter were carried out. Results show that failure occurs due to concrete crushing for thicker steel sections, while thinner steel sections fail due to sudden buckling. The shear connectors proved to have adequate strength to provide full shear connection and to transfer longitudinal shear force without failure. An increase of wire mesh layers resulted in greater strength capacity and improved crack formation. Finally, according to the analysis results, the ultimate strength capacity of these composite beams can be calculated by constitutive laws agreed by Eurocode 4.

Khadavi and Tahir [15] conducted research related to the bending resistance of encased composite beams composed of closed steel profiles filled with concrete. These closed profiles consist of two cold-formed C-sections oriented toe to toe and kept in place by a profiled sheeting installed on the beam top to increase stability and bearing performance. Shear connection is obtained using bent reinforcement bars. The proposed solution results show that using closed shapes of steel profiles can improve the bending strength as the confinement contributed by the concrete decreases the local buckling of the steel section. At the same time, the use of reinforcement as shear connectors further increases the bending and shear resistance and additionally stiffens the proposed beam. Similar composite cross-sections were researched by Salih et al. [16]. The results showed that filled beams have a higher bearing capacity than typical slender sections by reaching the yield stress. The proposed methods used in research proved to be applicable for calculating the ultimate limit state of the composite system according to the method from EC4. The continuous shear connection helped to move the neutral axis upward in the composite cross-section, which ensures that the steel top flange does not buckle under compression. Finally, research has shown that a partially encased steel beam increases flexural capacity.

Leal and Batista [17,18] present the investigation on the behaviour of structural composite floor system composed of cold-formed steel trusses and partially pre-cast concrete slab, which are interconnected using innovative shear connector solutions. In one case, the thin-walled channel connector is comprised of a lipped channel, single angle and reinforcing plate, which are altogether connected using self-drilling screws to the top chord, while in the other case, the thin-walled Perfobond rib connectors are composed of cold-formed plates, which are crossed by reinforcing steel bars and connected over a reinforcing plate to the top chord using self-drilling screws. The results of both specimen types showed that shear connectors were able to provide full interaction and adequate shear capacity up to a failure point. Analytical models based on the assumption of full yielding of the composite cross-section for the calculation of bending resistance is in good agreement with the experimental results.

Another interesting and innovative solution is presented in the paper [19]. The proposed lightweight cold-formed steel floor system is based on cold-formed steel trusses with a composite mortar. The trusses were formed with a U-shaped cross-section bottom chord, C-shaped cross-section web members and hat-shaped cross-section top chord. The paper presents the results of the bending tests on full-scale composite floor prototypes and the development of FE models and proposals of a theoretical method for determining the deflection and bearing capacity. The results show that the theoretical solutions were in good agreement with both the experimental and FE model results.

In order to improve weak interfaces that occur between the slab and the lower part (beam) and to achieve better composite behaviour, Liu et al. [20] proposed a new composite steel–concrete system composed of cold-formed U-profiles and rib truss, which is referred to as RCUCB. The system is characterised by two top varus flanges and rebar trusses in the opening sections. The study results showed that three possible failure modes could occur in such systems. In addition, tensile longitudinal reinforcement plays a significant role in controlling cracks and deformations of the encased concrete, and a proper shear stud placement can significantly improve ductility. Finally, the EC4 procedure can be used to calculate bending capacity with the additional application of three modification factors.

In addition to enhancements of the individual parts comprising composite steel–concrete structural systems by applying more adequate cross-section types or materials with more favourable mechanical properties, shear connection providing interaction of individual parts is another key parameter highly influencing the behaviour of composite systems. Its behaviour is characterised by its strength which depends on the realised degree of the shear connection. The degree of shear connection is defined as the ratio between the shear connection capacity provided by the interconnecting element used and the capacity of the weakest component of the composite cross-section (either steel beam or concrete slab) [3]. Although welded headed studs are still the most widely used shear connectors, there are numerous other solutions available to realise adequate shear connection in composite beams. Therefore, the research on shear connection solutions is still very popular to improve existing solutions or develop new ones. Among these solutions, the most interesting ones allow composite systems to be easily disassembled in case of structure demounting during deconstruction or when modification or repair is needed. Another reason to research demountable shear connection solutions is the previously mentioned goal to obtain sustainable structural systems. In other words, the ability to disassemble and reuse structural systems or their parts at the end of the building service life is of great importance to our environment.

Recently, Jakovljević et al. [21] provided a detailed chronological overview of the available demountable shear connection solutions that have been considered for use in composite structural systems. For better understanding, the shear connectors were classified according to their geometry and load-bearing mechanism, and the main results obtained from the push-out and large-scale beam bending test were presented and compared. In [22], a series of push-out tests on profiled steel sheeting and beams with demountable shear connectors were carried out. Comparison to the beam test results with welded shear connectors showed that the beam with demountable shear connectors has similar stiffness and superior ductility. A few examples of demountable headed shear stud connectors are shown in Figure 1. Solutions using friction-grip bolts are the ones that have been greatly investigated, but mostly as post-installed solutions in bridge rehabilitation. It is worth noting that complete code regulations are still lacking, and further research is needed. Bolted solutions using shear connectors with or without nuts embedded in the concrete slab have also been investigated. Due to tolerance of the hole for the bolts, the results exhibit lower initial stiffness, which can cause serviceability issues due to more considerable composite beam deflections.

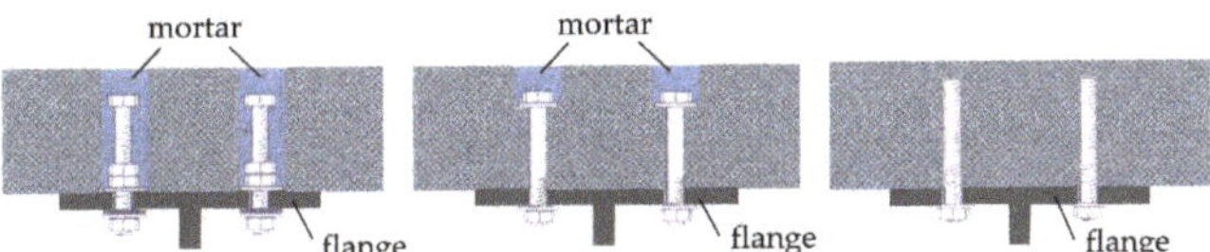

Figure 1. Variations of headed shear stud connectors.

To some extent, preloading can also be applied to avoid this, but only to the solutions that use embedded nuts. Such a solution leads to ultimate resistances close to the resistance of welded studs, but with failure at a very low slip rate due to brittle failure of the bolts [23]. In addition to many other solutions for shear connectors described in [21], there are still some significant issues regarding the resistance and ductility of the proposed solutions that need to be addressed. Firstly, the resistance of bolted shear connectors cannot be calculated according to Eurocode 3. The additional resistance provided through the nut to flange friction and catenary effects in bolts themselves has been observed [23]. Secondly, propositions of Eurocode 4 [24] for the determination of shear connector ductility cannot be implemented for bolted shear connectors as they exhibit large slips before reaching ultimate strength and therefore need to be modified [25].

On the other hand, research of the shear connection solutions between thin cold-formed steel profiles and concrete slabs has not been extensive. First, shear connection experiments were conducted by Hanaor [8], where two types of connectors, embedded and dry, were tested. The embedded connectors used channel sections that were either connected to the beam section using 6 mm diameter screws or were welded to it. The obtained capacities were well above the values computed for fasteners in the cold-formed section using standards, pointing out that appropriate code provisions can be used in the absence of test data. Lakkavalli and You [11] conducted research on four shear transfer mechanisms to join cold-formed C-sections and concrete. Namely, these included surface bonds, prefabricated bent-up tabs, pre-drilled holes, and self-drilling screws. As expected, the surface bond specimens showed the lowest strength and stiffness, and the ones using bent-up tabs performed best at both strength and serviceability limit states. Of the three specimens with shear transfer enhancements, the ones using self-drilling screws resulted in the lowest strength increase. Later, Irwan et al. [26] conducted experiments on shear connections using bent-up triangular tabs. Wehbe et al. [27] examined the behaviour of composite beams of concrete and cold-formed steel using stand-off screws 8 mm in diameter as shear connectors. The results showed that such a solution is feasible for providing composite action, where composite beams with such connections can be designed for ductile flexural failure. Alhajri et al. [13] also used bolts of 12 mm diameter to ensure composite action between the lipped C-channels to Ferro cement slab. Such connection proved to have adequate strength to transfer forces from slab to slab beam section. Recently, Bamaga et al. [28] proposed three different demountable shear connection solutions between the cold-formed steel beam and concrete slab. Push-out tests of all three solutions showed extremely ductile behaviour with adequate strength capacity.

In addition to the above presented discrete shear connection solutions, innovative and promising shear connection solutions based on composite dowel rib connectors [29], as shown in Figure 2, have also been developed. Such type of shear connector in cold-formed composite beams is titled a continuous furring channel and has been presented in papers [30,31]. The shear connection solution comprised a continuous cold-formed furring shear connector connected through the metal deck on top of the steel beam using self-drilling fasteners, providing horizontal and vertical interlock between steel–concrete parts of the cross-section.

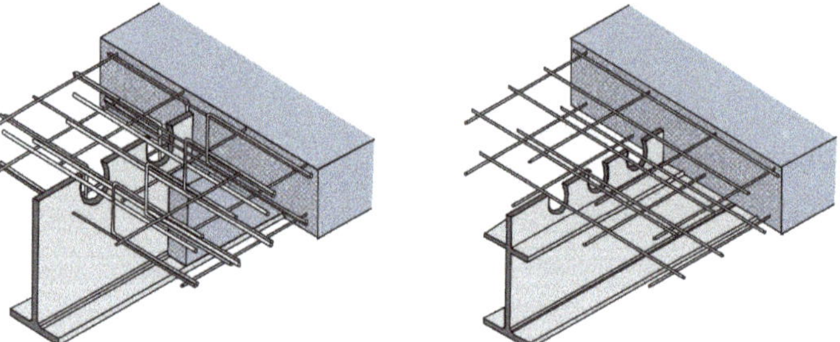

Figure 2. Possible solutions with composite dowel rib connectors.

From all the above results, it is clear that cold-formed steel sections and concrete slabs can act compositely. However, the data on the behaviour and performance of cold-formed sections in composite construction are still lacking. In addition, as one of the main reasons for using composite sections and applying CFS built-up profiles are environmental and sustainability concerns, the preference in selecting the shear connection type should be given to the demountable solutions. The environmental impacts of the demountable and conventional shear connections can be quantified using the Lifecycle Assessment (LCA) method. The evaluation of environmental benefits obtained using the LCA method comparing demountable systems, promoting disassembly and reuse, to the conventional monolithic ones, destined for demolition and recycling, is presented [32]. As expected, the results are quite encouraging for the demountable shear connector solutions. In papers [33,34], research on such demountable headed shear stud connectors together with the use of ultra-high performance concrete was carried out. The proposed connectors show excellent ductility in comparison to traditional solutions. The push-out tests were conducted for various types of fasteners, changing their shape, diameter and bolt class, and the obtained results showed that the average slip was close to the value required by the EN 1994-1-1 [24]. However, when designing for demountability, it is necessary to keep in mind that one needs to limit the fasteners' deformations during use and avoid the plastic response of the element to allow the reuse of parts of the structure. Some guidelines on this issue have been analysed for bolted joints with embedded nuts, considering the optimal ratio of beam span and depth of the demountable steel–concrete beam [35].

Developing a new type of shear connection requires experimental testing of the proposed connection type. Examples of the push-out tests on headed studs and composite dowel rib connectors are shown in Figure 3a,b, respectively.

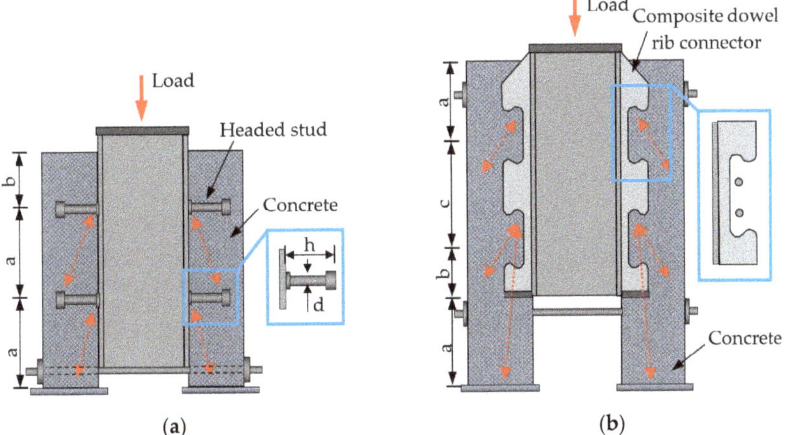

Figure 3. Push-out tests examples: (**a**) headed studs; (**b**) composite dowel rib connectors.

The presented overview of the application of cold-formed sections in steel–concrete floor systems leads to some advantages such as [8]:

- The possibility of reducing overall slab depth by using lighter sections at closer spacing.
- Easy variation of the cross-section for irregular layouts.
- Freedom in the design of cross-sections, i.e., cold-formed sections are made from flat sheets and can be designed and manufactured to order. It is relatively easy to produce built-up members from sections and flat strips screwed or spot welded together.
- Flexibility in assembling the sections and attached components in the workshop and/or on-site.
- The technology for manufacturing cold-formed sections is available and straightforward in regions and countries where a large selection of hot-rolled profiles is not available on short order, particularly for small or medium size projects.

As in the case of the shear connection for the system's overall behaviour, the connecting technique related to built-up cold-formed steel elements also has a vital aspect. A comprehensive experimental investigation on laser-welded connections based on lap-shear and tension tests is presented in [36]. In paper [37], a comparative resistance study was conducted of self-pierce riveting, resistance spot welding, and spot friction joining, identifying the resistance of spot welding as the most favourable option. Guenfoud et al. [38] tested welded specimens fabricated through one, two or four layers of thin steel sheets using the shear resistance and tension resistance of multi-layer arc spot welds.

2.2. Background to the LWT-FLOOR Structural System

Corrugated web beams represent a relatively new structural system that has emerged in the past two decades and was developed for various applications, i.e., the mainframes of single-storey steel buildings, secondary beams of multi-storey buildings, etc. Due to the thin webs, from 1.5 mm to 3 mm, corrugated web beams allow significant weight reduction compared to hot-rolled profiles or welded I-sections. The main benefits of this type of beam are that the corrugated webs increase the beam's stability against local and lateral-torsional buckling and against web crippling, which may result in a more effective design from both technical and economic points of view.

The corrugated web does not participate in the longitudinal transfer of bending stresses. Therefore, in static terms, the corrugated web beam behaves like a lattice girder, in which the bending moments and applied forces are transferred only through the flanges. In contrast, the shear forces are transferred only through the diagonals and verticals of the lattice girder. Consequently, the girder's flanges provide the flexural strength of the girder with no contribution from the corrugated web, which provides the girder's shear capacity. Furthermore, the use of thinner webs without stiffeners results in lower material cost, with an estimated cost savings of 10–30% compared to conventionally fabricated sections and more than 30% compared to standard hot-rolled beams [39,40]. For instance, the buckling resistance of 1 mm thick sinusoidal corrugated sheet web corresponds to buckling resistance of a plain flat web of 12 mm thickness or even more.

In developed solutions from the literature, already available on the construction market, the flanges are flat plates welded to the sinusoidal web sheet, requiring a specific welding technology [41,42]. Almost all the research on this kind of girder was devoted to studying bending and shear capacity [43–47]. The dimensioning of corrugated web beams is ruled by Annex D of the EN 1993-1-5 [48–50].

The main benefits of build-up corrugated web elements using fixed supports have already been summarised and demonstrated through research investigation on such types of beams connected with screws [40,51,52], spot welding (SW) and Cold Metal Transfer (CMT) techniques [53–55]. CMT, mainly developed in the last years, is a new connection technology for building steel structures. Since the technical solution of such types of beams enables standardisation of design detailing and fabrication, both SW and CMT welding techniques are appropriate for application in automated fabrication. Therefore, besides the structural advantages of such a system, the potential for automated series fabrication of the new technological solutions is another great advantage.

In research [53–55], for the purpose of full-scale tests, two beam specimens were built-up applying the SW connection technique, i.e., CWB SW-1 and CWB SW-2. The components of the built-up beam specimens, which were 5157 mm long and 600 mm high, are shown in Figure 4.

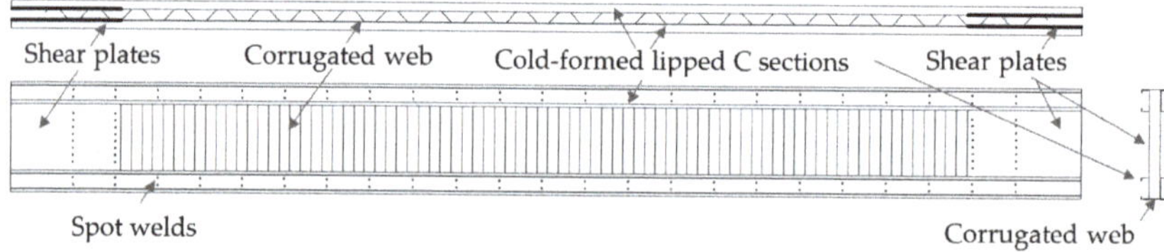

Figure 4. Components of the SW built-up beams.

Compared to the solution with built-up beams using self-drilling screws [40,51,52], it can be seen that the beams connected using the spot weld technique possess higher stiffness and higher load-bearing capacity. The intermittent spot welding technique avoids the difficulties of traditional arc welding and specific technologies, thereby significantly increasing the fabrication speed. The numerical parametric research on the calibration and validation of the numerical models with additional experimental tests considering the influence of web openings have resulted in several papers [56–58]. These studies confirmed the excellent behaviour of SW built-up corrugated web beams.

2.3. LWT-FLOOR Project

Based on built-up cold corrugated web cold-formed steel beams and other previously mentioned facts, it is possible to develop the new composite floor system. LWT-FLOOR system represents such innovation that maximises the values of each building component and used materials. The system is being investigated within the research project LWT-FLOOR at the University of Zagreb, Faculty of Civil Engineering, Croatia.

The LWT-FLOOR project integrates state-of-the-art knowledge in new, fast and productive spot welding technology and innovative cold-formed steel–concrete composite solutions proposing a new construction method as a combination of built-up cold-formed steel members and cast-in-place concrete slab. This potentially cost-effective and sustainable floor system could offer vital benefits in a high degree of prefabrication, reusability, and suitability for larger spans.

In order to investigate and validate components and the proposed system, extensive experimental, numerical and theoretical research is planned. Within the research, a particular focus is given to spot welding connections and innovative types of shear connections with the possibility for demountability and the potential for reuse or recycling at the end of the product design life, which are additionally evaluated through the application of lifecycle analyses. Calibrated and validated numerical models based on experimental tests of the system and its components allow evaluation of the system suitability for larger spans. All the accumulated knowledge, including the support of probabilistic methods, will be the basis for further structural detailing and the possibility of mass production. Analyses and performance evaluation of the proposed floor system solution will be crucial for establishing the first analytical proposal for design recommendations of this new system within the European standards.

The proposed solution is partially made of steel, which is highly recyclable, and cold-formed products use a high percentage of recyclable steel [59]. Because of the design for demountability in one of the proposed shear connection types, the system has the potential for future reuse and recycling of products and components. Spot welding technology proved its economic effectiveness, compared to laser welding and conventional techniques, and sustainable performance in terms of consumption of welding materials, energy, workers' health and safety issues, and elimination of inert shielding gases and particle emissions in the environment. The system will be durable under climate change effects and in an aggressive environment due to high protection from corrosion and the fact that all components are galvanised. It is easy to manipulate, transport and erect components of the system because of the reduced weight and safe connection technology. This can contribute to the safety of workers in the shops and building sites, cleaner activity on working sites, reduced energy consumption and lower emissions of greenhouse gasses. In the case of inner-city projects, construction speed and lack of on-site storage require a high level of prefabrication, which the proposed system can provide [60]. The proposed structural system falls into the category of adaptive building technology, following environmental and climate change requirements.

For the final applicability of the system in buildings, more knowledge about the behaviour in fire situations is required as well. Additionally, the system can also be applied in bridges; however, a detailed analysis of the fatigue behaviour is necessary.

Mentioned benefits of the system are investigated through experimental and numerical research with the support of probabilistic methods and lifecycle analyses by means of a holistic approach combining Lifecycle Assessment (LCA), Lifecycle Costs (LCC) and Lifecycle Performance (LCP) analyses.

The technical (mandatory O1 and O6) and scientific (O2 to O5) objectives with expected results (ER1 to ER6) of the LWT-FLOOR project are presented in Figure 5.

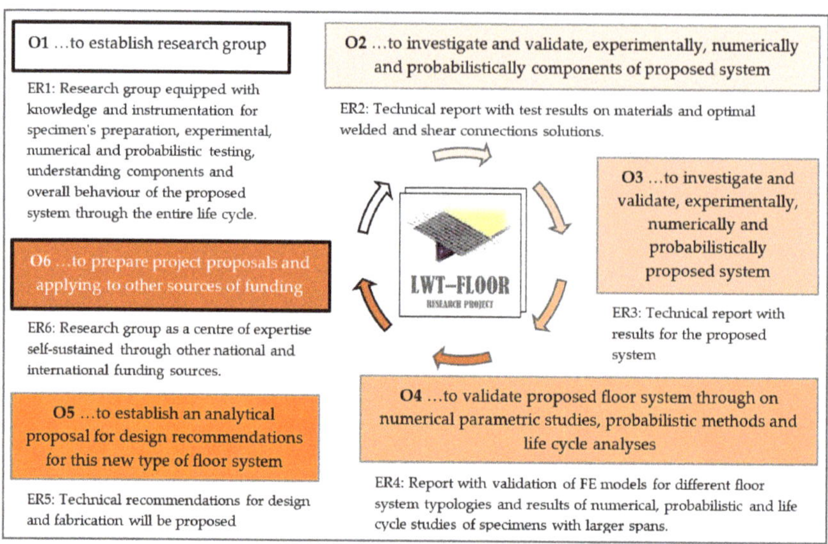

Figure 5. Technical and scientific objectives of the LWT-FLOOR project.

Experimental research is divided into five phases. Experimental research of LWT-FLOOR system materials and spot welds between different cold-formed sheet thicknesses are performed in the first two phases. Because of the great importance of the shear connection between the steel and concrete part of the system, the third phase includes experimental tests on the shear connection.

The results from the push-out test are implemented in the numerical models for their calibration, and the obtained models are used as the input before the experimental bending test of large-scale specimens of the overall system are conducted.

In Figure 6, two proposed solutions for shear connections are presented.

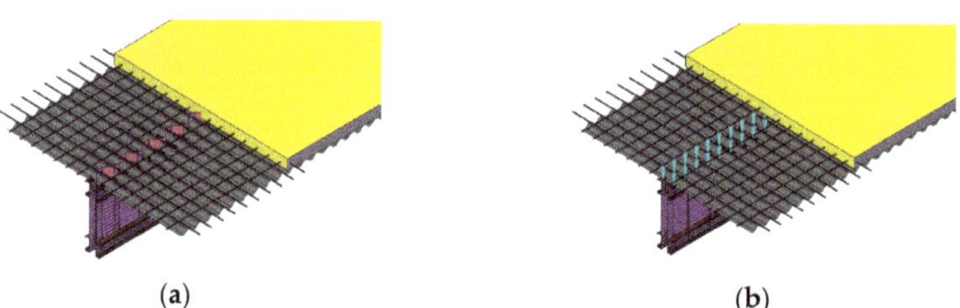

(a)   (b)

Figure 6. Proposed solutions for shear connection: (**a**) composite dowel rib connectors; (**b**) demountable headed shear stud connectors.

The solution in Figure 6a is based on extending the corrugated web of steel beam to form a shear connection. The solution in Figure 6b uses demountable shear connectors, which should ensure the demountability of the floor system at the end of its service life.

Within phases 4 and 5 of the experimental research full-scale LWT-FLOOR system is tested. The span of tested elements are approximately 6 m, corresponding to typical spans of floor systems in multi-storey buildings. In the first phase of experimental research, to understand the behaviour of corrugated built-up cold-formed girders without and with web openings, tests of the girders without concrete slab are conducted, Figure 7.

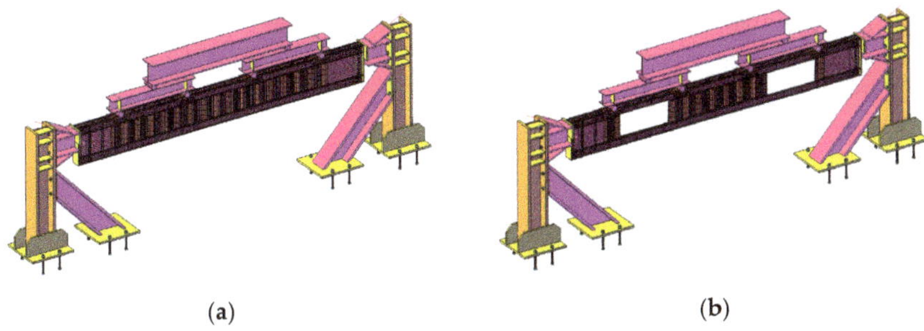

(a)   (b)

Figure 7. Built-up corrugated web girder: (**a**) without web openings; (**b**) with web openings.

Based on obtained results for optimal shear connection and steel girder solution, the experimental tests of the composite LWT-FLOOR system are performed. Figure 8 shows the proposed test set-up for the LWT-FLOOR beam.

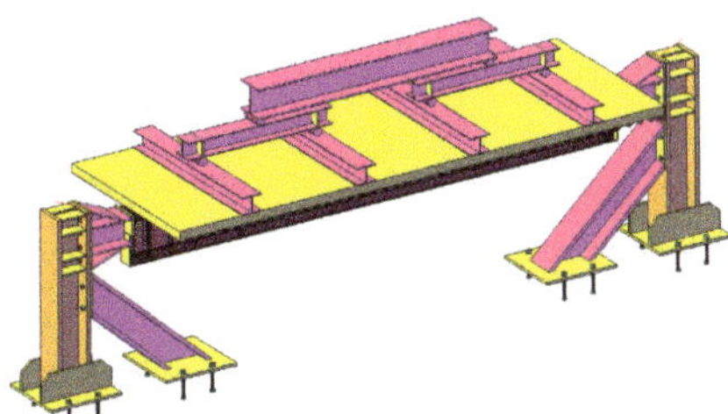

Figure 8. Proposal for test set-up for LWT-FLOOR system.

Based on calibrated numerical models, the numerical part of the research allows parametric research of the proposed solution. In the first phase, the numerical models of shear connection in combination with probabilistic analyses provide a selection of the optimal solution of the shear connection. The second phase of numerical research is numerical modelling of built-up corrugated web beams without and with web openings. Calibrated numerical models are the base for parametric studies to find the optimal solution for larger spans and different configurations of web openings.

Finally, calibration of a numerical model for composite LWT-FLOOR system and numerical parametric studies in combination with probabilistic analyses ensure the validation of the suitability of such system for larger spans and determine optimal dimensions of particular system components by taking into account different configurations of the shear connection and boundary conditions.

Based on probabilistic analyses and lifecycle analyses, the analytical proposal is be evaluated for analysed types of shear connections, steel girders with and without web openings, and composite LWT-FLOOR systems without and with web openings.

2.4. Analytical Model

Since cold-formed steel cross-sections are usually class 3 and 4, their plastic resistance in bending cannot be achieved. However, plastic resistance can be achieved in a composite beam with a full shear connection, but the cross-section's neutral axis position must be considered.

The neutral axis can be located in the concrete flange, at the interface between the concrete flange and the steel beam, or within the steel beam. Within steel beam means either in the flange or in the web of the steel beam.

Thus, taking into account the position of the neutral axis in the cross-section of the composite beam, and when the neutral axis is located in the concrete flange or at the

177

interface between the concrete flange and the steel beam, the steel cross-section will be completely or mainly in tension. Therefore, the bending resistance of the composite cross-sections, which consists of either class 3 or class 4 steel cross-section, can be calculated in the same way as the resistance for the class 1 or class 2 cross-section.

Investigated cross-section geometry is presented in Figure 9. The thickness of the concrete flange is 90 mm, and its effective width is 1000 mm with the concrete class C25/30. Built-up cold-formed steel beam consisted of four C120 sections with a thickness of 2.5 mm, flange width of 47 mm and height of 120 mm, and corrugated web and shear plates of 1.2 mm thickness with an overall height of 400 mm. All steel parts were considered to be made of steel grade S350. The span of the beam is 6 m. For the presented geometry and in the case of a full shear connection, the plastic neutral axis is located in a concrete flange.

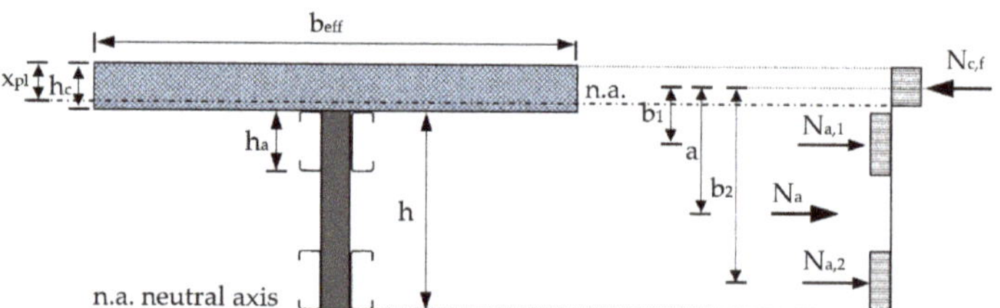

Figure 9. Cross-section of the analysed LWT-FLOOR beam.

The plastic bending resistance $M_{pl,Rd}$ in the case of full shear connection is calculated using Equations (1) and (2) or Equations (3) and (4), which depends on what part of the cross-section is considered, i.e., the concrete or the steel part, see Figure 9.

$$M_{pl,Rd} = N_{c,f} \times a, \tag{1}$$

$$M_{pl,Rd} = b_{eff} \times x_{pl} \times 0.85 \times f_{cd} \times \left(\frac{h}{2} + h_c - \frac{x_{pl}}{2} \right), \tag{2}$$

$$M_{pl,Rd} = 2 \times A_a \times f_{yd} \times b_1 + 2 \times A_a \times f_{yd} \times b_2, \tag{3}$$

$$M_{pl,Rd} = 2 \times A_a \times f_{yd} \times \left(\frac{h_a}{2} + h_c - \frac{x_{pl}}{2} \right) + 2 \times A_a \times f_{yd} \times \left(h - \frac{h_a}{2} + h_c - \frac{x_{pl}}{2} \right), \tag{4}$$

where: $N_{c,f}$ is the design value of the normal compressive force in the concrete flange with full shear connection, a is the distance according to Figure 9, b_{eff} is the total effective width of concrete flange, x_{pl} is the distance of the neutral axis from the upper edge of the concrete slab, f_{cd} is the design value of concrete compressive strength, h is the depth of the corrugated web structural steel section, h_c is the thickness of the concrete flange, A_a is the cross-sectional area of the cold-formed structural steel C-section, f_{yd} is the design value of the yield strength of structural steel, b_1 and b_2 are the centroids distances according to Figure 9, ha is the depth of the cold-formed structural steel C-section.

In the case of partial shear connection, the bending resistance M_{Rd} in the case of plastic structural steel cross-section can be calculated using Equation (5) according to EN 1994-1-1 [24].

$$M_{Rd} = M_{pl,a,Rd} + \left(M_{pl,Rd} - M_{pl,a,Rd} \right) \times \eta, \tag{5}$$

where: $M_{pl,a,Rd}$ is the plastic bending resistance of the structural steel section, and η is the degree of shear connection calculated according to Equation (6).

$$\eta = \frac{n \times 2 \times P_{Rd,min}}{b_{eff} \times 0.85 \times f_{cd} \times x_{pl}}, \tag{6}$$

where: n is the number of shear connector pairs and $P_{Rd,min}$ is the minimum design value of the shear resistance of a single connector. Bolts with 12 mm diameter and 8.8 steel grade were considered in this paper for calculation of $P_{Rd,min}$.

When using cold-formed corrugated web structural steel cross-sections, the following modified expression given in Equation (7) is proposed for the calculation of the composite cross-section bending resistance:

$$M_{Rd} = M_{el,a,Rd} + \left(M_{pl,Rd} - M_{el,a,Rd} \right) \times \eta, \tag{7}$$

where: $M_{el,a,Rd}$ is the elastic bending resistance of the cold-formed corrugated web structural steel cross-section.

According to Equation (7), the bending resistance of the analysed system will be equal to the elastic bending resistance of the cold-formed corrugated web structural steel cross-section $M_{el,a,Rd}$ when the degree of shear connection $\eta = 0$. In the case of full shear connection, $\eta = 1$, the plastic bending resistance of composite cross-section, $M_{pl,Rd}$ can be achieved.

Calculated bending resistance for full shear connection and solutions with the shear connectors distance of 300 mm ($\eta = 1$) and 600 mm ($\eta = 0.55$) were compared with the results of FE simulations in Section 3.

2.5. Numerical Model

2.5.1. Boundary Conditions and Finite Element Mesh

The numerical FE model is based on a static system with four concentrated forces as load, as presented in Figure 10. Figure 10a shows the static system of the beam with pinned supports, while Figure 10b shows an alternative solution with fixed supports. Because of the system's nonlinear behaviour and redistribution of moments at the supports, the analytically calculated bending resistance is compared with the resistance of the composite cross-section in the beam midspan. The bending moment from FE models is assumed to be based on a simply supported beam, Figure 10a. This assumption means that the bending moment in the beam midspan is equal to 4F multiplied by 0.75 for both cases, pinned and fixed solution.

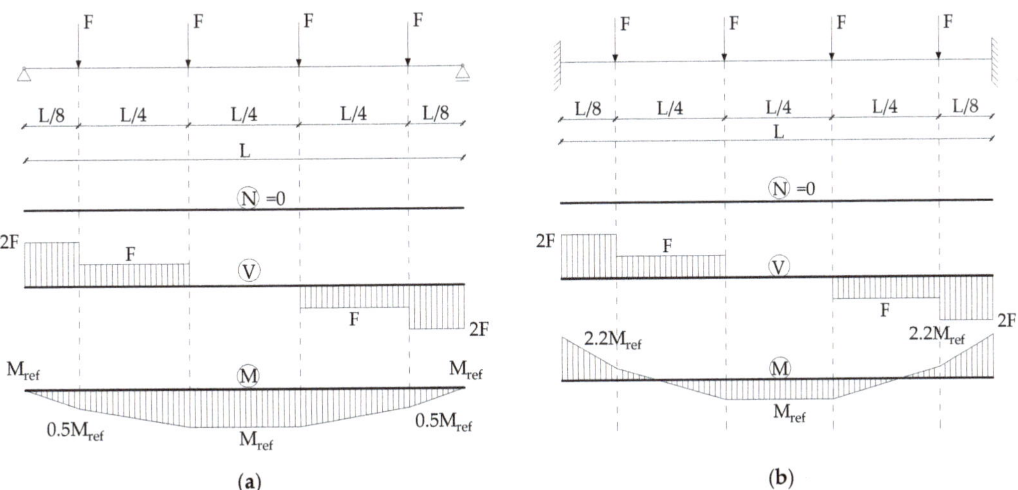

Figure 10. Static systems of analysed solutions: (a) pinned supported beam; (b) fixed supported beam.

The model of the analysed composite beam in Abaqus/CAE software [61] is presented in Figure 11. Figure 11 also shows two analysed solutions of spot weld configurations, i.e., solutions with 2 and 3 spot welds (SW2 and SW3, respectively).

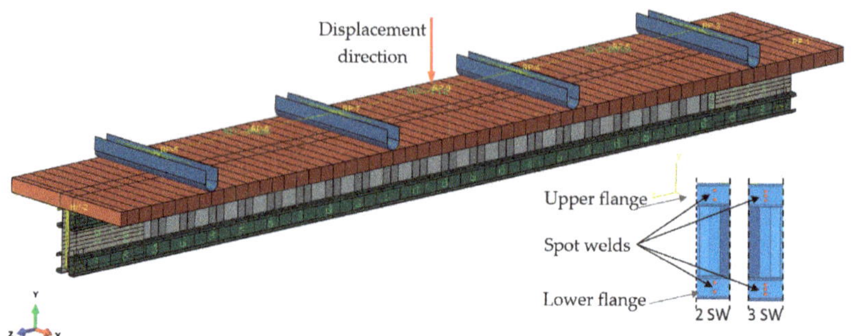

Figure 11. FE model in ABAQUS/CAE with the definition of 2 SW and 3 SW configurations.

According to [56,58], the beam loading was defined as a vertical displacement in a set of multipoint constraints (MPC) that forms a leverage system to transmit the deflection to the four loading points, see Figure 11. A kinematic coupling constraint defined the link between the control points and the pressure elements for all DOFs. RB3D2 elements were used as a rigid body for load transfer and multipoint constraint beam (MPC beam) for DOF coupling between groups of specified nodes. The total force, 4F, is obtained by monitoring the reaction force on the node where vertical displacement is applied.

Due to the complicated geometry and multiple contact interactions between different components, the "explicit" solving method was utilised to avoid the numerical convergence difficulties, which is often encountered in the "implicit" solving method. The loading speed is defined by ramping up the velocity smoothly over the analysis step. To increase the speed of performed analyses, loading velocity is increased, ensuring minimal inertia effects. This is ensured by maintaining kinetic energy at a small value throughout the analyses (mainly below 5% of all energy) and by limiting loading velocity to less than 1% of the wave speed of the used materials. However, several drops in the resulting curves of moments and forces can be observed. The authors believe that such drops result from spot weld failures and buckling of the steel beam components.

General contact with the following parameters between all elements in the model was used, i.e., normal direction—hard contact, transverse direction—a friction coefficient of 0.1 and separation was allowed after the general contact takes place.

Geometric and material nonlinear analyses, including the effects of initial imperfections (GMNIA), were performed. The numerical modelling consists of two steps. In the first step, the initial imperfections are modelled by performing static analysis, which results in desired imperfection shape—with the magnitude of the shell thickness. In the second step, the dynamic, explicit analysis is used to run the load-displacement analysis of the beam, including imperfect geometry obtained from previous analysis and all contacts and material nonlinearities.

In order to carry out dynamic analysis, the density of the material needs to be specified. The density of steel components is set to 7850 kg/m^3, while the density of concrete is 2400 kg/m^3.

Each part of the built-up steel beam was defined as a 3D shell element S4R, while solid C3D8R elements were used to model concrete flange. The FE mesh density is 20 mm for steel sections and 30 mm for the concrete flange.

2.5.2. Material Model for Spot Welds, Steel Beam and Headed Shear Studs

The material and spot weld properties from the research of Ungureanu et al. [53,54] were used for corrugated beam components and their assembly with spot welds.

The spot welds (SW) between different parts of the built-up beams were defined in function of the tensile-shear tests of the simple specimens according to [58]. Attachment points were defined on each part where SW was applied. The connection between the attachment points was defined using Point-Based Fasteners with the connector response [61] of the SW initially calibrated from the tensile-shear test results from [53,54]. The connector was considered using the elasticity, plasticity, damage and failure parameters. Bushing connector elements were used to model the SW. This type of element provides a connection between two nodes that allows independent behaviour in three local Cartesian directions that follow the system at both nodes and that allows different behaviour in two flexural rotations and one torsional rotation [61].

In order to obtain realistic results from the finite element nonlinear analyses, plastic strains were included in the material definition, according to EN 1993-1-5, Annex C [45]. The measured stress–strain curves based on tensile tests on coupons cut from the cross-section of component elements were included in the model. The static engineering stress–strain curves obtained from tensile coupon tests were converted to true stress vs. logarithmic true plastic strain curves in the plastic analysis. The true stress σ_{true} and true plastic strain ε_{true} were calculated using Equations (8) and (9) as follows:

$$\sigma_{true} = \sigma_{engineering}\left(1 + \varepsilon_{engineering}\right), \tag{8}$$

$$\varepsilon_{true} = \ln\left(1 + \varepsilon_{engineering}\right), \tag{9}$$

In order to include plasticity, the stress–strain points past yield must be input in the form of true stress and logarithmic plastic strain. The logarithmic plastic strain was calculated with Equation (10) as follows:

$$\varepsilon_{ln}^{plastic} = \varepsilon_{true} - \frac{\sigma_{true}}{E}, \tag{10}$$

where σ_{true}/E is the elastic strain and E is Young's modulus.

The shear connection is taken into account using three different solutions, i.e., full tie connection between concrete flange and steel beam, shear connectors modelled by wire elements spaced at 300 mm (degree of shear connection is 1) and 600 mm (degree of shear connection is 0.55). The connectors section was defined using the bushing connection type with defined elasticity, plasticity damage and failure parameters based on calculated resistances of the analysed shear stud connectors, i.e., bolts with 12 mm diameter and 8.8 steel grade. After obtaining experimental results, more detailed definitions of shear connectors will be considered. Examples of shear connectors definitions with local interactions and damage initiation can be found in the literature [62–64].

2.5.3. Material Model for Concrete

The concrete damaged plasticity model (CDP) was used to model the concrete slab. This model is primarily based on two main failure mechanisms of tensile cracking and compressive crushing of concrete [64]. The yield (or failure) surface evolution is controlled by two hardening variables linked to failure mechanisms under tension and compression loading, respectively [61]. This paper took the dilation angle as 40° according to [64], while default values were assumed for all other plasticity parameters.

The modulus of elasticity of concrete, E_{cm}, was calculated according to EN 1992-1-1: 2004 [65] as:

$$E_{cm} = 22\left[\frac{f_{cm}}{10}\right]^{0.3}, \tag{11}$$

$$f_{cm} = f_{ck} + 8, \tag{12}$$

where, f_{cm} is the mean value of concrete cylinder compressive strength, and f_{ck} is the characteristic compressive cylinder strength of concrete at 28 days.

In EN 1992-1-1 [65] the relationship between compressive stress, σ_c and shortening strain, ε_c for short term uniaxial loading is described by the Equation (13):

$$\sigma_c = f_{cm} \times \frac{k\eta - \eta^2}{1 + (k-2)\eta}, \quad \eta \leq \varepsilon_{cu1}/\varepsilon_c, \tag{13}$$

with

$$k = 1.05 \times E_{cm} \times \frac{\varepsilon_{c1}}{f_{cm}}, \tag{14}$$

$$\eta = \frac{\varepsilon_c}{\varepsilon_{c1}}, \tag{15}$$

where ε_c is compressive strain, ε_{c1} is the compressive strain in the concrete at the peak stress f_c, $\varepsilon_{c1} = 0.7 \cdot f_{cm}^{0.31} \leq 2.8 \times 10^{-3}$. Equation (13) is valid for compressive strain region $0 < |\varepsilon_c| < |\varepsilon_{cu1}|$ where ε_{cu1} is the nominal ultimate strain. The strain at peak stress $\varepsilon_{c1} = 2.25 \times 10^{-3}$, and nominal ultimate strain $\varepsilon_{cu1} = 3.5 \times 10^{-3}$ were adopted from EN 1992-1-1 [65].

The stiffness degradation dc on account of crushing of concrete was calculated with Equation (16):

$$d_c = 1 - \frac{\sigma_c}{f_{cm}}, \tag{16}$$

Stress–strain behaviour of plain concrete in uniaxial compression defined in the EN 1992-1-1 [65] with concrete stiffness degradation in compression dc is presented in Figure 12.

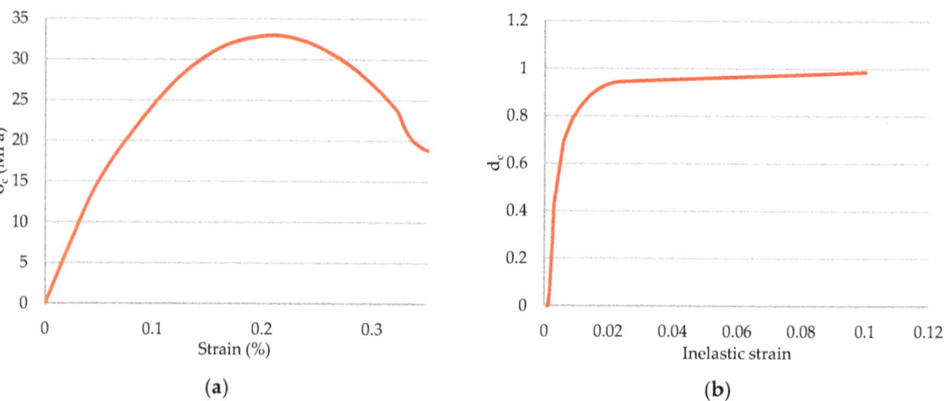

Figure 12. Parameters of concrete in compression: (**a**) stress–strain relation of concrete for structural analysis based on EN 1992-1-1; (**b**) concrete stiffness degradation in compression.

As presented with Equation (13) and Figure 12, the plasticity curve in EN 1992-1-1 is defined only up to nominal ultimate strain $\varepsilon cu1$. In cases when high crushing strains are expected, the extensions of the Eurocode model can be used as presented in the paper [23]:

$$\sigma_c(\varepsilon_c) = \begin{cases} f_{cm}\left[\dfrac{1}{\beta} - \dfrac{\sin(\mu^{\alpha_{tD}} \times \alpha_{tE}\pi/2)}{\beta \times \sin(\alpha_{tE}\pi/2)} + \dfrac{\mu}{\alpha}\right], & \varepsilon_{cuD} < \varepsilon_c \leq \varepsilon_{cuE} \\ [f_{cuE}(\varepsilon_{cuF} - \varepsilon_c) + f_{cuF}(\varepsilon_c - \varepsilon_{cuE})]/(\varepsilon_{cuF} - \varepsilon_{cuE}), & \varepsilon_c > \varepsilon_{cuE} \end{cases} \tag{17}$$

where $\mu = (\varepsilon_c - \varepsilon_{cuD})/(\varepsilon_{cuE} - \varepsilon_{cuD})$ is relative coordinate between points D–E and $\beta = f_{cm}/f_{cu1}$. Point D is defined as $\varepsilon_{cuD} = \varepsilon c_{u1}$ and $f_{cuD} = f_{cu1} = \sigma_c(\varepsilon_{cu1})$ from Equation (13). Point E is the end of the sinusoidal descending part at strain ε_{cuE} with concrete strength reduced to f_{cuE} by factor $\alpha = f_{cm}/f_{cuE}$. The linear descending part (residual branch) ends in point F at strain ε_{cuF} with the final residual strength of concrete f_{cuF}. In the paper [23] strain $\varepsilon_{cuF} = 0.10$ was chosen large enough not to be achieved in the analyses. Compression plasticity curve defined dependent on inelastic strain assuming that the concrete acts elastically up to $0.4 \times f_{cm}$ is presented in Figure 13.

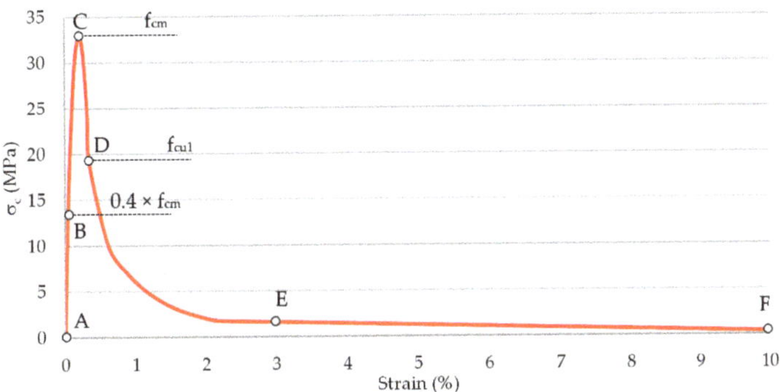

Figure 13. Parameters of concrete in compression, an extension of EN 1992-1-1 model.

The extension of the EN 1992-1-1 model does not change the presented models' results, which means that the compression strains are below strain ε_{cu1}.

The tension plasticity curve is defined as the function of cracking strain and tensile stress. Tensile stress increases linearly along with modulus of elasticity, up to the peak value f_t. In this paper, the post-failure tensile behaviour is defined by using the exponential function proposed by Cornelissen et al. [66]:

$$\frac{\sigma}{f_t} = f(w) - \frac{w}{w_c} f(w_c), \tag{18}$$

$$f(w) = \left[1 + \left(\frac{c_1 \times w}{w_c} \right)^3 \right] \exp\left(-\frac{c_2 \times w}{w_c} \right), \tag{19}$$

where w is the crack opening displacement, w_c is the crack opening displacement at which stress can no longer be transferred $w_c = 5.14 \times G_f/f_t$, c_1 is a material constant and $c_1 = 3.0$ for normal density concrete, c_2 is a material constant and $c_2 = 6.93$ for normal density concrete.

The value of fracture energy, the energy required to develop a unit area of a crack in (N/mm), is [67,68]:

$$G_f = 73 \times f_{cm}^{0.18}, \tag{20}$$

The degradation in tension d_t was calculated with Equation (21):

$$d_t = 1 - \frac{\sigma_t}{f_t}, \tag{21}$$

The tensile stress versus cracking displacement relation using Equation (18) and tensile damage versus cracking displacement relation, Equation (21), are shown in Figure 14.

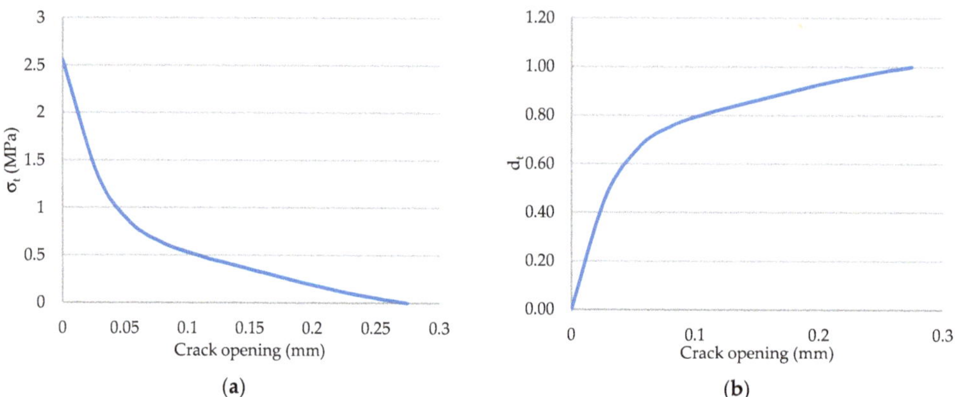

Figure 14. Parameters of concrete in tension: (**a**) tensile stress versus cracking displacement; (**b**) tensile damage versus cracking displacement.

3. Results and Discussion

Figure 15 shows analytically calculated bending moment resistance based on Equations (5) and (7) concerning the degree of shear connection for beam geometry from Section 2.4.

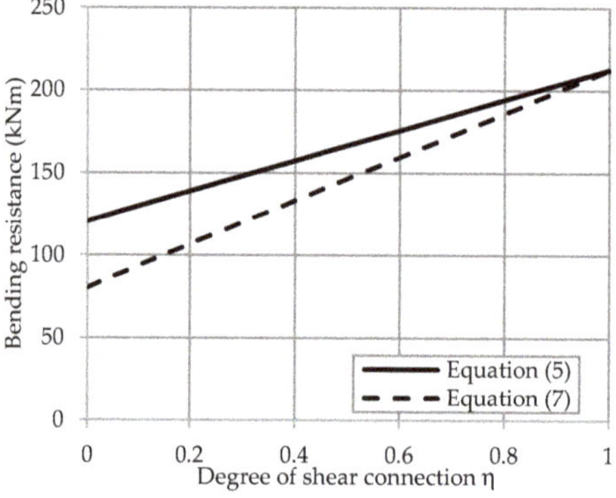

Figure 15. Analytically calculated bending moment resistance based on Equations (5) and (7) concerning the degree of shear connection.

Results of the numerical study are divided based on the beam supports definition. The results of the beam model with pinned supports and beam model with fixed supports are presented in the following subsections and compared with analytically obtained results from Figure 15.

3.1. Beam with Pinned Supports

Figure 16 shows the differences in end slip between three analysed solutions depending on the shear connection, i.e., solutions with fully tied concrete flange to the steel beam and solutions with shear connectors longitudinally spaced at 300 mm and 600 mm. The 300 mm shear connector spacing represents a full shear connection with the degree of shear connection around 1. In contrast, the 600 mm shear connector spacing represents a partial shear connection with the degree of the shear connection of 0.55. Figure 16a shows that the end slip is completely prevented, while on the other hand, in Figure 16b,c the existence of some end slip is observed.

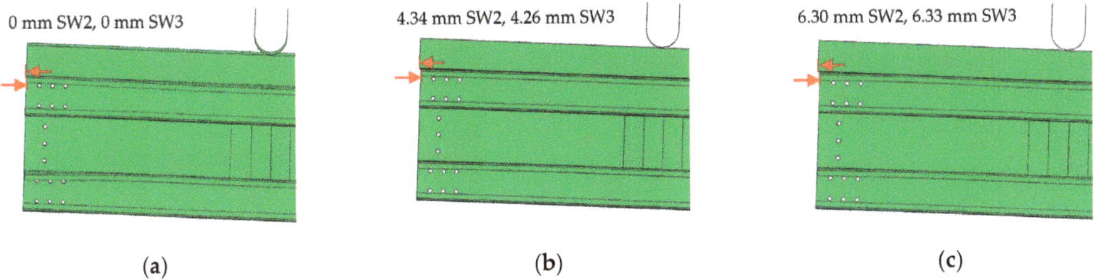

Figure 16. Differences in end slip between three analysed solutions of pinned beam: (**a**) full tie shear connection; (**b**) shear connectors spaced at 300 mm; (**c**) shear connectors spaced at 600 mm.

The bending resistance of the beam in the cases of fully tied shear connection and analysed spot weld solutions, i.e., SW2 and SW3, are presented in Figure 17. Both curves in Figure 17a,b shows that the numerically obtained bending resistance is higher than the analytically calculated bending resistance presented with a dashed horizontal line.

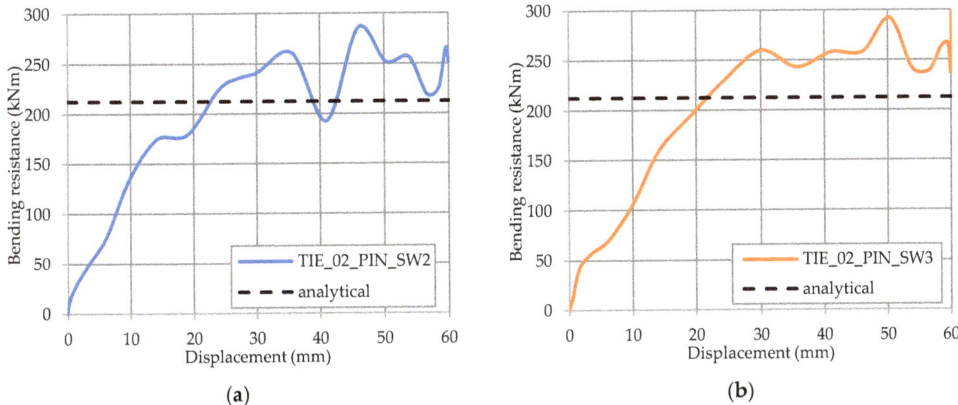

Figure 17. Bending resistance of the pinned beam for a fully tied shear connection: (**a**) SW2; (**b**) SW3.

In the case of beams with shear connectors longitudinally spaced at 300 mm, the reduced stiffness and bending capacity are observed compared to the full tie shear connection results, as shown in Figure 18. Figure 18a shows the bending resistance of simply supported beam in the cases when shear connectors are spaced at 300 mm with SW2 spot weld configuration while Figure 18b shows bending resistance when SW3 spot weld configuration is used.

In the case when shear connectors are spaced at 600 mm, similar results are obtained as the ones when the spacing of 300 mm is applied. The obtained results are presented with curves in Figure 19. From the curves in Figures 17–19, it can be observed that the solutions with SW2 and SW3 spot weld configurations are in good correlation. The number of SW does not govern the bending capacity of pinned beam solutions to a great extent.

3.2. Beam with Fixed Supports

Figure 20 shows end slip behaviour in the case of fixed end supports between three analysed solutions of shear connection. From the comparison with the solutions of the beam with pined supports, it can be observed that slip is prevented in the fixed beam, but some separation between the concrete flange and steel cross-section exists.

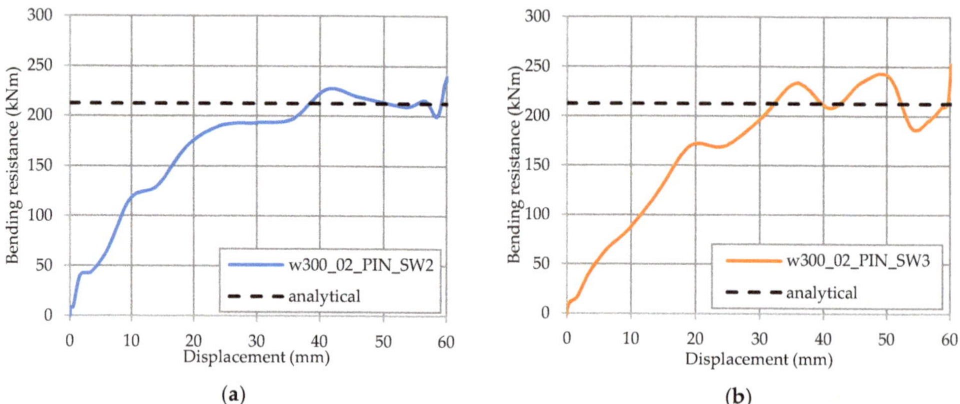

Figure 18. Bending resistance of the pinned beam for shear connectors spaced at 300 mm: (**a**) SW2; (**b**) SW3.

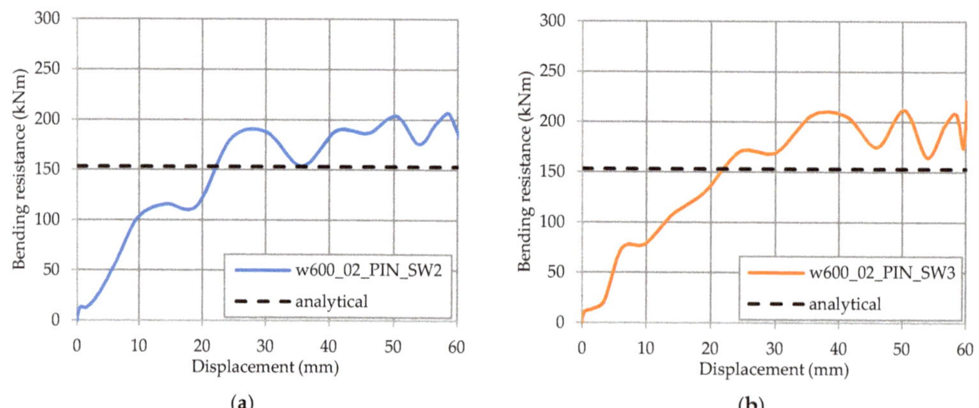

Figure 19. Bending resistance of the pinned beam for shear connectors spaced at 600 mm: (**a**) SW2; (**b**) SW3.

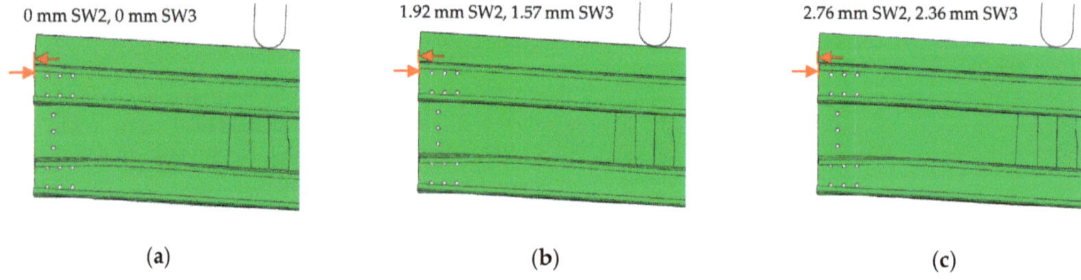

Figure 20. Differences in end slip between three analysed solutions of fixed beam: (**a**) full tie shear connection; (**b**) shear connectors spaced at 300 mm; (**c**) shear connectors spaced at 600 mm.

In the case of fixed end supports, the stiffness of the beam is generally increased. The behaviour in the cases of different degrees of shear connections is similar to the pinned solution. Figure 21 shows the bending resistance of the beam with fixed supports and a full tie shear connection. Both curves in Figure 21a,b shows that the numerically obtained bending resistance is higher than the analytically calculated bending resistance presented with a dashed horizontal line.

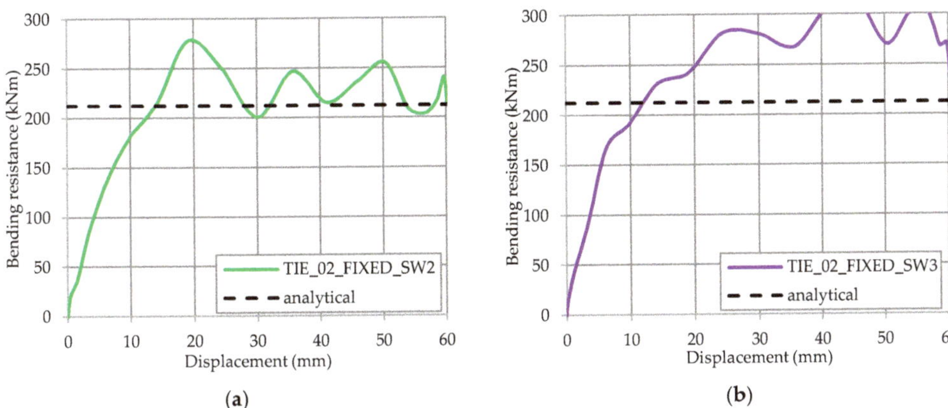

Figure 21. Bending resistance of the fixed beam for a fully tied shear connection: (**a**) SW2; (**b**) SW3.

Figure 22 shows the results in the 300 mm spacing between shear connectors, and again, bending resistance is higher than analytically calculated resistance. Contrary to the pinned beam case, the solution with SW3 configuration increases the bending capacity.

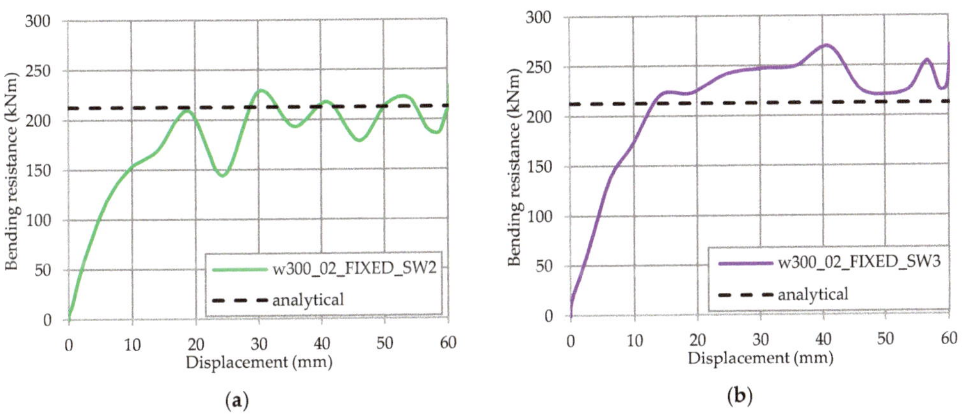

Figure 22. Bending resistance of the fixed beam for shear connectors spaced at 300 mm: (**a**) SW2; (**b**) SW3.

In the case of shear connectors being spaced at 600 mm, the results are presented in Figure 23. As was the case in the previous solutions, the bending resistance is higher than the analytically calculated resistance. In the cases of shear connectors being spaced at 300 mm and 600 mm, the solutions with SW3 and SW2 are different, which means that this solution with fixed supports is governed by the number of SW, which was not the case for the pinned beam solution.

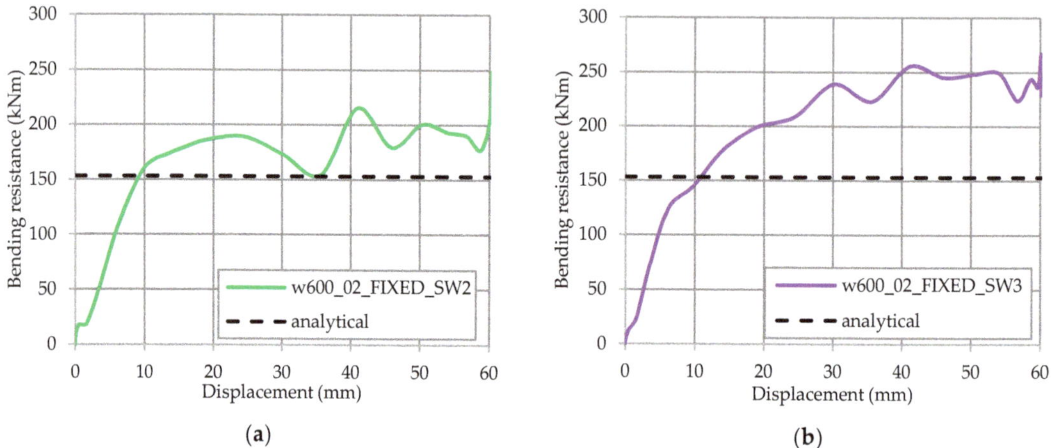

(a)   (b)

Figure 23. Bending resistance of the fixed beam for shear connectors spaced at 600 mm: (**a**) SW2; (**b**) SW3.

3.3. Comparison of the Pinned and Fixed Solution

Figures 24–26 show comparisons of the force/displacement diagrams for the solutions with SW2 and SW3 configurations. These comparisons clearly show that SW3 solutions have a small influence compared to SW2 for the pinned solution of the beam. Additionally, the stiffness and decrease of capacity are also evident from fully tied to partial shear connection. If we compare pinned and fixed solutions of beam end supports, we can conclude the following. The stiffness and the bending capacity are increased from pinned to fixed solutions. Such behaviour, as already discussed, can also be captured from the partial to fully tied shear connection.

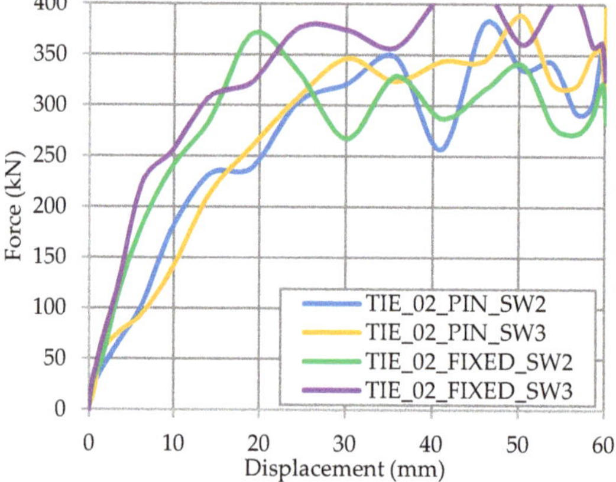

Figure 24. Force/displacement diagrams for 2 SW and 3 SW configurations of the pinned and fixed beam with full tie shear connection.

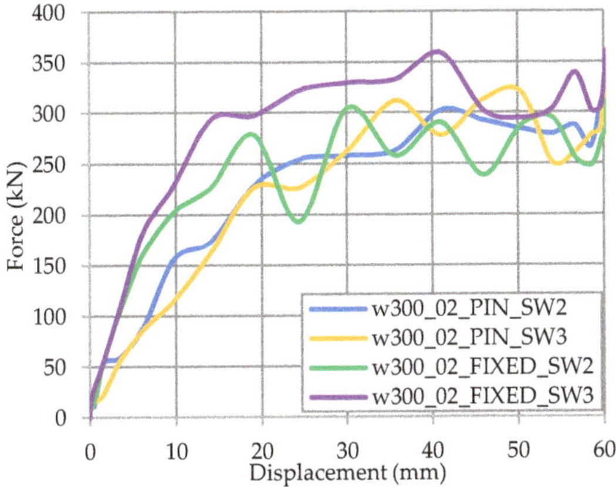

Figure 25. Force/displacement diagrams for 2 SW and 3 SW configurations of the pinned and fixed beam with shear connectors spaced at 300 mm.

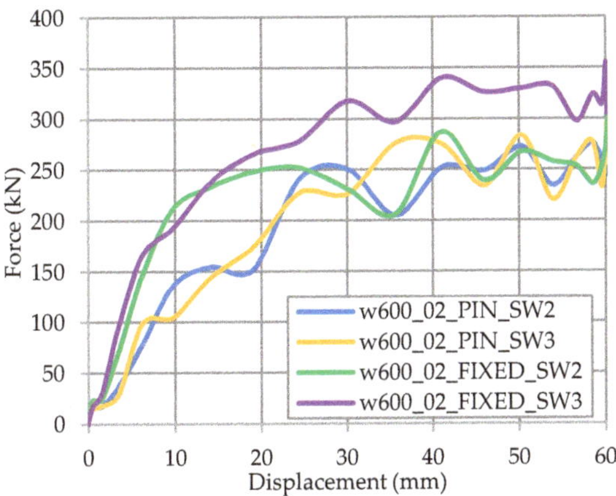

Figure 26. Force/displacement diagrams for 2 SW and 3 SW configurations of the pinned and fixed beam with shear connectors spaced at 600 mm.

4. Conclusions

Due to their efficiency, steel–concrete structural systems are increasingly used in the construction industry and are still the subject of intensive research. This results from their efficient material usage, especially with cold-formed built-up sections and innovative shear connections. The popularity of such systems arose from their flexibility, adaptability, and capacity for dismantling and reuse. This paper summarises an overview of cold-formed steel–concrete composite floor systems developments. Additionally, it gives the background, planned activities and recent results related to developing the new innovative solution within research project LWT-FLOOR, which is ongoing at the University of Zagreb, Faculty of Civil Engineering, Croatia. The LWT-FLOOR project investigates innovative lightweight solutions that can result in new sustainable and resilient construction products due to the holistic design approach integrating manufacturing, operation, maintenance and reuse stages up to the end-of-life.

From the literature survey, the following aspects can be highlighted.

Shear connection mechanisms and mechanisms of the connections between steel elements in built-up cold-formed steel beams greatly influence the behaviour of composite floor systems. These mechanisms are of vital interest to optimise components and produce more economical and efficient floor systems. Integrating the floor system related to its connection to the rest of the structure is another important issue. Simple and effective connections of the floor system will result in carrying the vertical load and providing a shear diaphragm, which is crucial for the performance of the overall structure.

The presented LWT-FLOOR project integrates state-of-the-art knowledge in a new and productive spot welding technology and innovative cold-formed steel–concrete composite solutions to develop a cost-effective and competitive floor system. The LWT-FLOOR system can give vital benefits in terms of a high degree of prefabrication, reusability and suitability for larger spans to maximise the values of building components and materials.

The proposed analytical solution for bending resistance of composite cross-section is based on the elastic bending resistance of steel beams. The numerical parametric FE analyses show that the bending capacity of the LWT-FLOOR beams made of cold-formed steel and the cast-in-place concrete slab can be compared with analytical results.

Results of the parametric FE study show that the bending capacity of the LWT-FLOOR beams made of cold-formed steel and the cast-in-place concrete slab is affected by the resistance of the cold-formed steel components and by the degree of shear connection. In the case of a fixed beam solution, the resistance of the LWT-FLOOR beam is governed by the resistance of the steel beam, i.e., spot welds. In the cases with pinned beam solution, the results show that the resistance of the LWT-FLOOR beam is governed by the degree of shear connection. At the same time, the increased number of SW from two to three between channel flanges and corrugated web does not influence the beam resistance. This means that the beam resistance in the pinned solution is rather more affected by the stability of the cold-formed steel components than the number of spot welds, with a significant influence on the type and degree of shear connection.

Based on previous studies on steel built-up corrugated web beams, it is observed that the resistance is dependent on the thicknesses of the corrugated web and the thickness of the shear plate. Therefore, such solutions will be investigated in the extension of this research.

Author Contributions: Conceptualisation, I.L.; methodology, I.L. and I.Ć.; software, I.L. and A.R.; validation, I.L., I.Ć. and M.B.; formal analysis, A.R.; investigation, I.L., I.Ć. and A.R.; resources, I.L., I.Ć. and M.B.; data curation, I.L., I.Ć. and A.R.; writing—original draft preparation, I.L.; writing—review and editing, all authors; visualisation, I.L. and A.R.; supervision, I.L., I.Ć. and M.B.; project administration, I.L., I.Ć. and A.R.; funding acquisition, I.L. All authors have read and agreed to the published version of the manuscript.

Funding: This research was partially funded by the Croatian Science Foundation, grant number UIP-2020-02-2964 (LWT-FLOOR project—Innovative lightweight cold-formed steel–concrete composite floor system), project leader: Ivan Lukačević.

Institutional Review Board Statement: Not applicable.

Informed Consent Statement: Not applicable.

Data Availability Statement: The data presented in this study are available on request from the corresponding author.

Conflicts of Interest: The authors declare no conflict of interest. The funders had no role in the design of the study; in the collection, analyses, or interpretation of data; in the writing of the manuscript, or in the decision to publish the results.

References

1. Dujmović, D.; Androić, B.; Lukačević, I. *Composite Structures According to Eurocode 4: Worked Examples*; John Wiley and Sons: Berlin, Germany, 2015.
2. Ahmed, I.M.; Tsavdaridis, K.D. The evolution of composite flooring systems: Applications, testing, modelling and eurocode design approaches. *J. Constr. Steel Res.* **2019**, *155*, 286–300. [CrossRef]
3. Androić, B.; Dujmović, D.; Lukačević, I. *Design of Composite Structures According to Eurocode 4*; IA Projektiranje: Zagreb, Croatia, 2012.

4. Hicks, S. EN 1994-Eurocode 4: Design of composite steel and concrete structures-Composite slabs. In Proceedings of the EUROCODES Background and Applications, "Dissemination of Information for Training" Workshop, Brussels, Belgium, 18–20 February 2008; pp. 1–24.

5. Nguyen, R.P. Thin-Walled, Cold-Formed Steel Composite Beams. *J. Struct. Eng.* **1991**, *117*, 2936–2952. [CrossRef]

6. Anwar Hossain, K.M. Designing thin-walled composite-filled beams. *Proc. Inst. Civ. Eng. Struct. Build.* **2005**, *158*, 267–278. [CrossRef]

7. Abdel-Sayed, G. Composite Cold-Formed Steel-Concrete Structural System. In Proceedings of the 6th International Specialty Conference on Cold-Formed Steel Structures, St. Louis, MO, USA, 16–17 November 1982; University of Missouri: Rolla, MO, USA, 1982; pp. 485–510.

8. Hanaor, A. Tests of composite beams with cold-formed sections. *J. Constr. Steel Res.* **2000**, *54*, 245–264. [CrossRef]

9. Přivřelová, V. Modelling of Composite Steel and Concrete Beam with the Lightweight Concrete Slab. *Int. J. Civ. Environ. Eng.* **2014**, *8*, 1130–1134. [CrossRef]

10. Zhou, X.; He, Y.; Jia, Z.; Nie, S. Experimental study on vibration behavior of cold-form steel concrete composite floor. *Adv. Steel Constr.* **2011**, *7*, 302–312.

11. Lakkavalli, B.S.; Liu, Y. Experimental study of composite cold-formed steel C-section floor joists. *J. Constr. Steel Res.* **2006**, *62*, 995–1006. [CrossRef]

12. Hsu, C.-T.T.; Punurai, S.; Punurai, W.; Majdi, Y. New composite beams having cold-formed steel joists and concrete slab. *Eng. Struct.* **2014**, *71*, 187–200. [CrossRef]

13. Alhajri, T.M.; Tahir, M.M.; Azimi, M.; Mirza, J.; Lawan, M.M.; Alenezi, K.K.; Ragaee, M.B. Behavior of pre-cast U-Shaped Composite Beam integrating cold-formed steel with ferro-cement slab. *Thin Walled Struct.* **2016**, *102*, 18–29. [CrossRef]

14. Saggaff, A.; Tahir, M.M.; Azimi, M.; Alhajri, T.M. Structural aspects of cold-formed steel section designed as U-shape composite beam. *AIP Conf. Proc.* **2017**, *1903*, 020025.

15. Khadavi; Tahir, M.M. Prediction on flexural strength of encased composite beam with cold-formed steel section. *AIP Conf. Proc.* **2017**, *1903*, 020016.

16. Salih, M.N.A.; Md Tahir, M.; Mohammad, S.; Ahmad, Y.; Sulaiman, A.; Shek, P.N.; Abraham, A.; Firdaus, M.; Aminuddin, K.M. Khadavi Experimental study on flexural behaviour of partially encased cold-formed steel composite beams using rebar as shear connector. *IOP Conf. Ser. Mater. Sci. Eng.* **2019**, *513*, 012038. [CrossRef]

17. de Seixas Leal, L.A.A.; de Miranda Batista, E. Composite floor system with CFS trussed beams, concrete slab and innovative shear connectors. *Rev. Esc. Minas* **2020**, *73*, 23–31.

18. de Seixas Leal, L.A.A.; de Miranda Batista, E. Composite floor system with cold-formed trussed beams and prefabricated concrete slab. *Steel Constr.* **2020**, *13*, 12–21. [CrossRef]

19. Tian, L.M.; Kou, Y.F.; Hao, J.P.; Zhao, L.W. Flexural performance of a lightweight composite floor comprising cold-formed steel trusses and a composite mortar slab. *Thin-Walled Struct.* **2019**, *144*, 106361. [CrossRef]

20. Liu, J.; Zhao, Y.; Chen, Y.F.; Xu, S.; Yang, Y. Flexural behavior of rebar truss stiffened cold-formed U-shaped steel-concrete composite beams. *J. Constr. Steel Res.* **2018**, *150*, 175–185. [CrossRef]

21. Jakovljević, I.; Spremić, M.; Marković, Z. Demountable composite steel-concrete floors: A state-of-the-art review. *J. Croat. Assoc. Civ. Eng.* **2021**, *73*, 249–263. [CrossRef]

22. Lam, D.; Dai, X.; Ashour, A.; Rehman, N. Recent research on composite beams with demountable shear connectors. *Steel Constr.* **2017**, *10*, 125–134. [CrossRef]

23. Pavlović, M.; Marković, Z.; Veljković, M.; Buđevac, D. Bolted shear connectors vs. headed studs behaviour in push-out tests. *J. Constr. Steel Res.* **2013**, *88*, 134–149. [CrossRef]

24. *EN 1994-1-1: Eurocode 4: Design of Composite Steel and Concrete Structures—Part 1-1: General Rules and Rules for Buildings*; CEN: Brussels, Belgium, 2004.

25. Kozma, A.; Odenbreit, C.; Braun, M.V.; Veljkovic, M.; Nijgh, M.P. Push-out tests on demountable shear connectors of steel-concrete composite structures. *Structures* **2019**, *21*, 45–54. [CrossRef]

26. Irwan, J.M.; Hanizah, A.H.; Azmi, I. Test of shear transfer enhancement in symmetric cold-formed steel–concrete composite beams. *J. Constr. Steel Res.* **2009**, *65*, 2087–2098. [CrossRef]

27. Wehbe, N.; Bahmani, P.; Wehbe, A. Behavior of Concrete/Cold Formed Steel Composite Beams: Experimental Development of a Novel Structural System. *Int. J. Concr. Struct. Mater.* **2013**, *7*, 51–59. [CrossRef]

28. Bamaga, S.O.; Tahir, M.M.; Tan, C.S.; Shek, P.N.; Aghlara, R. Push-out tests on three innovative shear connectors for composite cold-formed steel concrete beams. *Constr. Build. Mater.* **2019**, *223*, 288–298. [CrossRef]

29. Kopp, M.; Wolters, K.; Claßen, M.; Hegger, J.; Gündel, M.; Gallwoszus, J.; Heinemeyer, S.; Feldmann, M. Composite dowels as shear connectors for composite beams—Background to the design concept for static loading. *J. Constr. Steel Res.* **2018**, *147*, 488–503. [CrossRef]

30. Majdi, Y.; Hsu, C.T.T.; Punurai, S. Local bond-slip behavior between cold-formed metal and concrete. *Eng. Struct.* **2014**, *69*, 271–284. [CrossRef]

31. Majdi, Y.; Hsu, C.T.T.; Zarei, M. Finite element analysis of new composite floors having cold-formed steel and concrete slab. *Eng. Struct.* **2014**, *77*, 65–83. [CrossRef]

32. Brambilla, G.; Lavagna, M.; Vasdravellis, G.; Castiglioni, C.A. Environmental benefits arising from demountable steel-concrete composite floor systems in buildings. *Resour. Conserv. Recycl.* **2019**, *141*, 133–142. [CrossRef]

33. Wang, J.-Y.; Guo, J.-Y.; Jia, L.-J.; Chen, S.-M.; Dong, Y. Push-out tests of demountable headed stud shear connectors in steel-UHPC composite structures. *Compos. Struct.* **2017**, *170*, 69–79. [CrossRef]

34. Yang, F.; Liu, Y.; Jiang, Z.; Xin, H. Shear performance of a novel demountable steel-concrete bolted connector under static push-out tests. *Eng. Struct.* **2018**, *160*, 133–146. [CrossRef]

35. Girão Coelho, A.M.; Lawson, R.M.; Aggelopoulos, E.S. Optimum use of composite structures for demountable construction. *Structures* **2019**, *20*, 116–133. [CrossRef]

36. Landolfo, R.; Mammana, O.; Portioli, F.; Di Lorenzo, G.; Guerrieri, M.R. Laser welded built-up cold-formed steel beams: Experimental investigations. *Thin-Walled Struct.* **2008**, *46*, 781–791. [CrossRef]

37. Briskham, P.; Blundell, N.; Han, L.; Hewitt, R.; Young, K.; Boomer, D. Comparison of Self-Pierce Riveting, Resistance Spot Welding and Spot Friction Joining for Aluminium Automotive Sheet. In Proceedings of the SAE 2006 World Congress & Exhibition, Detroit, MI, USA, 3–6 April 2006; p. 15.

38. Guenfoud, N.; Tremblay, R.; Rogers, C.A. Arc-Spot Welds for Multi-Overlap Roof Deck Panels. In Proceedings of the Twentieth International Specialty Conference on Cold-Formed Steel Structures, St. Louis, MO, USA, 3–4 November 2010; pp. 535–549.

39. Ungureanu, V.; Dubina, D. Influence of Corrugation Depth on Lateral Stability of Cold-Formed Steel Beams of Corrugated Webs. *Acta Mech. Autom.* **2016**, *10*, 104–111. [CrossRef]

40. Dubina, D.; Ungureanu, V.; Gîlia, L. Experimental investigations of cold-formed steel beams of corrugated web and built-up section for flanges. *Thin-Walled Struct.* **2015**, *90*, 159–170. [CrossRef]

41. Hamada, M.; Nakayama, K.; Kakihara, M.; Saloh, K.; Ohtake, F. Development of welded I-beam with corrugated web. *Sumitomo Search* **1984**, *29*, 75–90.

42. Zeman & Co GmbH. *Corrugated Web Beam, Technical Document*; Zeman & Co GmbH: Wien, Austria, 1993.

43. *EN1993-1-1 Eurocode 3: Design of Steel Structures—Part 1-1: General Rules and Rules for Buildings*; Including; CEN: Brussels, Belgium, 2005.

44. *EN1993-1-3 Eurocode 3: Design of Steel Structures. Part 1-3: General Rules. Supplementary Rules for Cold-Formed Thin Gauge Members and Sheeting*; CEN: Brussels, Belgium, 2006.

45. *EN1993-1-5 Eurocode 3: Design of Steel Structures—Part 1-5: Plated Structural Elements*; CEN: Brussels, Belgium, 2006.

46. Elgaaly, M.; Dagher, H. Beams and Girders with Corrugated Webs. In Proceedings of the SSRC Annual Technical Session, St. Louis, MO, USA, 10–11 April 1990; pp. 37–53.

47. Lindner, J. Lateral-torsional buckling of beams with trapezoidally corrugated webs. In Proceedings of the 4th International Colloquium on Stability of Steel Structures, Budapest, Hungary, 25–27 April 1990; pp. 79–82.

48. Moon, J.; Yi, J.-W.; Choi, B.H.; Lee, H.-E. Lateral–torsional buckling of I-girder with corrugated webs under uniform bending. *Thin-Walled Struct.* **2009**, *47*, 21–30. [CrossRef]

49. Pasternak, H.; Robra, J.; Kubieniec, G. New proposals for EN 1993-1-5, Annex D: Plate girders with corrugated webs. In Proceedings of the Codes in Structural Engineering, Joint IABSE-fib Conference, Dubrovnik, Croatia, 3–5 May 2010.

50. Elgaaly, M.; Seshadri, A.; Hamilton, R.W. Bending Strength of Steel Beams with Corrugated Webs. *J. Struct. Eng.* **1997**, *123*, 772–782. [CrossRef]

51. Dubina, D.; Ungureanu, V.; Gîlia, L. Cold-formed steel beams with corrugated web and discrete web-to-flange fasteners. *Steel Constr.* **2013**, *6*, 74–81. [CrossRef]

52. Dubina, D.; Ungureanu, V.; Dogariu, A. Lightweight Footbridges of Cold Formed Steel Corrugated Web Beams: Technical Solution and Evaluation. In Proceedings of the The 8th International Symposium on Steel Bridges: Innovation & New Challenges 2015 (SBIC-2015), Istanbul, Turkey, 14–16 September 2015.

53. Ungureanu, V.; Both, I.; Burca, M.; Tunea, D.; Grosan, M.; Neagu, C.; Dubina, D. Welding technologies for built-up cold-formed steel beams: Experimental investigations. In Proceedings of the Ninth International Conference on Advances in Steel Structures (ICASS'2018), Hong Kong, China, 5–7 December 2018.

54. Ungureanu, V.; Both, I.; Burca, M.; Grosan, M.; Neagu, C.D. Built-up cold-formed steel beams using resistance spot welding: Experimental investigations. In Proceedings of the Eighth International Conference on Thin-Walled Structures (ICTWS 2018), Lisbon, Portugal, 24–27 July 2018.

55. Benzar, Ş.; Ungureanu, V.; Dubină, D.; Burcă, M. Built-Up Cold-Formed Steel Beams with Corrugated Webs Connected with Spot Welding. *Adv. Mater. Res.* **2015**, *1111*, 157–162. [CrossRef]

56. Ungureanu, V.; Lukačević, I.; Both, I.; Burca, M.; Dubina, D. Built-Up Cold-Formed Steel Beams with Corrugated Webs Connected by Spot Welding—Numerical Investigations. In Proceedings of the International Colloquia on Stability and Ductility of Steel Structures (SDSS 2019), Prague, Czech Republic, 11–13 September 2019.

57. Lukačević, I.; Ungureanu, V.; Valčić, A.; Pedišić, M. Bending resistance of cold-formed back-to-back built-up steel sections. In Proceedings of the 18th International Symposium of MASE, Ohrid, North Macedonia, 2–5 October 2019; Cvetkovska, M., Ed.; MASE: Skopje, North Macedonia, 2019; pp. 1080–1089.

58. Ungureanu, V.; Lukačević, I.; Both, I.; Burca, M. Numerical investigation of built-up cold-formed steel beams connected by spot welding. In Proceedings of the Evolving Metropolis, 2019 IABSE Congress New York City, New York, NY, USA, 4–6 September 2019.

59. Pauliuk, S.; Kondo, Y.; Nakamura, S.; Nakajima, K. Regional distribution and losses of end-of-life steel throughout multiple product life cycles—Insights from the global multiregional MaTrace model. *Resour. Conserv. Recycl.* **2017**, *116*, 84–93. [CrossRef]
60. Skejić, D.; Lukačević, I.; Ćurković, I.; Čudina, I. Application of steel in refurbishment of earthquake-prone buildings. *J. Croat. Assoc. Civ. Eng.* **2020**, *72*, 955–965. [CrossRef]
61. Dassault Systèmes Simulia Corp. *ABAQUS, User's Manual, Version 6.12*; Dassault Systèmes Simulia Corp.: Providence, RI, USA, 2012.
62. Qureshi, J.; Lam, D.; Ye, J. Effect of shear connector spacing and layout on the shear connector capacity in composite beams. *J Constr Steel Res* **2011**, *67*, 706–719. [CrossRef]
63. Amadio, C.; Bedon, C.; Fasan, M.; Pecce, M.R. Refined numerical modelling for the structural assessment of steel-concrete composite beam-to-column joints under seismic loads. *Eng. Struct.* **2017**, *138*, 394–409. [CrossRef]
64. Qureshi, J.; Lam, D. Behaviour of Headed Shear Stud in Composite Beams with Profiled Metal Decking. *Adv. Struct. Eng.* **2012**, *15*, 1547–1558. [CrossRef]
65. *EN 1992-1-1: Eurocode 2: Design of Concrete Structures—Part 1-1: General Rules and Rules for Buildings*; CEN: Brussels, Belgium, 2004.
66. Cornelissen, H.; Hordijk, D.; Heron, H.R.-; 1986, U. Experimental determination of crack softening characteristics of normalweight and lightweight. *HERON* **1986**, *31*, 45–56.
67. Kordina, K.R.; Mancini, G.; Schäfer, K.; Schießl, A.; Zilch, K. *Structural Concrete Textbook on Behaviour, Design and Performance*, 2nd ed.; Fib Bulletin No. 54; Balázs, G.L., Ed.; The International Federation for Structural Concrete: Washington, DC, USA, 2010; Volume 4, ISBN 9782883940949.
68. Tao, Y.; Chen, J.-F. Closure to Concrete Damage Plasticity Model for Modeling FRP-to-Concrete Bond Behavior. *J. Compos. Constr.* **2015**, *19*, 07015003. [CrossRef]

Article

An Efficient Reliability-Based Approach for Evaluating Safe Scaled Distance of Steel Columns under Dynamic Blast Loads

Mohammad Momeni [1,2], Chiara Bedon [3,*], Mohammad Ali Hadianfard [1] and Abdolhossein Baghlani [1]

[1] Department of Civil and Environmental Engineering, Shiraz University of Technology, Shiraz 7155713876, Iran; m.momeni@sutech.ac.ir (M.M.); hadianfard@sutech.ac.ir (M.A.H.); baghlani@sutech.ac.ir (A.B.)

[2] Department of Civil Engineering, Shahrekord Branch, Islamic Azad University, Shahrekord 8813733395, Iran

[3] Department of Engineering and Architecture, University of Trieste, 34127 Trieste, Italy

* Correspondence: chiara.bedon@dia.units.it

Abstract: Damage to building load-bearing members (especially columns) under explosions and impact are critical issues for structures, given that they may cause a progressive collapse and remarkably increase the number of potential victims. One of the best ways to deal with this issue is to provide values of safe protective distance (SPD) for the structural members to verify, so that the amount of damage (probability of exceedance low damage) cannot exceed a specified target. Such an approach takes the form of the so-called safe scaled distance (SSD), which can be calculated for general structural members but requires dedicated and expensive studies. This paper presents an improved calculation method, based on structural reliability analysis, to evaluate the minimum SSD for steel columns under dynamic blast loads. An explicit finite element (FE) approach is used with the Monte Carlo simulation (MCS) method to obtain the SSD, as a result of damage probability. The uncertainties associated with blast and material properties are considered using statistical distributions. A parametric study is thus carried out to obtain curves of probability of low damage for a range of H-shaped steel columns with different size and boundaries. Finally, SSD values are detected and used as an extensive databank to propose a practical empirical formulation for evaluating the SSD of blast loaded steel columns with good level of accuracy and high calculation efficiency.

Keywords: safe protective distance; safe scaled distance; steel beam-column; dynamic blast load; reliability analysis; Monte Carlo simulation

Citation: Momeni, M.; Bedon, C.; Hadianfard, M.A.; Baghlani, A. An Efficient Reliability-Based Approach for Evaluating Safe Scaled Distance of Steel Columns under Dynamic Blast Loads. *Buildings* **2021**, *11*, 606. https://doi.org/10.3390/buildings11120606

Academic Editor: Paulo Santos

Received: 21 October 2021
Accepted: 29 November 2021
Published: 2 December 2021

Publisher's Note: MDPI stays neutral with regard to jurisdictional claims in published maps and institutional affiliations.

1. Introduction

Crowded buildings such as schools, shopping venues, stadiums, transportation infrastructure and public locations are well-known attractive targets for terrorist attacks. The disruption of such places has irreversible consequences, including severe casualties and fatalities and negative impact on society [1]. There is a need to identify areas that may be potentially at risk and to take preventive measures to improve their safety and security. In this regard, securing the perimeter of structures or buildings using landscaping or barrier methods is one of the valid risk reduction options recommended in the literature for protecting buildings against terrorist attacks, including vehicle-borne improvised explosive devices [2]. These secure barriers must be installed at the minimum required stand-off distance from a structure, in order to minimize the damage probability of primary structural elements and consequently the risk of progressive collapse. In order to design a blast-resistant building, the design engineer first has to determine blast loads on the building and its structural components. To determine the characteristics and intensity of blast loading, the parameters for explosive charge weight (W) and stand-off distance (R) must be necessarily known. There are several formulas and graphs that can be used to determine blast load parameters, as a function of the scaled distance parameter (Z). The Z parameter, also known as $R/W^{1/3}$ [3], indicates that two charges with similar geometry, ambient conditions, explosive composition, but different size (weight) will produce self-similar blast waves as far as their distances $R = Z \times W^{1/3}$ are identical. A much more complete discussion on features and applicability of the scaling law is given in [4]. In [5], the scaled distance parameter is used to assess the safety and resistance of structures under air blast loads. As an example, for un-strengthened buildings, an SSD of 4.46 m/kg$^{1/3}$ is specified from suffering a damage of "approaching to destruction" [5,6]. The SSD parameter, in this context, represents a guide to determine the explosive weight that can be used at a given distance, without exceeding the safe limit states of the structure (allowable support rotation values or damage index, for low damage). It should be noted that the SSD parameter is derived so that probability of failure is lower than an acceptance criterion. The probability of failure, as explained later in Section 3, is a function of capacity and demand called state function. In some cases, the state function can be expressed mathematically, but in most cases it does not have an explicit mathematical closed-form and must be defined by other methods such as FE analysis. When the state function is defined in mathematical form, it is possible to calculate the SSD parameter directly, otherwise an iteration-based method should be used to meet the acceptance criterion (Section 4). The SSD values presented in standards and regulations are usually obtained from blast tests on simple structural models and the effects of structural configuration or material properties are usually disregarded. As such, guidelines can be used for a quick safety assessment of structures, but do not provide clear damage scenarios [6]. Some studies have been performed to also investigate the SPD and SSD of structural elements under blast loads.

The blast performance assessment of structural systems is one of the critical issues for research. Accordingly, the need of empirical but accurate tolls in support of design optimization is an ongoing challenge. Among others, Byfield and Paramasivam [7] developed an iterative method to establish the minimum SSD of Reinforced Concrete (RC) columns for a given charge weight, column geometry and material. The iterative process must be repeated until the strength of the column is equal to the dynamic force in it. Thomas et al. [8] implemented MCS method for the reliability analysis of circular RC columns subjected to sequential vehicular impact and blast. Given that the stand-off distance has marked effects on reliability predictions, minimum SPD values have been proposed for selected configurations. Hadianfard and Malekpour [9] evaluated safe explosion distances of a steel column with IPBv220 and length of 3.6 m under different blast scenarios by utilizing the Single Degree of Freedom (SDOF) and FE methods via MCS method. Zhai et al. [10] investigated the blast effects on reticulated domes, proposing a method to determine the SSD based on the intersections of W-R charts and Pressure–Impulse (P–I) curves. Wu and Hao [6,11] numerically derived the SSD for masonry infilled RC frame structures. The presented SSD values for different damage levels (RC frame collapse, side wall collapse, front wall collapse and excessive damage) were compared with the corresponding estimates by the US DoD [5] and ASCE [12] technical documents. A simple approximate method was proposed by Dorofeev for unconfined hydrogen explosions in three hypothetical obstructed areas with different congestion levels. Based on [13], a number of different safety distance relationships were stipulated depending upon the receptor under consideration, comprising storage distances, process building distances and public building and traffic distances.

To provide a robust background and comprehensive feedback for civil engineering applications, experimental and theoretical investigations on the effects of blast loads on steel structural members have been also reported in [14–17]. Bao and Li [18] focused on the residual axial capacity of square RC columns, while the study in [19] was dedicated to H-section steel columns. A number of numerical investigations used equivalent SDOF systems and FE for primary members [16,20–32]. Besides, the uncertainty of input variables for blast load parameters and material properties (but also geometrical parameters and FE modelling errors [33]) can severely affect the predicted structural response. As such, probabilistic methods are preferred to support a more holistic risk-based approach [34–37]. Several studies have been focused on the reliability analysis of selected structures, such as RC buildings [38,39], steel structures [40], RC slabs [36,41], RC columns [35,42–44], RC wall panels [33,45], RC beams [46], composite walls [47], masonry walls [48], profiled wall structures [49], clamped aluminum plate [50] and steel columns [51,52], by considering the uncertainties of input variables related to material properties and blast load parameters. Stewart et al. explored the reliability analysis of structures under blast [33,35,43,44] and supported the definition of a general framework for quantitative probabilistic risk assessment of structures subjected to blast [53–56]. Most of those studies have been developed based on MCS method along with SDOF, Multi Degree of Freedom (MDOF) and full 3D FE models. In [52], a methodology based on structural reliability analysis using MCS and explicit FE modelling (shell element formulation) was proposed for determining the damage probability of H-shape steel columns (IPBv 200 section) under various blast scenarios. It was shown that the time required in a probabilistic analysis for iterations of 1000 and 300 can be expected to be about 100 and 30 h. Such a run time may be acceptable for a single reliability analysis, but it is not suitable for SSD calculations that require a trial-and-error process with several reliability analyses. In this regard, a parametric analysis was performed in [57] to capture the effect of several FE modelling techniques (based on solid, shell or beam elements), blast intensity (medium and high levels) and supports (pinned or fixed ends), on damage evaluation assessment. It was proved that the beam formulations can offer good results for the calculation of the residual axial capacity of blast loaded steel columns, with high computational efficiency.

As mentioned above, recommending the minimum SSD is of high practical interest, especially for the design of structures in congested urban areas. Once SSD is known, the corresponding SPD can be easily calculated as a function of SSD and W. Although the blast dynamic behavior of structures has been largely investigated, the SSD of axially preloaded columns has been rarely considered and, to date, no comprehensive studies have been conducted. In this regard, this paper represents an effort toward the definition of a reliable and efficient methodology based on reliability analysis along with explicit FE approach (using beam element formulation) to determine the SPD and SSD for blast loaded steel columns. The proposed strategy, as shown, can be extended to different structural members (or assemblies) under the effect of a given explosion. In more detail, reliability analyses are carried out to obtain the curves of probability of low damage for a set of H-shape steel columns with different cross sections (IPB180 to IPB500), lengths (2.8, 3.2, 3.6 and 4.0 m) and boundary conditions (pinned or fixed ends), under different explosive charge weights. Afterward, using the obtained curves of probability of low damage, the SSD are extracted for the selected configurations, to present a correlation between SSD and several input parameters (such as the explosive charge weight and the initial axial capacity of a given column) and derive some useful empirical formulas for practical design. An illustrative calculation example is finally discussed, in order to highlight the applicability of the proposed equations for calculating SSD and SPD of steel columns under blast loads.

2. FE Numerical Analysis and Failure Assessment

2.1. Steel Columns

A set of explicit FE models is developed using LS-DYNA software [58], to examine the blast loaded behavior of steel columns with different boundary conditions. As a reference, the limit pinned and fixed ends are considered. To characterize steel, MAT_PLASTIC_KINE MATIC material model is used. This constitutive model can adequately describe the isotropic and kinematic hardening plasticity, with the inclusion of strain rate effects based on the Cooper–Simonds relationship, that is [59–61]:

$$\text{DIF} = 1 + \left(\frac{\dot{\varepsilon}}{C}\right)^{\frac{1}{P}} \tag{1}$$

where $\dot{\varepsilon}$ is the material strain rate, DIF is the dynamic increase factor and C and P are constant coefficients that were set to 40.4 and 5 for mild steel [61,62]. The stress–strain curve provided by MAT_PLASTIC_KINEMATIC material model is shown in Figure 1, where L_0 and L_1 are undeformed and deformed lengths of uniaxial tension specimen, respectively. Furthermore, E_s, F_y, E_t, σ_t and ε_t are the modulus of elasticity, yield stress, the slope of the bilinear stress strain curve in strain hardening region, true stress and true strain, respectively. Furthermore, kinematic, isotropic, or a combination of kinematic and isotropic hardening may be specified by varying β' between 0 and 1 as shown in Figure 1 [58].

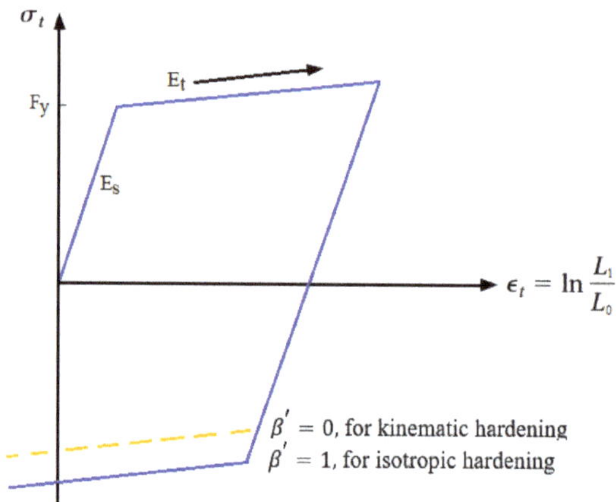

Figure 1. Stress–strain curve of MAT_PLASTIC_KINEMATIC model used for steel material [58].

It should be noted that strain hardening was not considered in this study and the value of E_t was set to zero ($E_t = 0$). Such a choice was derived from earlier preliminary sensitivity studies where the results showed that strain hardening has no significant effect on residual axial capacity of a steel column [63]. For sake of conciseness, the aforementioned results are not included in the discussion herein reported.

The Hughes–Liu beam element formulation is used for the FE modelling of the selected steel columns [57,64]. There is an integration refinement factor in Hughes–Liu beam element formulation to determine integration points throughout a cross section. The number of integration points can vary depending on the desired accuracy required. A greater number of integration points can also more accurately represent the structural response. In this study k was set to 5 ($k = 5$) following carried out sensitivity analyses that are not presented in the discussion herein for sake of brevity. A schematic drawing of the typical FE modelling of steel columns (as columns of a building (not columns of a boundary wall)) with H-shape cross section and $k = 2$ is shown in Figure 2.

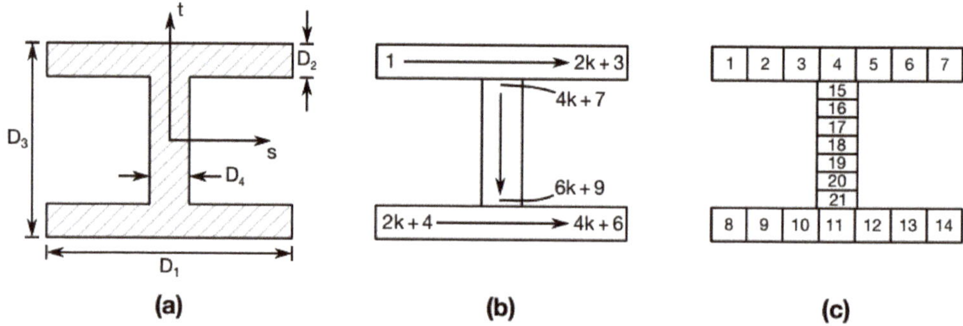

Figure 2. FE numerical modelling of H-section steel column with Hughes–Liu beam elements: (**a**) cross section geometry; (**b**) integration point numbering; (**c**) section example for $k = 2$.

Although the use of beam elements is notoriously deficient for simulating the effects of local buckling phenomena or shear damage mechanisms induced by blast loads, this choice can significantly improve the calculation efficiency of simulations (compared to shell and solid elements) when the global behavior prediction prevails on local behaviors. This advantage is further appreciated when the time reduction is a relevant issue, like in the case of reliability analysis.

2.2. Nature of Blast Loading

The magnitude of blast waves due to terrorist attacks can be generally classified in the number of explosive packs (portable by humans) and different types of used vehicles (such as automobiles, vans and trucks), based on the amount of W and the distance of detonation R [52,65].

In this research study, three types of surface burst explosive packs with 55, 275 and 555 kg of trinitrotoluene (TNT) are considered and can be reasonably assumed to be carried by an automobile, a van or a truck, respectively. The stand-off distance is also considered to modify in each blast scenario, in order to find the minimum required SPD, using the reliability analysis. It should be noted that human-made explosions generally occur on the vicinity of the ground surface. Due to this closeness, an immediate interaction initiates between the blast wave and the ground, which forms the hemispherical surface burst [66]. The incident waves are quickly reflected from the ground surface and lead to higher pressure values [67]. Based on [38,52,66,68–70], the parameters of the hemispherical surface burst (reflected pressure (P_r) and the positive time duration (t_d)) can be defined using the conventional relationships for free air burst, by replacing W with the effective charge weight ($W_{eff} = 1.8 \times W$). Finally, the blast load time history is defined based on the calculated surface burst parameters and is considered as a simplified equivalent triangular pulse, for all the FE modelled configurations. For the sake of conciseness and to avoid lengthening, no major details about definition of time history of blast loading are reported in this paper. More details can be found in [52,57].

2.3. Damage Evaluation Assessment Based on Damage Index Criterion

To numerically assess the expected damage of blast loaded steel columns after explosion, the damage index (DI) based on residual axial carrying capacity is taken into account in this study. According to Shi et al. [25], this index is given by:

$$DI = 1 - \frac{P_{residual}}{P_{initial}} \tag{2}$$

where $P_{residual}$ is the post-blast residual axial capacity of the damaged column and $P_{initial}$ is the maximum axial load-carrying capacity of the undamaged column. The degrees of damage are thus categorized into four levels [25], namely corresponding to:

(a) DI = 0–0.2 low damage;
(b) DI = 0.2–0.5 medium damage;
(c) DI = 0.5–0.8 high damage;
(d) DI = 0.8–1.0 collapse.

It should be noted that for vertical load bearing components belonging to high-class buildings that are sensitive to lateral deformations and must be designed for maximum lateral ductility ratio 1, no relevant damage is allowed and consequently DI must be selected in such a way that this limitation is satisfied. From a computational point of view, several calculation steps must be generally carried out to find the expected $P_{residual}$ value. The sequence of required steps, however, is not reported in this paper for the sake of conciseness. Additional details can be found in [52,57]. In blast-resistant design of structures, it is often stated that the damage caused in a structure due to blast loads would be reduced if the structure is well designed against seismic loadings. This is not true in all cases and it should not be assumed that a structure designed to withstand seismic loads is sufficient to resist the prescribed blast loading or prevent subsequent progressive collapse. Despite the similarities between seismic and blast loadings, the global response of buildings subjected to blast loading is not usually critical. In the other words, for a structure that is affected by an explosion, only its critical members (i.e., closer to detonation) are individually assessed and designed by means of different methods (SDOF and FE models) and damage criteria (support rotation and damage index), while for building structures under the effects of earthquake loadings, the global deformations (inter-story drifts) must be evaluated based on the desired performance level (life safety) as the most important response parameter [71,72]. In the design of structures under seismic and blast loading, the desirable features of design—that is, the provision for ductility in member response and increasing the ability to redistribute extreme loads to lesser-loaded elements—must be satisfied.

3. Random Variables and Reliability Analysis Using MCS

3.1. Random Variables

The variability of blast loading parameters is one of the key variables of the problem explored herein. A number of documents [34,36,73] have reported constant coefficient of variation (COV) values for the variability of blast loads at various scaled distances. On the other hand, the observed statistics obtained from blast tests and empirical formulations confirm the basic variability. To overcome this major limit, additional studies were performed in [35,37,73]. Among others, the proposals by Hao et al. [35] and Netherton and Stewart [37] are of general application and thus often used in the reliability analysis of blast loaded structures. Although the cited strategies are different, the shared feature is the blast load variability, which is expressed in terms of Z. Furthermore, the same strategies are validated for a wide range of scaled distance values ($0.24 \, \mathrm{m/kg^{1/3}} \leq Z \leq 40 \, \mathrm{m/kg^{1/3}}$ in [35] and $0.59 \, \mathrm{m/kg^{1/3}} \leq Z \leq 40 \, \mathrm{m/kg^{1/3}}$ in [37]). As a final result of the formulations provided in [35] and [37], for a blast scenario (with specified charge weight and stand-off distance) the mean, standard deviation (σ) and COV of wave parameters can be estimated as a function of Z. In this paper, P_r and t_d are selected as random variables for blast loading. It is also assumed that the uncertainties are defined based on [35], that is:

$$\log P_{r(mean)} = \begin{aligned}&3.651 - 3.018 \times \log Z + 0.1967 \times (\log Z)^2 + 0.8873 \times (\log Z)^3 \\ &-0.3795 \times (\log Z)^4\end{aligned} \tag{3}$$

$$\log \sigma_{P_r} = \begin{aligned}&3.03 - 3.533 \times \log Z + 0.4534 \times (\log Z)^2 + 0.3248 \times (\log Z)^3 \\ &+0.07896 \times (\log Z)^4\end{aligned} \tag{4}$$

$$\log COV_{P_r} = \begin{aligned}&-0.6239 - 0.5726 \times \log Z + 0.3203 \times (\log Z)^2 - 0.3538 \times (\log Z)^3 \\ &+0.2973 \times (\log Z)^4\end{aligned} \tag{5}$$

$$\log\left(\frac{t_{d(mean)}}{w^{1/3}}\right) = \begin{aligned}&-0.00307 + 1.2186 \times \log Z - 0.5207 \times (\log Z)^2 - 0.2835 \times (\log Z)^3 \\ &+0.2132 \times (\log Z)^4\end{aligned} \tag{6}$$

$$\log \sigma_{t_d/w^{1/3}} = \begin{aligned}&-0.8433 + 1.0982 \times \log Z - 0.8127 \times (\log Z)^2 + 0.4214 \times (\log Z)^3 \\ &-0.1046 \times (\log Z)^4\end{aligned} \tag{7}$$

$$\log COV_{(t_d/w^{1/3})} = \begin{aligned}&-0.8411 - 0.1186 \times \log Z - 0.2868 \times (\log Z)^2 + 0.6955 \times (\log Z)^3 \\ &-0.3141 \times (\log Z)^4\end{aligned} \tag{8}$$

Equations (3)–(8), in more detail, represent the statistical characteristics (mean, σ and COV) of the P_r and t_d variables, as a function of Z, in the range of $0.24 \, \mathrm{m/kg^{1/3}} \leq Z \leq 40 \, \mathrm{m/kg^{1/3}}$. It should be noted that the log in these equations is the logarithm to the base 10. As reported in [35], the proposed formulas are valid only for an open field explosion and large enough flat reflection surface. For a complex explosion scenario, such as an explosion in a complex city environment, more significant variations are expected because of blast wave interactions with surrounding structures.

It should be noted that t_d is the positive time duration of an idealized triangular blast loading history with sufficient accuracy instead an exponentially decayed loading history of a real explosion. The assumption of using triangular pressure-time history for blast loading originates from past research studies, such as [35,37,38,42,43,55,74–79]. There, the variation of the waveform coefficient for the positive pressure phase has been generally disregarded for the reliability analysis, due to lack of information. This issue depends on blast load databanks that have been used to propose analytical formulas to calculate the variation of blast load parameters as functions of scaled-distance. Among others, the parameter corresponding to the waveform coefficient has been considered probabilistic in [48,73], but the intended scaled distance was set between 1.62 and 2.78 m/kg$^{1/3}$, and an explicit relation was not presented for the calculation of statistical properties (i.e., mean and standard deviation) of the waveform coefficient based on scaled distance. Although the linear assumption of blast load imposes some unwanted approximation in the problem [35], such an assumption might cause an error up to 10% for the final results. In this paper, following former dedicated research [35,37,38,42,43,55,74–79], a linear function is thus used to define the input blast.

Normal probability density function (PDF) is used for all input random parameters including loading parameters (P_r and t_d) and steel material properties (F_y and E_s), see Table 1 [35,42,80–82].

Table 1. Statistical properties of input random variables.

Random Variable	Mean	σ	COV	PDF
P_r	Equation (3)	Equation (4)	Equation (5)	Normal
t_d	Equation (6)	Equation (7)	Equation (8)	Normal
F_y	240 × 1.15 MPa	16.56 MPa	0.06	Normal
E_s	210 GPa	8.40 GPa	0.04	Normal

3.2. MCS Method

The MCS method is a well-known technique for estimating statistical properties of structural systems under stochastic uncertainties of input parameters [83,84] and is used in this paper to carry out the reliability analyses. MCS is one of the simplest and relatively most accurate methods which provides a feasible way to determine the reliability index, where the limit state function is more complicated. Most of the literature studies on the reliability of structures under blast loading have been performed using MCS. The probability of failure based on MCS equals to $P_f = N_f/N$, where N is the number of total simulations and N_f is the number of trials for which limit state function, $g(\mathbf{X}) = r{-}q$, falls in the failure region or has negative value. In the definition of the state function, $\mathbf{X}$ is the vector of input random variables, r is the capacity or resistance, q is the demand or loading. The probability of failure can also be written as follows:

$$P_f = P[g(\mathbf{X}) \leq 0] = \int_{g(\mathbf{X}) \leq 0} f_{\mathbf{X}}(\mathbf{X})\, d\mathbf{X} = \frac{\sum_{i=1}^{N} I_F(\mathbf{X}_i)}{N} \tag{9}$$

where $f_{\mathbf{X}}(\mathbf{X})$ is the joint probability density function and I_F is the failure indicator which equals 1 if $g(\mathbf{X}) \leq 0$ and 0 if $g(\mathbf{X}) > 0$.

The accuracy and precision of MCS in damage estimation directly depends on the N value. The higher the N value, the more precise the MCS. On the other hand, by increasing the number of simulations, the computational effort is also increased, which is the main disadvantage of the MCS method. In this regard, there are many procedures in the literature to find the minimum number of iterations required for MCS for a certain level of accuracy. The equation proposed by Broding et al. [85] is taken into account in this paper:

$$N > \frac{-\ln(1 - C_L)}{P_f} \tag{10}$$

where N is the minimum number of required random samples, P_f is the probability of failure and C_L is the confidence level. In this paper, the value N = 300 is taken into account for reliability analyses, which corresponds to 95% confidence (C_L = 0.95) and 0.99 reliability (P_f = 0.01).

4. Methodology of Calculating SSD Using Reliability

4.1. SSD Definition

SPD is defined as the minimum required stand-off distance where the probability of low damage based on Equation (2) is at least 95% [86–88] or the damage probability is lower than an acceptance criterion 5%. Figure 3 shows schematically an instance of probability of low damage diagram for a given charge weight, based on stand-off distance, that can be obtained from the results of reliability analysis for any blast loaded member. The concept in Figure 3 is shown for the specific case of probability of low damage 95%. The philosophy is that the structure object of analysis is examined for different blast scenarios (under constant charge weight and a variable stand-off value) and the probability of low damage in each case is calculated and drawn in Figure 3. In the other words, each point in Figure 3 corresponds to probability of low damage for the selected configuration under a blast scenario.

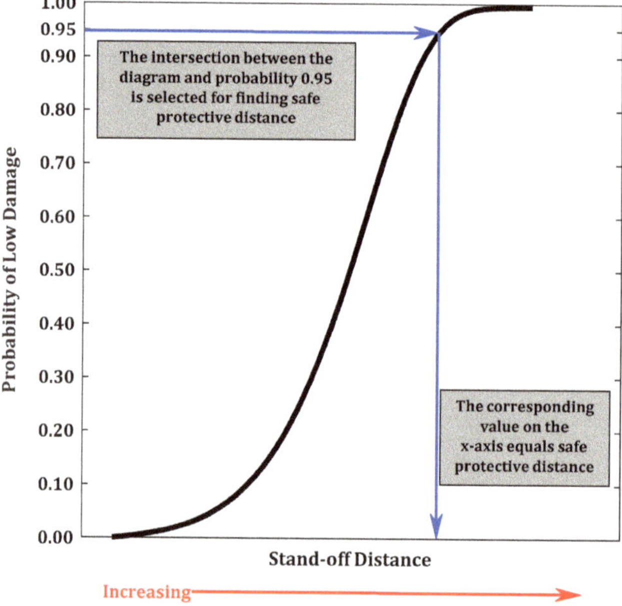

Figure 3. The concept of calculating SPD.

The desired output is obtained by a complete probabilistic analysis using MCS with 300 different simulations. The final result agrees with Figure 3 and is the basis for the SPD determination. This is in fact calculated as the stand-off distance corresponding to the intersection point between the diagram in Figure 3 and the probability of low damage 0.95. Once the SPD is known, the SSD can be easily determined, given that $SSD = SPD/W^{1/3}$.

4.2. Application of Reliability Analysis Based on MCS in Calculating SSD

Key steps to perform the SSD by implementing the concept of reliability analysis with MCS method and FE method (based on LS-Dyna software simulations) are summarized in this section. The full calculation process takes advantage of a set of LS-PrePost, MATLAB, LS-DYNA and C# coding for each FE model, thus importing the models into LS-DYNA and extracting and post-processing the results of interest. As shown in Figure 4, the general procedure can be basically described as follows:

1. Definition of boundary conditions, section properties and length for the examined steel column.
2. Generation of the initial LS-DYNA model (input file) for the considered steel column. At this stage, hypothetical (average) values are used for input random parameters. The same values are then updated in the following calculation steps, based on the real values of generated samples, for each random parameter. The aim of step 2 is only to create a .k file format for the column that will be object of the probabilistic analysis.
3. Selection of a blast scenario, by defining corresponding values for charge weight and stand-off distance.
4. Calculation of the mean values and standard deviations for the input random variables. In this paper, the attention is focused on blast load parameters (P_r and t_d) and material properties (F_y and E_s), according to Table 1.
5. Choice of appropriate probability density functions for the selected input random variables.
6. Generation of random variables (MATLAB code) according to the selected PDF (step 5).
7. Update of the initially generated LS-DYNA model (input file, see step 2), for the number of generated random variables (step 6), using MATLAB.
8. Analysis of all the FE models (by automatically running LS-Dyna software with C# coding) and extracting all the damage indices (MATLAB).
9. Derivation of histogram, PDF and Cumulative Distribution Function (CDF) for the calculated DI (from step 8).
10. Calculation of the probability of low damage, or P[DI $\leq$ 0.2].
11. And in conclusion, a double check must be necessarily carried out, given that:
    (a) If the probability of low damage from step 10 is approximately 95%, the selected stand-off distance (step 3) coincides with SPD and consequently the required SSD can be calculated.
    (b) Otherwise, if the probability of low damage is less or more than 95%, the selected stand-off distance (step 3) must be increased or decreased, respectively. The full algorithm must be thus repeated (from step 3), until the probability of low damage reaches 95%.

4.3. Verification of Reliability Analysis Based on MCS Using Beam Element Formulation

In order to verify the current MCS results based on beam element formulation, major outcomes from the reliability analyses presented in [52] are compared in this study. In more detail, the numerical results of two loading cases (Case 1 and Case 2) are considered, as obtained for a steel column with section type IPBv200, nominal length of 3.6 m and pinned ends. The column from [52] is made of ST37 steel, with density of 7850 kg/m^3. Yield strength, elastic modulus, Poisson's ratio and failure strain are set equal to 240 MPa, 210 GPa, 0.3 and 0.2 respectively. In Case 1, the column is subjected to W = 55 kg of TNT and R = 6 m. In Case 2, the explosive charge weight and the stand-off distance are set to 55 kg and 8 m. For both configurations, the number of simulations is set to 300. The comparison of past [52] and current numerical results is shown in Figure 5a,b, for Cases 1 and 2, respectively, in terms of CDFs for DI.

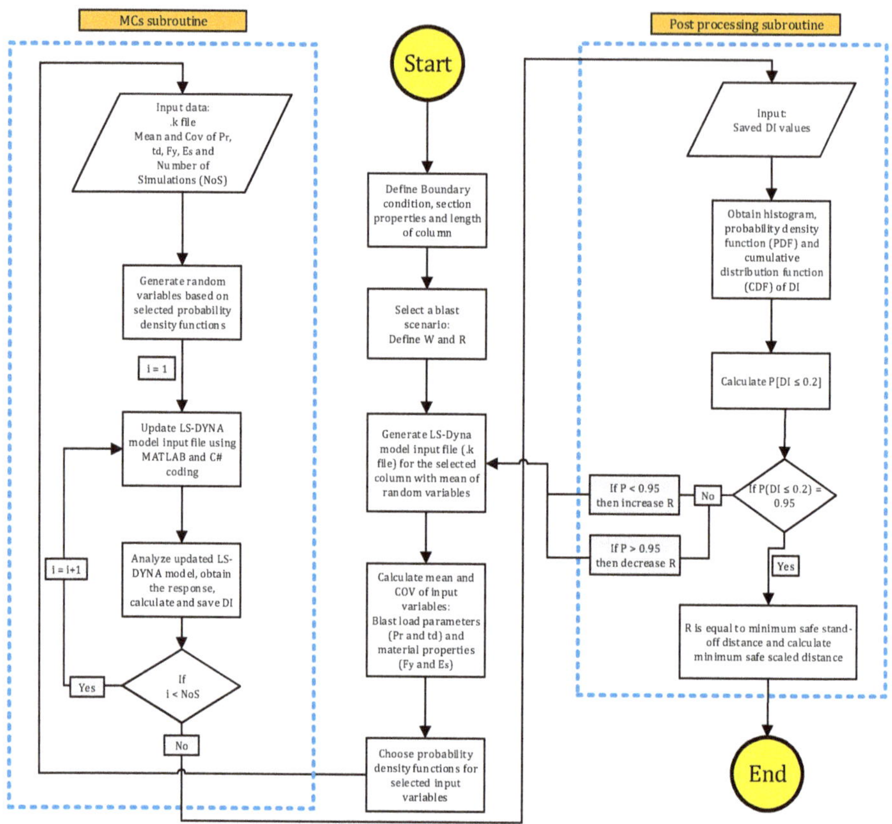

Figure 4. Procedures for SSD derivation based on structural reliability approach.

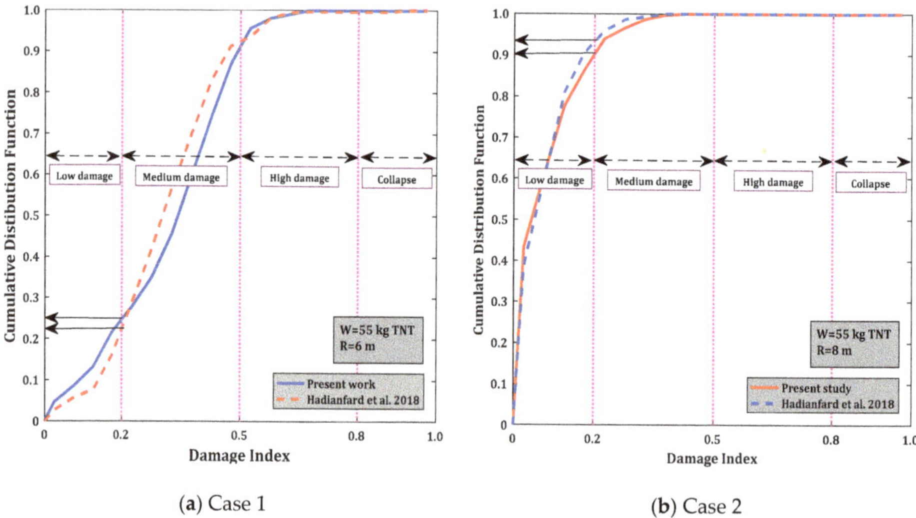

(**a**) Case 1

(**b**) Case 2

Figure 5. Comparison of the results of MCS based on beam (present study) and shell element types [52] for CDFs of DI:
(**a**) Case 1; (**b**) Case 2.

Based on Figure 5a,b, it can be clearly noticed that there is a rather close correlation between the collected results, even in the presence of different accuracy levels for the FE models in use (beam and shell elements, respectively). For Case 1, according to Figure 5a, the probabilities of low damage (DI < 0.2) are found to be 0.25 (present study) and 0.22 [52], with a scatter of 12%. In Case 2, see Figure 5b, the probabilities of low damage (DI < 0.2) are calculated at 0.90 (present study) and 0.93 [52], respectively, with a 3.22% scatter that further confirms the good agreement between the compared MCS results. In order to better clarify the performance of beam elements in probabilistic analysis, the required analysis durations for reliability analyses of the so-called Case 1 with beam and shell element types are thus presented in this paper. For beam elements, the typical required time was approximately 0.90 h for 300 MCS iterations. For shell elements [52], the required time was in the order of 30 h for the same number of iterations (that is, 33.33 times higher), which shows that implementing beam elements speeds up the procedure, especially in finding SPD and SSD values which need many separate reliability analyses.

4.4. Selected Columns

Given the potential of beam element formulation, a set of H-section steel columns with different geometrical properties in cross sections (IPB180 to IPB500) and lengths (2.8, 3.2, 3.6 and 4.0 m) are considered in the FE parametric investigation. The reference cross-sectional parameters are shown in Table 2, where A_g is the cross-sectional area, I_x is the moment of inertia about the strong axis (x-axis), I_y is the moment of inertia about the weak axis (y-axis).

Table 2. Geometrical properties of selected H-sections for parametric FE simulations.

Identification	b (mm)	h (mm)	s (mm)	t (mm)	A_g (cm^2)	I_x (cm^4)	I_y (cm^4)
IPB 180	180	180	8.5	14.0	65.3	3831	1363
IPB 220	220	220	9.5	16.0	91.0	8091	2843
IPB 260	260	260	10.0	17.5	118.4	14,920	5135
IPB 300	300	300	11.0	19.0	149.1	25,170	8563
IPB 340	300	340	12.0	21.5	170.9	36,660	9690
IPB 400	300	400	13.5	24.0	197.8	57,680	10,820
IPB 500	300	500	14.5	28.0	238.6	107,200	12,620

Reference cross section

5. Results and Discussions

5.1. Curves of Probability of Low Damage

The curves of probability of low damage are obtained in this paper for all the steel sections presented in Table 2 via reliability analysis, for both the pinned and fixed ends under different blast scenarios (with TNT charges of 55, 275 and 555 kg). For example, Figure 6a–c illustrate the numerical curves for IPB220 with pinned ends under explosive charge weights of 55, 275 and 555 kg, respectively. Each plot corresponds to a column with a specified length and explosive charge. Furthermore, it consists of some observed points that are obtained from reliability analyses. Each observed point and its probability of low damage is in fact the result of a reliability analysis based on the MCS method with 300 simulations. In total, 15,900 simulations were performed to extract Figure 6a–c, indicating the high amount of computational effort, which was around 47.4 h. Additionally, the normal CDFs have been fitted to the observed points for each case related to a specified column length, to convey a better understanding of presented concepts in Section 4 and also calculating the probability of low damage for points other than the observed points, if necessary. Given that the stand-off distance corresponding to the 95% probability of low damage is considered for finding safe distance (see Figure 4), then the fitting operation was used to find the stand-off distance corresponding to exactly 95% probability of low damage. In this case, lower tails are not important and will not affect the calculation and extraction of SPD and SSD.

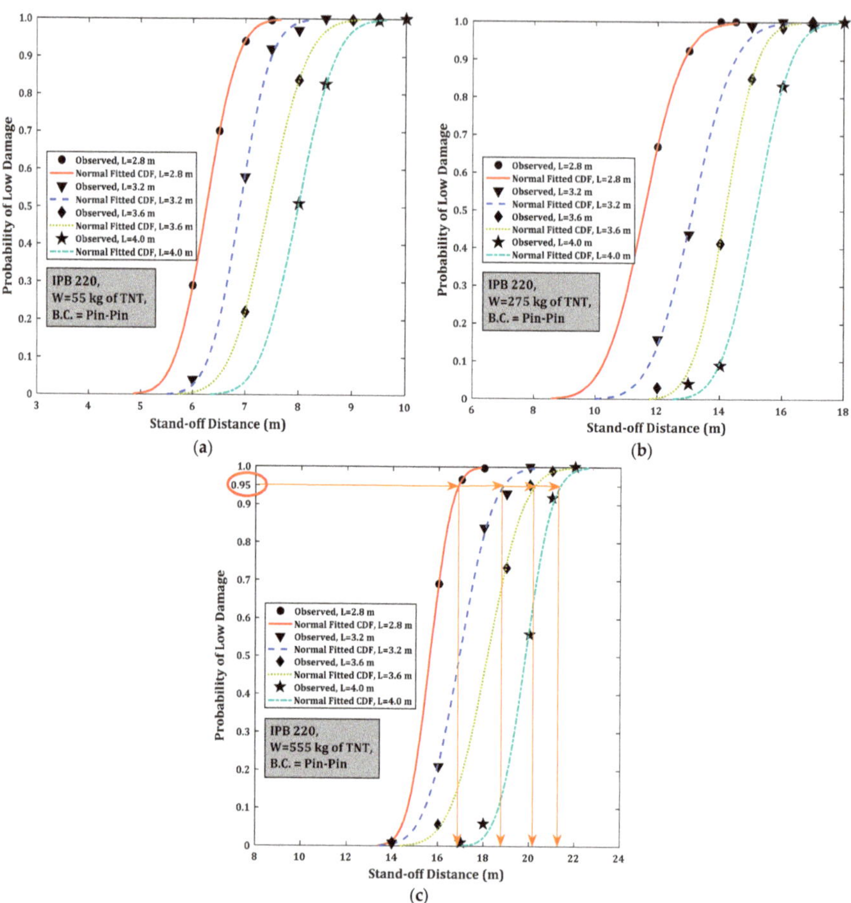

Figure 6. Curves of probability of low damage for IPB220 with pinned ends and different lengths subjected to: (**a**) 55, (**b**) 275 and (**c**) 555 kg of TNT.

According to Figure 6c, the SPD value for IPB220 with pinned ends and 3.6 m of length, subjected to 555 kg of TNT, is approximately calculated as 20.2 m by considering the 95% probability of low damage as a criterion and using orange arrows connected to normal fitted CDF for L = 3.6 m which eventually shows the value of SPD on the horizontal axis. As can be estimated from the figure, by changing the criterion from 95% to 99%, the SPD increases to 21.0 m. The important aspect, in this regard, is that the exclusive calculation of the SPD value for a column is not a sufficient way to provide protection against catastrophic events or major releases. In other words, finding the SPD value of 21.0 m by selecting 99% probability of low damage instead of 20.2 m (with 95% probability) does not mean that the safety of the examined steel column (or generally the whole structure) is ensured at this distance, without considering additional special arrangements. As such, the SPD values can be thus used as a valuable guidance to design and provide special arrangements around the building (such as appropriate access control and security guards), so as to reduce the frequency and/or the possible consequences to an acceptable level. The reliable prediction of adequate distances or separation zones around the building is thus one of the fundamental considerations for safe layout and can be designed according to SPD values. In some cases, it is worth mentioning that providing SPD for a structure to protect from all possible events is not practicable and this is especially the case in urban places, due to the lack of sufficient space between buildings and access roads. The assessment of the frequency of the expected event and its potential consequences is thus necessary to understand which risks can be reasonably mitigated by an SPD. For cases in which the obtained SPD value is too large, of course, additional mitigating or prevention measures should be considered. In a nutshell, the 95% criterion is internationally recognized to represent a rational choice for finding practical SPD that causes no serious damage for the structure (if special arrangements are provided) and the designer's judgment along with SPD values should lead to better decisions for ensuring safety of the structure under such events.

Similarly, Figure 7a–c present the curves of probability of low damage for IPB220 with fixed ends, subjected to explosive charge weights 55, 275 and 555 kg, respectively. As also explained in Section 4, using the obtained curves, the required SPD and the corresponding SSD can be calculated for each column by using orange arrows connected to each fitted curve. For instance, as Figure 6c reveals, the SPD values for IPB220 with different column lengths (2.8, 3.2, 3.6 and 4.0 m) and pinned ends are calculated in 16.9, 18.8, 20.2 and 21.2 m, respectively. Consequently, the SSD for lengths 2.8, 3.2, 3.6 and 4.0 m (and an explosive charge weight of 555 kg of TNT) are 2.06, 2.29, 2.46 and 2.58 m/kg$^{1/3}$, respectively.

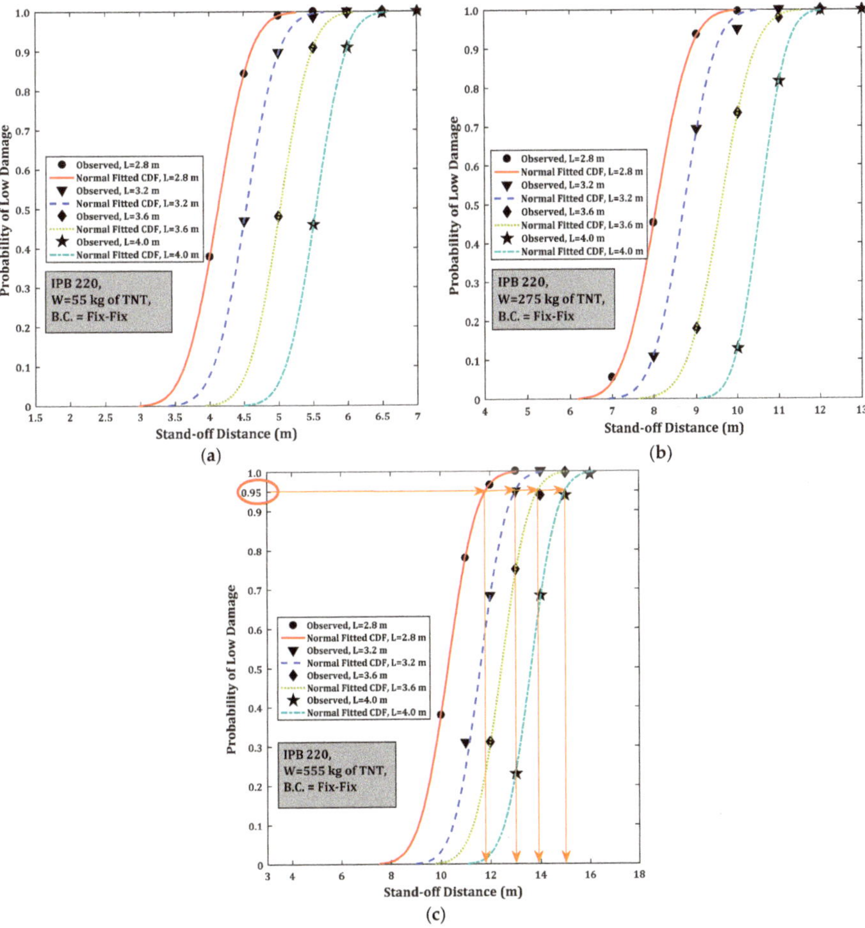

Figure 7. Curves of probability of low damage for IPB220 with fixed ends and different s lengths, subjected to: (**a**) 55, (**b**) 275 and (**c**) 555 kg of TNT.

As shown, when increasing the column length, the SSD also increases. The reason is that the longer the column length (and thus slenderness), the more it is exposed to premature buckling which reduces the overall residual axial load capacity and further results in more severe damage and thus higher SPD value. In the case of fixed ends, according to Figure 7c, the SPD values are obtained in 11.8, 13.0, 13.9 and 15.0 m, which correspond to SSD values of 1.44, 1.58, 1.69 and 1.83 $m/kg^{1/3}$ for 555 kg of TNT and column lengths of 2.8, 3.2, 3.6 and 4.0 m, respectively. As evidenced, by changing the support condition from pinned to fixed ends, the calculated SSD decreases. The actual boundary condition has thus a significant effect on the final performances and results. Moreover, considering the fact that the end conditions for real columns are neither fully pinned nor fixed, it is preferable for designers to take into account an actual value between the two limit conditions of perfectly pinned or fixed column models, which ultimately leads to choosing an SPD in between [89].

It should be noted that the variability of stand-off distance is highly dependent on the position of the explosive, given that the location of a terrorist device is not a certain parameter. When the target is known, the minimum stand-off distance from a facility (building, bridge, etc.) is obtained from the knowledge of the site (roads, parking, etc.), the access control (security gates, bollards, etc.) and the perimeter security [78]. Generally, for a critical building that may represent a target for terrorist attacks with variable and portable explosive weights (i.e., by human and different vehicles), the minimum stand-off distance can be easily found using the proposed strategy. Such a minimum stand-off distance can be thus used to provide appropriate access control and security guards around the building, thus ensuring that the risk of damage for the structural members in the first stage and the progressive collapse can be reduced.

Further, the residual axial load carrying capacity of a column (which is used to obtain DI) alternates between minimum (zero) and maximum ($P_{initial}$) values. This variation depends on geometrical properties and boundaries of the column, as well as on some uncertainties associated with blast loading and material properties. For all the cases in which the SPD is calculated, due to the fact that 95% probability of low damage is considered as decision criterion, the residual capacity of a given column under a selected blast scenario is expected to approach $P_{initial}$, or equivalently, the DI values are expected to approach zero. To better clarify the given explanations, Figure 8 shows the DI histograms obtained from MCS for a given IPB260 steel column with length of 3.6 m and pinned ends, subjected to 55, 275 and 555 kg of TNT. In all these cases, the probabilities of low damage are approximately calculated as 0.95, and, consequently, the proposed stand-off distances on the top of each figure are related to SPD value for the selected charge weights. As Figure 8 reveals, in all cases, the frequency of the DI obtained from MCS tends to low DI values (almost between DI = 0 and DI = 0.5), while the possibility of high damage (0.5 < DI < 0.8) and collapse (DI > 0.8) is really rare.

5.2. Empirical Relationship for Calculating SSD

The curves of probability of low damage were extracted similar to the approach presented in Section 5.1 to find the SSD values for all selected configurations (Table 2). Both pinned and fixed end conditions are examined, including different column lengths and explosive charge weights (55, 275 and 555 kg of TNT). A large number of FE simulations (approximately 252,000) were conducted for reliability analyses based on MCS, which took nearly 756 h of run time. The collected data were further investigated to find a practical relationship which could support designers in the predicting the SSD for steel columns under blast loads. By examining the results after a lot of trial-and-error process, it was finally found that the SSD can be expressed as:

$$SSD(W, P_{initial}) = \alpha_0 + \alpha_1 W + \alpha_2 P_{initial} + \alpha_3 W^2 + \alpha_4 W.P_{initial} + \alpha_5 P_{initial}^{\,2} \tag{11}$$

where the parameters α_0 to α_5 are constant coefficients. The final values of these α_i coefficients, as well as the coefficient of determination (R^2) values, goodness of fit (GoF) and root-mean-square error (RMSE), are shown in Table 3 for different columns, as obtained from curve fitting. It is clear that the R^2 values are higher than 97% for all the selected configurations, hence indicating a very satisfactory accuracy of the proposed formula for SSD predictions (usually, R^2 values higher than 80% are considered satisfactory).

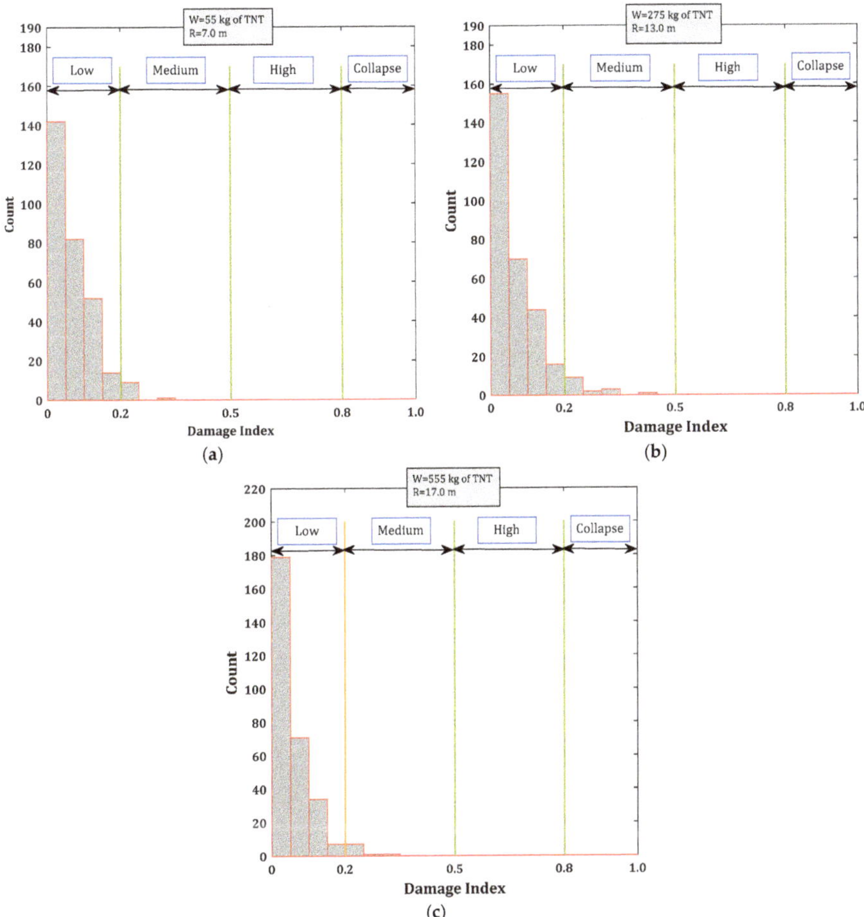

Figure 8. DI histograms, as obtained from MCS for IPB260 steel column with length of 3.6 m and pinned ends under: (**a**) 55, (**b**) 275 and (**c**) 555 kg of TNT.

Table 3. Constant coefficients of Equation (11) and corresponding R^2, GoF and RMSE for each selected case (BC = boundary condition).

BC	Length (m)	α_0	α_1	α_2	α_3	α_4	α_5	R^2	GoF	RMSE
Pinned	2.80	+2.600	$+1.131 \times 10^{-3}$	-4.101×10^{-4}	-1.351×10^{-6}	$+3.199 \times 10^{-8}$	$+1.502 \times 10^{-8}$	0.9943	0.0199	0.0364
	3.20	+2.765	$+1.383 \times 10^{-3}$	-4.219×10^{-4}	-1.528×10^{-6}	$+5.498 \times 10^{-9}$	$+1.356 \times 10^{-8}$	0.9958	0.0177	0.0343
	3.60	+2.959	$+1.269 \times 10^{-3}$	-4.348×10^{-4}	-1.527×10^{-6}	$+4.701 \times 10^{-8}$	$+9.666 \times 10^{-9}$	0.9930	0.0338	0.0475
	4.00	+3.056	$+1.173 \times 10^{-3}$	-4.018×10^{-4}	-1.399×10^{-6}	$+6.757 \times 10^{-8}$	-2.381×10^{-10}	0.9945	0.0287	0.0437
Fixed	2.80	+1.852	$+6.884 \times 10^{-4}$	-2.723×10^{-4}	-5.899×10^{-7}	$+4.833 \times 10^{-9}$	$+9.599 \times 10^{-9}$	0.9825	0.0298	0.0446
	3.20	+2.039	$+5.398 \times 10^{-4}$	-2.825×10^{-4}	-3.748×10^{-7}	-1.978×10^{-9}	$+9.041 \times 10^{-9}$	0.9739	0.0513	0.0585
	3.60	+2.105	$+1.046 \times 10^{-3}$	-2.847×10^{-4}	-9.854×10^{-7}	-1.322×10^{-8}	$+7.141 \times 10^{-9}$	0.9771	0.0538	0.0599
	4.00	+2.266	$+6.718 \times 10^{-4}$	-2.705×10^{-4}	-5.376×10^{-7}	$+1.577 \times 10^{-8}$	$+1.872 \times 10^{-9}$	0.9782	0.0562	0.0612

There are two conventional ways to calculate $P_{initial}$: one based on FE modelling and another based on the empirical relationships that are presented in several regulations. In this study, the second method is used, which is easier to apply, is efficient and can be extended to each column, without the need of any complex calculation. The final result is that, even disregarding sophisticated FE methods, the $P_{initial}$ prediction can be used in Equation (11) and it can be consequently assessed (for a specific explosive charge weight) whether the column is in a safe condition or not. According to the regulations, in more detail, $P_{initial}$ for members under compression without slender elements can be calculated as:

$$P_{initial} = F_{cr} \times A_g$$

$$\left(\frac{k_e L}{r_g}\right)_{max} \leq 4.71\sqrt{\frac{E_s}{F_y}} \qquad \rightarrow \qquad F_{cr} = \left[0.658^{\frac{F_y}{F_e}}\right] F_y \qquad (12)$$

$$\left(\frac{k_e L}{r_g}\right)_{max} > 4.71\sqrt{\frac{E_s}{F_y}} \qquad \rightarrow \qquad F_{cr} = 0.877\ F_e$$

where F_{cr} is the critical stress due to flexural buckling and L, r_g and k_e are, respectively, the column length, the radius of gyration and the effective length factor. For a column with pinned ends or fixed ends, it is assumed $k_e = 1$ or $k_e = 0.5$, respectively. Finally, F_e is the elastic buckling stress which can be calculated as:

$$F_e = \frac{\pi^2 E_s}{\left(\frac{k_e L}{r_g}\right)^2} \qquad (13)$$

Figures 9 and 10 show the predicted SSD values and the corresponding fitted planes (according to Equation (11)) for pinned and fixed ends and for different column lengths. In these figures, the numerical data collected from reliability analyses for different steel configurations and blast scenarios are represented by 21 points which are used for curve fitting (Equation (11)).

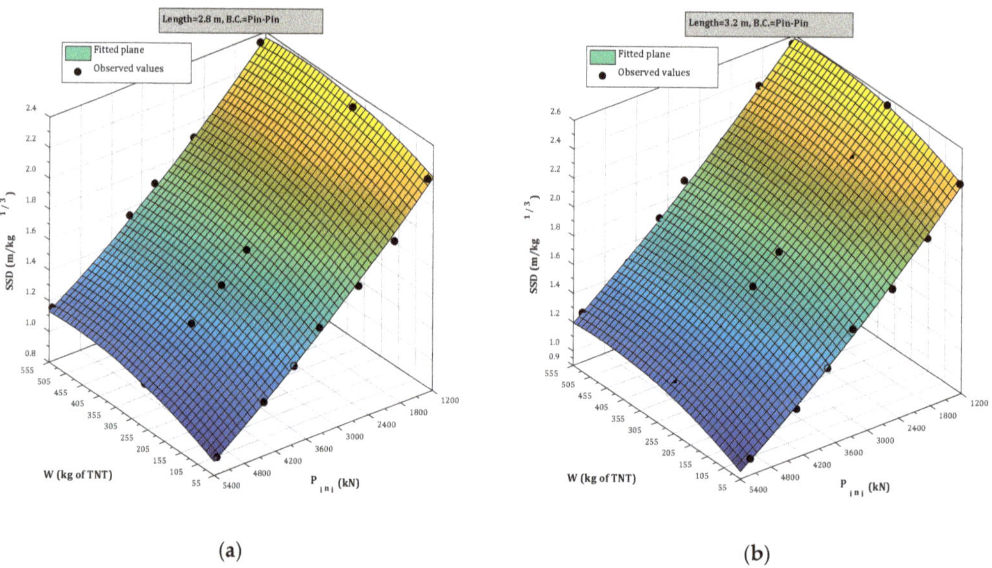

(a)　　　　　　　　　　　　　　　　(b)

Figure 9. *Cont.*

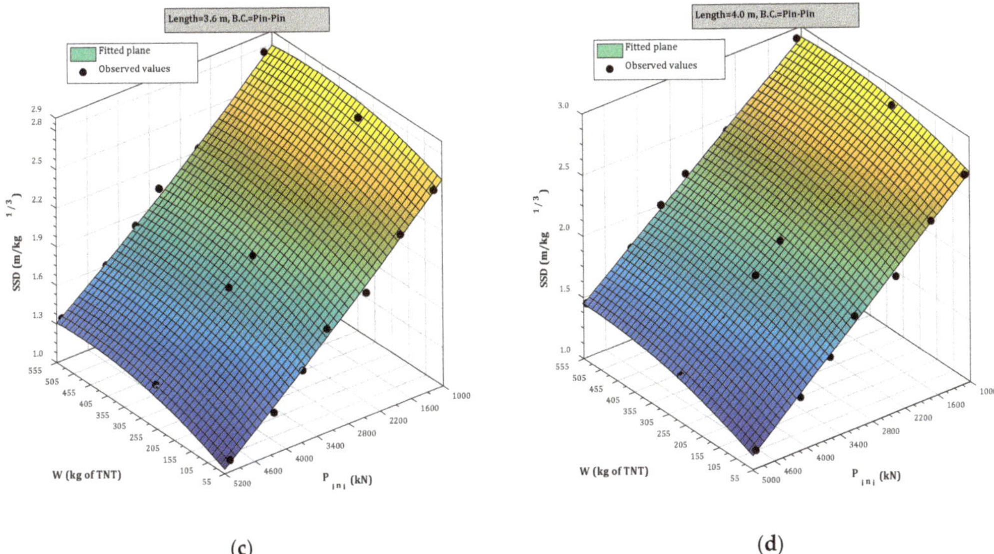

(**c**)　　　　　　　　(**d**)

Figure 9. SSD values for steel columns with pinned ends and different lengths: (**a**) L = 2.8 m, (**b**) L = 3.2 m, (**c**) L = 3.6 m and (**d**) L = 4.0 m.

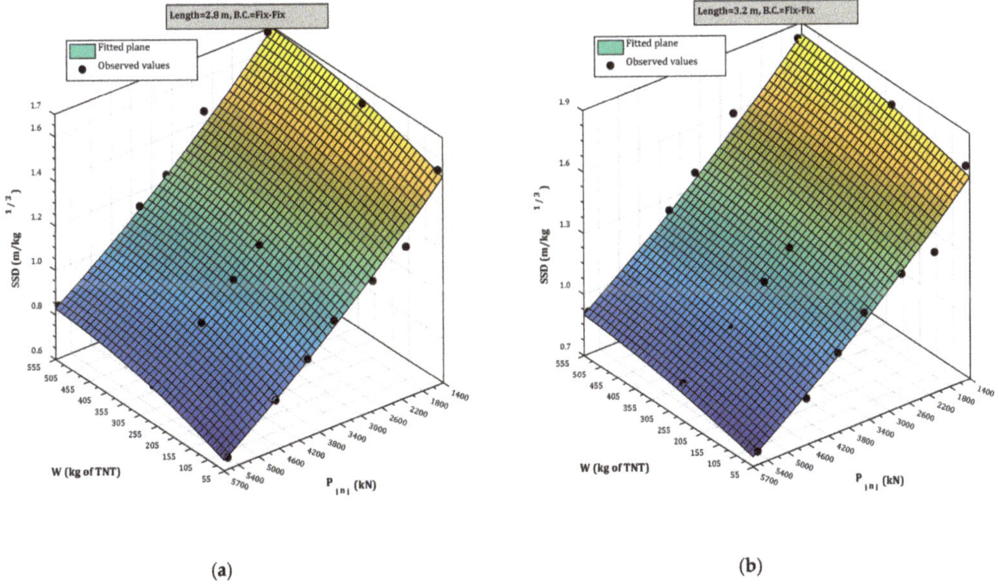

(**a**)　　　　　　　　(**b**)

Figure 10. *Cont.*

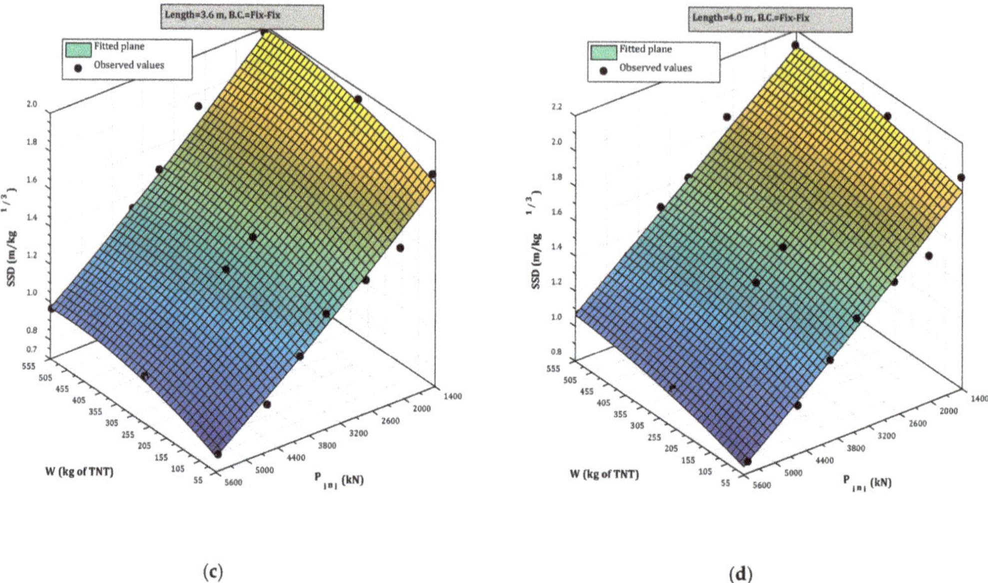

(c)   (d)

Figure 10. SSD values for steel columns with fixed ends and different lengths: (**a**) L = 2.8 m, (**b**) L = 3.2 m, (**c**) L = 3.6 m and (**d**) L = 4.0 m.

As shown, the SSD value typically increases with increasing explosive charge weight and decreases with increasing initial axial load capacity. For a given column length, by changing the support condition from pinned to fixed, the SSD decreases. Additionally, the effect of the explosive charge weight on SSD is minimum, compared to $P_{initial}$. For W-values higher than or equal to 275 kg, the effect of W is almost negligible and the values obtained for SSD are almost identical. In other words, by maintaining $P_{initial}$ constant, the obtained SSD values for W = 275 kg of TNT can be still considered for W > 275 kg, with an acceptable level of accuracy (Equation (14)). For a given $P_{initial}$ value and a variable W, the SPD can be in fact calculated as follows:

$$SSD_{(W=275)} \cong SSD_{(W>275)} \tag{14}$$

$$\left. \begin{aligned} SSD_{(W=275)} &= \frac{SPD_{(W=275)}}{275^{1/3}} \\ SSD_{(W>275)} &= \frac{SPD_{(W>275)}}{W^{1/3}} \end{aligned} \right\} \xrightarrow[\text{with acceptable estimation}]{\text{Substituting Eq. 14}} SPD_{(W>275)} = SPD_{(W=275)} \times \left(\frac{W}{275} \right)^{1/3} \tag{15}$$

where $SSD_{(W=275)}$ and $SSD_{(W>275)}$ are the safe scaled distances for W = 275 and W > 275 kg of TNT, respectively. Similarly, $SPD_{(W=275)}$ and $SPD_{(W>275)}$ are the SPDs for W = 275 kg and W > 275 kg of TNT, respectively. Based on such an equation, the SPD for W values higher than 275 kg of TNT can be thus easily obtained, without the need for any further simulation. As an instance, the predicted SSD value for a steel column with IPB 220 section and length of 3.6 m is 2.22 and 2.15, as obtained for TNT explosive charge weights equal to 555 or 275 kg. This results in SPD values of 18.24 and 13.98 m, respectively. Using Equations (14) and (15), the $SPD_{(W>275)}$ for the aforementioned column is calculated as 17.66 m, thus with a minimum percentage scatter (3.15%) compared to the expected value.

5.3. Verification of the Proposed Formula

The proposed analytical correlation for calculating the SSD was finally further discussed and verified. To this end, the numerical results from [52] were taken into account.

For a pinned steel column with IPBv200 cross section (ST37 steel) and a total length of 3.6 m, subjected to different charge weights (55, 275 and 555 kg of TNT), the SPD values in [52] were predicted to be 8, 15 and 19 m, thus resulting in SSD values of 2.10, 2.30 and 2.31 m/kg$^{1/3}$, respectively. The input steel yield stress, density, elastic modulus and Poisson's ratio were set to 240 MPa, 7850 kg/m^3, 210 GPa and 0.3, respectively. Furthermore, the geometrical characteristics of the resisting section (i.e., b, h, s and t, see Table 2), were set to 206, 220, 15 and 25 mm, respectively. It should be noted that the IPBv200 cross section is different from the intended steel sections presented in Table 2.

In order to analytically predict the SSD value based on Equation (11), the initial axial capacity of the column must first be calculated. Based on Equations (12) and (13), such a value can be estimated as 2518 kN. Consequently, by using Equation (14) and the constant coefficients presented in Table 2 (pinned ends and column length of 3.6 m), the corresponding SSD are easily calculated for selected explosive charges. The related values of SSD and SPD from Equations (14) and (15), together with those obtained from the previous study [52], are thus compared in Table 4. As shown, the percentage scatter Δ is also calculated for each case, giving evidence of the accuracy of the proposed method.

Table 4. Verification results of the proposed formula, with respect to previous study [52].

W (kg of TNT)	SSD Value (m/kg$^{1/3}$)			SPD (m)		
	Hadianfard et al. [52]	Present Study	Δ (%)	Hadianfard et al. [52]	Present Study	Δ (%)
55	2.10	1.99	5.24	8	7.57	5.37
275	2.30	2.18	5.21	15	14.17	5.53
555	2.31	2.21	4.33	19	18.16	4.42

As Table 4 reveals, there is in fact a rather close correlation between the current proposed formula for the SSD calculation and the past numerical study reported in [52]. SSD values obtained by Equation (11) were equal to 1.99, 2.18 and 2.21 m/kg$^{1/3}$, respectively, for explosive charge weights of 55, 275 and 555 kg, and thus corresponding to a negligible scatter (5.24%, 5.21% and 4.33%) compared to the SSD values from [52].

6. Calculation Examples

In conclusion, to emphasize the applicability and usefulness of the proposed relationship, some calculation examples are presented. The objective of these examples is to find the SPD and SSD of:

(i) A steel column with IPB 240 cross section and L = 3.4 m (Section 1), and

(ii) A box shape steel column with L = 3.6 m and given geometrical properties in Table 5 (Section 2).

Table 5. Sectional properties of selected box section steel column for calculation examples.

	Section Properties							Reference cross section
Identification	b (mm)	h (mm)	f (mm)	w(mm)	A (cm^2)	I$_x$ (cm^4)	I$_y$ (cm^4)	
BOX	300	300	19.6	6.2	149.9	24,986	15,799	

Section 1 differs from the intended columns presented in Table 2. Its length of 3.4 m, in particular, is further modified and differs with considered lengths (i.e., 2.8, 3.2, 3.6 and 4.0 m) in the reliability analyses, in order to create a databank for the relationship proposal. Section 2 is selected from [57], in which its sectional properties about bending axis is the same as for a IPB300 steel column, while its cross-sectional shape is totally different. Again, the material yield stress, density, elastic modulus and Poisson's ratio are set equal to 240 MPa, 7850 kg/m^3, 210 GPa and 0.3, respectively. Finally, Section 1 is subjected to six different explosive charge weights (55, 200, 275, 350, 555 and 1000 kg of TNT), while Section 2 is subjected to 275 kg of TNT, under the assumption of two idealized boundary conditions, (a) pinned and (b) fixed ends. The reason for selecting an explosive weight of 1000 kg of TNT, in this regard, is to verify the accuracy of the proposed Equations (14) and (15) for calculating the SPD and SSD for explosive charge weights higher than 555 kg of TNT.

To determine the SSD of the selected configurations, the proposed relationship (Equation (11)) and the methodology earlier presented are employed. The results are collected in Table 6, for both the pinned and fixed ends. It is worth mentioning that the SSD values calculated by Equation (11) are obtained by an interpolation approach. This means that, for a column with a length other than those reported in Figures 9 and 10, for both pinned and fixed end conditions, the interpolation method should be always used to calculate the corresponding SSD. For the present calculation example, the SSD values for column lengths of 3.2 and 3.6 m are thus first calculated by Equation (11) and then, using the interpolation, the required SSD values are estimated for the assigned column length of 3.4 m.

Table 6. SSD and SPD calculations with different methods MCS and proposed practical equations.

Column	B.C	W (kg of TNT)	SSD ($kg/m^{1/3}$)					SPD (m)				
			Equation (11)	Equation (14)	MCS	Δ_1 (%)	Δ_2 (%)	Equation (11)	Equation (15)	MCS	Δ_1 (%)	Δ_3 (%)
IPB 240	Pinned	55	2.052	–	2.039	0.63	–	7.80	–	7.75	0.64	–
		200	2.192	–	2.077	5.24	–	12.82	–	12.15	5.23	–
		275	2.244	2.244	2.214	1.34	1.33	14.59	14.59	14.40	1.30	1.30
		350	2.254	2.244	2.172	3.63	3.20	15.88	15.81	15.31	3.58	3.16
		555	2.274	2.244	2.265	0.39	0.93	18.69	18.44	18.61	0.43	0.91
		1000	–	2.244	2.278	–	1.49	–	22.44	22.78	–	1.49
	Fixed	55	1.466	–	1.416	3.41	–	5.58	–	5.38	3.58	–
		200	1.532	–	1.438	6.14	–	8.96	–	8.41	6.13	–
		275	1.587	1.587	1.533	3.40	3.40	10.32	10.32	9.97	3.39	3.39
		350	1.614	1.587	1.558	3.46	1.83	11.37	11.18	10.98	3.43	1.78
		555	1.646	1.587	1.657	0.66	4.22	13.53	13.04	13.62	0.66	4.26
		1000	–	1.587	1.598	–	2.92	–	15.87	15.98	–	0.69
BOX	Pinned	275	1.874	1.874	1.811	3.36	3.36	12.18	12.18	11.78	3.28	3.28
	Fixed	275	1.386	1.386	1.396	0.71	0.71	9.01	9.01	9.08	0.77	0.77

From Table 6, see Section 1, it is possible to notice that the proposed equation is able to provide a good level of accuracy for the estimation of the expected SSD under different explosive charge weights. Such an outcome is confirmed by the calculated percentage scatters, obtained between Equation (11) and MCS as Δ_1, and between Equation (14) and a MCS as Δ_2, and between Equation (15) and MCS as Δ_3. It should be noted that the symbol dash (–) in Table 6 shows that input value of W is out of range for corresponding equation. Furthermore, for Section 2, it can be seen that the proposed relationship accurately predicts the required SPD and SSD values, in comparison to the FE-based MCS results. This is also consistent with the results presented in [57], where for steel columns with different cross-sectional shapes (but similar section properties) subjected to the same loading/boundary conditions, it was proved that the cross section shape has mostly null effects on the global response. In case of pinned ends, in more detail, the cross-sectional shape has little effect on the response of a given column, while in case of fixed ends the results may change up to 20% [57].

In this paper, the results obtained for Section 2 show that, for a non H-shape steel column that can be equaled to an H-section (within the examined range and with the same loading/boundary conditions), the proposed formula can be used with accuracy to determine the SPD and SSD parameters. It should be kept in mind, however, that the proposed relationship is generally based on interpolation within the range of the intended columns. In general, it is hence recognized that the use of the proposed formula for other steel columns can provide a preliminary estimation of the required SPD and SSD values, but the accuracy of these results should be examined through engineering judgment and further investigations.

It is also clear, in this regard, that considering the high computational cost of the rigorous approach (almost 0.9 h for 300 simulations, and almost 4.5 h for each curve of probability of low damage, consisting of 5 points), the proposed approximate relationships can be efficiently used to obtain practical and reliable estimates.

The illustrative calculation examples, in conclusion, proved that the proposed formula is capable of properly interpolating the available data, and thus finding the SSD of various types of columns, which may differ in length and/or cross-sectional properties. Furthermore, such a relationship could be further extended to find the SSD of blast loaded steel columns with semi-fixed boundary conditions. Recently, a research study was in fact reported in [89] to investigate the effect of semi-fixed supports on the response of flexural members under impact loads. The equivalent SDOF system was used and the transformation factors including load, mass, stiffness and ultimate resistance factors were obtained for different fixity values in the elastic, elastic–plastic and plastic regions. It was thus shown that the semi-fixed condition can severely affect the blast response of a given column, compared to the two ideal assumptions of fully pinned and fully fixed support conditions. Accordingly, it is recognized that the current research outcomes—based on the SSD results for two perfectly pinned and fixed conditions—can support the designer with their engineering judgment in the choice of the SSD value (in between two limit support conditions) which is closest to the real boundary condition.

7. Conclusions

Safe scaled distance (SSD) is of practical interest, especially for design purposes of structural elements or assemblies, in order to minimize the damage probability and consequently the risk of progressive collapse against terrorist attacks in congested urban areas. In this paper, an improved methodology based on reliability analysis and implementing the beam element formulation was presented for calculating the SSD and the safe protective distance (SPD) for steel beam columns subjected to blast loads. To obtain the probability of low damage, the Monte Carlo simulation (MCS) method was used, so as to account for the uncertainties of blast loading parameters and material properties. The proposed methodology was extended to steel columns with different cross sections (IPB180 to IPB500), lengths (2.8, 3.2, 3.6 and 4.0 m) and boundary conditions (pinned or fixed ends). The collected data were thus further investigated to find a practical relationship to predict the SSD of steel columns under blast loads. From the comparative discussion, the following conclusions were obtained:

- The results showed that the improved methodology, based on the beam element formulation, has good efficiency and accuracy in predicting the damage probability of blast loaded steel columns and further remarkably reduces the run time of probabilistic analyses.
- A practical relationship was proposed and verified against numerical studies in the literature, to relate the SSD of blast loaded steel columns to the initial axial capacity and explosive charge weight.
- The proposed equation has a very good agreement with FE results based on MCS, which indicates its very high level of accuracy in predicting the SSD and thus its efficiency in obtaining practical and reliable estimates.
- The results showed that for both pinned and fixed end conditions, by increasing the initial axial carrying capacity of a given column and the amount of explosive charge weight, the SSD decreases and increases, respectively. The variation of the explosive

charge weight, however, has minimum effects on the calculated SSD, compared to variations in the initial axial capacity of the column.

- The discussed results proved that upon changing the support condition from pinned to fixed ends, the corresponding SSD decreases significantly. This indicates that the actual boundary condition has substantial effects on the SSD and the designer should consequently select an SSD value between two perfectly pinned and fixed models to account for real support conditions.

- For explosive charge weights (W) higher than or equal to 275 kg of TNT, by keeping constant the initial axial capacity, the effects of W variations on SSD are almost negligible. In a nutshell, the SSD obtained for W = 275 kg of TNT can be rationally taken into account, with an acceptable level of accuracy, for W values higher than 275 kg of TNT ($SSD_{(W=275)} \cong SSD_{(W>275)}$).

- Similarly, the SPD of a given steel column subjected to explosive charge weights higher than or equal to 275 kg of TNT can be easily obtained by calculating the SPD for W = 275 kg of TNT using the proposed equation.

Author Contributions: Conceptualization, M.M., C.B., M.A.H. and A.B.; methodology, M.M., C.B., M.A.H. and A.B.; software, M.M., C.B., M.A.H. and A.B.; validation, M.M., C.B., M.A.H. and A.B.; formal analysis, M.M., C.B., M.A.H. and A.B.; investigation, M.M., C.B., M.A.H. and A.B.; writing—original draft preparation, M.M., C.B., M.A.H. and A.B.; writing—review and editing, M.M., C.B., M.A.H. and A.B.; visualization, M.M., C.B., M.A.H. and A.B.; supervision, C.B., M.A.H. and A.B. All authors have read and agreed to the published version of the manuscript.

Funding: This research received no external funding.

Institutional Review Board Statement: Not applicable.

Informed Consent Statement: Not applicable.

Data Availability Statement: Supporting data will be made available upon request.

Acknowledgments: The authors wish to acknowledge and express their special thanks to Sina Malekpour and Fateme Hajari from Department of Civil and Environmental Engineering, Shiraz University of Technology, Shiraz, Iran.

Conflicts of Interest: The authors declare no conflict of interest.

Abbreviations

MCS	Monte Carlo Simulation
SPD	Safe Protective Distance
SSD	Safe Scaled Distance
FE	Finite Element
RC	Reinforced Concrete
SDOF	Single Degree of Freedom
MDOF	Multi Degree of Freedom
DIF	Dynamic Increase Factor
DI	Damage Index
GoF	Goodness of Fit
RMSE	Root-Mean-Square Error
R^2	Coefficient of determination
PDF	Probability Density Function
CDF	Cumulative Distribution Function
COV	Coefficient of Variation
TNT	Trinitrotoluene
BC	Boundary Condition
W	Explosive charge weight
W_{eff}	Effective charge weight
R	Stand-off distance

Z	Scaled distance
C and P	Constant coefficients of Cooper-Simonds relationship
k	Integration refinement factor
σ	Standard deviation
ε_t	True stress
ε_t	True strain
$\dot{\varepsilon}$	Strain rate
L_1	Deformed length of uniaxial tension member
L_0	Undeformed length of uniaxial tension member
P_r	Reflected pressure
$P_{r(mean)}$	Mean value of reflected pressure
σ_{Pr}	Standard deviation of reflected pressure
COV_{pr}	Coefficient of variation of reflected pressure
t_d	Positive time duration
$t_{d(mean)}$	Mean value of positive time duration
σ_{td}	Standard deviation of positive time duration
COV_{td}	Coefficient of variation of positive time duration
$P_{residual}$	Post-blast residual axial capacity of the damaged column
$P_{initial}$	Maximum axial load-carrying capacity of the undamaged column
F_y	Yield stress
E_s	Modulus of elasticity
E_t	Slope of the bilinear stress strain curve in strain hardening region
P_f	Probability of failure
N_f	Number of trials for which limit state function falls in the failure region
N	Number of total simulations
$\mathbf{X}$	Vector of input random variables
$g(\mathbf{X})$	Limit state function
r	Capacity
q	Demand
$f_x(\mathbf{X})$	Joint probability density function
I_F	Failure indicator
C_L	Confidence level
I_x	Moment of inertia about the strong axis
I_y	Moment of inertia about the weak axis
α_0 to α_5	Constant coefficients
F_{cr}	Critical stress due to flexural buckling of members without slender elements
A_g	Total cross-sectional area
L	Column length
r_g	Radius of gyration
k_e	Effective length factor
F_e	Elastic buckling stress
Δ, Δ_1, Δ_2 and Δ_3	Percentage scatters

References

1. *Soft Targets and Crowded Places Security Plan Overview*; U.S. Department of Homeland Security: Washington, DC, USA, 2018. Available online: https://www.cisa.gov/sites/default/files/publications/DHS-Soft-Target-Crowded-Place-Security-Plan-Overview-052018-508_0.pdf (accessed on 25 November 2021).
2. Hinman, E. Primer for design of commercial buildings to mitigate terrorist attacks. *FEMA* **2003**, *427*, 40–41.
3. Command, U.A.M. *Engineering Design Handbook: Explosions in Air Part One*; AD/A-003 817 (AMC Pamphlet AMCP 706-181); AMC Pamphlet: Alexandria, VA, USA, 1974.
4. Baker, W.E.; Westine, P.S.; Dodge, F.T. *Similarity Methods in Engineering Dynamics: Theory and Practice of Scale Modeling*; Hayden Book Co.: Indianapolis, IN, USA, 1973; Available online: https://www.elsevier.com/books/similarity-methods-in-engineering-dynamics/westine/978-0-444-88156-4 (accessed on 25 November 2021).
5. U.S. DoD. *Ammunition and Explosives Safety Standards*; U.S. DoD: Washington, DC, USA, 2004.
6. Wu, C.; Hao, H. Safe scaled distance for masonry infilled RC frame structures subjected to airblast loads. *J. Perform. Constr. Facil.* **2007**, *21*, 422–431. [CrossRef]
7. Byfield, M.; Paramasivam, S. Estimating safe scaled distances for columns subjected to blast. *Eng. Comput. Mech.* **2014**, *167*, 23–29. [CrossRef]

8. Thomas, R.; Steel, K.; Sorensen, A.D. Reliability analysis of circular reinforced concrete columns subject to sequential vehicular impact and blast loading. *Eng. Struct.* **2018**, *168*, 838–851. [CrossRef]

9. Hadianfard, M.A.; Malekpour, S. Evaluation of explosion safe distance of steel column via structural reliability analysis. *ADST J.* **2017**, *8*, 349–359.

10. Zhai, X.; Wang, Y.; Sun, Z. Damage model and damage assessment for single-layer reticulated domes under exterior blast load. *Mech. Based Des. Struct. Mach.* **2019**, *47*, 319–338. [CrossRef]

11. Hao, H.; Wu, C. Numerical simulation of damage of low-rise RC frame structures with infilled masonry walls to explosive loads. *Aust. J. Struct. Eng.* **2006**, *7*, 13–22. [CrossRef]

12. Bounds, W.L. *Design of Blast-Resistant Buildings in Petrochemical Facilities*; ASCE Publications: Reston, VA, USA, 2010.

13. Jarrett, D. Derivation of the British explosives safety distances. *Ann. N. Y. Acad. Sci.* **1968**, *152*, 18–35. [CrossRef]

14. Magallanes, J.M.; Martinez, R.; Koenig, J.W. *Experimental Results of the AISC Full-Scale Column Blast Test*; The American Institute of Steel Construction: Chicago, IL, USA, 2006.

15. Nassr, A.A.; Razaqpur, A.G.; Tait, M.J.; Campidelli, M.; Foo, S. Dynamic response of steel columns subjected to blast loading. *J. Struct. Eng.* **2013**, *140*, 04014036. [CrossRef]

16. Nassr, A.A.; Razaqpur, A.G.; Tait, M.J.; Campidelli, M.; Foo, S. Strength and stability of steel beam columns under blast load. *Int. J. Impact Eng.* **2013**, *55*, 34–48. [CrossRef]

17. Nassr, A.A.; Razaqpur, A.G.; Tait, M.J.; Campidelli, M.; Foo, S. Single and multi degree of freedom analysis of steel beams under blast loading. *Nucl. Eng. Des.* **2012**, *242*, 63–77. [CrossRef]

18. Bao, X.; Li, B. Residual strength of blast damaged reinforced concrete columns. *Int. J. Impact Eng.* **2010**, *37*, 295–308. [CrossRef]

19. Momeni, M.; Hadianfard, M.A.; Bedon, C.; Baghlani, A. Damage evaluation of H-section steel columns under impulsive blast loads via gene expression programming. *Eng. Struct.* **2020**, *219*, 110909. [CrossRef]

20. Rong, H.-C.; Li, B. Probabilistic response evaluation for RC flexural members subjected to blast loadings. *Struct. Saf.* **2007**, *29*, 146–163. [CrossRef]

21. Yokoyama, T. Limits to Deflected Shape Assumptions of the SDOF Methodology for Analyzing Structural Components Subject to Blast Loading. *J. Perform. Constr. Facil.* **2014**, *29*, B4014008. [CrossRef]

22. Crawford, J.E.; Magallanes, J.M. The effects of modeling choices on the response of structural components to blast effects. *Int. J. Prot. Struct.* **2011**, *2*, 231–266. [CrossRef]

23. Al-Thairy, H. A modified single degree of freedom method for the analysis of building steel columns subjected to explosion induced blast load. *Int. J. Impact Eng.* **2016**, *94*, 120–133. [CrossRef]

24. Lee, K.; Kim, T.; Kim, J. Local response of W-shaped steel columns under blast loading. *Struct. Eng. Mech.* **2009**, *31*, 25–38. [CrossRef]

25. Shi, Y.; Hao, H.; Li, Z.-X. Numerical derivation of pressure–impulse diagrams for prediction of RC column damage to blast loads. *Int. J. Impact Eng.* **2008**, *35*, 1213–1227. [CrossRef]

26. Hadianfard, M.A.; Farahani, A. On the effect of steel columns cross sectional properties on the behaviours when subjected to blast loading. *Struct. Eng. Mech.* **2012**, *44*, 449–463. [CrossRef]

27. Hadianfard, M.A.; Nemati, A.; Johari, A. Investigation of Steel Column Behavior with Different Cross Section under Blast Loading. *Modares Civ. Eng. J.* **2016**, *16*, 265–278.

28. Ibrahim, Y.E.; Nabil, M. Assessment of structural response of an existing structure under blast load using finite element analysis. *Alex. Eng. J.* **2019**, *58*, 1327–1338. [CrossRef]

29. Amadio, C.; Bedon, C. Blast analysis of laminated glass curtain walls equipped by viscoelastic dissipative devices. *Buildings* **2012**, *2*, 359–383. [CrossRef]

30. Momeni, M.; Bedon, C. Uncertainty Assessment for the Buckling Analysis of Glass Columns with Random Parameters. *Int. J. Struct. Glass Adv. Mater. Res.* **2020**, *4*, 254–275. [CrossRef]

31. Figuli, L.; Cekerevac, D.; Bedon, C.; Leitner, B. Numerical analysis of the blast wave propagation due to various explosive charges. *Adv. Civ. Eng.* **2020**, *2020*, 8871412. [CrossRef]

32. Goel, M.D.; Thimmesh, T.; Shirbhate, P.; Bedon, C. Enhanced Single-Degree-of-Freedom Analysis of Thin Elastic Plates Subjected to Blast Loading Using an Energy-Based Approach. *Adv. Civ. Eng.* **2020**, *2020*, 8825072. [CrossRef]

33. Shi, Y.; Stewart, M.G. Damage and risk assessment for reinforced concrete wall panels subjected to explosive blast loading. *Int. J. Impact Eng.* **2015**, *85*, 5–19. [CrossRef]

34. Bogosian, D.; Ferritto, J.; Shi, Y. *Measuring Uncertainty and Conservatism in Simplified Blast Models*; Karagozian And Case Glendale: Glendale, CA, USA, 2002.

35. Hao, H.; Stewart, M.G.; Li, Z.-X.; Shi, Y. RC column failure probabilities to blast loads. *Int. J. Prot. Struct.* **2010**, *1*, 571–591. [CrossRef]

36. Low, H.Y.; Hao, H. Reliability analysis of reinforced concrete slabs under explosive loading. *Struct. Saf.* **2001**, *23*, 157–178. [CrossRef]

37. Netherton, M.D.; Stewart, M.G. Blast load variability and accuracy of blast load prediction models. *Int. J. Prot. Struct.* **2010**, *1*, 543–570. [CrossRef]

38. Kelliher, D.; Sutton-Swaby, K. Stochastic representation of blast load damage in a reinforced concrete building. *Struct. Saf.* **2012**, *34*, 407–417. [CrossRef]

39. Olmati, P.; Sagaseta, J.; Cormie, D.; Jones, A. Simplified reliability analysis of punching in reinforced concrete flat slab buildings under accidental actions. *Eng. Struct.* **2017**, *130*, 83–98. [CrossRef]
40. Ding, Y.; Song, X.; Zhu, H.-T. Probabilistic progressive collapse analysis of steel frame structures against blast loads. *Eng. Struct.* **2017**, *147*, 679–691. [CrossRef]
41. Low, H.Y.; Hao, H. Reliability analysis of direct shear and flexural failure modes of RC slabs under explosive loading. *Eng. Struct.* **2002**, *24*, 189–198. [CrossRef]
42. Hao, H.; Li, Z.-X.; Shi, Y. Reliability analysis of RC columns and frame with FRP strengthening subjected to explosive loads. *J. Perform. Constr. Facil.* **2015**, *30*, 04015017. [CrossRef]
43. Shi, Y.; Stewart, M.G. Spatial reliability analysis of explosive blast load damage to reinforced concrete columns. *Struct. Saf.* **2015**, *53*, 13–25. [CrossRef]
44. Stewart, M.G. Reliability-based load factors for airblast and structural reliability of reinforced concrete columns for protective structures. *Struct. Infrastruct. Eng.* **2019**, *15*, 634–646. [CrossRef]
45. Olmati, P.; Petrini, F.; Gkoumas, K. Fragility analysis for the Performance-Based Design of cladding wall panels subjected to blast load. *Eng. Struct.* **2014**, *78*, 112–120. [CrossRef]
46. Stochino, F. RC beams under blast load: Reliability and sensitivity analysis. *Eng. Fail. Anal.* **2016**, *66*, 544–565. [CrossRef]
47. Hussein, A.; Mahmoud, H.; Heyliger, P. Probabilistic analysis of a simple composite blast protection wall system. *Eng. Struct.* **2020**, *203*, 109836. [CrossRef]
48. Campidelli, M.; El-Dakhakhni, W.; Tait, M.; Mekky, W. Blast design-basis threat uncertainty and its effects on probabilistic risk assessment. *ASCE-ASME J. Risk Uncertain. Eng. Syst. Part A Civ. Eng.* **2015**, *1*, 04015012. [CrossRef]
49. Hedayati, M.H.; Sriramula, S.; Neilson, R.D. Reliability of Profiled Blast Wall Structures. In *Numerical Methods for Reliability and Safety Assessment*; Springer: Berlin/Heidelberg, Germany, 2015; pp. 387–405.
50. Borenstein, E.; Benaroya, H. Sensitivity analysis of blast loading parameters and their trends as uncertainty increases. *J. Sound Vib.* **2009**, *321*, 762–785. [CrossRef]
51. Momeni, M.; Hadianfard, M.A.; Baghlani, A. Implementation of Weighted Uniform SimulationMethod in Failure Probability Analysisof Steel Columns under Blast Load. In Proceedings of the 11th International Congress on Civil Engineering, University of Tehran, Tehran, Iran, 8–10 May 2018.
52. Hadianfard, M.A.; Malekpour, S.; Momeni, M. Reliability analysis of H-section steel columns under blast loading. *Struct. Saf.* **2018**, *75*, 45–56. [CrossRef]
53. Thöns, S.; Stewart, M.G. On decision optimality of terrorism risk mitigation measures for iconic bridges. *Reliab. Eng. Syst. Saf.* **2019**, *188*, 574–583. [CrossRef]
54. Stewart, M.G.; Netherton, M.D. A Probabilistic Risk-Acceptance Model for Assessing Blast and Fragmentation Safety Hazards. *Reliab. Eng. Syst. Saf.* **2019**, *191*, 106492. [CrossRef]
55. Stewart, M.G.; Netherton, M.D. Security risks and probabilistic risk assessment of glazing subject to explosive blast loading. *Reliab. Eng. Syst. Saf.* **2008**, *93*, 627–638. [CrossRef]
56. Netherton, M.D.; Stewart, M.G. Risk-based blast-load modelling: Techniques, models and benefits. *Int. J. Prot. Struct.* **2016**, *7*, 430–451. [CrossRef]
57. Momeni, M.; Hadianfard, M.A.; Bedon, C.; Baghlani, A. Numerical damage evaluation assessment of blast loaded steel columns with similar section properties. *Structures* **2019**, *20*, 189–203. [CrossRef]
58. Hallquist, J.O. LS-DYNA Theory Manual. *Livermore Softw. Technol. Corp.* **2006**, *3*, 25–31.
59. Hallquist, J. *LS-Dyna Theory Manual, March 2006*; Livermore Software Technology Corporation (LSTC): Livermore, CA, USA, 2012.
60. LSTC. *Keyword User's Manual Volume II*; Livermore Software Technology Corporation (LSTC): Livermore, CA, USA, 2007.
61. Cowper, G.R.; Symonds, P.S.; United States Office of Naval Research; Brown University Division of Applied Mathematics. *Strain-Hardening and Strain-Rate Effects in the Impact Loading of Cantilever Beams*; Brown University: Providence, RI, USA, 1957.
62. Jones, N. *Structural Impact*; Cambridge University Press: Cambridge, UK, 2011.
63. Momeni, M. Damage Evaluation of Steel Beam-Columns Subjected to Impulsive Loads Using Reliability Approach. Ph.D. Thesis, Shiraz University of Technology, Shiraz, Iran, 2021. (In Persian)
64. Rackauskaite, E.; Kotsovinos, P.; Rein, G. Model parameter sensitivity and benchmarking of the explicit dynamic solver of LS-DYNA for structural analysis in case of fire. *Fire Saf. J.* **2017**, *90*, 123–138. [CrossRef]
65. *FEMA 426: Reference Manual to Mitigate Potential Terrorist Attacks against Buildings—Buildings and Infrastructure Protection Series*; U.S. Department of Homeland Security: Washington, DC, USA, 2003.
66. Karlos, V.; Solomos, G. *Calculation of Blast Loads for Application to Structural Components*; European Commission's science and knowledge service, Joint Research Centre: Ispra, Italy, 2013.
67. U.S. DoD. *Structures to Resist the Effects of Accidental Explosions*; UFC 3-340-02; U.S. DoD: Washington, DC, USA, 2008.
68. Beshara, F. Modelling of blast loading on aboveground structures—I. General phenomenology and external blast. *Comput. Struct.* **1994**, *51*, 585–596. [CrossRef]
69. *Suppressive Shields Structural Design and Analysis Handbook*; US Army Corps of Engineers, Huntsville Division: Huntsville, AL, USA, 1977; HNDM-1110-1-2.
70. Krauthammer, T. *Blast Effects and Related Threats*; Pennsylvania State University: State College, PA, USA, 1999.

71. Ruggieri, S.; Porco, F.; Uva, G.; Vamvatsikos, D. Two frugal options to assess class fragility and seismic safety for low-rise. *Bull. Earthq. Eng.* **2021**, *19*, 1415–1439. [CrossRef]
72. Miranda, E.; Reyes, C.J. Approximate lateral drift demands in multistory buildings with nonuniform stiffness. *J. Struct. Eng.* **2002**, *128*, 840–849. [CrossRef]
73. Campidelli, M.; Tait, M.; El-Dakhakhni, W.; Mekky, W. Inference of blast wavefront parameter uncertainty for probabilistic risk assessment. *J. Struct. Eng.* **2015**, *141*, 04015062. [CrossRef]
74. Asprone, D.; Jalayer, F.; Prota, A.; Manfredi, G. Proposal of a probabilistic model for multi-hazard risk assessment of structures in seismic zones subjected to blast for the limit state of collapse. *Struct. Saf.* **2010**, *32*, 25–34. [CrossRef]
75. Netherton, M.D.; Stewart, M.G. The effects of explosive blast load variability on safety hazard and damage risks for monolithic window glazing. *Int. J. Impact Eng.* **2009**, *36*, 1346–1354. [CrossRef]
76. Olmati, P.; Vamvatsikos, D.; Stewart, M.G. Safety factor for structural elements subjected to impulsive blast loads. *Int. J. Impact Eng.* **2017**, *106*, 249–258. [CrossRef]
77. Stewart, M.; Netherton, M.; Shi, Y.; Grant, M.; Mueller, J. Probabilistic terrorism risk assessment and risk acceptability for infrastructure protection. *Aust. J. Struct. Eng.* **2012**, *13*, 1–17. [CrossRef]
78. Stewart, M.G. Reliability-based load factor design model for explosive blast loading. *Struct. Saf.* **2018**, *71*, 13–23. [CrossRef]
79. Stewart, M.G.; Mueller, J. Terror, Security, and Money: Balancing the Risks, Benefits, and Costs of Critical Infrastructure Protection. In Proceedings of the Reliability Engineering Computing REC 2012, Brno, Czech Republic, June 2012; pp. 513–533.
80. Bartlett, F.; Dexter, R.; Graeser, M.; Jelinek, J.; Schmidt, B.; Galambos, T. *Updating Standard Shape Material Properties Database for Design and Reliability (k-Area 4)*; Technical Report for American Institute of Steel Construction; American Institute of Steel Construction: Chicago, IL, USA, 2001.
81. Hadianfard, M.; Razani, R. Effects of semi-rigid behavior of connections in the reliability of steel frames. *Struct. Saf.* **2003**, *25*, 123–138. [CrossRef]
82. Thai, H.-T.; Uy, B.; Kang, W.-H.; Hicks, S. System reliability evaluation of steel frames with semi-rigid connections. *J. Constr. Steel Res.* **2016**, *121*, 29–39. [CrossRef]
83. Rubinstein, R. *Simulation and the Monte Carlo Method*; John Wiley&Sons: New York, NY, USA, 1981.
84. Melchers, R.E.; Beck, A.T. *Structural Reliability: Analysis and Prediction*; John Wiley & Sons Ltd.: Chichester, UK, 2018.
85. Broding, W.; Diederich, F.; Parker, P. Structural optimization and design based on a reliability design criterion. *J. Spacecr. Rocket.* **1964**, *1*, 56–61. [CrossRef]
86. Lees, F. *Lees' Loss Prevention in the Process Industries: Explosion*; Butterworth-Heinemann: Oxford, UK, 2012.
87. UNEP Industry and Environment Programme Activity Centre. *Paris, FR. Hazard Identification and Evaluation in a Local Community*; UNEP Industry and Environment Programme Activity Centre: Paris, France, 1992; Volume 12.
88. Center for Chemical Process Safety. *Guidelines for Chemical Process Quantitative Risk Analysis*; Center for Chemical Process Safety of the American Institute of Chemical: New York, NY, USA, 2000.
89. Hadianfard, M.A.; Shekari, M. An Equivalent Single-Degree-of-Freedom System to Estimate Nonlinear Response of Semi-fixed Flexural Members Under Impact Load. *Iran. J. Sci. Technol. Trans. Civ. Eng.* **2019**, *43*, 343–355. [CrossRef]

 *buildings*

Article

FRP Pedestrian Bridges—Analysis of Different Infill Configurations

Lucija Stepinac [1,*], Ana Skender [2], Domagoj Damjanović [2] and Josip Galić [1,*]

[1] Faculty of Architecture, University of Zagreb, Andrije Kačića Miošića 26, 10000 Zagreb, Croatia

[2] Faculty of Civil Engineering, University of Zagreb, Andrije Kačića Miošića 26, 10000 Zagreb, Croatia; ana.skender@grad.unizg.hr (A.S.); domagoj.damjanovic@grad.unizg.hr (D.D.)

* Correspondence: lstepinac@arhitekt.hr (L.S.); jgalic@arhitekt.hr (J.G.)

Abstract: The main aim of this study is to analyze fiber-reinforced polymer (FRP) bridge decks according to their material, cross-section, and shape geometry. Infill cell configurations of the decks (rectangular, triangular, trapezoidal, and honeycomb) were tested based on the FRP cell units available in the market. A comparison was made for each cell configuration in flat and curved bridge shapes. Another comparison was made between the material properties. Each model was computed for a composite layup material and a quasi-isotropic material. The quasi-isotropic material represents chopped fibers within a matrix. FE (finite element) analysis was performed on a total of 24 models using Abaqus software. The results show that the bridge shape geometry and infill configuration play an important role in increasing the stiffness, more so than improving the material properties. The arch shape of the bridge deck with quasi-isotropic material and chopped fibers was compared to the cross-ply laminate material in a flat bridge deck. The results show that the arch shape of the bridge deck contributed to the overall stiffness by reducing the deformation by an average of 30–40%. The results of this preliminary study will provide a basis for future research into form finding and laboratory testing.

Keywords: FRP deck; pedestrian bridges; pultruded deck; sandwich deck; cell configuration; laminate; quasi-isotropic; geometry optimization; 3D printing

Citation: Stepinac, L.; Skender, A.; Damjanović, D.; Galić, J. FRP Pedestrian Bridges—Analysis of Different Infill Configurations. *Buildings* **2021**, *11*, 564. https://doi.org/10.3390/buildings11110564

Academic Editors: Chiara Bedon, Mislav Stepinac and Flavio Stochino

Received: 15 October 2021
Accepted: 17 November 2021
Published: 22 November 2021

Publisher's Note: MDPI stays neutral with regard to jurisdictional claims in published maps and institutional affiliations.

1. Introduction

Henry Ford, a great innovator in the auto industry, introduced FRP material to the world under the motto "Ten times stronger than steel" [1]. However, even though it is not a novelty in developed countries, it is still not widely used in developing countries.

FRP is a two-component material whose volume consists of 30–70% fiber, with 50% of its total weight incorporated into a polymer matrix. With regard to the mechanical characteristics of FRP, fibers are the ones in charge of carrying the load and providing strength, stiffness, and thermal stability. The matrix wraps the fibers and protects them during the production process and during the exploitation time, ensuring an even load distribution to each fiber. It is also crucial in providing composite durability [2].

At this point, the market is full of various fibers. Glass fibers (GFRP) are the most common choice for investors, architects, and structural engineers due to their good mechanical and physical properties and low price compared to carbon fibers (CFRP), aramid fibers (AFRP), and basalt fibers (BFRP).

There are many applications of FRP in bridge engineering, mostly with pedestrian bridges. FRP material is a good alternative to traditional materials that will extend the life of the structure or enable the usage of pedestrian bridges for a longer period. The replacement of parts in the existing bridges, such as concrete decks, with FRP decks, or even replacing the whole structure, can be performed without significantly disturbing the structure. Composites excel because of their high modulus of elasticity, ultimate load capacity, and low density, making them ideal for strengthening the existing girders and decks and using them in a new, fully composite or hybrid bridge structure [3]. Some of this material's other beneficial properties include rapid construction, anticorrosive properties, water resistance, pleasant appearance, lasting color, overloading resistance, good dynamic performance [4], and the potential to build bridges with a greater span [5]. In addition, research by Mara [6] shows that less energy is needed for the production and maintenance of FRP bridges than for construction based on other materials.

FRP is promising in terms of its durability, but its high price makes it difficult for it to compete with traditional materials. Insufficient building codes and poor knowledge of the benefits of anisotropic materials are the reasons why architects and structural engineers often choose traditional materials over FRP.

The research from M. Gunaydin et al. [7] on the Halgavor bridge details the use of three different materials: steel, GFRP, and a steel–GFRP combination. The dynamic analysis results are positive for when only GFRP is used in a static representation of the Halgavor suspended bridge. Comparing the total weight, GFRP has only half of the weight of the steel bridge and almost a fifth of the weight of the concrete bridge. Structural and physical appearances are of great value, and architectural design should be taken into consideration more often when designing with FRP. Each new material must go through a process of finding the optimal structural system and the shape that suits it best, but that involves copying other traditional materials at the beginning. There are numerous examples of when FRP decks are used on steel and concrete girders [8–11]. FRP composites make a good alternative to traditional plate systems due to their strength-to-weight ratio and their pleasant appearance [12].

In this research, several FRP decks were numerically investigated to show how geometry modification can influence the global behavior of the panels—in this case, FRP bridges.

The main aim of this research is to see how the inclination and shape of the bridge, its infill geometry, and its material can affect stiffness. We assume, in our models, that the type of material is known. Twenty-four models were made. Twelve of them have been modeled with an anisotropic GFRP material and a layup orientation in two orthogonal directions (0/90/0/90). Another 12 models have quasi-isotropic material properties, including apropos matrix mixes with chopped fibers. Quasi-isotropic material will result in a higher deformation and a lower load-bearing capacity due to its lower modulus of elasticity and strength. A curved bridge geometry will result in lower deformation and higher stiffness. The main aim of this preliminary research is to quantify the level of this reduction according to material and geometry. The reason for the use of quasi-isotropic material (even though its modulus of elasticity is much lower than laminated alternatives) is because of the application of FRP in more advanced production processes, such as 3D printing. The configuration of infill cells in a 3D printed bridge is used to define a self-standing bridge deck. No additional girders under the deck are required, as suggested by [13], while 2.7 m is the standard span length for currently available sandwich panels on the market. The "sandwich deck" is the main load-bearing structure and is supported on two edges. Smart formworks can also be used, as well as 3D printing. Thus, this research could be applicable to laminate, too.

2. FRP Bridge Decks

2.1. Structural Systems

After defining the material properties, it is necessary to define a cross-section of the element. The infill geometry is based on established plate systems that can be found on the market. Four of them have longitudinally oriented cells with different web inclinations, and two have orthogonally oriented cells, known as sandwich panels. Designing a bridge in the form of an arch enabled us to quantify an increase in stiffness and, consequently, a deflection reduction.

Plate bridge deck systems have not changed much since 1991 [14]. The most common systems are sandwich panels and pultruded profiles that are adhesively connected. The technology of pultrusion is the most commonly used in the FRP production process, and for those reasons, the most common plate systems available on the market are EZSpan (Atlantic Research, Gainesville, VA, USA), Superdeck (Creative Pultrusions, Alum Bank, PA, USA), DuraSpan (Martin Marietta Materials, Raleigh, NC, USA), and Strongwell. Sandwich panels consist of top and bottom flanges that carry the load and a low-weight infill that connects them and transmits the load but does not contribute to the panel's stiffness [15].

In a recent review paper [16], a summary of FRP pultruded deck and sandwich panel failure modes has been provided [13,17–19]. It can be concluded that FRP decks demonstrate linear elastic behavior up until failure. Some pseudo-ductility could be achieved through the structural system and cell configuration [16].

An example of a temporary bridge structure [20] shows that after eight years of service, some visible damages were: flange crushing, longitudinal cracks, visible fibers due to top-surface blooming, and some local damages. Despite these flaws, structural stability was not yet jeopardized, and all of the damages were easily repaired, for the connections stayed in perfect condition.

It has been reported [21] that for triangular cell configurations in a pultruded deck (EZ Span and Asset deck), only around 20% of the material compression strength is utilized. This is caused by local buckling and delamination between the web and flanges, which are the symptoms of deck failure. Local deformation or cracking of the top surface or wear surface will occur under patch load, well before the compression strength limit of the material is reached. For this reason, many parameters of FRP decks should be considered during the design to obtain a more optimal, cost-acceptable, attractive, and sustainable solution.

The pultruded plate systems that are currently available on the market vary in depth from 80 to 225 mm depending on the production process (see Table 1). Sandwich panels, on the other hand, are more flexible in terms of production dimensions, and can also be inclined. An FRP composite structural system with FRP sandwich decking laying on a U-box girder with a bridge span of 15–25 m has recently been proposed and examined in Poland [22,23].

Table 1. Physical properties of plate systems from adhesively bonded pultruded profiles [6].

System	Production Process	Thickness (mm)	Weight per m^2 (kN/m^2)
Superdeck	Pultrusion	203	1.0
DuraSpan	Pultrusion	195	1.05
EZSpan	Pultrusion	216	0.96
Strongwell	Pultrusion	170	-

The FRP's load-bearing shell structure can be flat or curved. A shell in arch has a higher stiffness and a smaller deformation. The problem with forming the elements in an arch is that additional material is required at the area where the bending moment is significant. A bridge made in cooperation with the FiberCore company [24] is a good example of where each bridge structure's design fulfills the required, ultimate serviceability limit states for every member within the allocated costs. Lately, optimization with additional software, such as Grasshopper, Karamba, and Kangaroo, has been of great help [25].

Many aspects of FRP decks affect the final results, such as material (e.g., GFRP or CFRP), the configuration and inclination angle of the internal cavities, and thicknesses, etc. It is shown [9,26] that a triangular configuration in pultruded decks has a higher in-plane shear stiffness (G_{xz}) compared to a rectangular configuration. In sandwich panels, the density of the cavities will influence stiffness more than geometry. Sandwich panels can be filled with polymeric foam in order to increase stiffness [6]. Sandwich decks usually have sufficient shear stiffness to transfer these strains, whereas pultruded FRP decks have lower shear stiffness.

2.2. Production

Several technological processes have been used to produce FRP elements depending on the required shape and structural parameters. One of the first was the hand-lamination process, which is fully manual, as indicated by the name. The number of products that can be made is limited, but the form and shape of the products is limitless, depending on the formwork. Pultrusion is when fibers soaked in a matrix go through a heated mold where the element is shaped. This is a fully automated process, and after the element leaves the mold, it is cut at the required length. Although the size of continuous cross-section limits the process, the quality of the produced members is of the highest standard among all currently established production processes. It is also one of the most economical processes.

Technology is leading us to smarter and more cost-effective solutions. An important recent innovation in the production of curved-plate elements has been the use of smart formworks, especially when these are used not for a series of the same product, but for the creation of a unique product. At the moment, these formworks are well known in the production of concrete [27] and glass panels, but it could also be applicable to FRP. Greater freedom of architectural design is one of the most important advantages that this process offers.

Another growing industry in pedestrian bridge production is 3D printing technology and production assisted by robotic arms. There are few examples of 3D-printed steel bridges, but also some promising projects that involve printing with FRP. An example of a 3D-printed pedestrian bridge can be found in China from 2020 [28], which has total length-width dimensions of 15.5 m × 3.8 m (see Figure 1). The total load capacity is estimated to be 250 kg/m^2, and it has an expected lifetime of 30 years. The material is ASA (acrylonitrile styrene acrylate), reinforced with glass fibers. The total weight of the bridge is 5.8 t, and it has a fiber content of 12.5%. Total production took 30 days on a 3D printer, which had the capacity to print 8 kg/h and a maximum volume of 24 m × 4 m × 1.5 m. The main problem is that 3D printing is still a relatively slow process compared to the construction of bridges with traditional materials and techniques.

Figure 1. First 3D-printed bridge in China, Shanghai (**left**); 3D printing infill pattern (**middle**); 3D-printed cross section (**right**) [28].

The pultrusion production process limits the FRP bridge deck to unidirectional decks with a straight and constant depth. Unlike pultruded decks, sandwich decks can be produced with varying depths and larger sizes. Sandwich panels include two stiff and strong face plates adhesively connected to a core [16]. The core material can vary in configuration (e.g., honeycomb, triangular, sinusoidal) and density. A denser raster will result in greater stiffness. The optimization of the core material within the deck could be obtained if the denser core would appear in areas of higher stresses. Moreover, novel materials for higher stiffness and strength could be utilized, resulting in a more elegant structure [29,30].

The most cost-effective production process will likely govern the final bridge design. In this study, longer life cost–benefit is considered. The totality of the labor could be automatized without sacrificing another aspect of the project, such as time and money saving, or the appearance of the structure. Production processes assisted by robotic arms, 3D printers, and smart formworks are some of the possible solutions. These are costly in the beginning but easily adaptable to various project solutions. More sophisticated technologies would allow variations in slope, depth, and 3D core configuration to be integrated into the FRP bridge deck. Thus, additional material could be added in places where higher local stiffness is needed. A subject for future research will be the performance of 3D cell configurations [31].

2.3. Connections

The serviceability limit state will most likely govern the design in the structural analysis unless there are mechanical connections. In this case, connection location becomes critical, rather than the element itself. In adhesively connected elements, the load is transferred uniformly and, for that reason, this feature is considered to be more appropriate in FRP structures.

During the design of such FRP plate systems, the main purpose of these connections is to transfer bending moment and shear force between two systems without significant deflections. Laboratory tests showed that the collapse of the component would occur before a connection failure occurs. Laboratory examinations for more significant projects, where unreasonably high partial safety factors are applied, lead to oversized structures [32]. The study and testing of FRP material [33] will reduce the time required for laboratory testing and facilitate the design of structures with this material.

The connections and adhesive bonds between elements in pultruded decks can significantly influence the global stiffness [26,34]. The force is transferred more uniformly when adhesive connections are used, making them more appropriate for FRP structures. Mechanical connectors are mandatory when an element exceeds limitations in the production process or during transportation.

According to the Eurocomp standard and manual [35], there are three types of connections: connections that should last a whole lifetime and whose collapse would be devastating for the whole structure; connections whose collapse would affect the structure locally and would not have a serious impact on the structure as a whole; and non-structural connections which basically connect secondary elements, such as a fence.

3. Materials and Methods

In the previous section, a general overview of existing research on FRP decks was presented. Based on the literature review on FRP deck systems, two general groups of products can be distinguished: panels made of adhesively pultruded profiles and sandwich panels [36]. Models of bridge decks consist of three adhesively bonded components (the top surface, web, and bottom surface).

In this study, a bridge deck with two hinge supports and a span of 5 m was used. The depth of all of the decks was uniform at 200 mm, while the width was 1.8 m. Uniform wall thicknesses of 6 mm were used in all cross-sections (four layers of 1.5 mm in two orthogonal directions (0/90/0/90)). The main aim of this study was to have a similar depth and mass for all decks. Thus, the stiffness would be primarily influenced by the cell configuration (Figure 2).

Different infill geometries and web inclinations were analyzed using Abaqus software. Panel depth was taken as 200 mm according to the available standard panel systems (Figure 3): (a,b) EZ-Span (49.91 kg/m^2), (c,d) Superdeck (50.48 kg/m^2), (e,f) DuraSpan (50.74 kg/m^2), and (g,h) Strongwell (41.08 kg/m^2), which are made of pultruded profiles, and sandwich panels: (i,j) with a profiled triangular infill (74.84 kg/m^2) and (k,l) honeycomb sandwich panels (54.19 kg/m^2). The geometry of the deck cross-sections (rectangular, triangular, trapezoidal, and honeycomb) was taken from the different FRP cell units available in the market. Each plate system weighs approximately 50 kg/m^2. Future research will be performed for 3D cell configuration [37].

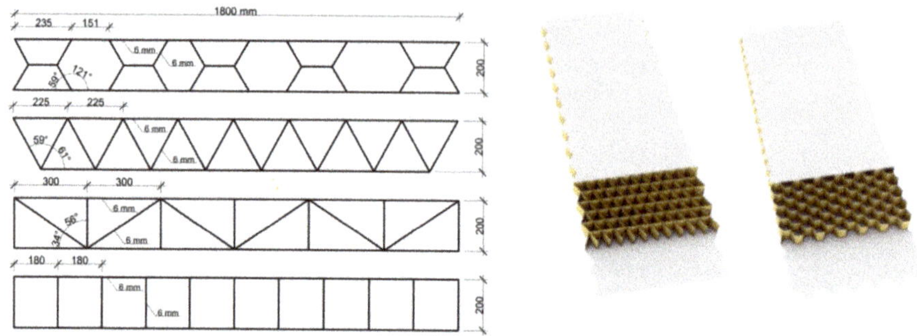

Figure 2. Cross-section of pultruded deck panels in mm (**left**); sandwich panels (**right**).

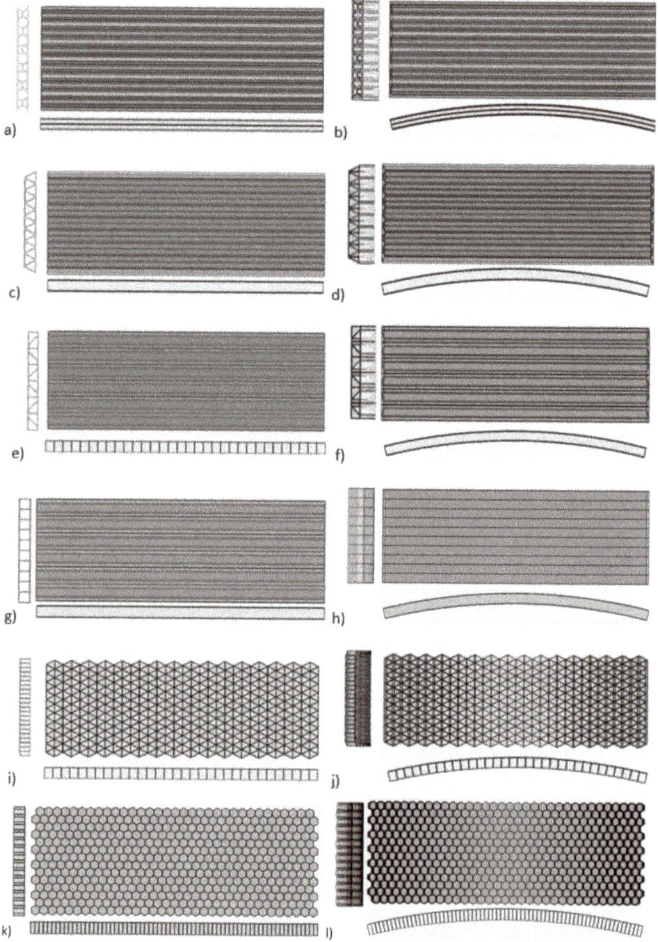

Figure 3. GFRP plate, adhesively bonded pultruded profiles: (**a**) Superdeck; (**b**) Superdeck in an arch; (**c**) EZSpan; (**d**) EZSpan in an arch; (**e**) DuraSpan; (**f**) DuraSpan in an arch; (**g**) Strongwell; (**h**) Strongwell in an arch. Sandwich panels: (**i**) triangular-shaped infill; (**j**) triangular-shaped infill in an arch; (**k**) honeycomb-shaped infill; (**l**) honeycomb-shaped infill in an arch.

The geometry was defined using the Rhinoceros software package. The serviceability load combination was applied to the top of each deck according to the British code [38]. A distributed live load of 5 kN/m^2 and a dead load of 1 kN/m^2 was applied (in case of wearing, the surface is applied).

In this study, a simple arch with a height of 30 cm at the middle of the span was developed. With innovative production processes, such as 3D printing, more design freedom is possible. Optimal structures, based on internal forces, could be created. The selection of bridge geometry would not be determined by the location of the bridge.

In addition, a comparison was made between a composite fiber layup (0/90/0/90) material and a quasi-isotropic material. The composite layup consists of four orthogonally positioned fiber layers, each 1.5 mm thick, with their material properties shown in Table 2. On the other hand, one modulus of elasticity is defined for the quasi-isotropic material. The modulus of elasticity for composites reinforced with glass fibers can vary from 5 GPa to 50 GPa [39]. For a resin reinforced with 40% of chopped strand mat by weight, a modulus of elasticity of 10 GPa was chosen (see Figure 4). Three-dimensional models were imported from Rhinoceros software into Abaqus software as surface shell parts. Tie constraints connected the top and bottom surfaces to infill surfaces, thus simulating the rigid behavior of all composite panel board instances. The pedestrian bridge was analyzed as a clamped supported beam and a clamped arch. Finite element modeling was performed with quadratic mesh size elements of 50 mm $\times$ 50 mm.

Table 2. Composite layup material properties for glass fiber epoxy resin [40].

E_1 (MPa)	E_2 (MPa)	ν_{12}	G_{12} (MPa)	G_{13} (MPa)	G_{23} (MPa)	ρ (kg/m^3)
34,412	6531	0.217	2433	1698	2433	2000

E_1: Young's modulus in longitudinal (fiber) direction. E_2: Young's modulus in transverse. ν_{12}: Minor Poisson ratio. G_{12}: In-plane shear modulus. G_{13}: Out of plane shear modulus. G_{23}: Out of plane shear modulus. ρ: Density.

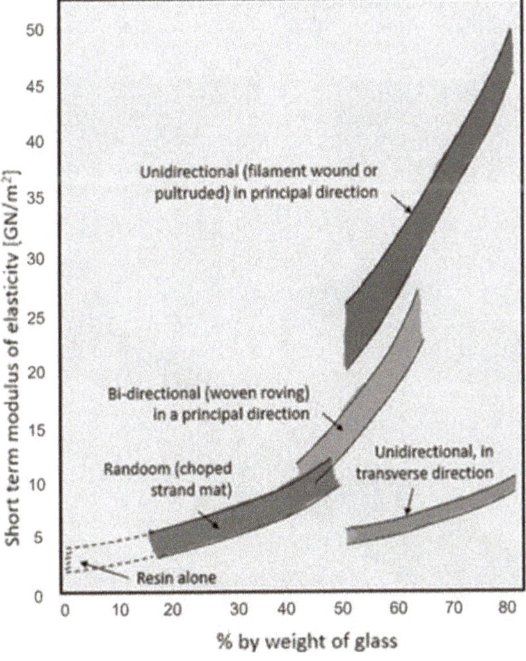

Figure 4. Modulus of elasticity for glass fiber reinforced polymers [39].

4. Results

The most effective cell unit configuration was determined based on the serviceability design criteria as governed by deflections. According to the British code [38], a deformation limitation of L/300 (16.6 mm) was compared to the maximum deformation at the middle of the span. To obtain the deformations, the complex numerical modeling of several FRP decks (see Figure 2) was performed in Abaqus. The results are shown in Figures 5–10.

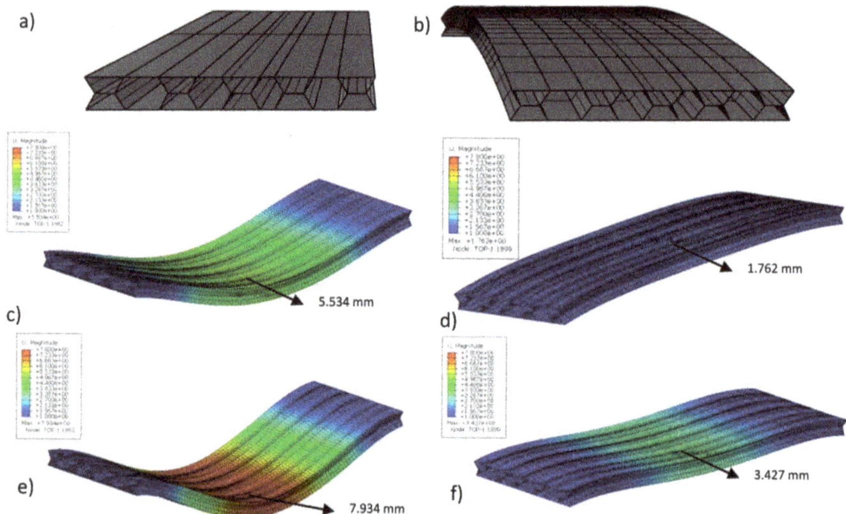

Figure 5. Geometry of: (**a**) Superdeck and (**b**) Superdeck in an arch. Deformation for: (**c**) Superdeck (laminate); (**d**) Superdeck (laminate) in an arch; (**e**) Superdeck (quasi-isotropic); (**f**) Superdeck (quasi-isotropic) in an arch.

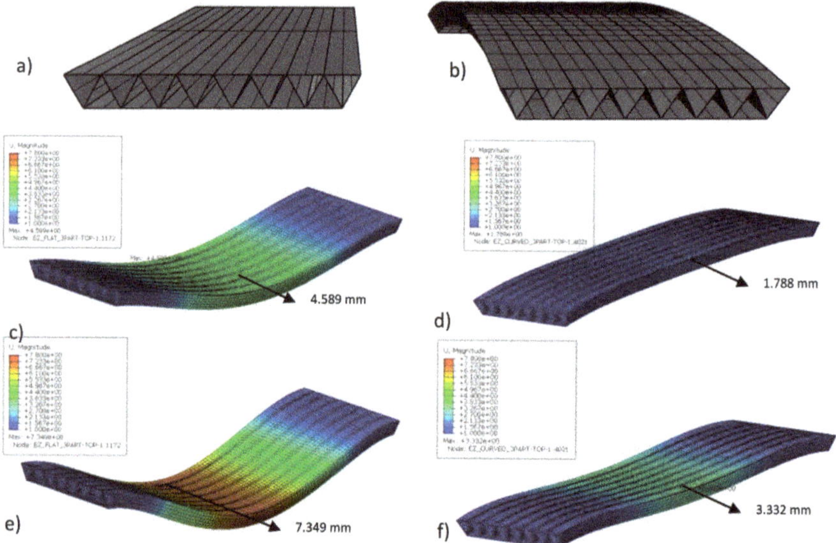

Figure 6. Geometry of: (**a**) EZSpan and (**b**) EZSpan in an arch. Deformation for: (**c**) EZSpan (laminate); (**d**) EZspan (laminate) in an arch; (**e**) EZspan (quasi-isotropic); (**f**) EZspan (quasi-isotropic) in an arch.

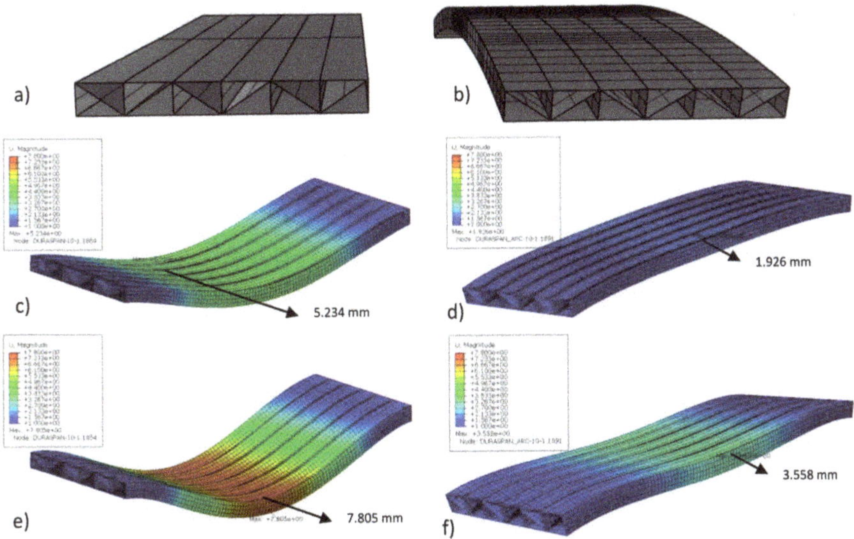

Figure 7. Geometry of: (**a**) Duraspan and (**b**) Duraspan in an arch. Deformation for: (**c**) Duraspan (laminate); (**d**) Duraspan (laminate) in an arch; (**e**) Duraspan (quasi-isotropic); (**f**) Duraspan (quasi-isotropic) in an arch.

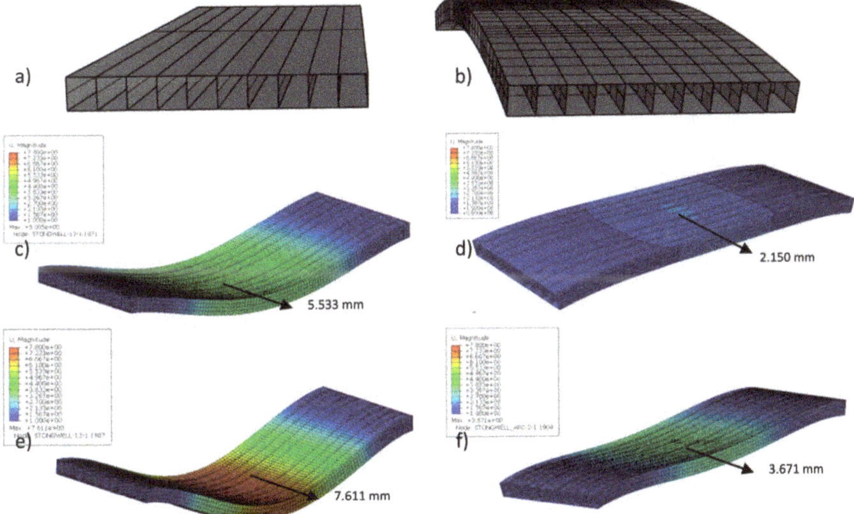

Figure 8. Geometry of: (**a**) Strongwell and (**b**) Strongwell in an arch. Deformation for: (**c**) Strongwell (laminate); (**d**) Strongwell (laminate) in an arch; (**e**) Strongwell (quasi-isotropic); (**f**) Strongwell (quasi-isotropic) in an arch.

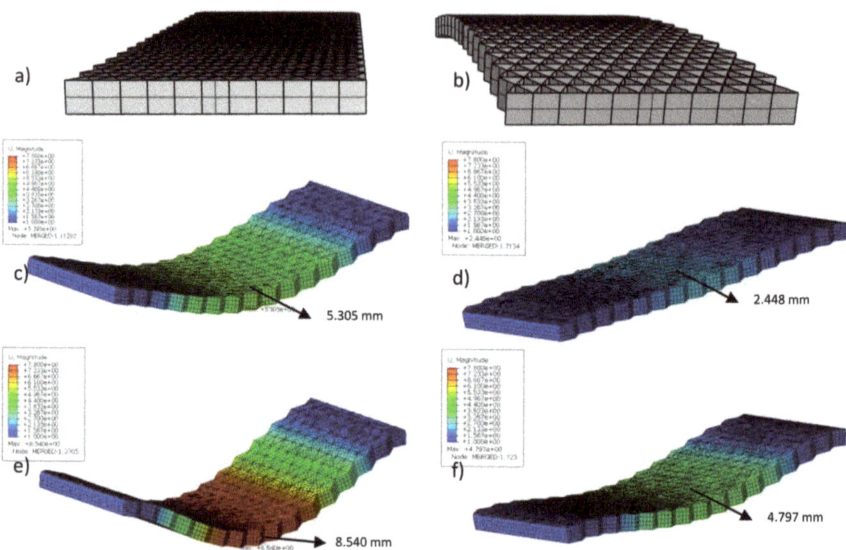

Figure 9. Geometry of sandwich panel with: (**a**) triangular cell configuration and (**b**) triangular cell configuration in an arch. Deformation for sandwich panel with: (**c**) triangular cell configuration (laminate); (**d**) triangular cell configuration (laminate) in an arch; (**e**) triangular cell configuration (quasi-isotropic); (**f**) triangular cell configuration (quasi-isotropic) in an arch.

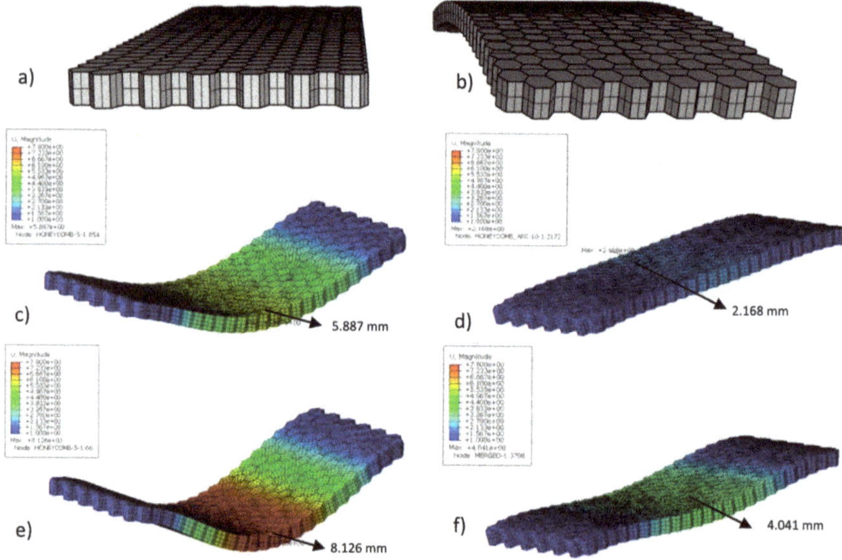

Figure 10. Geometry of sandwich panel with: (**a**) honeycomb cell configuration and (**b**) honeycomb cell configuration in an arch. Deformation for sandwich panel with: (**c**) honeycomb cell configuration (laminate); (**d**) honeycomb cell configuration (laminate) in an arch; (**e**) honeycomb cell configuration (quasi-isotropic); (**f**) honeycomb cell configuration (quasi-isotropic) in an arch.

The comparison of the results is shown in Figure 11. Since the serviceability limit states govern the design of these types of bridges, the displacements are more relevant for the analysis.

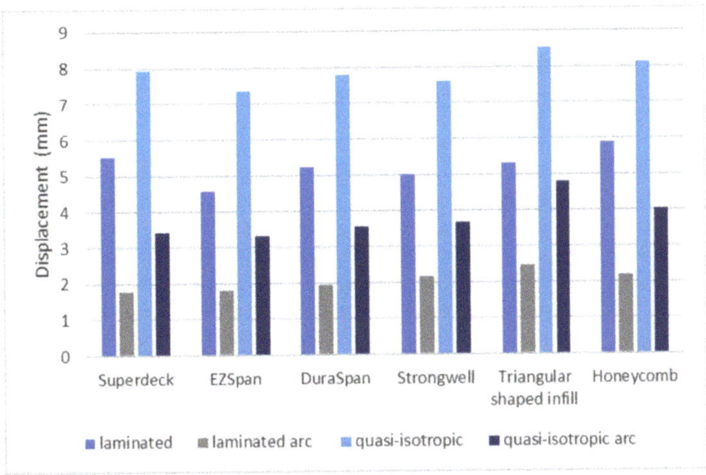

Figure 11. Results for displacements.

The study showed that the lowest deflections of about 1.75 mm were obtained in the arched Superdeck (which has a hexagonal cell configuration) and EZ Span (with a triangular cell configuration) bridge decks. The decks made of pultruded profiles were stiffer overall than the sandwich panels. In previous research by [41], the authors found that Superdeck (hexagonal cell) and ASSET (triangular cell, similar to EZ Span) were stiffer than others, which is in accordance with the results of this study.

Bilim et al. [41] investigated the influence of the additional concrete layer on the top face and concluded that the displacements of the bridge deck are reduced by an additional 60%. Several studies have been conducted showing an increase in local stiffness after the application of the wear layer [11,21]. In this study, no additional wear layer was applied that could influence the stiffness, but it could be part of future research.

The midspan displacement values for decks with a quasi-isotropic material exceed the midspan displacement values for decks with the laminate layup (0/90/0/90) by 32% on average for flat decks and by 46% on average for arched decks. Thus, the calculated mass per m^2 and midspan displacement values are given in Table 3 for arched and flat decks with the laminate layup. For example, the arched Superdeck system has a smaller deflection than the Strongwell system (by 18%), but a higher value in mass per m^2 (by 23%) than the Strongwell system.

Table 3. Results for composite layup (0/90/0/90) plate systems.

Plate System	Flat	Arch	Mass Per m^2
	Displacement (mm)		(kg/m^2)
Superdeck (laminated)	5.53	1.76	50.48
EZSpan (laminated)	4.59	1.79	49.91
DuraSpan (laminated)	5.23	1.93	50.74
Strongwell (laminated)	5.01	2.15	41.08
Triangular (laminated)	5.31	2.45	74.87
Honeycomb (laminated)	5.89	2.17	54.19

The results show that the maximum deflection occurred in the flat bridge with quasi-isotropic material properties, reaching 8.5 mm in the sandwich panel with triangular infill configuration.

The main conclusion can be drawn by comparing the calculated displacements in the flat laminated decks with the quasi-isotropic decks in an arch. Even though the quasi-isotropic material with the chopped strands has lower material properties than the cross-ply laminate, the arch shape of the bridge deck contributed to the overall stiffness by reducing the deformation by around 30–40%.

5. Discussion and Conclusions

FRP as a material is still under investigation and is waiting to fulfill its potential in the construction industry. All aspects, from geometry, to material parameters, to connections, to its production, are important for a successful bridge design.

In this study, finite element modeling of FRP bridge decks made of pultruded profiles and sandwich panels was performed in Abaqus. Typical infill geometries (rectangular, triangular, trapezoidal, and honeycomb) were used, with two different materials for each infill, namely a quasi-isotropic material and a cross-ply laminate. The objective of this research was to analyze the effect of the arch shape of the bridge deck on the overall stiffness when compared to a flat bridge deck. An increase in deflection was observed for the quasi-isotropic material in all of the bridge decks considered here when compared to the laminated decks. On the other hand, there is also an obvious reduction in deformation for the plates formed in an arch. Nevertheless, there is a 30–40% stiffness increase in curved bridges with quasi-isotropic material properties compared to laminate bridge decks in a flat shape. In pultruded decks, hexagonal and triangular infill configurations show smaller deformations than rectangular, while in sandwich panels, honeycomb is the best configuration. When comparing pultruded decks and sandwich panels under a uniform load, pultruded profiles with longitudinal cavities are stiffer. There are research possibilities for finding new infill configurations, especially in 3D infill geometries that could vary in density and cross-sectional height. The results justify the use of 3D printing technology and polymeric filament that allow variations in the shape and infill cell configuration of the bridge structure.

Form finding could reduce the material cost of retaining the structural integrity. In future research, the investigation of and laboratory tests for material properties are mandatory. Bridge self-weight plays an important role in decision making. The main idea is the definition of curved shapes and tridimensional infill configurations to find out which would give the lightest possible bridge structures.

Analytical approaches through software such as Grasshopper, Karamba and Kangaroo will help in designing architecturally appealing bridge solutions. Material properties will depend on the requirements (economical), but since each project will be individual, the material properties, infill geometry, and shape of the structure will vary. Future bridge design should be unlimited in all its segments, from modeling and design to its production.

Author Contributions: Conceptualization, L.S., A.S., J.G. and D.D.; methodology, L.S. and A.S.; software, L.S.; validation, L.S., A.S., J.G. and D.D.; formal analysis, L.S.; investigation, L.S.; resources, L.S., A.S., J.G. and D.D.; data curation, L.S., A.S., J.G. and D.D.; writing—original draft preparation, L.S. and A.S.; writing—review and editing, L.S., A.S., J.G. and D.D.; visualization, L.S.; supervision, A.S., J.G. and D.D.; project administration, J.G. and D.D.; funding acquisition, A.S., J.G. and D.D. All authors have read and agreed to the published version of the manuscript.

Funding: This research received no external funding.

Institutional Review Board Statement: Not applicable.

Informed Consent Statement: Not applicable.

Data Availability Statement: Data available on request due to restrictions, e.g., privacy or ethical. The data presented in this study are available on request from the corresponding author.

Conflicts of Interest: The authors declare no conflict of interest.

References

1. Smits, J. Fiber-Reinforced Polymer Bridge Design in the Netherlands: Architectural Challenges toward Innovative, Sustainable, and Durable Bridges. *Engineering* **2016**, *2*, 518–527. [CrossRef]
2. Sonnenschein, R.; Gajdosova, K.; Holly, I. FRP Composites and their Using in the Construction of Bridges. *Procedia Eng.* **2016**, *161*, 477–482. [CrossRef]
3. Ascione, L.; Caron, J.-F.; Godonou, P.; van Ijselmuijden, K.; Knippers, J.; Mottram, J.; Oppe, M.; Gantriis Sorensen, M.; Taby, J.; Tromp, L. *Prospect for New Guidance in the Design of FRP: Support to the Implementation, Harmonization and Further Development of the Eurocodes*; Joint Research Centre Institute for the Protection and the Security of the Citizen: Ispra, Italy, 2016; ISBN 9789279542251.
4. Kišiček, T.; Stepinac, M.; Renić, T.; Hafner, I.; Lulić, L. Strengthening of masonry walls with FRP or TRM. *Gradjevinar* **2020**, *72*, 937–953. [CrossRef]
5. Jin, F.; Feng, P.; Ye, L. Study on dynamic characteristics of light-weight FRP footbridge. In *Advances in FRP Composites in Civil Engineering, Proceedings of the 5th International Conference on FRP Composites in Civil Engineering, CICE 2010, Beijing, China, 27–29 September 2010*; Tsinghua University Press: Beijing, China, 2011; pp. 173–176.
6. Mara, V. Fibre Reinforced Polymer Bridge Decks: Sustainability and a Novel Panel-Level Connection. Bachelor's Thesis, Chalmers University of Technology, Göteborg, Sweden, 2014.
7. Gunaydin, M.; Adanur, S.; Altunisik, A.C.; Sevim, B. Static and dynamic responses of Halgavor footbridge using steel and FRP materials. *Steel Compos. Struct.* **2015**, *18*, 51–69. [CrossRef]
8. Davalos, J.F.; Chen, A.; Zou, B. Stiffness and Strength Evaluations of a Shear Connection System for FRP Bridge Decks to Steel Girders. *J. Compos. Constr.* **2011**, *15*, 441–450. [CrossRef]
9. Keller, T.; Gürtler, H. Quasi-static and fatigue performance of a cellular FRP bridge deck adhesively bonded to steel girders. *Compos. Struct.* **2005**, *70*, 484–496. [CrossRef]
10. Solomon, G.; Godwin, G. Expanded use of composite deck projects in USA. *Struct. Eng. Int.* **2002**, *12*, 102–104. [CrossRef]
11. Sebastian, W.M.; Keller, T.; Ross, J. Influences of polymer concrete surfacing and localised load distribution on behaviour up to failure of an orthotropic FRP bridge deck. *Compos. Part. B Eng.* **2013**, *45*, 1234–1250. [CrossRef]
12. Hollaway, L.C. A review of the present and future utilisation of FRP composites in the civil infrastructure with reference to their important in-service properties. *Constr. Build. Mater.* **2010**, *24*, 2419–2445. [CrossRef]
13. Keller, T.; Schollmayer, M. Plate bending behavior of a pultruded GFRP bridge deck system. *Compos. Struct.* **2004**, *64*, 285–295. [CrossRef]
14. Gürtler, H.W. *Composite Action of Frp Bridge Decks Adhesively Bonded To Steel Main Girders Par*; EPFL: Lausanne, Switzerland, 2004; Volume 3135.
15. Bakis, C.E.; Bank, L.C.; Brown, V.L.; Cosenza, E.; Davalos, J.F.; Lesko, J.J.; Machida, A.; Rizkalla, S.H.; Triantafillou, T.C. Fiber-Reinforced Polymer Composites for Construction—State-of-the-Art Review. *J. Compos. Constr.* **2003**, *6*, 369–383. [CrossRef]
16. Mara, V.; Haghani, R. Review of FRP decks: Structural and in-service performance. In *Proceedings of the Institution of Civil Engineers-Bridge Engineering*; Thomas Telford Ltd.: London, UK, 2015; Volume 168, pp. 308–329. [CrossRef]
17. Alagusundaramoorthy, P.; Harik, I.E.; Asce, M.; Choo, C.C. Structural Behavior of FRP Composite Bridge Deck Panels. *J. Bridge Eng.* **2006**, *11*, 384–393. [CrossRef]
18. Camata, G.; Shing, P.B. Static and fatigue load performance of a gfrp honeycomb bridge deck. *Compos. Part B Eng.* **2010**, *41*, 299–307. [CrossRef]
19. Majumdar, P.K.; Liu, Z.; Lesko, J.J.; Cousins, T.E. Performance Evaluation of FRP Composite Deck Considering for Local Deformation Effects. *J. Compos. Constr.* **2009**, *13*, 332–338. [CrossRef]
20. Keller, T.; Bai, Y.; Valle, T. Long-term performance of a glass fiber-reinforced polymer truss bridge. *J. Compos. Constr.* **2007**, *11*, 99–108. [CrossRef]
21. Gabler, M.; Knippers, J. Improving fail-safety of road bridges built with non-ductile fibre composites. *Constr. Build. Mater.* **2013**, *49*, 1054–1063. [CrossRef]
22. Kulpa, M.; Siwowski, T.; Rajchel, M.; Wlasak, L. Design and experimental verification of a novel fibre-reinforced polymer sandwich decking system for bridge application. *J. Sandw. Struct. Mater.* **2021**, *23*, 2326–2357. [CrossRef]
23. Siwowski, T.; Kulpa, M.; Rajchel, M.; Poneta, P. Design, manufacturing and structural testing of all-composite FRP bridge girder. *Compos. Struct.* **2018**, *206*, 814–827. [CrossRef]
24. FiberCore Europe. Available online: https://www.fibercore-europe.com/en/ (accessed on 2 February 2021).
25. Smits, J. *The Art of Bridge Design*; TU Delft Open: Delft, The Netherlands, 2019; ISBN 978-94-6366-164-5.
26. Keller, T.; Gürtler, H. Design of hybrid bridge girders with adhesively bonded and compositely acting FRP deck. *Compos. Struct.* **2006**, *74*, 202–212. [CrossRef]
27. Schipper, R.; Janssen, B. *Curving Concrete: A Method for Manufacturing Double Curved Precast Concrete Panels Using a Flexible Mould*; Hemming Group Ltd: London, UK, 2011.
28. Polymaker. 3D Printed Bridge. Available online: https://polymaker.com/3d-printed-bridge/ (accessed on 12 January 2021).
29. Chen, Y.; Ye, L.; Fu, K. Progressive failure of CFRP tubes reinforced with composite sandwich panels: Numerical analysis and energy absorption. *Compos. Struct.* **2021**, *263*, 113674. [CrossRef]

30. Chen, Y.; Cheng, X.; Fu, K.; Ye, L. Failure characteristics and multi-objective optimisation of CF/EP composite sandwich panels under edgewise crushing. *Int. J. Mech. Sci.* **2020**, *183*. [CrossRef]
31. Kreja, I. A literature review on computational models for laminated composite and sandwich panels. *Cent. Eur. J. Eng.* **2011**, *1*, 59–80. [CrossRef]
32. Canning, L.; Luke, S. Development of FRP bridges in the UK: An overview. *Adv. Struct. Eng.* **2010**, *13*, 823–835. [CrossRef]
33. Kulpa, M.; Wiater, A.; Rajchel, M.; Siwowski, T. Comparison of material properties of multilayered laminates determined by testing and micromechanics. *Materials* **2021**, *14*, 761. [CrossRef]
34. Park, S.Z.; Hong, K.J.; Lee, S.W. Behavior of an adhesive joint under weak-axis bending in a pultruded GFRP bridge deck. *Compos. Part. B Eng.* **2014**, *63*, 123–140. [CrossRef]
35. Clarke, J.L. *European Structural Polymeric Composites Group. Structural Design of Polymer Composites: EUROCOMP Design Code and Handbook*; E & FN Spon: London, UK, 2005; Volume 148, ISBN 0203475135.
36. Zhou, A.; Keller, T. Joining techniques for fiber reinforced polymer composite bridge deck systems. *Compos. Struct.* **2005**, *69*, 336–345. [CrossRef]
37. Kreja, I. On geometrically non-linear FEA of laminated FRP composite panels. In *Shell Structures: Theory and Applications*; CRC Press: Boca Raton, FL, USA, 2014; Volume 3, pp. 33–42. [CrossRef]
38. *Design of FRP Bridges and Highway Structures*; The Stationery Office: London, UK, 2005; Volume 1, p. 21.
39. Leggatt, A.J. *GRP and buildings: A Design Guide for Architects and Engineers*; Butterworths: London, UK, 1984; Volume 1.
40. Supeni, E.E.; Epaarachchi, J.A.; Islam, M.M.; Lau, K.T. Design of Smart Structures for Wind Turbine Blades. In Proceedings of the 2nd Malaysian Postgraduate Conference, Bond University, Gold Coast, Queensland, Australia, 7–9 July 2012; Noor, M.M., Rahman, M.M., Ismail, J., Eds.; Education Malaysia: Kuala Lumpur, Malaysia, 2012.
41. Bilim, C.; Kara, İ.F.; Ashour, A.F. Flexural Behavior of Hybrid Frp-Concrete Bridge Decks. *Turk. J. Eng.* **2019**, *3*, 206–217. [CrossRef]

 *buildings*

Review

An Abridged Review of Buckling Analysis of Compression Members in Construction

Manmohan Dass Goel [1], Chiara Bedon [2,*], Adesh Singh [1], Ashish Premkishor Khatri [1] and Laxmikant Madanmanohar Gupta [1]

[1] Department of Applied Mechanics, Visvesvaraya National Institute of Technology, Nagpur 440 010, India; mdgoel@apm.vnit.ac.in (M.D.G.); adeshsingh3108@gmail.com (A.S.); ashishkhatri@apm.vnit.ac.in (A.P.K.); lmgupta@apm.vnit.ac.in (L.M.G.)

[2] Department of Engineering and Architecture, University of Trieste, 34127 Trieste, Italy

* Correspondence: chiara.bedon@dia.units.it; Tel.: +39-040-5583837

Abstract: The column buckling problem was first investigated by Leonhard Euler in 1757. Since then, numerous efforts have been made to enhance the buckling capacity of slender columns, because of their importance in structural, mechanical, aeronautical, biomedical, and several other engineering fields. Buckling analysis has become a critical aspect, especially in the safety engineering design since, at the time of failure, the actual stress at the point of failure is significantly lower than the material capability to withstand the imposed loads. With the recent advancement in materials and composites, the load-carrying capacity of columns has been remarkably increased, without any significant increase in their size, thus resulting in even more slender compressive members that can be susceptible to buckling collapse. Thus, nonuniformity in columns can be achieved in two ways—either by varying the material properties or by varying the cross section (i.e., shape and size). Both these methods are preferred because they actually inherited the advantage of the reduction in the dead load of the column. Hence, an attempt is made herein to present an abridged review on the buckling analysis of the columns with major emphasis on the buckling of nonuniform and functionally graded columns. Moreover, the paper provides a concise discussion on references that could be helpful for researchers and designers to understand and address the relevant buckling parameters.

Keywords: buckling; compression members; Euler's load; nonprismatic sections; imperfections; slenderness

Citation: Goel, M.D.; Bedon, C.; Singh, A.; Khatri, A.P.; Gupta, L.M. An Abridged Review of Buckling Analysis of Compression Members in Construction. *Buildings* **2021**, *11*, 211. https://doi.org/10.3390/buildings11050211

Academic Editor: Francisco López Almansa

Received: 12 March 2021
Accepted: 14 May 2021
Published: 18 May 2021

Publisher's Note: MDPI stays neutral with regard to jurisdictional claims in published maps and institutional affiliations.

1. Introduction

Compression members are an integral part of the structures, and unlike other load-bearing members, their capacity to carry loads is governed by the different sets of influencing parameters. This difference in behaviour questions their structural integrity and necessitates the analysis of compression members with numerical models that could offer a minimum deviation from the reality and thus ensure a fairly close estimation of the actual buckling load.

While the stability issue was first pointed out in 1675 by Hooke [1], several other formulations followed especially during the 18th century, and even further important developments in the support of design have been obtained in the last few decades. Currently, the development of novel design applications, materials, and composites solutions enforces a further need for dedicated calculation tools. In the last decades, the column buckling issue has become relevant for traditional constructional applications but especially for innovative material solutions, as in Figure 1, in which selected examples can be seen for FRP-reinforced concrete columns [2], repaired timber columns [3] and even hollow square glass columns [4].

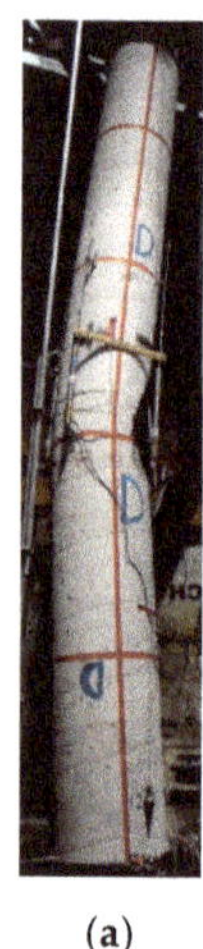

(a)   (b)   (c)

Figure 1. Examples of column buckling for constructional members: (**a**) FRP-reinforced concrete columns (reproduced from [2] with permission from Elsevier®, license n. 5026030791841); (**b**) repaired timber columns (reproduced from [3] with permission from Elsevier®, license n. 5026030922470); and (**c**) hollow square glass columns (reproduced from [4] with permission from Elsevier®, license n. 5026031011542).

In Section 2, some basics concepts and background theories are first presented. Section 3 provides a brief overview of the methods and critical issues on the buckling failure of short columns, while slender columns are discussed in Section 4. Finally, Section 4.5 presents a subdiscussion on compressed members with variable stiffness due to thermal gradients and constructional materials that can be remarkably sensitive to degradation and hence to the premature column buckling collapse. It is important to mention that this work is primarily focused on the global buckling of the compression member.

2. Basics

The necessary preliminary analysis of the stability problem was proposed by Hooke in 1675, wherein it was shown that the displacement in any structural body is directly proportional to the load causing the displacement. This law can be applied to spring bodies, stone, wood, metal, etc., and it is commonly known as Hooke's law [1]. Further, Bernoulli studied the curvature and deflection of a cantilever beam using Hooke's law in 1705. It was Euler who was credited with the first systematic study of the stability problem in equilibrium. In his first publications, Euler investigated the stability of a hinged bar, having flexural rigidity (EI), in equilibrium, subjected to an axially compressive force (p) and uniformly distributed load (q) along the longitudinal axis (z) by two different approaches [5–8]. It is interesting to note that Euler has defined all his formulations in terms of Ek^2 instead of EI, with E defined as strength property and k^2 as a dimensional property of the column. Further, the transformation from Ek^2 to EI requires the knowledge of Hooke's law, and it was Coulomb, who, for the first time, applied Hooke's law and equation of static equilibrium to develop the bending moment and normal stress due to the elastic bending in cantilever column as follows [9]:

$$EI\frac{d^3v}{dz^3} + qz\frac{dv}{dz} + p\frac{dv}{dz} = 0 \tag{1}$$

As is evident from Equation (1), its solution will contain only three constants, and the equation has failed to satisfy four boundary conditions. Euler identified this error and presented a corrected differential equation in his third paper by including the presence of a horizontal force N [5]. However, it is interesting to note that Euler did a numerical mistake, and calculated the second eigenvalue instead of the first, which was later corrected by [10–12]. Thus, the equation of static equilibrium (Equation (2)) to develop bending moment and normal stress due to the elastic bending in a cantilever column is:

$$EI\frac{d^3v}{dz^3} + qz\frac{dv}{dz} + p\frac{dv}{dz} = N \tag{2}$$

Euler's analytical conclusion supported the experimental results obtained by Musschenbroek [13] for slender wooden columns. However, Coulomb discarded the result of Musschenbroek and concluded that the breaking strength was independent of length, based on experiments on masonry columns [9]. Duleau, Hodgkinson, Considère, and Engesser discussed Euler's formulation and its exclusive validity for "slender" columns [13–17]. Moreover, Hodgkinson proposed an empirical formula for the design of short columns based on the experimental investigations on cast-iron columns. In the year 1845, Lamarle proposed a critical load expression in terms of the critical stress and stated that Euler formulation is applicable when the critical stress (σ_{cr}) is less than the elastic limit (σ_0) for the constructional material in use. In other words, it is applicable for the struts whose slenderness ratio (l/h) is greater than the limit value given as follows i.e., Equation (3) [18]:

$$\left(\frac{l}{r}\right)^2 = \frac{\pi^2 E}{\sigma_0} \tag{3}$$

where r is the radius of gyration about the weaker axis of the column. Although there is no record of whether Lamarle's suggestion was used anywhere practically, the formula suggested by Gordon provides the same result as Lamarle's model, and this is verified both for large and small slenderness ratios [19]. Figure 2 shows some typical design curves, as conventionally obtained in terms of stress and slenderness ratio, based on Lamarle and Gordon models.

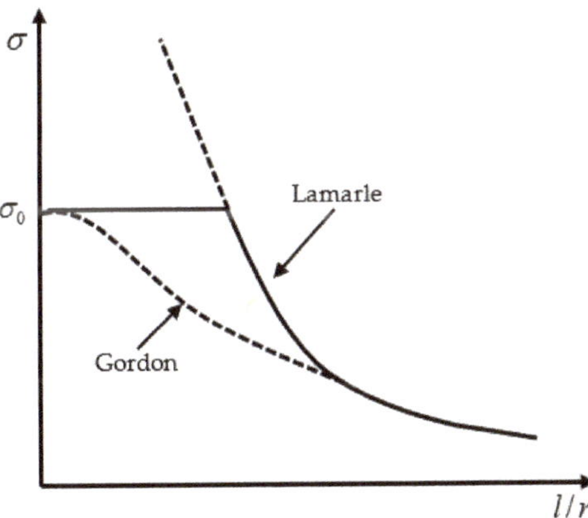

Figure 2. Comparison of design curves for compression members.

The proportionality between the stress and the strain was proposed by Young [20]. Johnson et al. [21] suggested using Euler's formula by incorporating modifying constant, which is similar to the use of equivalent length coefficient, k.

It is to be noted that, despite considering all the assumptions to transform a real column into an ideal column, the existence of perfectly clamped or pinned boundary conditions at either end and no demand of flexural strength from compression members are hard to achieve. In real problems, these assumptions rarely meet since columns in framed structures are supposed to have sufficient flexural rigidity and restrain. Due to this gap, the use of interaction equations is favoured which is based on Ayrton-Perry's approach [22]. They first related the concept of the elastic critical stress to the failure stress, which was later simplified further in [23]. Herein, the average compressive stress (f_c), the allowable compressive stress in an axially loaded strut (p_c), the resultant compressive stress due to bending about the rectangular axis (f_{bc}), and the allowable compressive stress for a member subjected to bending (p_{bc}) are related as per Equation (4) using the well-known beam-column interaction:

$$\frac{f_c}{p_c} + \frac{f_{bc}}{p_{bc}} < 1 \tag{4}$$

3. Buckling Failures

3.1. Self-Buckling

Self-buckling is a phenomenon wherein a column buckles under its own weight; these columns are commonly known as heavy columns. Generally, self-buckling is not considered since it is assumed that the weight of the column is small, compared to the applied axial loads. However, there may be cases in which self-buckling may govern and hence need attention. Self-buckling was first investigated in 1881 by Greenhill [24], and based on his analysis, he proposed that a vertical column may buckle under its own weight if its length exceeds, as given in the following (Equation (5)):

$$l \approx 7.8373 \left(\frac{EI}{\rho g A} \right)^{1/3} \tag{5}$$

where ρ is the density of column material, E is Young's modulus, I is the moment of inertia of column, g is gravitational constant, and A is the cross-sectional area of the column.

Duan and Wang [25] considered buckling of heavy columns and presented an analytical solution in terms of hypergeometric function. They highlighted the fact that buckling capacities were not only dependent on end support condition, shape, size, material but also on the weight. They suggested using a fourth-order differential equation instead of second order. Later on, Darbandi et al. [26] presented closed-form solutions for variable section columns subjected to distributed axial force. Herein, the column was modelled using the Euler's-Bernoulli theory, and solutions were presented using the singular perturbation method of Wentzel-Kramers-Berilloui (WKB), see [26].

In the year 2010, Wei et al. [27] outlined a procedure to compute the buckling load of prismatic and nonprismatic columns under self-weight and tip force. This method did not use Bessel's function as others [25], which strongly depends on the form of an ordinary differential equation with a variable coefficient [27]. Huang and Li [28] studied the column with a nonuniform section using Fredholm's integral equation and presented closed-form solutions. Fredholm's equation transformed the exercise of finding solutions of differential equations to simple algebraic expressions [28]. Later on, Riahi et al. computed the buckling capacity of columns with variable moment of inertia through the slope-deflection method, and dimensionless charts were proposed [29]. On the same line of study, columns with variable inertia (trigonometric-varied inertia column) iteration-perturbation method was applied and obtained results were compared with the result obtained by modelling the same column in ANSYS by Afsharfard and Farshidianfar [30]. Later on, detailed work was reported by Nikolić and Šalinić [31], wherein they assumed that the column is doubly symmetric to apply the method of rigid elements in order to perform buckling analysis of columns with continuously varying cross section and multistepped columns under different boundary conditions.

The described method removes the limitation of the existing rigid body element approach. This method also serves an additional advantage that the boundary condition can be introduced without any extra calculation. However, the limitation of this method lies in the discretisation of elastic segments with rigid segments [31].

3.2. Failure of Inelastic or Short Columns

Duleau, Hodgkinson, Considère, and Engesser, while working independently, suggested that Euler's formula is valid only for slender columns. It is to be noted here that Hodgkinson had already suggested an empirical formula which was used for the design of short columns. However, there was a need to develop a theory which can govern the failure of short columns or columns with a smaller slenderness ratio [14–17]. Considering this, Engesser suggested tangent modulus theory, wherein he assumed that axial load was increasing during the transition from straight to the bent position and presented the value of critical stress in terms of tangent modulus (E_t) as follows as given by Equation (6):

$$\sigma_{cr} = \frac{\pi^2 E_t I}{\lambda^2} \tag{6}$$

In the same year, Considère suggested that, if an ideal column is subjected to load greater than the proportional load, the column begins to bend, and stresses on the concave side increase according to tangent modulus theory, whereas, on the convex side, stress peaks decrease according to Hooke's law. He defined critical load by employing E, which is a function of average stress in the column. He also suggested that the E value should lie in between the modulus of elasticity and the tangent modulus. Later on, in 1995, the error in tangent modulus theory was put forward by Jasinski, and he pointed out that determination of function which describe E was impossible to find theoretically [32]. After this, a double modulus theory was developed by Karman and proposed the actual evaluation of E for rectangular cross section and idealised H-section consisting of infinitely thin flange and negligible web. The general expression for the critical stress σ_{cr} was thus defined in terms of reduced modulus, E_r [33] by Equations (7) and (8), respectively as:

$$\sigma_{cr} = \frac{\pi^2 E_r I}{\lambda^2} \tag{7}$$

$$E_r = \frac{E I_1 + E I_2}{I} \tag{8}$$

where I_1 and I_2 are the moment of inertia of either side of the section about the neutral axis. Since then, the value of E has been evaluated by several authors.

In 1947, using an imaginary column, Shanley concluded that there will be bending once the tangent modulus load is exceeded following which axial load increases and reaches a maximum value which lies in between the tangent modulus load and reduced modulus load, and there will be stress reversal once the bending deformation becomes finite. In other words, Shanley's analysis clearly described that the first bifurcation will occur at tangent load and a sequence of equilibrium can be constructed in between two limiting loads, i.e., tangent load and double modulus load. Shanley thus proposed an interaction curve to link eccentricities to the tangent modulus theory in order to apply his theory for practical problem and design calculations [34,35]. In this regard, Figure 3 compares the average stress for different slenderness ratios, as collected from the experimental investigation of a specimen (aluminium solid round rod with 0.72 cm diameter having flat ends) discussed in [36].

Later on, a model similar to Shanley was analysed by Johnston by replacing the two-area element with a solid rectangular segment and determined the magnitude of stress distribution for various loads above the tangent modulus load across the section [37]. With the advancement in computer technology, computer programs were written by Batterman [38] to find the maximum load for aluminium alloy H-section with finite web areas about weak as well as the strong axis of bending, in both initially straight position and with initial curvature [38]. In 1987, Groper and Kenig proposed the inelastic stability of stepped columns with the help of Newton's method or bisection method [39]. In general, the Engesser-Shanley definition for the critical load of a column in an inelastic range is widely acceptable. The same concept is extended for structural steel columns having initial stress due to differential cooling, although the material is in an elastic range.

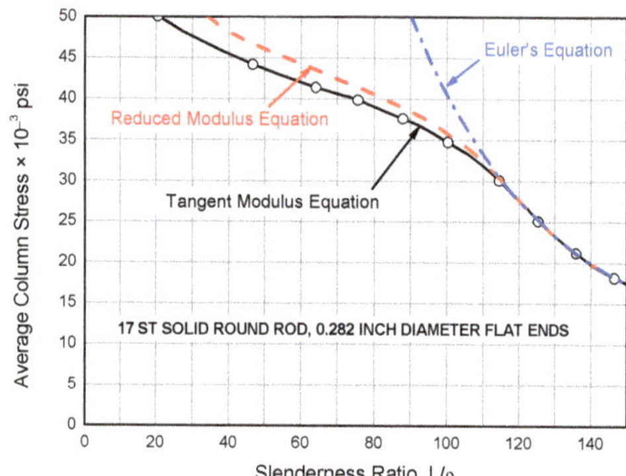

Figure 3. Comparison of experimental data and column theories.

3.3. Failure of Imperfect Long Columns

Long columns, more than short ones, are notoriously sensitive to initial imperfections, defects, etc. Hence, they necessitate careful investigations since a minor change in the loading and geometrical parameters may lead to their sudden failure. It is well accepted that perfect columns are theoretical identities, and in practice, their behaviour is altogether different. One of the important examples of such columns is a walking stick which is subjected to a large amount of eccentricity.

There exists a wide scatter of results for long columns, due to many reasons, and some of the motivations include nonideal supports, plastic behaviour, the interaction of buckling modes (wherein local buckling of columns is more important), along with possible residual stresses. Due to these imperfections, column behaviour is altogether different practically, in comparison with its theoretical treatment, and the reason for this may be attributed to the treatment of these imperfections. Thus, these columns primarily fail due to elastic instability. Section 4 reports further details about the failure of imperfect long columns.

4. Imperfections in Long Columns

4.1. Imperfections Due to Large Deformations

It is well understood that Euler's original formulation was based on some defined assumptions, and hence, he modelled the behaviour of ideal columns which hardly exist in the reality of the structures. In order to apply his theories to practical problems of engineering, it becomes important to understand the difference in the behaviour of a real and an ideal column. This result can be achieved by removing the various assumptions, one by one, and then analysing the column response. One of the prominent assumptions in Euler's theory, for example, is that all deformations are considered as "small". This results in curvature $(1/R)$ of deflected shape of a column of length (L) with flexural rigidity (EI), subjected to axial load (P), with pinned boundary condition on either edge becomes equal to double differentiation of deflection (y''), thus neglecting (y'), as in the following [40] given by Equations (9) and (10) as:

$$\frac{\delta}{L} = \frac{2p}{\pi\sqrt{\frac{P}{P_{cr}}}} \tag{9}$$

$$p = sin(\alpha/2) \tag{10}$$

where α is the slope of deflected shape at support, and Equation (9) represents the solution in terms of mid-height deflection, δ, applied load, P, and Euler load, P_{cr}.

According to [40], Figure 4 shows the variation of P/P_{cr} with δ/L and highlights that the estimation of the expected critical load by linear theory is valid for a considerable range of deformations. The reason for such a behaviour is attributed to the fact that for most of the columns, a combination of bending and axial stresses reaches the proportionality limit long before the difference between linear and nonlinear theory becomes notable.

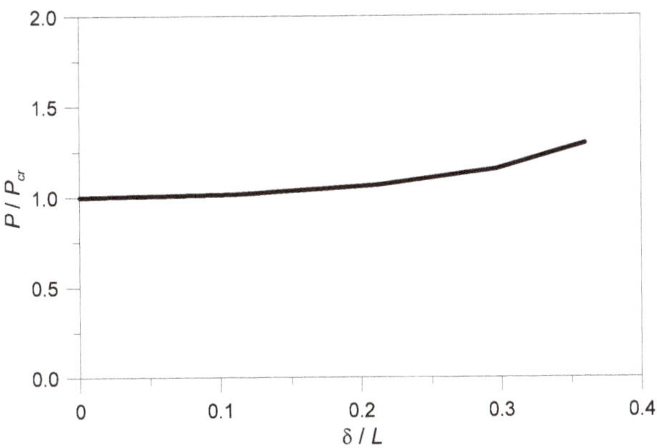

Figure 4. Normalised load-deflection curve.

4.2. Imperfections Due to Initial Curvature and Eccentric Loading

It was Young who, in 1807, tried to find out the effect of eccentricity (e) and initial curvature on the load-carrying capacity of a given column [41]. However, his original research results were not presented in usable form. Later on, during the year 1858, Scheffler [42] presented the complete solution for eccentrically loaded columns, by taking into account the effect of direct stress and bending stress. This solution is now commonly known as the "Secant Formula" (SF).

It is important to highlight, in this context, that the SF is accurate until the predicted stresses are within the elastic limit of the constructional material in use. The behaviour of a column with a given initial curvature or a column subjected to eccentric load is more or less the same, considering the fact that, in either case, the behaviour of the column is the same. Further, if the initial imperfections are small, the original Euler's formula results in a fairly accurate estimation of the total compressive load which a straight slender member can support.

$$\frac{P}{P_{cr}} = \left[\frac{2}{\pi}\cos^{-1}\left\{\frac{1}{1+\delta/e}\right\}\right]^2 \tag{11}$$

$$\frac{P}{P_{cr}} = 1 - \frac{a}{\delta} \tag{12}$$

Equations (11) and (12), in this regard, describe the correlation between the Euler's load for an ideal column (P_{cr}) and the critical load (P) for a column either having a certain initial curvature (a) or subjected to eccentric loading (e). Figure 5 shows the graphical interpretation of Equations (11) and (12) for different values of eccentricity and initial curvature. The example calculations are carried out by taking into account Equations (11) and (12) by assuming the different values of δ, along with a and e to consider the initial curvature or eccentricity, respectively.

Along with that, the graph also shows that it does not matter how the initial imperfection is introduced in a perfect column, given that the critical load for an imperfect column will always be smaller than the critical load of the perfect one.

Table 1. *Cont.*

Ref.	Variable/Method	Column	Remarks/Findings
[53]	Finite-difference approach	Nonuniform	• Method to compute approximate lower bound buckling load • Recursion relations developed for the coefficient of characteristic equation from which an approximate lower bound buckling load was calculated
[54]	Finite-difference method (FDM) through matrix iteration approach	Nonuniform; tapered	• Finite-difference form used to write the differential equation for the equilibrium at a number of points with small lateral deflection • Set of homogeneous simultaneous linear equations and the lowest value of eigenvalue gave the required buckling load • Simple model formulation and concise nature of its solution • The method is preferred over the Rayleigh-Ritz energy approach
[55]	Bessel's function	Tapered	• Computed the exact Bernoulli-Euler's static load using Bessel's function
[39]	Continuum approach	Tapered	• Investigation on inelastic buckling of nonprismatic columns
[44]	ODE approach	Tapered	• Study limited to concentrated load
[56]	Energy approach	Fixed-free; square pyramid; truncated cone	• Cross section written as function of axial coordinate by assuming the deflected shape corresponding to the first mode of buckling • Analytical solutions were obtained • The method can be also extended to other boundary conditions
[57]	FEM approach (i.e., power series solution of differential equation with variable coefficients to generate the stiffness matrix)	Variable stiffness	• Computed the stiffness of columns with varying cross-sectional bending stiffness, as well as varying axial load along their length, in the form of a polynomial expression • It can easily be incorporated into FEM software
[58]	Semi-analytical approach	Nonprismatic	• Step varying column which can be extended to incorporate continuously varying column • Step changes in the profile represented by distribution, and finally solved by polynomial functions • Accuracy of method dependent on the number of assumed segments • The method did not gain popularity due to the very lengthy formulation, even for simple variations of the basic cross section
[59]	Four normalised fundamental equations	Constant width and tapered depth; constant depth and tapered width; double tapered	• Nonuniform column approximated as stepped uniform column • Normalised approximated fundamental solution found using the recurrence formula • Buckling load easily obtained after substituting the fundamental solutions into the characteristic equations • The significant advantage of this method was that it does not require any computer-based technique, thus saves computational time

Table 1. *Cont.*

Ref.	Variable/Method	Column	Remarks/Findings
[60]	New numerical method (i.e., eigenvalue problem transformed into a boundary value problem which can be solved using the numerical integration)	Nonprismatic; self-weight	• The problem with consideration of distributed axial force will leave the governing differential equation with variable coefficient • For column with variable distributed axial force or varying cross section the governing differential equation cannot be converted into Bessel's equation • Numerical method such as energy method, FEM, Finite Difference Method, etc. are required to arrive at solutions
[61]	Semi-analytical procedure	Nonprismatic	• The method worked well with step discontinuity but for continuously varying profile minimum 30 segments must be considered to obtain correct solution • The procedure can be used to generate geometric stiffness matrix for variable beam-column element which can be used in FEM
[62]	Power function or exponential function and distribution of flexural stiffness along with Bessel's function	Nonprismatic	• Obtained general solution using the mentioned functions • The general solution can be used to solve the problem discussed by [32,61,63–65]
[66]	ODE approach	Variable moment of inertia	• Predicted exact mode shape along with their closed-form solution • Since then, till 1999, no closed-form solutions were reported for columns with variable moment of inertia subjected to axial load (until [67])
[67]	Fixed polynomial variation of flexural rigidity	Variable moment of inertia	• Solutions similar to [66] considering the fact that buckling mode shape was employed as polynomial function • Method suggested to generalise solution by [66]
[68]	Transcendental equations, Bessel's or Lommel's function	Variable stiffness; self-weight	• Exact closed-form solutions • Results useful for columns wherein the variation of elasticity can be constructed
[69]	Continuum approach	Variable stiffness	• exact solutions for the buckling analysis of nonuniform columns subjected to concentrated axial force at different point along the longitudinal axis • method exact, simple and efficient • method limited only to very special buckling mode, and thus not able to solve a column with general heterogeneity

Table 1. *Cont.*

Ref.	Variable/Method	Column	Remarks/Findings
[70]	Arbitrary distribution of flexural stiffness	Variable stiffness; axially distributed load	• differential equation reduced to Bessel's equation • distribution of axial loading expressed as a functional related with the distribution of flexural stiffness
[71]	Eigenvalue approach	Variable stiffness	• closed-form solutions for simple shapes only
[72]	ODE approach	Tapered (parabolic and sinusoidal); polygonal cross section	• In order to derive buckled shape of linear elastic columns, relationship between buckled shape and load in free vibration was utilised • governing differential equations solved by Runge-Kutta method and determinant search method combined with Regula-Falsi method
[73]	ODE approach with Green's function		• Differential equation whose solution was obtained by Green's Function to give buckling load of heterogeneous column by Functional Perturbation Method (FPM) • In order to find the material around which Optimised Differential Functional Perturbation Method (ODFPM), solution more accurate • Second-order Perturbation term in Frechet's series minimised, which yielded nonlinear differential equation and related material property to bending stiffness
[29]	Modified vibrational mode shape (MVM) and energy method	Multistep	• buckling capacity of multistep column using modified vibrational mode

* Additionally, Weinberger (unpublished research).

4.4. Imperfections Due to Functionally Graded Material

From the basic equation of Euler's [5] one can directly infer that the buckling capacity of columns can also be varied by varying the modulus of elasticity. This method was not preferred until the technology advances to a level that variation of modulus of elasticity either in the axial or longitudinal direction was feasible. Based on the literature, it is observed that it is still a relatively unexplored area. In order to increase the buckling capacity of the column, it was suggested to vary the modulus of elasticity, but the solution of instability becomes difficult to compute. Signer [74] investigated the buckling solution for columns with continuous monotonic variations of flexural rigidity along the column. Fixing the origin at one end and x coordinate running along the centre line, variation of modulus of elasticity (E) and moment of inertia (I) were assumed as follows by Equation (13):

$$E(x)I(x)\eta(x) = E(0)I(0) = E_\circ I_\circ, \eta(x) = 1 + \beta(x), \beta \in R \tag{13}$$

It is to be noted that compressed members used in the civil structure are supported at the intermediate point (bracing). Considering the importance of the intermediate restraints, the study in [75] approximated a column with spatial variation of flexural stiffness due to material gradation or nonisoperimetric shape by an equivalent column with piecewise constant geometrical and material properties (Figure 7). This method uses a transcendental function that results in a closed-form solution of uniform columns. The suggested method was unique because the mathematical model preserves the properties of a continuous system by containing the infinite eigenvalues corresponding to all higher buckling modes [75]. The buckling analysis of axially graded columns was conducted by Huang and Li [28]. They transformed the governing differential equation with variable coefficients to Fredholm's integral equation which were further reduced to a system of algebraic equations. The accuracy of the suggested procedure was confirmed by comparing the obtained result with the available closed-form and numerical solutions. The significant role of their work was that, unlike other research, it was not restricted only to suitable buckling mode. Through this method, one can successfully solve the problem of buckling, if the variation of flexural rigidity was polynomial, trigonometric, or exponential function.

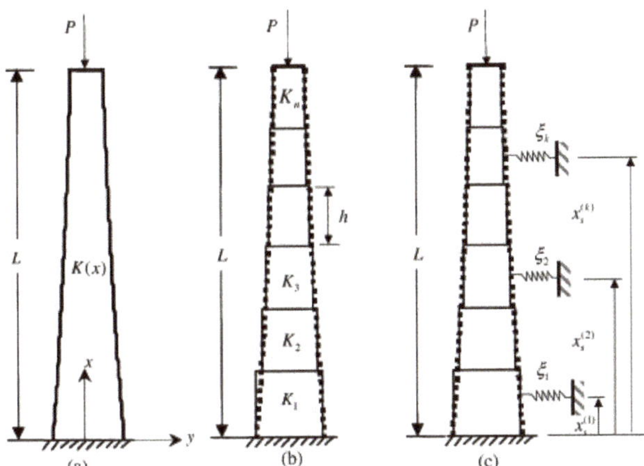

Figure 7. Examples of (**a**) nonuniform and nonhomogeneous axially graded columns, (**b**) piecewise continuous approximated model (of order *n*), and (**c**) axially graded column (order *n*) with multiple intermediate elastic restraints. Reproduced from [75] with permission from Elsevier®, license n. 5026040108100.

Recently, Elishakoff [76] studied the buckling of columns made from functionally graded material in an axial direction. The study was limited to find the polynomial variation of modulus of elasticity, E such that the buckling value exceeds in case of cantilever column whose cross-sectional area was kept constant. In another recent investigation, Rychlewska [77] presented buckling solutions for a beam with clamped-clamped, hinged-hinged, and hinged-clamped boundary conditions with an exponential variation of material properties in the axial direction and subjected to distributed load in exponential form.

4.5. Imperfections Due to Elevated Temperature or Fire Exposure

Specific attention can be paid to the buckling analysis of columns with variable stiffness, as in Section 4.3, but with a focus on stiffness variations due to the use of materials that are remarkably sensitive to temperature variations, as well as for resisting cross section that may suffer for long-term temperature exposure. This is the case of load-bearing members that are susceptible to elevated temperature exposure, and even fire, or any kind of phenomena that can be represented by a "thermal gradient" for the resisting cross section to analyse.

Most of the research studies, in this regard, are relatively recent and specifically focused on columns composed of steel [78–80], reinforced concrete [81–84], timber [85–87].

Developments in building technology and design strategies are even more frequently focused on innovative laminated glass solutions that are bonded by thermoplastic layers [88–91] or even composite-laminated insulated panels in which both mechanical and climatic loads can severely affect the overall column buckling performance. In this last case, thermal exposure effects do necessarily coincide with extreme accidents since fire loading can have marked effects on the overall mechanical performance, given the typically small thickness that is of common use in structural glass applications. Besides different materials and characteristics that are used for these members, the common aspect of the above documents is represented by the progressive bending weakness deriving from the degradation of the constituent materials. Therefore, the total compressive load acts on a resisting section and member that prematurely collapses due to its lack of load-bearing capacity. From a practical point of view, the shared feature for the cited literature studies is the basic trend to define standardised design buckling curves for columns made of mostly different constructional materials and thus to collect, in a simplified and univocal formulation, all the possible uncertainties and effects due to material behaviours, eccentricities, imperfections [92].

Even more attention is indeed required for load-bearing members in general that can be subjected to scattered thermal patterns and are thus potentially characterised by a number of critical cross sections.

Additionally, in this latter case, the first efforts are certainly related to the classical material for buildings, thus steel members. Alpsten [93] showed, for example, that residual nonuniform thermal stresses can severely affect the column buckling performance of a given member and result in even more pronounced degradation than geometrical initial imperfections. Culver [94] also focused on the analysis of pinned columns with thermal exposure. The study proved that severe thermal gradients in the mid-span region of columns are typically associated with a remarkable loss of global buckling capacity. While such a concept can be intuitive—due to stiffness reduction—this is in contrast with the discussion by Hoffend [95]. The reason is in the idealisation of the thermal gradient profile. The generally recognised idea, finally, is that thermal gradient effects can be generally schematised in the form of an equivalent initial imperfection. Therefore, the overall buckling performance of an axially loaded member in compression can be severely compromised. On the other hand, this issue can be efficiently addressed for safe design by means of conventional calculation methods that include a given initial geometrical imperfection.

5. Conclusions

The problem investigated by Euler was much simpler since it did not involve finding the solution of differential equation with varying coefficient because neither the material properties nor the cross-sectional dimensions were changing. However, in an attempt to maximise the buckling capacity of the column, modifications were performed in the column due to which the differential equation governing the mathematical model is left with varying coefficients. This review paper provides a complete synopsis of the development of various theories related to column buckling. A significant number of methods were recalled to obtain close-form solutions, but providing evidence for each one of them had certain intrinsic restrictions—either the buckling shape was assumed to be governed by a specific function or the distribution of flexural stiffness was not random. Moreover, all the discussed methods in the literature showed rather good agreement with some experimental results available in the literature. However, which one of them is more suitable to find the solution for a given arbitrary variation of coefficients still remains an unanswered question. In the self-buckling of columns, more research emphasis is required for the proper discretisation of elastic segments with rigid segments. For short column analysis, detailed experiments with various materials are one of the areas wherein research still needs to be carried out. In long columns analysis, more emphasis shall be put on the development of closed-form solution with variable moment of inertia with emphasis on varying material properties along the length of the column. Further, different functions can be developed to investigate the variation of modulus of elasticity and its effect on the buckling strength of columns. Furthermore, attention is indeed required for load-bearing members in general that can be subjected to scattered thermal patterns and are thus potentially characterised by a number of critical cross sections. This can be achieved by incorporating the effect of thermal gradient in terms of initial imperfections.

Author Contributions: The reported work is carried out authors as a team wherein Conceptualization, M.D.G.; methodology, M.D.G. and C.B.; software, A.S.; validation, A.S., A.P.K. and L.M.G.; formal analysis, M.D.G., C.B. and A.S.; investigation, M.D.G., C.B., A.S., A.P.K. and L.M.G.; resources, A.S. and M.D.G.; data curation, M.D.G., C.B. and A.S.; writing—original draft preparation, M.D.G., C.B. and A.S.; writing—review and editing, M.D.G., C.B., A.P.K. and L.M.G.; visualization, M.D.G. and C.B.; supervision, M.D.G.; project administration, L.M.G.; funding acquisition. All authors have read and agreed to the published version of the manuscript.

Funding: This research received no external funding.

Institutional Review Board Statement: Not applicable.

Informed Consent Statement: Not applicable.

Data Availability Statement: Data will be available upon request.

Acknowledgments: Authors acknowledge the support and encouragement provided by Director, VNIT Nagpur, India. MDPI is also acknowledged for the invited free-of-charge submission of this manuscript in the special issue "Innovation in Structural Analysis and Dynamics for Constructions" (C.B. Guest Editor vouchers).

Conflicts of Interest: The authors declare no conflict of interest.

References

1. Hooke, R. *Lectures de Potentia Restitutiva, or of Spring Explaining the Power of Springing Bodies*; John Martyn: London, UK, 1678; pp. 331–388, reprinted, of R. T.Günther, Early Sciences in Oxford 8, Oxford, 1931.
2. Hales, T.A.; Pantelides, C.P.; Reavele, L.D. Analytical buckling model for slender FRP-reinforced concrete columns. *Compos. Struct.* **2017**, *176*, 33–42. [CrossRef]
3. Cheng, W.-S. Repair and reinforcement of timber columns and shear walls–A review. *Constr. Build. Mater.* **2015**, *97*, 14–24. [CrossRef]
4. Bedon, C.; Kalamar, R.; Eliasova, M. Low velocity impact performance investigation on square hollow glass columns via full-scale experiments and Finite Element analyses. *Compos. Struct.* **2017**, *182*, 311–325. [CrossRef]
5. Euler, L. Euler's calculation of buckling loads for columns of non uniform section. In *The Rational Mechanics of Flexible or Elastic Bodies 1638–1788*; Orell FüssliTurici, Societatis Scientiarum Naturalium Helveticae: Zurich, Switzerland, 1960; pp. 345–347, Originally published in 1757.
6. Euler, L. De altitudinecolumnarum sub proprioponderecorruentium. *Acta Acad. Sci. Petropolitanae* **1780**, *1*, 163–193. (In Latin)
7. Euler, L. Determinatioonerum, quaecolumnaegestarevalent. *Acta Acad. Sci. Petropolitanae* **1947**, *1*, 121–145. (In Latin)
8. Euler, L. Exameninsignispuradoxi in theoria column arumoccurentis. *Acta Acad. Sci. Petropolitanae* **1780**, *1*, 146–162. (In Latin)
9. Coulomb, C.A. *Essai sur une Application des Regles de Maximis et Minimis a Quelques Problemes de Statiquerelatifs a L'architecture*; Academie Royale Des Sciences: Paris, France, 1773; Volume 7, pp. 343–382.
10. Dinnik, A.N. Buckling under own weight. *Proceeding Don Polytech.* **1912**, *1*, 19. (In Russian)
11. Dinnik, A.N. Design of column of varying cross section" Transactions of American Society of Mechanical Engineering. *Appl. Mech.* **1929**, *51*, 105–114.
12. Dinnik, A.N. Design of column of varying cross section. *Trans. Am. Soc. Mech. Eng.* **1932**, *54*, 105–109.
13. van Musschenbroek, P. *Physicaeexperimentale Set Geometricae de Magnete, Tuborum Capillarium Vitreorum Quespeculorum Attractione, Magnitudine terrae, Cohaerentia Corporum Firmorum Dissertationes: Ut et Ephemerides Meteorologicae Ultrajectinae*; Luchtmans, Nova Editio: Viennae, Austria, 1756.
14. Duleau, A.J.C. *Essaitheorique et Experimental sur la Resistance du Fer Forge*; Mme.Ve. Courcier, Nabu Press: Paris, France, 1820.
15. Hodgkinson, E. On the Transverse Strain and Strength of Materials. In *Memoirs of the Literary and Philosophical Society of Manchester*; 2nd Series; Literary and Philosophical Society of Mancheste: Manchester, UK, 1824; Volume 2–4.
16. Engesser, F. Ueber die KnickfestigkeitgeraderSUibe. In *ZeitschriftfiirArchitektur und Ingenieurwesen*; W. Ernst & Sohn: Berlin, Germany, 1889; Volume 35.
17. Considère, A. Résistance des piècescomprimées. In *Congrès International des Procédés de Construction*; Libraire Polytechnique: Paris, France, 1891; pp. 3, 371.
18. Lamarle, E. Memoire sur la flexion du bois. In *Annales des Travaux Publics de Belgique*; Vandooren, B.J., Ed.; Imprimeur Des Annales Des Travaus Publication: Bruxelles, Belgium, 1845; Volume III, pp. 1–64, Volume IV, pp. 1–36.
19. Rankine, W.J.M. *A Manual of Civil. Engineering*; Charles Griffin: London, UK, 1862.
20. Young, D.H. Rational design of steel columns. *Am. Soc. Civ. Eng. Transl.* **1936**, *101*, 442–500.
21. Johnson, J.B.; Bryan, C.W.; Turneaure, F.E. *The Theory and Practice of Modern Framed Structures: Designed for the Use of Schools, and for Engineers in Professional Practice*; John Wiley & Sons, Inc.: New York, NY, USA, 1905; p. 561.
22. Ayrton, W.E.; Perry, J. On struts. *Engineer* **1886**, *62*, 464–465.
23. Robertson, A. The strength of struts. *Inst. Civ. Eng.* **1925**, *1*. [CrossRef]

24. Greenhill, A.G. Determination of the greatest height consistent with stability that a vertical pole or mast can be made, and the greatest height to which a tree of given proportions can grow. *Proc. Camb. Philos. Soc.* **1881**, *4*, 65–73.
25. Duan, W.H.; Wang, C.M. Exact solution for buckling of columns including self-weight. *J. Eng. Mech.* **2008**, *134*, 116–119. [CrossRef]
26. Darbandi, S.M.; Firouz-Abadi, R.D.; Haddadpour, H. Buckling of variable section column under axial loading. *J. Eng. Mech.* **2010**, *136*, 472–476. [CrossRef]
27. Wei, D.J.; Yan, S.X.; Zhang, Z.P.; Li, X.-F. Critical load for buckling of non-prismatic columns under self-weight and tip force. *Mech. Res. Commun.* **2010**, *37*, 554–558. [CrossRef]
28. Huang, Y.; Li, X.F. Buckling analysis of nonuniform and axially graded columns with varying flexural rigidity. *J. Eng. Mech.* **2010**, *137*, 73–81. [CrossRef]
29. Rahai, A.R.; Kazemi, S. Buckling analysis of non-prismatic columns based on modified vibration modes. *Commun. Nonlinear Sci. Numer. Simul.* **2008**, *13*, 1721–1735. [CrossRef]
30. Afsharfard, A.; Farshidianfar, A. Finding the buckling load of non-uniform columns using the iteration perturbation method. *Theor. Appl. Mech. Lett.* **2014**, *4*. [CrossRef]
31. Nikolić, A.; Šalinić, S. Buckling analysis of non-prismatic column: A rigid multibody approach. *Eng. Struct.* **2017**, *143*, 511–521. [CrossRef]
32. Jasinski, F. NocheinWortzu den Knickfragen. *Schweiz. Bauztg.* **1895**, *25*, 172.
33. von Karman, T. *Untersuchungen Liber Knickfestigkeit*; Mitteilungeniiber Forsclulflgsarbeitenaufdem Gebiete des Ingenieul'lvesens: Berlin, Germany, 1910.
34. Shanley, F.R. The column paradox. *J. Aeronaut. Sci.* **1946**, *13*, 618–679. [CrossRef]
35. Shanley, F.R. Inelastic column theory. *J. Aeronaut. Sci.* **1947**, *14*, 261–268. [CrossRef]
36. Templin, R.L.; Strum, R.G.; Hartmann, E.C.; Holt, M. Column strength of various aluminium alloys. In *Aluminium Research Laboratories Technical Paper 1*; Aluminium Company of America: Pittsburgh, PA, USA, 1938.
37. Johnston, B.G. Buckling behavior above the tangent modulus load. *J. Eng. Mech. Div. ASCE* **1961**, *87*, 79–100. [CrossRef]
38. Batterman, R.H.; Johnston, B.G. Behavior and maximum strength of metal columns. *J. Struct. Div. ASCE* **1967**, *93*, 205–230. [CrossRef]
39. Groper, M.; Kenig, M.J. Inelastic buckling of nonprismatic column. *J. Eng. Mech.* **1987**, *113*, 1233–1239. [CrossRef]
40. Chajes, A. *Principles of Structural Stability*; Prentice Hall: Hoboken, NJ, USA, 1974.
41. Young, T. *A Course of Lectures on Natural Philosophy and the Mechanical Arts*; Royal Society: London, UK, 1807; Volume 1, 2, pp. 135–156.
42. Salmon, E.H. *Columns by, Frowde*; Hodder & Stoughton: London, UK, 1920.
43. Murphy, G.M. *Ordinary Differential Equations and Their Solutions*; D. Van Nostrand: New York, NY, USA, 1960.
44. Gere, J.M.; Carter, W.O. Critical buckling loads for tapered columns. *J. Struct. Div. Proc.* **1962**, *1*, 1–12.
45. Lagrange, J.-L. Sur la figure des colonnes. In *Oeuvres de Lagrange (Publ. de M.J.-A. Serret)*; Gauthier-Villars: Paris, France, 1868; Volume 2, pp. 125–170.
46. Clausen, T. Über die Form architektonischerSäulen. *Bull. Phys. Math. De L'academie St. Petersburg* **1851**, *9*, 368–379.
47. Keller, J.B. The shape of the strongest column. *Arch. Ration. Mech. Anal.* **1960**, *5*, 275–285. [CrossRef]
48. Tadjbakhsh, I.; Keller, J.B. Strongest columns and isoperimetric inequalities for eigenvalues. *J. Appl. Mech.* **1962**, *29*, 159–164. [CrossRef]
49. Taylor, J.E. The strongest column: An energy approach. *J. Appl. Mech.* **1967**, *34*, 486–487. [CrossRef]
50. Martin, G.H. A procedure for determining the critical load for a column of varying section. *J. Aeronaut. Sci.* **1946**, *13*, 135–140. [CrossRef]
51. Harris, C.O. A suggestion for columns of varying section. *J. Aeronaut. Sci.* **1942**, *9*. [CrossRef]
52. Wilson, J.F.; Holloway, D.M.; Biggers, S.B. Stability experiments on the strongest columns and circular arches. *Exp. Mech.* **1971**, *11*, 303–308. [CrossRef]
53. O'Rourke, M.; Zebrowski, T. Buckling load for nonuniform columns. *Comput. Struct.* **1977**, *7*, 717–720. [CrossRef]
54. Iremonger, M.J. Finite difference buckling analysis of non-uniform columns. *Comput. Struct.* **1980**, *12*, 741–748. [CrossRef]
55. Banerjee, J.R.; Williams, F.W. Exact Bernoulli-Euler static stiffness matrix for a range of tapered beam-columns. *Int. J. Numer. Methods Eng.* **1985**, *21*, 2289–2302. [CrossRef]
56. Smith, W.G. Analytic solutions for tapered column buckling. *Comput. Struct.* **1988**, *28*, 677–681. [CrossRef]
57. Chen, Y.Z.; Cheung, Y.K.; Xie, J.R. Buckling loads of columns with varying cross-section. *J. Eng. Mech.* **1989**, *115*, 662–667. [CrossRef]
58. Eisenberger, M. Buckling loads for variable cross-section members with variable axial forces. *Int. J. Solids Struct.* **1991**, *27*, 135–143. [CrossRef]
59. Lee, S.Y.; Kuo, Y.H. Elastic stability of non-uniform column. *J. Sound Vib.* **1991**, *148*, 11–24. [CrossRef]
60. Vaziri, H.H.; Xie, J. Buckling of columns under variably distributed axial loads. *Comput. Struct.* **1992**, *45*, 505–509. [CrossRef]
61. Arbabi, F.; Li, F. Buckling of variable cross-section columns: Integral-equation approach. *J. Struct. Eng.* **1991**, *117*, 2426–2441. [CrossRef]
62. Qiusheng, L.; Hong, C.; Guiqing, L. Stability analysis of bars with varying cross-section. *Int. J. Solids Struct.* **1995**, *32*, 3217–3228. [CrossRef]

63. Timoshenko, S.P.; Gere, J.M. *Theory of Elastic Stability*; McGraw Hill: New York, NY, USA, 1961.
64. Genik, A.N. *Scientific Paper of Genik*; AN: Moscow, Russia, 1950.
65. von Kármán, T.; Biot, M. *Mathematical Methods in Engineering*; McGraw-Hill: New York, NY, USA, 1940.
66. Duncan, W.J. *Galerkin's Method in Mechanics and Differential Equations*; Aeronautical Research Committee Reports and Memoranda; H.M.S.O.: London, UK, 1937; No. 1798.
67. Elishakoff, I.; Rollot, O. New closed-form solutions for buckling of a variable stiffness column by mathematica. *J. Sound Vib.* **1999**, *224*, 172–182. [CrossRef]
68. Elishakoff, I. A closed form solution for the generalized Euler problem. *Proc. R. Soc. Lond. A Math. Phys. Eng. Sci.* **2000**, *456*. [CrossRef]
69. Li, Q.S. Buckling analysis of multi-step non-uniform columns. *Adv. Struct. Eng.* **2000**, *3*, 139–144. [CrossRef]
70. Li, Q.S. Exact solutions for buckling of non-uniform column under axial concentrated and distributed loading. *Eur. J. Mech.-A/Solids* **2001**, *20*, 485–500. [CrossRef]
71. Wang, C.M.; Wang, C.Y.; Reddy, J.N. *Exact Solutions for Buckling of Structural Members*; CRC: Boca Raton, FL, USA, 2005.
72. Lee, B.K.; Carr, A.J.; Lee, T.E.; Kim, J. Buckling loads of columns with constant volume. *J. Sound Vib.* **2006**, *294*, 381–387. [CrossRef]
73. Totry, E.M.; Altus, E.; Proskura, A. Buckling of non-uniform beams by a direct functional perturbation method. *Probabilistic Eng. Mech.* **2007**, *22*, 88–89. [CrossRef]
74. Siginer, A. Buckling of columns of variable flexural rigidity. *J. Eng. Mech.* **1992**, *118*, 640–643. [CrossRef]
75. Singh, K.V.; Li, G. Buckling of functionally graded and elastically restrained non-uniform columns. *Compos. Part B Eng.* **2009**, *40*, 393–403. [CrossRef]
76. Elishakoff, I. Buckling of a column made of functionally graded material. *Arch. Appl. Mech.* **2012**, *82*, 1355–1360. [CrossRef]
77. Rychlewska, J. Buckling analysis of axially functionally graded beams. *J. Appl. Math. Comput. Mech.* **2014**, *13*, 103–108. [CrossRef]
78. Janss, J.; Minne, R. Buckling of steel columns in fire conditions. *Fire Saf. J.* **1981**, *4*, 227–235. [CrossRef]
79. Ng, K.T.; Gardner, L. Buckling of stainless steel columns and beams in fire. *Eng. Struct.* **2007**, *29*, 717–730. [CrossRef]
80. Gomes, F.C.T.; Providencia e Costa, P.M.; Rodrigues, J.P.C.; Neves, I.C. Buckling length of a steel column for fire design. *Eng. Struct.* **2007**, *29*, 2497–2502. [CrossRef]
81. Tie, L.L. Fire Resistance of Reinforced Concrete Columns: A Parametric Study. *J. Fire Prot. Eng.* **1989**, *1*, 121–129.
82. Rodrigues, J.P.C.; Laím, L.; Correia, A.M. Behaviour of fiber reinforced concrete columns in fire. *Compos. Struct.* **2010**, *92*, 1263–1268. [CrossRef]
83. Han, L.H.; Tan, Q.-H.; Song, T.-Y. Fire Performance of Steel Reinforced Concrete Columns. *J. Struct. Eng.* **2015**, *141*. [CrossRef]
84. Gernay, T. Fire resistance and burnout resistance of reinforced concrete columns. *Fire Saf. J.* **2019**, *104*, 67–78. [CrossRef]
85. Malhotra, H.L.; Rogowski, B.F. Fire resistance of laminated timber columns. *Fire Res. Notes* **1967**, *671*, 1–60.
86. Schnabl, S.; Turk, G.; Planinc, I. Buckling of timber columns exposed to fire. *Fire Saf. J.* **2011**, *46*, 431–439. [CrossRef]
87. Wiesner, F.; Bisby, L. The structural capacity of laminated timber compression elements in fire: A meta-analysis. *Fire Saf. J.* **2019**, *107*, 114–125. [CrossRef]
88. Amadio, C.; Bedon, C. Buckling of Laminated Glass Elements in Compression. *J. Struct. Eng.* **2011**, *137*. [CrossRef]
89. Oikonomopoulou, F.; van den Broek, E.A.M.; Bristogianni, T.; Veer, F.A.; Nijsse, R. Design and experimental testing of the bundled glass column. *Glass Struct. Eng.* **2017**, *2*, 183–200. [CrossRef]
90. Aiello, S.; Campione, G.; Minafo, G.; Scibilia, N. Compressive behaviour of laminated structural glass members. *Eng. Struct.* **2011**, *33*, 3402–3408. [CrossRef]
91. Bedon, C.; Amadio, C. Buckling analysis and design proposal for 2-side supported double Insulated Glass Units (IGUs) in compression. *Eng. Struct.* **2018**, *168*, 23–34. [CrossRef]
92. Bedon, C.; Amadio, C. Design buckling curves forglass columns and beams. *Struct. Build.* **2015**, *168*, 514–526. [CrossRef]
93. Alpsten, G.A. On numerisk simulering av barformagan hos isolerade stalpelareutsatta for brandpaverkan. In *Nordiskafrosknings-dagar for Stalbygguad*; ETH, Zurich: Stockholm, Sweden, 1970. (In German)
94. Culver, C.G. Steel column buckling under thermal gradient. *ASCE J. Struct. Div.* **1972**, *92*, 1853–1865. [CrossRef]
95. Hoffend, F. Brandverhalten von Stahlstutzenbeiausmittiger Lasteintung. In *Dehnbehindeung ode teilweiser Beckleidun*; Sonderforschungsbereich 148, Brandverhalten von Bauteilen, Arbeitsbericht, Teil; VS Verlag für Sozialwissenschaften, Wiesbaden Technical University: Wiesbaden, Germany, 1978–1980. (In German)

 *buildings*

Article

Facial Expression-Based Experimental Analysis of Human Reactions and Psychological Comfort on Glass Structures in Buildings

Chiara Bedon * and Silvana Mattei

Department of Engineering and Architecture, University of Trieste, 34127 Trieste, Italy;
silvana.mattei@phd.units.it
* Correspondence: chiara.bedon@dia.units.it; Tel.: +39-040-558-3837

Abstract: For engineering applications, human comfort in the built environment depends on several objective aspects that can be mathematically controlled and limited to reference performance indicators. Typical examples include structural, energy and thermal issues, and others. Human reactions, however, are also sensitive to a multitude of aspects that can be associated with design concepts of the so-called "emotional architecture", through which subjective feelings, nervous states and emotions of end-users are evoked by constructional details. The interactions of several objective and subjective parameters can make the "optimal" building design challenging, and this is especially the case for new technical concepts, constructional materials and techniques. In this paper, a remote experimental methodology is proposed to explore and quantify the prevailing human reactions and psychological comfort trends for building occupants, with a focus on end-users exposed to structural glass environments. Major advantages were taken from the use of virtual visual stimuli and facial expression automatic recognition analysis, and from the active support of 30 volunteers. As shown, while glass is often used in constructions, several intrinsic features (transparency, brittleness, etc.) are responsible for subjective feelings that can affect the overall psychological comfort of users. In this regard, the use of virtual built environments and facial expression analysis to quantify human reactions can represent an efficient system to support the building design process.

Keywords: structural glass; building design; human reactions; psychological comfort; experiments; virtual reality (VR)

Citation: Bedon, C.; Mattei, S. Facial Expression-Based Experimental Analysis of Human Reactions and Psychological Comfort on Glass Structures in Buildings. *Buildings* **2021**, *11*, 204. https://doi.org/10.3390/buildings11050204

Academic Editor: Geun Young Yun

Received: 9 April 2021
Accepted: 10 May 2021
Published: 14 May 2021

1. Introduction

In building design, the end-user's comfort is a target for a multitude of applications. These include, for example, thermal comfort, indoor air quality, visual comfort, noise nuisance, ergonomics and vibrations. With around 90% of our lives spent in buildings, there is a strong link between comfort and the built environment [1,2]. For decades, various researchers have sought to understand how the characteristics of the built environment can impact the emotions, behaviors and physical well-being of end-users [2]. However, comfort itself depends on a great number of factors which can, if not addressed properly, lead to annoyance. Further, it is known that such analysis requires convergent teams from the humanities, arts, social sciences, technology, engineering and medicine. With a more explicit focus on engineering issues, comfort strictly relates to many topics of primary interest for building technology [3–5].

Structural issues and comfort can be, for example, considered in terms of vibration serviceability assessments. Usually, the evaluation of vibration issues in relation to possible annoyance risk is carried out in terms of acceleration peaks and recommended limit values. Similar methods can be used for the comfort analysis of floors under walking or standing conditions [6–8], and for tall buildings against wind [9–11]. As a matter of fact, human reactions and comfort levels strongly depend on engineering parameters, but also on the physiological perception of the frequency and amplitudes of vibrations. The operational context, including location (street, gym, office, home, etc.), time of day (morning or evening) and stimuli duration (seconds or hours) also have severe impacts on the degree of human tolerance [8].

The structural use of materials and design concepts in buildings is an additional factor that can implicitly result in possible discomfort for the occupants. The well-known psychological effect of architectural solutions can have both positive and negative effects, and can evoke subjective feelings that reduce to a minimum the mechanical efforts of structural designers [12]. This is the major challenge of architects when evoking emotions in the design of so-called "emotional buildings" [13–18]. In the context of the built environment, experimental measurements can be carried out to quantify emotions [19]. Smart sensors or virtual walks are proven to represent useful tools in support of the analysis of human reactions to visual stimuli [20,21].

Among other things, structural glass is known to represent an attractive material for construction [22,23]. Key aspects are its transparency and abilities to adapt to various configurations and replace/interact with more traditional constructional materials. Typical examples can be found in facades, roofs, walkways, bridges and balustrades, as in Figure 1a–c. Besides, glass is also recognized as one of the most vulnerable components in constructions, and thus to need special structural design efforts in order to ensure appropriate safety levels. Glass facades and windows, for example, are the first physical barrier for building occupants, both in presence of ordinary operational conditions but also during accidental events that could result in potential shards and injuries (Figure 1d,e). The high aesthetic impact of glass components and structures is thus often in contrast with a basic discomfort and lack of safe feelings for end-users.

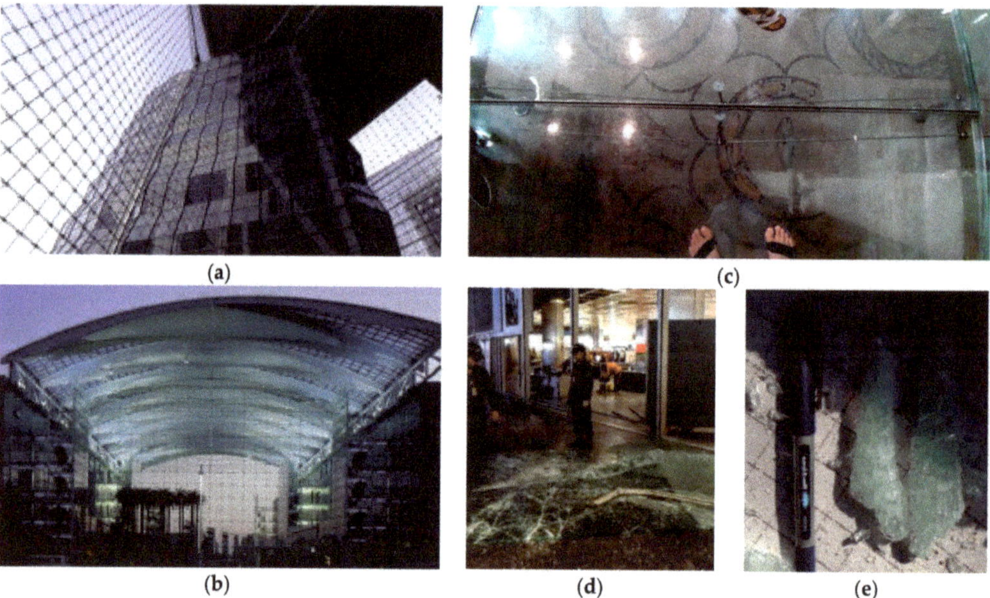

Figure 1. Examples of glass in buildings: (**a,b**) Facades and roofs for the Poly Plaza Cable-Net Wall (Beijing, CN) and the Kempinski Hotel (Munich, DE), (reprinted with permission from [24], Copyright 2021 Elsevier®, license number 5042040336444); (**c**) a pedestrian system (walkway in Aquileia, IT; reprinted from [8] under the terms and conditions of CC-BY license); (**d,e**) cracks and shards due to a hazard (reprinted with permission from [25], Copyright 2021 Elsevier®, license number 5042381253637).

Glass design can be highly demanding for pedestrian systems, due to the combination of dynamic mechanical parameters and complex human–structure interaction (HSI) phenomena that typically occur during a walk, but also due to subjective reactions [8].

In general terms, glass structures are often called "architectures of vertigo," given that transparent load-bearing systems are increasingly conceived as "spaces of visceral thrills with deep socio-spatial implications" [23]. Glass floors, in particular, can be recognized as intense psycho-physiological stimuli for pedestrians, as they induce unavoidable sensory experiences that go beyond the conventional HSI design procedures of analysis.

2. Research Goals and Approach

2.1. Glazed Stimuli

In this paper, a virtual experimental approach is presented in support of building design. The goal of the developed method is to explore and quantify the prevailing reactions and psychological comfort levels for end-users exposed to different visual built environments. At the time of analysis, careful attention was paid to the selection of visual stimuli characterized by a primary role of structural glass components in buildings and constructions (i.e., Figure 2a,b). The methodology can be adapted to different topics and design fields.

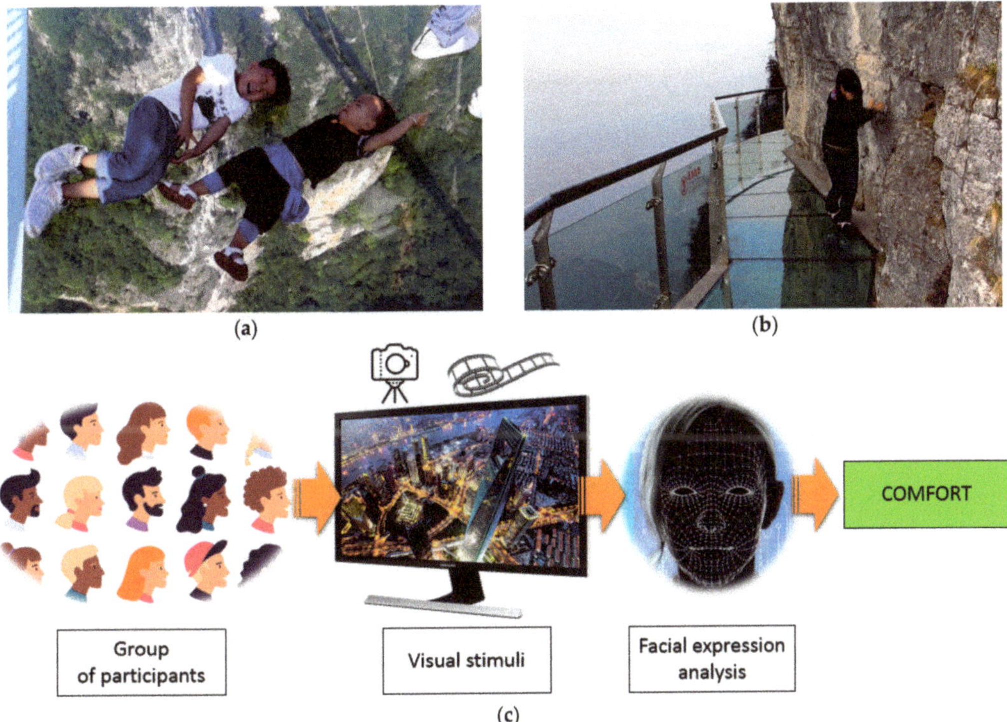

Figure 2. Examples of human reactions on glass structures: (**a**) Zhangjiajie bridge, Hunan, CN (adapted from [26], © Imagechina/REX/Shutterstock); (**b**) Cliffside skyway, CN (adapted from [27], © Visual China Group via Getty Images); and (**c**) a virtual experimental procedure.

While the use of glass in buildings has been increasing in the last few decades, with new challenges for structural designers [25], the impact on end-users is often severely affected by the material's transparency and brittleness, and also by the lack of technical knowledge regarding its mechanical properties [28,29]. In this regard, it is expected that the building design process could take advantage of subjective measurements and their possible combination with mathematical models and engineering calculations. As schematized in Figure 2c, the basic approach takes benefit from the quantitative measures of emotions and subjective feelings of individuals exposed to pre-selected visual stimuli. Based on the analysis of facial micro-expressions, comfort levels and trends can be analyzed by sub-groups of subjects of stimuli.

2.2. Experimental Procedure

The overall methodology was elaborated and designed at the University of Trieste (Department of Engineering and Architecture, Italy) in Winter 2020. The experimental study followed the procedural steps in Figure 3, and was carried out remotely.

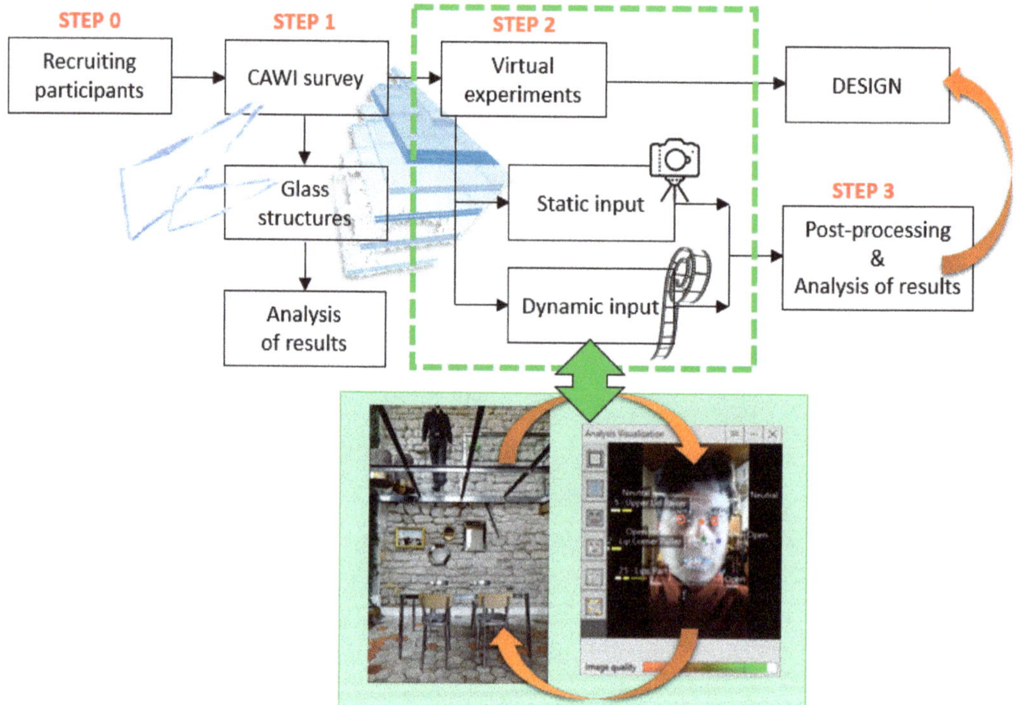

Figure 3. Flowchart of the virtual experimental analysis. A detailed example: a glazed apartment floor in Paris, FR (adapted from [30], © Jerry Jacobs Design) and facial expression analysis (© C. Bedon).

More precisely:

- STEP 0: the participants were first recruited. The group of volunteers (30 in total) included mostly students and researchers, both females and males (58% and 42% respectively), with an age range of 20–56 (28.8 years-old was the average, ± 8.4; 26 years-old the median value). Some volunteers were students or workers living in Trieste, or residing in the Friuli Venezia Giulia Region. The age demographics and geographic distribution of participants are shown in Figure 4. Their educational experience and background skills were characterized by different fields of study or activity. Most importantly, no preliminary technical knowledge on the use and features of structural glass in construction was required at the time of the experiment. Similarly, no direct experience on glass structures was needed to join the virtual investigation.

- STEP 1: A preliminary computer assisted Web interviewing (CAWI) survey was shared online and used to assess the multitude of subjective reactions for participants under different stimuli (Section 3).
- STEP 2: The virtual experimental analysis was carried out (Section 4) to study human behaviors in glass-involving scenarios, based on facial expression analysis. To that end, the FaceReader^{TM} automatic facial expression recognition software (version 8, Noldus Information Technology bv, Wageningen, Netherlands) [31] was used in support of the quantitative analysis of experimental measurements. Two different visual stimuli were designed to assess the reactions of volunteers, namely, consisting of a set of static input items (Section 4.2) and a dynamic virtual reality (VR) video clip of pre-recorded walks in glass environments (Section 4.3). In both cases, based on STEP 1, special care was taken in the selection and arrangement of stimuli, so as to capture different reactions and emotions of participants.
- STEP 3: the post-processing analysis of experimental measurements from STEP 2 was partly based on the automatic software analysis, and further elaborated as discussed in Sections 4.4 and 4.5. Detailed comparative results are presented in Sections 5 and 6 for static and dynamic stimuli respectively. As shown, the analysis of experimental measures proves that the use of structural glass in buildings is still affected by scattered human reactions. Moreover, the results from the proposed methodology suggest that the design of glass structures could benefit from subjective parameters that should be taken into account in the overall design process.

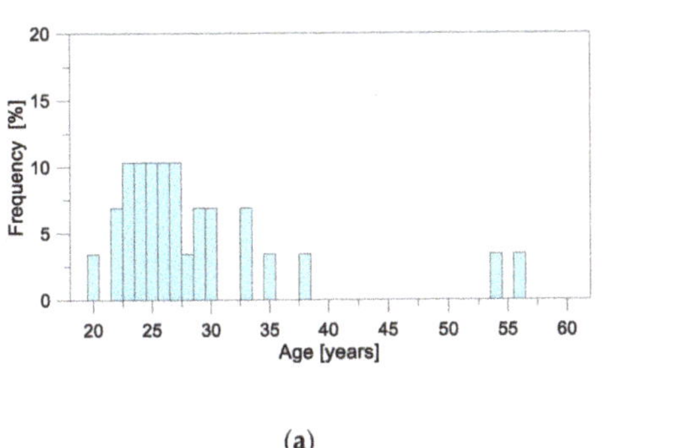
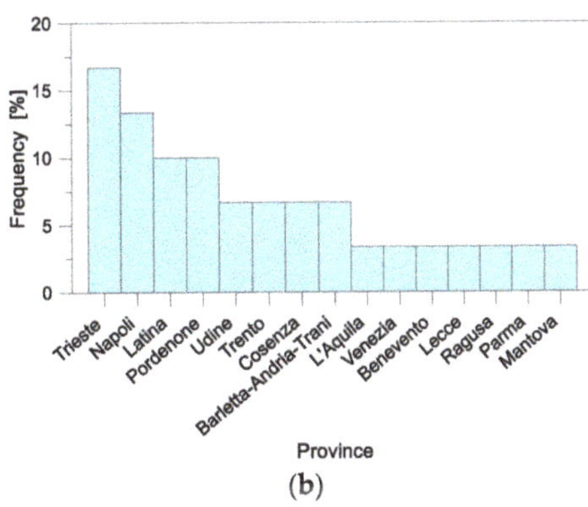

(a)   (b)

Figure 4. (a) Age demographics and (b) geographic distribution of participants.

2.3. Strengths and Limitations of the Study

The proposed experimental methodology took partial inspiration from past literature, which has proved that human emotions can be used as efficient guide for designers [32,33]. The study and measuring of human behaviors based on facial micro-expressions is in fact typical of many research and market fields, especially where understanding consumers feelings can help designers to establish an efficient emotional communication between consumers and products. It is in fact generally recognized that both verbal and non-verbal behaviors enable humans to communicate emotions [12,13]. Non-verbal communication includes physical behaviors that are commonly referred to as body language and gestures, including also facial expressions. Most importantly, facial expressions are essential for the quantitative measure of basic emotions, because they provide information about inner states of individuals.

For the present study, the use of a virtual experimental setup and its adaptation to the constructional field was suggested by several motivations. The first and most severe one was represented, at the time of the investigation, by severe movement limitations, remote teaching regulations and smart-working rules due to the COVID-19 pandemic emergency (i.e., obligations for "red zone"/maximum risk measures in the Italian territory). As such, the experimental analysis was carried out with the active support of volunteers, but respectfully of social distance rules and general temporary prohibitions. Furthermore, technical trends and considerations suggested the design of the experimental study. The sanitary emergency revealed in fact that the use of transparent materials for constructions is expected to increase in the next future [29,34,35], so as to facilitate the definition of new design concepts for the organization of private or public spaces in buildings. Differently, this increasing use of glass in structures can be associated with a further magnification of "architecture of vertigo" concepts [23] earlier described. Balanced technical solutions should be thus necessarily developed. Both the cited aspects were merged to support the design of the experimental strategy herein presented.

Besides, the investigation was carried out with a clear preliminary analysis of intrinsic limitations that should be taken into account for future extensions. The number of participants (30), for example, was selected on the base of individuals availability (i.e., to join remotely the study, with own devices and high-speed internet connection). A possible extension of group size, and demographic/geographic distribution, could facilitate a more generalized analysis of reactions for specific sub-groups of participants. Another limitation of the study was represented by the fully remote experimental procedure. Under COVID-19 restrictions, the volunteers were invited to take part from homes/different regions of Italy. As such, the measure of facial expressions was carried out with shared screens only, without additional smart sensors, wearable devices or VR glasses [36,37].

3. CAWI Survey

The preliminary STEP 1 took the form of a CAWI survey that was designed to include 40 questions. The survey was designed with both the simple multiple choice (SMC) approach and the open question (OQ) method. The goal was to expose the participants to well-defined built conditions characterized by the presence of structural glass at different levels (residential buildings, open spaces, public offices, etc.), and to capture their feelings as active occupants. In this context, the survey revealed some important aspects, and suggested the development of the virtual experimental investigation herein discussed. For the majority of participants, it was proved that:

- background technical knowledge regarding mechanical features of structural glass in buildings often lacks. This evokes discomfort and negative perceptions, especially for design scenarios characterized by contact of end-users with glass (partition walls, balustrades, roofs, etc.).
- The brittle behavior of glass represents the most influencing parameter, and is frequently associated to a common feeling of protection lack against accidental events (i.e., shards due to impact, etc.).
- The visual detection of minor damage or material degradation in glass structural components (delamination, etc.) evokes severely negative reactions, even in presence of safe mechanical performances.
- The use of structural glass in pedestrian systems generally evokes very positive feelings on the architectural side. When the same participants are asked to ideally walk on those systems, however, the presence of transparency and possible vibrations, spots or noise during walks results in discomfort and highly negative comments.

4. Static and Dynamic VR Experiments

4.1. Methods

The experimental investigation was based on a sequence of pre-selected input sources that were used to act as emotional visual stimuli for the involved participants. The so-called STEP 2 from the flowchart in Figure 3 was developed as schematized in Figure 5, and represented the core of remote analysis. STEP 2 as a whole consisted of three sub-steps for each participant. The process was repeated twice, first with a static stimulus (sequence of pictures described in Section 4.2) and later with a VR video clip of pre-recorded virtual walks (Section 4.3). The remote experiments were carried out with the support of personal computers for all the volunteers, a shared screen for the visual stimuli and a webcam for recording the facial expressions of participants.

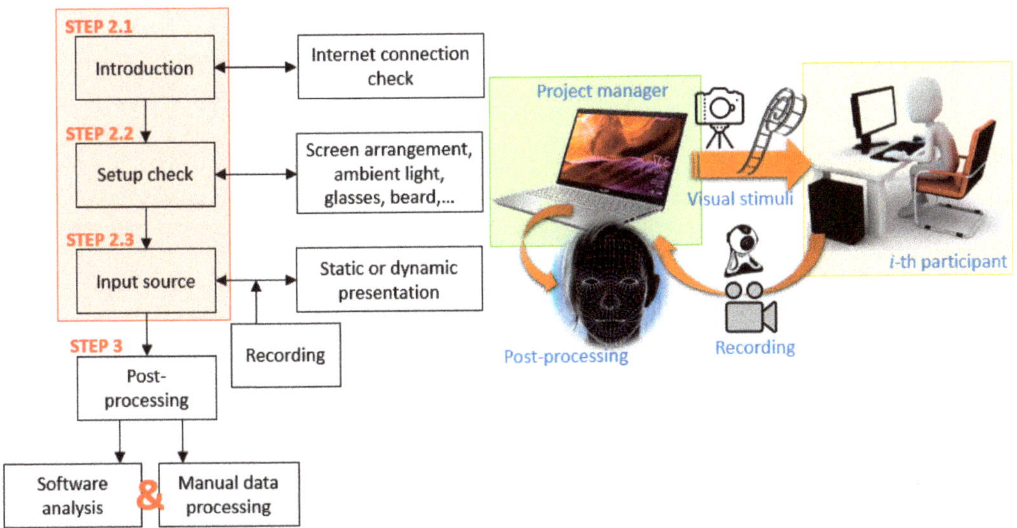

Figure 5. General procedure for the virtual experimental investigation (STEP 2).

Before watching the shared input source, the invited subjects were informed about the experimental procedure (STEP 2.1 in Figure 5). A special care was paid to verify the connection speed, and minimize possible delay in transmission and acquisition of shared data. Successively (STEP 2.2 in Figure 5), the visualization setup was also properly checked, so as to avoid disturbing effects for the quality level of signals, and thus for the post-processing analysis. More in detail, the attention was given to ensure an optimal and mostly uniform ambient illumination for the participants. The screen position was also assessed for each participant, in order to prevent severe facial distortions and to avoid introducing high head/eye inclinations through the experiments. Finally, the presence of participants with glasses or beard was separately noted, for a more refined post-processing analysis of measured data.

Once verified the basic operational conditions, the input source was shared separately for each volunteer (STEP 2.3 in Figure 5). No auditory stimuli were introduced. During the presentations, however, audio communication of volunteers with the project manager was allowed to provide technical feedback (when needed). During the presentations (STEP 2.3), the project manager webcam was used to record the facial expressions of participants, while looking at the shared screen. To this end, a preliminary analysis to capture the optimal webcam resolution and sampling rate was carried out. The final choice resulted in records with minimum 1280×720 (or 1920×1028) video resolution and around 4300 frames for each recorded signal (30–60 the range of frame rate; 12,432 kbits/s the average bitrate).

The analysis of recorded signals was carried out during STEP 3 of Figure 5. The interpretation of facial micro-expressions and emotions from the participants exposed to the shared input source was partly carried out with the use of the commercial software FaceReader^{TM} [31]. The automatic computational analysis was based on all the collected frames ($\approx$4300 the average) for each one of the available records (30 volunteers $\times$ 2 stimuli), and required approximately 10 min of analysis/each. The overall elaboration of signals and data analysis was then carried out as in Section 4.4. For the current study, it is thus important to note that the automatic software analysis was used to provide row quantitative data only of nervous states. The detection of human reactions and comfort trends for individuals exposed to glass environments was manually derived as in Section 4.5.

4.2. Static Input Source

The invited participants were first subjected to a shared input presentation that consisted in a selection of 27 pictures for various glass structures and scenarios. These pictures were equally spaced at time intervals of 5 s, so that the total duration of the visual stimulus could not exceed a maximum of 120 s, to avoid annoyance. Figure 6 shows an example of input pictures. A special care was spent for their selection (to capture emotions in the invited participants), but also for the definition of the optimal sequence. To that end, the pictures were selected from magazines, scientific journals, webpages of construction companies or newspapers. For all the participants, the sequence and duration of the static presentation was kept fix. Regarding the sequence, the choice was to alternate items possibly associated to comfort or negative feelings.

Figure 6. Selection of static input sources for the first stage of the static virtual experiment (examples from the set of 27 items): (**a**) private room (Off-grid itHouse, Pioneertown, USA; adapted from [38], © Airbnb); (**b**) Amsterdam RAI Hotel, NL (adapted from [39], © Egbert de Boer); (**c**) Hongyagu bridge, Hebei, CN (adapted from [40], © REUTERS/Stringer); (**d**) windows under a blast hazard (adapted from [29], © C. Bedon); (**e**) stairs of the Apple Cube, New York, USA (adapted from [41], © Sedak GmbH and Co KG); (**f**) facade and walls (adapted from [29], © C. Bedon); (**g**) window under vehicle impact (adapted from [42], © Times Colonist); (**h**) Space Needle tower (Seattle, WA, USA; adapted from [43], © Space Needle LLC and John Lok); (**i**) skydeck (Willis Tower, Chicago, USA; adapted from [44], © Ranvestel Photographic).

4.3. Dynamic Input Source

The setup schematized in Figure 5 was adapted and applied to the group of participants. The source consisted in a pre-recorded VR clip with total duration of 120 s, see Figure 7. The second stage of the investigation was in fact carried out to capture any kind of possible variation in the amplitude and trend of emotions of participants, when exposed to a virtual walk in a glazing environment. Major benefit was taken from the availability of a VR context representative of a case-study building located in Paris, and characterized by a large amount of glass components in facades, roofs, floors, balustrades. The VR context was used to pre-record a set of walks that could be used as stimulus. Based on preliminary studies, the clip was designed and divided onto three walks (40 s/each):

- At the top terrace level (W1),
- At the ground level of the building, from the main entrance (W2),
- At the first story level (W3).

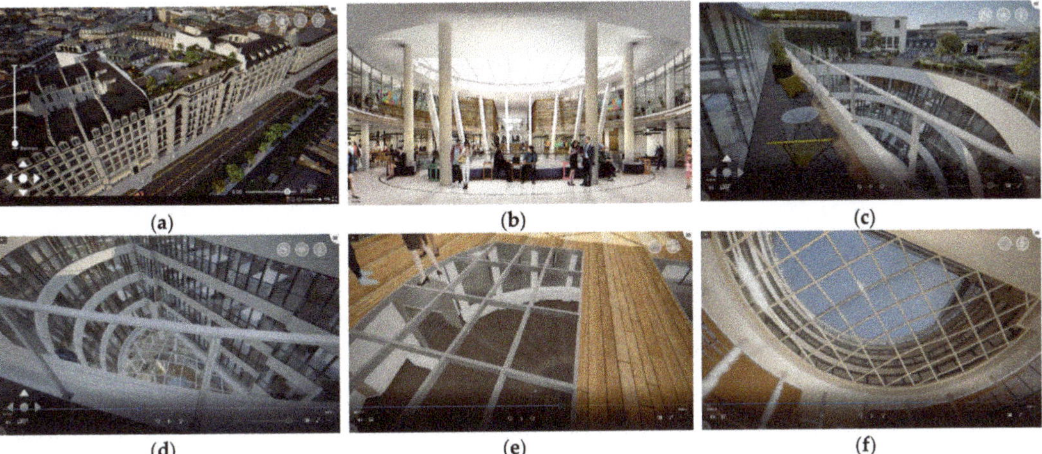

Figure 7. Selection of screenshot frames from the pre-recorded VR video clip inspired by the case-study palace in Paris, FR (adapted from [45], © Generali Real Estate French Branch): (**a**) general view of the building (from Rue Réaumur); (**b**) indoor open space; (**c**) top terrace; (**d**) glass conical envelope/facade; (**e**) glass floor (from the ground level); (**f**) glass roof (from the ground level).

Similarly to Section 5, no auditory stimuli were considered for the dynamic setup, and the participants were asked to watch the shared screen and join the virtual walks.

4.4. Experimental Measurements

The measurement of relevant data for comfort analysis was carried out in the post-processing stage of records for each participant, when exposed to both static and dynamic VR glazed environments.

The FaceReader^{TM} software, in particular, was chosen because it allows to model accurately the face of a given individual, by taking advantage of 500 key points for the description of facial movements and micro-expressions. The software is based on Artificial Intelligence tools and can be used to detect subjective emotions and human reactions from input video records or even pictures. As a whole, the software output takes the form of a series of quantitative charts and data that can be used for the interpretation of basic emotional states. In addition to the reference "neutral" condition, the standard classification carried out by the software includes basic definitions for individual states ($N = 6$ in total) targeted as "happy", "sad", "scared", "disgusted", "angry" or "surprised". From a practical point of view, the software offers an instantaneous evaluation of Action Units (AUs, ranging from 0 to 1) for all these measured emotions. Additional useful feedback can be obtained by the quantitative analysis of output parameters such as valence, arousal, gaze direction, head orientation, heart rate (based on remote skin conductance analysis) and few additional personal characteristics of participants (such as gender, average age, presence of glasses, beard).

For the present study, AU records were collected for each participant/virtual stimulus/basic emotion, as a function of the time of experiment (with $0 \leq t \leq 120$ s). AU data were exported from the software analysis with a sampling rate of 1 s. The typical example of AU records is shown in Figure 8 for one of the participants subjected to the dynamic VR setup. The chart, more in detail, gives evidence of specific AU values over time of virtual stimulus (i.e., STEP 2.3 in Figure 5), but omits the first instants of the experimental arrangement, when some preliminary discussion was carried out with the participant to provide instructions and create a comfortable condition.

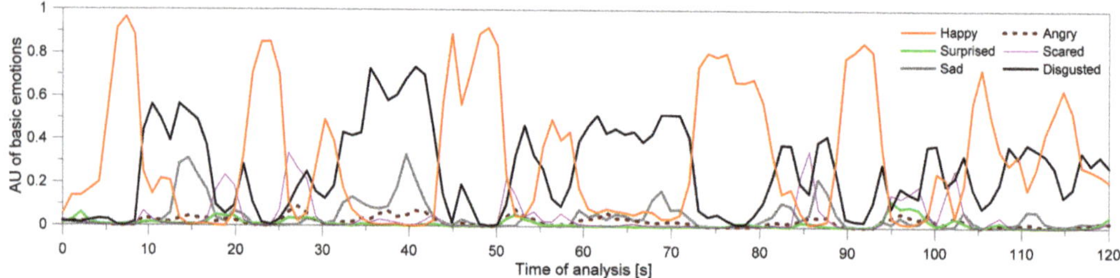

Figure 8. Example of AU records of basic emotional states, as obtained for one of the invited participants exposed to the shared VR input source.

In the overall procedure, moreover, it has to be noted that the "neutral" state computed by the software was excluded from manual elaboration of measured data, as it is shown in Figure 8. In presence of VR input sources and visual stimuli that do not involve active movements of participants, it is in fact generally recognized that the expected reactions and emotions cannot be quantitatively and qualitatively compared to feelings evoked by real field experiments. Accordingly, combination of "neutral" data with other basic emotional signals for the present study would have alter most of the measured subjective reactions and comfort trends.

4.5. Manual Post-Processing Strategy for AU of Basic Emotions

The quantitative and qualitative analysis of human comfort was based on the detection and measure of the herein called "positive" (POS, in the following) or "negative" (NEG) feelings from all the participants, when visually exposed to specific glazed environments.

Given that several AU records were obtained (as for 30 participants, $N = 6$ basic emotions, 2 virtual scenarios), the first approach of manual data elaboration consisted in the individual analysis of collected AU records. This stage was developed to explore the subjective feelings of single volunteers when exposed to a specific building scenario (i.e., glass floor, roof, balustrade, etc.). In addition, AU calculations were used to correlate the subjective response of different participants when exposed to the same VR input, and thus to find a prevailing emotional reaction/comfort trend for groups of built environment conditions. This means that for each VR scenario, basic emotions were subdivided onto two major categories representative of the herein called POS or NEG feelings and comfort trends. In doing so, the approach schematized in Figure 9 was take into account. The automatic software analysis was first carried out at STEP 3.1.

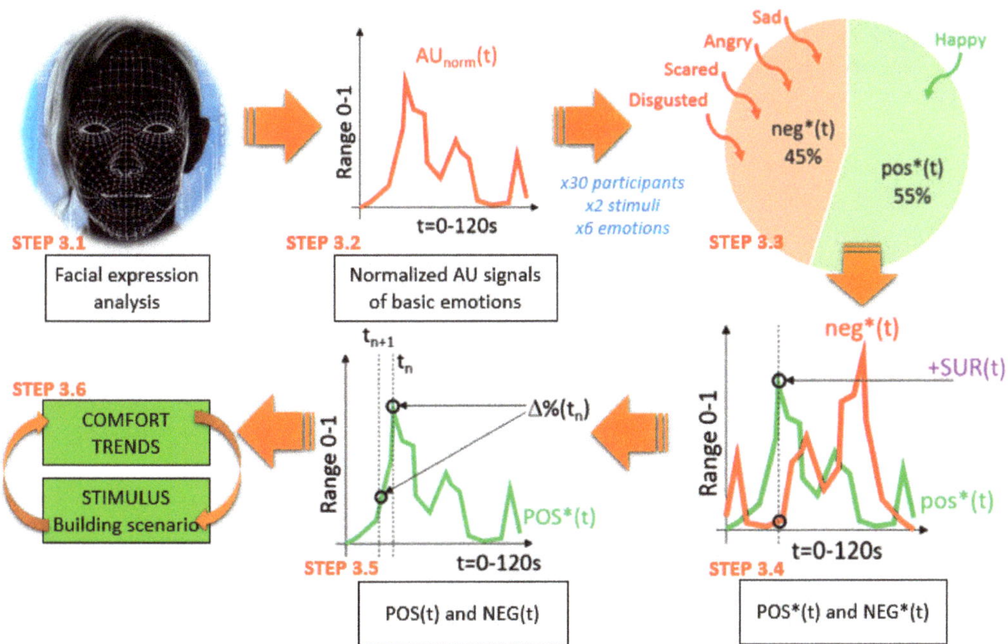

Figure 9. Schematic procedure for the post-processing analysis of AU of basic emotions (STEP 3) and detection of comfort trends.

According to AU charts as in Figure 8, the preliminary requisite of analysis consisted in the normalization of AU records of basic emotions in the time of experiment, so as to facilitate the objective comparison of multiple AU signals (STEP 3.2 in Figure 9). Starting from the exported original AU charts ($AU_{original}$), more in detail, the average AU for each emotion at the beginning of the individual experiment/participant ($AU_{initial,avg}$) was taken as a reference for calculating the variation of subsequent AU measurements in time, that is:

$$AU_{norm}(t) = AU_{original}(t) - AU_{initial,avg} \tag{1}$$

with $0 \leq t \leq 120$ s and:

- $AU_{norm}(t)$ the normalized record for each participant, emotion, VR scenario, at the time instant t;
- $AU_{original}(t)$ the original signal record for each participant, emotion, VR scenario, at the time instant t;
- $AU_{initial,avg}$ the mean value of each emotion, participant, VR scenario (preliminary stage).

Following Equation (1), the individual records of basic emotions, $AU_{norm}(t)$, were successively grouped in basic pos*(t) and neg*(t) plots (STEP 3.3 in Figure 9). In the pos*(t) set, the AU data in time for feelings marked as "happy" were considered only:

$$pos^*(t) = \sum_{i=1}^{N-1} AU_{norm}(t) = f(happy) \tag{2}$$

Similarly, the neg*(t) set included the input data from feelings marked as "sad", "angry", "scared", "disgusted":

$$neg^*(t) = \sum_{i=1}^{N-1} AU_{norm}(t) = f(sad, angry, scared, disgusted) \tag{3}$$

Note in STEP 3.4 of Figure 9 that the "surprised" AU data ("SUR(*t*)") were omitted from the so-calculated pos*(*t*) and neg*(*t*) signals, and analyzed separately. This because (depending on the individual subject, context and stimulus) "surprised" states can be typically associated with both positive or negative feelings for the participants. The final marking of emotional data was thus based on the analysis of prevailing AUs at a given time instant *t*. The SUR(*t*) input was in fact superimposed as:

$$\text{POS}^*(t) = \begin{cases} \text{pos}^*(t) + \text{SUR}(t) & if\ \text{pos}^*(t) > \text{neg}^*(t) \\ \text{pos}^*(t) & if\ \text{pos}^*(t) < \text{neg}^*(t) \end{cases} \tag{4}$$

or

$$\text{NEG}^*(t) = \begin{cases} \text{neg}^*(t) + \text{SUR}(t) & if\ \text{neg}^*(t) > \text{pos}^*(t) \\ \text{neg}^*(t) & if\ \text{neg}^*(t) < \text{pos}^*(t) \end{cases} \tag{5}$$

An example of calculation results is shown in Figure 10, in terms of POS*(*t*) and NEG*(*t*) measured states for one selected participant.

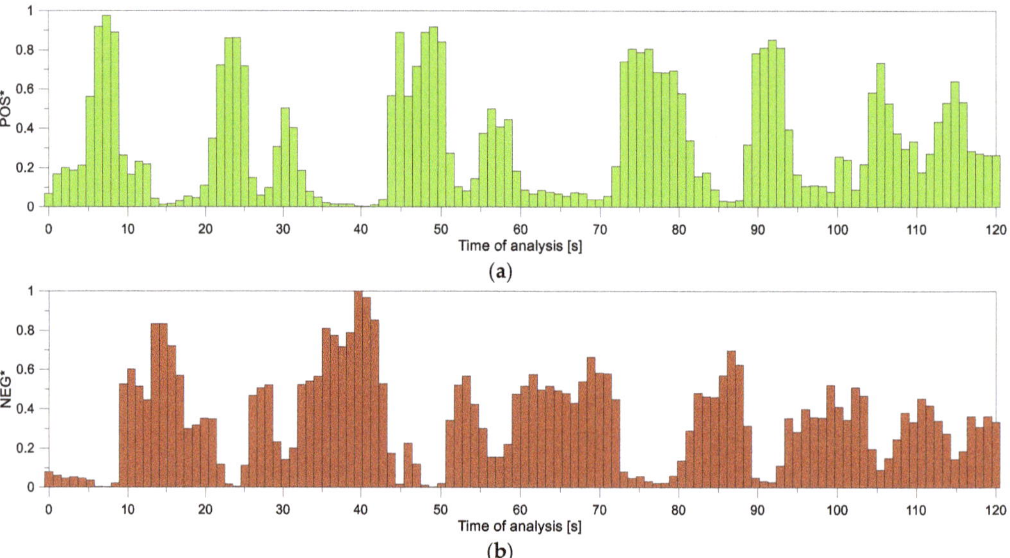

Figure 10. Example of calculated (**a**) POS* and (**b**) NEG* measurements of AU signals, as obtained for one of the participants under the VR input source, as a function of the time of analysis.

In conclusion, the final detection of comfort trends for a given visual stimulus was carried out as in STEP 3.5 of Figure 9. To that end, the percentage variation of elaborated signals from STEP 3.4 was measured over the time of analysis. At each increment *t* of the experiments, the emotional variation was calculated for both POS*(*t*) and NEG*(*t*) data, and defined as:

$$\Delta\% = 100\frac{\text{Y}(t_n) - \text{Y}(t_{n-1})}{\text{Y}(t_{n-1})} \tag{6}$$

with Y = POS*(*t*) or NEG*(*t*) respectively, as shown in the example of Figure 11.

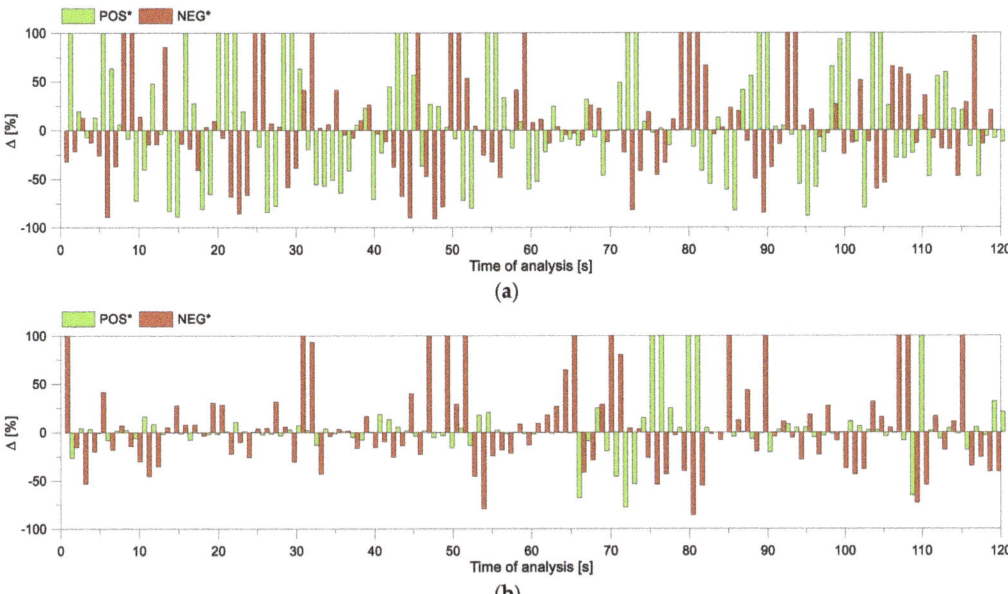

Figure 11. Examples of POS and NEG variation over the time of analysis, for a single participant exposed to (**a**) static and (**b**) dynamic VR stimuli.

The prevailing emotional state (and thus comfort trend) was established in terms of absolute maximum values of nervous feelings and max/min percentage variations in the time of analysis. Each time instant t was classified with a POS or NEG mark, uniquely detected from data elaboration. A "comfort weight" value $C(t) =$ POS or NEG (based on the prevailing nervous state) was assigned to each participant/instant/scenario, to facilitate the comparative analysis of multiple signals and stimuli.

The global correlation of comfort trends with the VR source was in fact carried out based on the synchronized comparison of elaborated data for the 30 participants subjected to an identical stimulus (STEP 3.6). More precisely:

- For the static experiment (Section 5), the prevailing comfort trend was calculated over time intervals of 5 s (corresponding to the duration of a single picture presentation), in terms of average comfort level;
- For the dynamic experiment (Section 6), the attention was focused on specific frames of pre-recorded video clips, and quantified in terms of average comfort levels for selected context scenarios.

5. Discussion of Results from Static Experiment

5.1. Prevailing Reactions

Figure 12 shows the analysis of prevailing reactions in the group of participants, as obtained by the investigation of elaborated records in the time of analysis (Figure 12a), and by the comparison of group reactions for a given static stimulus (Figure 12b). POS and NEG values are reported in percentage values for the whole group of volunteers. As such, the comparative data can be used to address the prevailing emotional state (in terms of average or maximum/minimum reaction of participants), towards the imposed picture/context. Figure 12a, more in detail, shows a clear modification of human reactions for the assigned stimuli, with marked fluctuations of POS and NEG data. It can be easily perceived that the calculated POS signal prevails on the NEG one for some time intervals, while the opposite condition can be noticed for other instants. In some other cases, rather balanced POS and NEG values can be also observed.

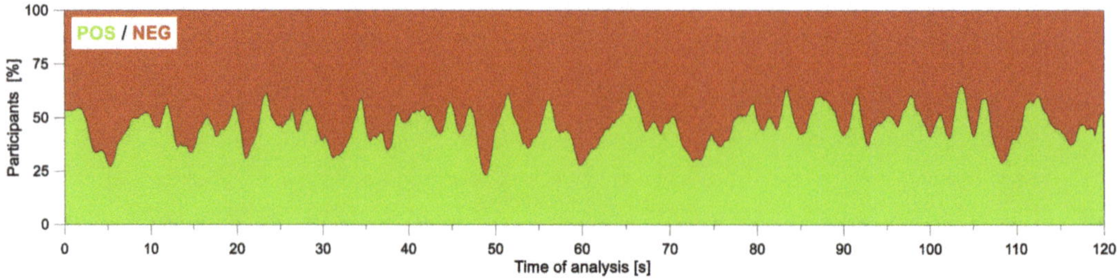

Figure 12. Detection of prevailing reactions (in percentage terms for the group of participants), as a function of (**a**) time of analysis or (**b**) static stimulus (27 pictures).

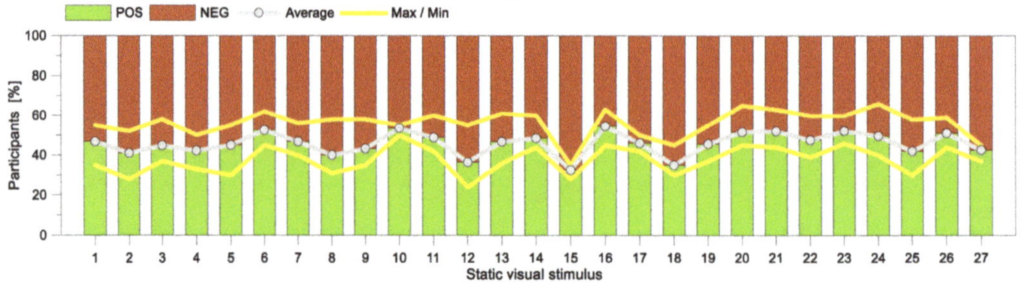

Most importantly, the POS and NEG trends were elaborated to derive the prevailing emotion for each input item, as shown in Figure 12b. For each time interval of 5 s, the average POS and NEG reactions were first calculated and associated with the 27 input pictures. Further, the maximum and minimum reaction peaks were also analyzed, as emphasized in the chart. This operation allowed us to derive a more precise analysis of human responses and nervous states, as a function of the imposed stimulus.

From the collected data, in particular, it was possible to assert that:

- The static virtual experience generally highlighted a strong "neutral" reaction for most of the participants. The average value of such a state was calculated in around 60% of measured AU signals. This finding can be easily justified by the static nature of a visual stimulus, and can be considered as an intrinsic limit of the overall experimental approach.
- A more detailed analysis of POS and NEG results, as in Figure 12, gave evidence, for most of the 27 pictures, of clear emotional trends calculated from the software analysis of minor facial micro-expressions.

The analysis of average reactions in Figure 12b highlighted in fact that:

- For most of the 27 items, the "average" measure of reactions can provide a reliable measure of experimental data and comfort trends.
- Differently, extreme reactions (quantified as max/min values for the fluctuation of average data) still suggest a rather variable subjective response for some of involved individuals.
- Rather few pictures can be associated with a mostly uniform reaction from the whole group of participants (i.e., with minimum max/min variations compared to the average response). This is the case for items #15 (Figure 6) and #27 (Figure 13). Furthermore, in both cases, the NEG reaction prevails on the POS.

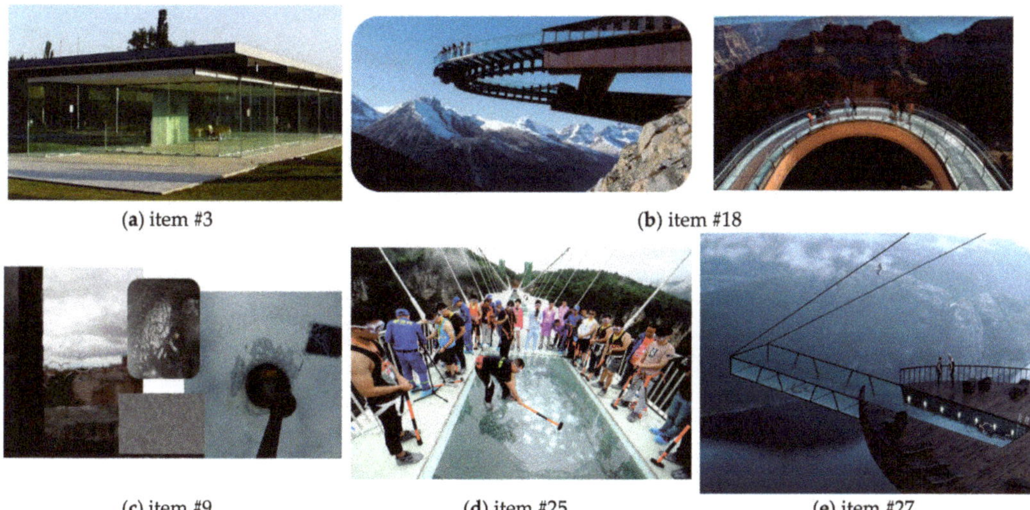

(**a**) item #3   (**b**) item #18

(**c**) item #9   (**d**) item #25   (**e**) item #27

Figure 13. Static items with prevailing NEG reactions: (**a**) Glass Pavilion Rheinbach, DE (adapted from [46], © F. Wellershoff); (**b**) skywalk (Jasper National Park, CA; (adapted from [47], © Getty Images); (**c**) delamination (adapted from [48] under the terms and permissions of a CC-BY license and adapted from [49], Copyright 2021 Elsevier®, license number 5067511162164); (**d**) Zhangjiajie bridge, Hunan, CN (adapted from [50], © VCG via Getty Images); (**e**) cliff concept boutique hotel (adapted from [51], © Hayri Atak).

The detailed analysis of any kind of discomfort revealed that:

- Average NEG peaks (>60% of participants) were found to prevail for POS states for items #8, #12, #15 (Figure 6) and #18 (Figure 13).
- In general, the minimum absolute NEG response was calculated for 46% of the group of participants. This suggests a constant presence of discomfort for the proposed stimuli, for several volunteers, and a weak prevalence of POS states.
- For some of the input items (as for example #12 and #15 in Figure 6), the average NEG outcome was found in line with preliminary expectations (i.e., possible discomfort due to hazard or risk of falling).
- For picture #2 in Figure 6, a marked NEG trend was also measured, even without a clear safety risk perception for end-users. This is also the case for item #3 in Figure 13. Both the outcomes could thus suggest some discomfort due to the use of structural glass in new solutions for constructions.
- A moderate average NEG reaction (>55% of participants) was also measured for items #2, #3, #4, #5, #9, #25 and #27. NEG and POS reactions were obtained for the residual items (see also Figure 14).

The analysis of major comfortable states, finally, can be summarized as follows:

- Absolute POS peaks were measured for item #1 (Figure 6), but also items #10 and #16 (Figure 15).
- For items #1 and #16, more in detail, this result could be justified by the presence of private/residential spaces and comfortable context conditions that do not involve direct contact with end-users. Regarding item #16, the POS peak could be justified by the "safe" presence of individuals on the floor mockup.
- Relevant POS reactions (>52% of participants) were also observed for items #20 (Figure 6) and #23 (Figure 15). Again, it is interesting to notice that the presence of end-users with "safe" behavior in the context of presented static stimuli can have positive effect on the emotional state of participants. This is in contrast with the structural typology and built environment (i.e., structures with risk of falling).

(**a**) item #5   (**b**) item #7   (**c**) item #13   (**d**) item #14

(**e**) item #17   (**f**) item #19   (**g**) item #21   (**h**) item #22

(**i**) item #24   (**j**) item #26

Figure 14. Static items with mostly balanced POS & NEG reactions: (**a**) private room (Off-grid itHouse, Pioneertown, USA; adapted from [38], © Airbnb); (**b**) glass bricked wall (Chanel Amsterdam Store, NL; adapted from [52], © MVRDV); (**c**) skywalk (Mahanakhon, Bangkok, THA; adapted from [53], © Tripadvisor); (**d**) roof (30 St Mary Axe Tower, London, UK; adapted from [54], © Nigel Young - Foster + Partners, Richard Bryant); (**e**) footbridge (Lisbon, PT; adapted from [55], © Schlaich Bergermann Partner); (**f**) Markthal, Rotterdam, NL (adapted from [56], © MVRDV) (**g**) walls for the 360 revolving restaurant Moon in Amsterdam, NL (adapted from [57], © A'DAM Lookout); (**h**) walkway (Coiling Dragon Cliff skywalk, CN; adapted from [58], © VCG via Getty Images); (**i**) swimming pool (Hubertus Hotel (Valdaora, IT; adapted from [59], © Design and Contract)); (**j**) underwater restaurant in Norway (adapted from [60], © Dezeen).

(**a**) item #10   (**b**) item #16   (**c**) item #23

Figure 15. Static items with prevailing POS reactions: (**a**) mock-up of glass floor system (adapted from [61], © Vitroplena bvba, BE); (**b**); roof (Botanical Garden, Amsterdam, NL); (**c**) observation deck (360 Chicago Tilt, Chicago, USA; adapted from [62], © Tripster).

5.2. Analysis of Reactions by Context of Pictures

Based on the experimental outcomes as in Figure 12, a special care was paid for the analysis of POS or NEG reactions for pictures grouped by context/building configuration. Two major categories of items were considered, namely represented by:

- Group-A: pedestrian systems or load-bearing elements characterized by risk of falling for the end-users (items #4, #8, #10, #13, #15, #17, #18, #20, #22, #23, #27).
- Group-B: elements characterized by the presence of damage and/or under hazard (items #6, #9, #12, #25).

In case of Group-A, a NEG reaction was typically found to prevail for the majority of pictures (8 of 11 items). For three pictures only (#10, #20, #23), the POS reaction minimally prevailed on the NEG measures. Comparative results are summarized in Figure 16a. Such an outcome suggests that pedestrian glass systems evoke the highest discomfort and emotional state for the selected stimuli, compared to other glazing solutions in buildings. The effect can be justified by the fact that, compared to other building conditions, the human interaction of pedestrians with the glass structure becomes predominant.

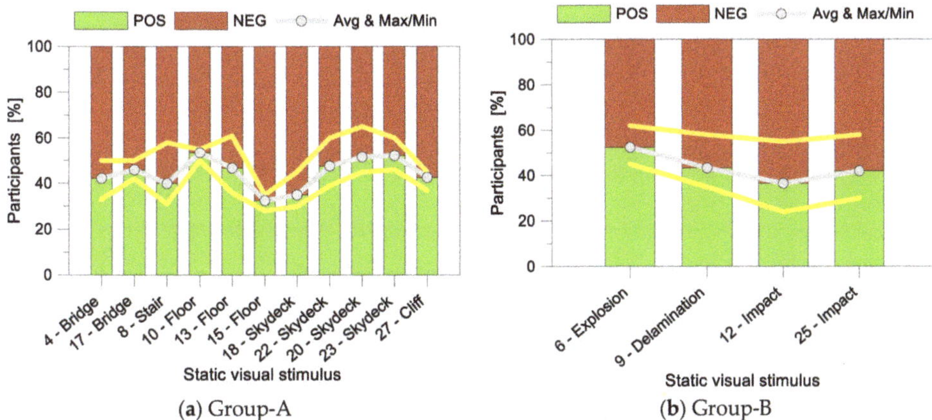

Figure 16. Detection of prevailing reactions (in percentage terms for the group of participants), as a function of static stimulus, for (**a**) Group-A or (**b**) Group-B.

Further, the few pictures from Group-A with a prevailing POS reaction were found associated with comfortable presence of individuals (i.e., item #10), and thus suggesting an appropriate safety level for the participants of the present experiment. Anyway, it is necessary to explore further this kind of scenario and extend the virtual experimental stage with field experiments, so as to capture the human reactions of active pedestrians and combine their emotional feelings with the structure vibrations and other mechanical parameters of primary interest for dynamic characterization.

Regarding the experimental results for Group-B of stimuli, NEG reactions were found to prevail on POS for most of the pictures (3 of 4 items), see Figure 16b. For the whole set, the presence of visual damage in glass proved to involve discomfort for the participants, thereby resulting in a rather good correlation of experimental predictions with the expectations of planned stimuli. The exception was represented by item #6 (Figure 6), characterized by a balanced POS/NEG average response.

5.3. Analysis of Reactions by Sub-Groups of Participants

The analysis of human reactions was also focused on the emotional state of sub-groups of participants, as, for example, obtained by gender or demographic sub-groups. As far as the POS comfort state is take into account, results according with Figures 17 and 18 can be observed. From Figure 17, more in detail, the total average POS response was found to prevail for the group of female volunteers, for most of items. This is the case for pictures #15 and #18 of Group-A and the whole Group-B. For some other items of Figure 17a, a more balanced reaction was indeed measured.

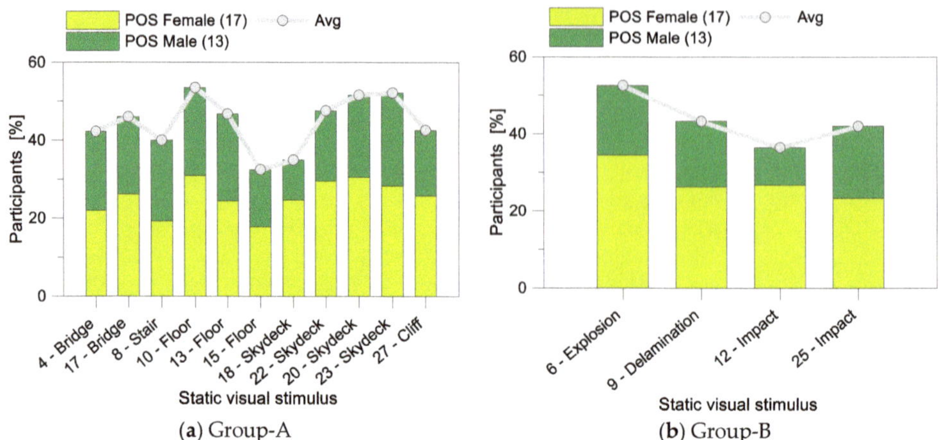

Figure 17. Detection of prevailing POS reactions (in percentage terms for participants), as a function of static stimulus, for (**a**) Group-A or (**b**) Group-B items. In evidence: the response for gender sub-groups of volunteers.

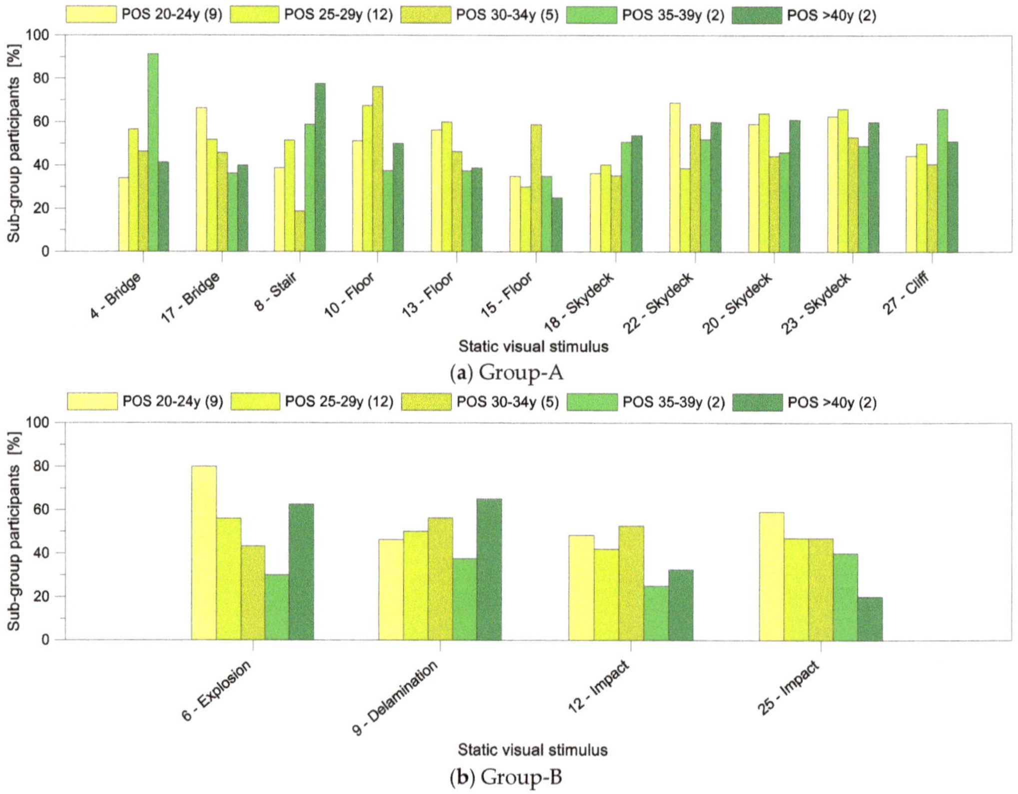

Figure 18. Detection of prevailing POS reactions (in percentage terms for participants), as a function of static stimulus, for (**a**) Group-A or (**b**) Group-B items. In evidence: the response for demographic sub-groups of volunteers.

Another aspect worth to discuss could be related to the emotional response for demographic sub-groups of volunteers, as in Figure 18. The distribution of volunteers onto age groups as in Figure 4 is not uniform and sufficiently extended to draw general conclusions. However, it is possible to notice that the trend of POS reactions modifies with age. Especially for Group-B with damage/hazard in Figure 18b, POS data decrease with age increase. Further investigations are hence needed in this direction.

6. Discussion of Results from the Dynamic Experiment

6.1. Analysis of Reactions by Context of Walks

For the dynamic setup, the post-processing stage of measured AUs was qualitatively carried out as in Section 4. However, major attention was focused on the analysis of absolute POS and NEG reaction peaks over the time of the experiment. This choice was suggested by the dynamic nature of the stimulus, with fast sequence of frames/scenarios and a different expected response compared to the static setup in Section 5.

Typical results can be seen in Figure 19, where the sequence of three different VR walks (40 s/each) is also emphasized. More precisely, the Wn labels are used to denote:

- W1: a walk outside the building, at the level of the roof/tope terrace, characterized by risk of falling (glass balustrades), but also by the presence of glass facades;
- W2: a walk inside the building (ground level), with a glass floor and a glass roof on the top, glazing walls and internal partitions;
- W3: a final walk inside the building (first story level), characterized by the risk of falling (glass balustrades), but also the floor in the below ground level (W2).

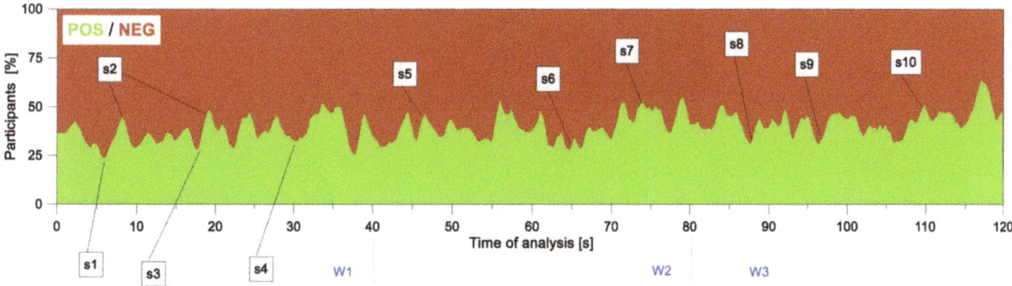

Figure 19. Detection of prevailing reactions (in percentage values for the group of participants), as a function of the time of analysis for the dynamic VR scenario.

Key labels for selected frames in Figure 19 are defined as:

- s1, s3, s4 = external balustrades, characterized by risk of falling for end-users (as in Figure 7c,d);
- s2 = external facades / walls, without contact with end-users (Figure 20a,b);
- s5, s7 = roofs, without contact with end-users (Figure 7f);
- s6 = floors with risk of falling (Figure 20c);
- s8, s9 = indoor balustrades, with risk of falling (Figure 20d);
- s10 = indoor walls, with possible contact of end-users (Figure 20e).

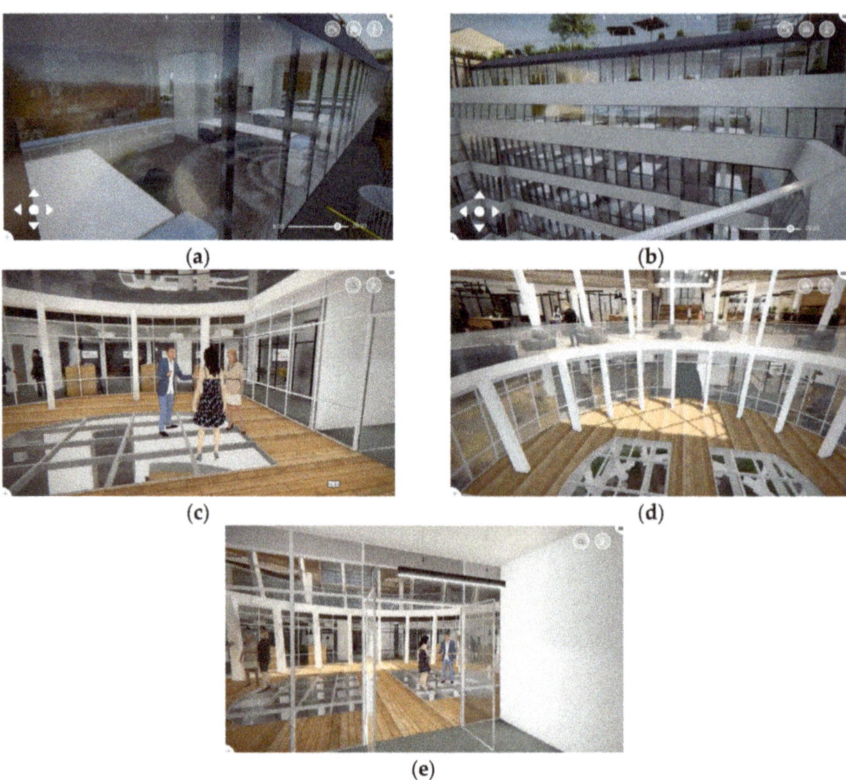

Figure 20. Selection of frames from the pre-recorded VR video clip with marked POS or NEG reactions (adapted from [45], © Generali Real Estate French Branch): (**a**)–(**b**) facade panels; (**c**) floor; (**d**) indoor balustrades; (**e**) indoor partition walls.

Worth of interest in Figure 19 is that most of the POS or NEG peaks can be associated with well-defined s1–s10 scenarios in which glazing structural components have a primary role in the building context.

More precisely, for the W1 walk:

- NEG reactions were measured in s1, s3 and s4 frames, when the pedestrian impacts the glass balustrades.
- The frame labelled as s2 is associated with glass facades and walls as in Figure 18a,b.
- During the W2 walk:
- POS reaction peaks were found, especially for s5 and s7 stimuli, to be characterized by the presence of the glass roof (Figure 7f), as well as the facade walls and the indoor partition walls for the building entrance (Figure 20c).
- The pedestrian glass floor (s6 and Figure 20d) still evoked negative perceptions for most of the participants.

Finally, for the W3 walk, the experimental measurements were still found in line with the earlier scenarios, given that:

- Major NEG peaks were related to the presence of glass balustrades with risk of falling for the end-users (Figure 20d).
- The glazing roof in s10, otherwise, reflected POS reactions peaks as previously observed for the W2 frames.

6.2. Analysis of Reactions under Static or Dynamic Stimuli

In conclusion, a comparative analysis was carried out in terms of subject/context of input stimuli for the group of volunteers, so as to capture any difference in the prevailing reactions due to a static or a dynamic setup. In Figure 21, results are shown in the form of average percentage values of measured POS and NEG reactions for selected categories of built environments and scenarios.

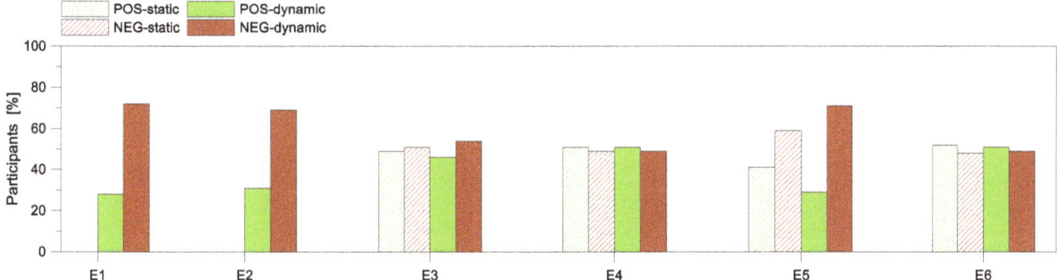

Figure 21. Analysis of prevailing POS or NEG reactions (average percentage results for the group of participants), as a function of context, as obtained from the static and dynamic VR experiments.

These are defined as:

- E1 = outdoor balustrades (with risk of falling);
- E2 = indoor balustrades (with risk of falling);
- E3 = outdoor facades/walls;
- E4 = indoor facades/walls;
- E5 = floors/walkways/pedestrian systems (with risk of falling);
- E6 = roofs.

For groups E1 and E2, the results are presented in Figure 21 for the dynamic input only, given that no direct correlation could be found with static items. Note, as also previously discussed, the remarkable NEG reaction of participants, both for outdoor or indoor balustrades. This reaction is typically observed when the end-user approaches the glass balustrades during virtual walks, and thus perceives a possible risk of falling.

Regarding the emotional states and reactions of participants towards glass facades and walls (E3 and E4 in Figure 21), the proposed experimental setup did not capture relevant modifications or prevailing nervous states. This effect can be quantified in mostly balanced POS/NEG reactions for both static and dynamic VR stimuli. Finally, it is possible to notice in Figure 21 that the configurations labelled as E5 (i.e., pedestrian systems) were mostly associated with prevailing NEG reactions. As shown, the discomfort and NEG reactions further increase with a dynamic stimulus, compared to the static one. This is not the case for glass roofs (E6) that do not involve direct contact with end-users and are characterized in Figure 21 by a slightly prevailing POS reaction, both under static and dynamic stimuli. In conclusion, it is worth noting that no items presenting a hazard or damage in the glass were taken into account for the comparative analysis of static and dynamic emotional states, due to the lack of appropriate comparative VR stimuli. Besides, data in Figure 21 still suggest a more concise emotional reaction of participants based on dynamic rather than static stimuli. As a further extension of the present analysis, it could be thus useful to take into account accidental scenarios and built environments in which glass walls are expected to act as physical barriers for the protection of occupants.

7. Conclusions

The optimization of human comfort in the built environment is a target for several design fields, and depends on a multitude of aspects. For engineering applications, some of these aspects can be mathematically controlled and limited to performance indicators, including structural, energy and thermal issues. Differently, human reactions are also sensitive to a various aspects that relate to subjective feelings, in the same way in which the so-called "emotional architecture" aims at evoking emotional responses from building users.

In this paper, an original experimental analysis was presented in support of the definition of new design strategies and tools. The goal was to assess the potential of virtual experimental methods and the possible combination of quantitative measures of subjective feelings with engineering targets, as well as architectural concepts and technology solutions for the constructional fields. Special attention was focused on glass structures and components in buildings, due to the material features, innovative use in constructions and emotional evocation for end-users. Major advantage was taken from the use of virtual experimental techniques and facial expression analyses able to capture the micro-reactions of the group of participants (30 volunteers).

By designing a visual stimulus composed of both static and dynamic virtual items for building configurations characterized by a prevailing role of structural glass elements at different context levels (roofs, floors, balustrades, etc.), the attention was focused on the quantification of "positive" or "negative" emotions and comfort feelings for the participants. As shown, the analysis of experimental measurements proved that the use of facial micro-expression recognition tools in support of comfort analyses can represent and efficient approach for optimal design.

Moreover, the proposed experimental methodology highlighted that the design of structural glass elements for buildings can be severely affected by subjective feelings and nervous states of end-users. This is especially the case for structural components that involve direct interaction and contact from end-users (such as floor or balustrades which present a risk of falling), but also the condition of structural components with visual damage. Differently, mostly positive feelings were experimentally measured for the presence of structural glass in walls, facades and roofs. In this regard, the experimental investigation could be extended to different contexts, and to groups of volunteers characterized by different age demographics, geographic distributions or background skills.

Author Contributions: This study resulted from a joint collaboration of the authors. C.B.: conceptualization; software; data curation; writing—original draft preparation; writing – review and editing; project administration. S.M.: conceptualization; data curation; investigation; writing – original draft preparation; writing—review and editing. All authors have read and agreed to the published version of the manuscript.

Funding: This research received no external funding.

Institutional Review Board Statement: Not applicable.

Informed Consent Statement: Informed consent was obtained from all subjects involved in the study.

Data Availability Statement: Data will be shared upon request.

Acknowledgments: Seretti Vetroarchitetture S.r.l. (www.seretti.it (accessed on 12 May 2021)) is acknowledged for sharing the dynamic VR video clip for the case-study building in Paris (courtesy of © Generali Real Estate French Branch, adapted from www.helloworldparis.com (accessed on 12 May 2021)). All the participants that voluntarily and actively contributed to the experimental investigation are warmly acknowledged. A special thanks is for Corine Tetteroo (Noldus Information Technology bv, NL). Finally, Alice Iurlaro is acknowledged for supporting part of the recording stage.

Conflicts of Interest: The authors declare no conflict of interest.

References

1. Levin, H. Designing for people: What do building occupants really want? In Proceedings of the Healthy Buildings 2003, Singapore, 7–11 December 2003.
2. Colenberg, S.; Jylha, T.; Arkesteijn, M. The relationship between interior office space and employee health and well-being—A literature review. *Build. Res. Inf.* **2020**, *49*, 35–366. [CrossRef]

3. Li, D.; Menassa, C.C.; Kamat, V.R. Personalized human comfort in indoor building environments under diverse conditioning modes. *Build. Environ.* **2017**, *126*, 304–317. [CrossRef]
4. Yun, G.Y.; Kim, J.T. Creating sustainable building through exploiting human comfort. *Energy Procedia* **2014**, *62*, 590–594. [CrossRef]
5. Abdel-Ghany, A.M.; Al-Helal, I.M.; Shady, M.R. Human thermal comfort and heat stress in an outdoor urban arid environment: A case study. *Adv. Meteorol.* **2013**. [CrossRef]
6. Busca, G.; Cappellini, A.; Manzoni, S.; Tarabini, M.; Vanali, M. Quantification of changes in modal parameters due to the presence of passive people on a slender structure. *J. Sound Vib.* **2014**, *333*, 5641–5652. [CrossRef]
7. Davis, B.; Avci, O. Simplified vibration serviceability evaluation of slender monumental stairs. *J. Struct. Eng.* **2015**, *141*, 04015017. [CrossRef]
8. Bedon, C.; Fasan, M. Reliability of field experiments, analytical methods and pedestrian's perception scales for the vibration serviceability assessment of an in-service glass walkway. *Appl. Sci.* **2019**, *9*, 1936. [CrossRef]
9. Burton, M.D.; Kwok, K.C.S.; Abdelrazaq, A. Wind-Induced motion of tall buildings: Designing for occupant comfort. *Int. J. High-Rise Build.* **2015**, *4*, 1–8.
10. Denoon, R.O.; Roberts, R.D.; Letchford, C.W.; Kwok, K.C.S. *Field Experiments to Investigate Occupant Perception and Tolerance of Wind-Induced Building Motion*; Research Report No. R803; Department of Civil Engineering, University of Sydney: Sydney, NWS, Australia, 2000.
11. Kwok, K.C.S.; Hitchcock, P.A. Occupant comfort test using a tall building motion simulator. In Proceedings of the Fourth International Conference on Advances in Wind and Structures, Jeju, Korea, 28–30 May 2008.
12. Bower, I.; Tucker, R.; Enticott, P.G. Impact of built environment design on emotion measured via neurophysiological correlates and subjective indicators: A systematic review. *J. Environ. Psychol.* **2019**, *66*, 101344. [CrossRef]
13. Juan-Vidal, F.; Inarra-Abad, S. Introducing emotions in the architectural design process. In Proceedings of the 1st International Conference on Higher Education Advances(HEAd'15), Universitat Politecnica de Valencia, Valencia, Spain, 24–26 June 2015. [CrossRef]
14. Shearcroft, G. The Joy of Architecture: Evoking Emotions through Building. *Archit. Des.* **2021**, *91*, 108–117.
15. Suri, C. Inside the Rise of Emotional Design. *Architectural Digest*. 18 May 2017. Available online: https://www.architecturaldigest.com/story/emotional-design (accessed on 12 May 2021).
16. Shahid, M. The Different Emotions Architecture (Buildings) Can Express. *Rethinking the Future*. Available online: https://www.re-thinkingthefuture.com/fresh-perspectives/a2586-the-different-emotions-architecture-buildings-can-express/ (accessed on 12 May 2021).
17. Kozlovsky, R. Architecture, emotions and the history of childhood. In *Childhood, Youth and Emotions in Modern History. Palgrave Studies in the History of Emotions*; Olsen, S., Ed.; Palgrave Macmillan: London, UK, 2015. [CrossRef]
18. Ricci, N. The Psychological Impact of Architectural Design. CMC Senior Theses. 2018. Available online: https://scholarship.claremont.edu/cmc_theses/1767 (accessed on 12 May 2021).
19. Canepa, E.; Scelsi, V.; Fassio, A.; Avanzino, L.; Lagravinese, G.; Chiorri, C. Atmospheres: Feeling Architecture by Emotions— Preliminary Neuroscientific Insights on Atmospheric Perception in Architecture. *Ambiances* **2019**. Available online: http://journals.openedition.org/ambiances/2907 (accessed on 12 May 2021). [CrossRef]
20. Shu, L.; Yu, Y.; Chen, W.; Hua, H.; Li, Q.; Jin, J.; Xu, X. Wearable emotion recognition using heart rate data from a smart bracelet. *Sensors* **2020**, *20*, 718. [CrossRef] [PubMed]
21. Geiser, M.; Walla, P. Objective measures of emotion during virtual walks through urban environments. *Appl. Sci.* **2011**, *1*, 1–11. [CrossRef]
22. Gulnick, J. The Psychology of Perception, Threshold, and Emotion in Interior Glass Design. *Glass on Web*. 2 December 2019. Available online: https://www.glassonweb.com/article/psychology-perception-threshold-and-emotion-interior-glass-design (accessed on 12 May 2021).
23. Deriu, D. Skywalking in the city: Glass platforms and the architecture of vertigo. *Emot. Space Soc.* **2018**, *28*, 94–103. [CrossRef]
24. Bedon, C.; Amadio, C. Exploratory numerical analysis of two-way straight cable-net facades subjected to air blast loads. *Eng. Struct.* **2014**, *79*, 276–289. [CrossRef]
25. Bedon, C.; Zhang, X.; Santos, F.; Honfi, D.; Kozłowski, M.; Arrigoni, M.; Figuli, L.; Lange, D. Performance of structural glass facades under extreme loads—Design methods, existing research, current issues and trends. *Constr. Build. Mater.* **2018**, *163*, 921–937. [CrossRef]
26. China Daily. World's Longest, Highest Glass Bridge Opens in Hunan. 2016. Available online: https://www.chinadaily.com.cn/travel/2016-08/21/content_26546587_7.htm (accessed on 12 May 2021).
27. Darral, S. Don't Look Down! The Terrifying See-Through Path Stuck to a Chinese Cliff-Face 4,000ft above a Rocky Ravine. 2011. Available online: https://www.dailymail.co.uk/news/article-2060023/Chinas-newest-tourist-attraction--glass-bottomed-walkway-cliff-face.html (accessed on 12 May 2021).
28. Bedon, C.; Kruszka, L. An insight on the mitigation of glass soft targets and design of protective facades. In *Critical Infrastructure Protection. Best Practices and Innovative Methods of Protection, NATO Science for Peace and Security Series D: Information and Communication Security*; Kruszka, L., Klósak, M., Muzolf, P., Eds.; IOS Press: Amsterdam, The Netherlands, 2019; Volume 52, pp. 107–117. ISBN 978-1-61499-963-8.

29. Bedon, C. Transparent materials and new design strategies in the COVID-19 era. In *Abstract Book of ICCECIP 2020, International Conference on Central European Critical Infrastructure Protection, Budapest, Hungary, 16–17 November 2020*; Nyikes, Z., Kovacs, T.A., Molnar, A., Eds.; Óbuda University: Budapest, Hungary, 2020; p. 39. ISBN 978-963-449-221-4.
30. Nothingam, S. Simply Breathtaking: Glass Floor Ideas for the Polished Modern Home. Available online: https://www.decoist.com/glass-floor-ideas/?firefox=1 (accessed on 12 May 2021).
31. Loijens, L.; Krips, O. *FaceReader Methodology*; Noldus Information Technology: Wageningen, The Netherlands, 2013.
32. Zaman, B.; Shrimpton-Smith, T. The FaceReader: Measuring instant fun of use. In Proceedings of the 4th Nordic Conference on Human-Computer Interaction (NordiCHI '06). Oslo, Norway, 14–18 October 2006.
33. Yu, C.Y.; Ko, C.H. Applying FaceReader to recognize consumer emotions in graphic styles. *Procedia CIRP* **2017**, *60*, 104–109. [CrossRef]
34. Mordor Intelligence. Europe Flat Glass Market—Growth, Trends, COVID-19 Impact, and Forecasts (2021–2026). Technical Report. 2020. Available online: https://www.mordorintelligence.com/industry-reports/europe-flat-glass-market (accessed on 12 May 2021).
35. Creating Safe, Sustainable Spaces: How Building Materials Will Evolve Post COVID-19. 2020. Available online: https://www.gauzy.com/creating-safe-sustainable-spaces-how-building-materials-will-evolve-post-covid-19/ (accessed on 12 May 2021).
36. Chan, Z.Y.S.; MacPhail, A.J.C.; Au, I.P.H.; Zhang, J.H.; Lam, B.M.F.; Ferber, R.; Cheung, R.T.H. Walking with head-mounted virtual and augmented reality devices: Effects on position control and gait biomechanics. *PLoS ONE* **2019**, *14*, e0225972. [CrossRef] [PubMed]
37. Yang, Y.R.; Tsai, M.P.; Chuang, T.Y.; Sung, W.H.; Wang, R.Y. Virtual reality-based training improves community ambulation in individuals with stroke: A randomized controlled trial. *Gait Posture* **2008**, *28*, 201–206. [CrossRef] [PubMed]
38. Airbnb. Off-Grid itHouse. Available online: https://www.airbnb.it/rooms/19606?source_impression_id=p3_1620923233_mkY0whaLq1nXgbul (accessed on 12 May 2021).
39. AASA Architecture. Nhow-Amsterdam-RAI-Hotel-by-OMA-01. 2020. Available online: https://aasarchitecture.com/2020/02/cod-and-being-development-delivered-the-nhow-amsterdam-rai-hotel-by-oma.html/nhow-amsterdam-rai-hotel-by-oma-01/ (accessed on 12 May 2021).
40. ABC News. China Closes "Scary" Glass Bridges Due to "Safety Problems". 2019. Available online: https://www.abc.net.au/news/2019-11-01/china-closes-glass-bridges-due-to-safety-problems/11662372 (accessed on 12 May 2021).
41. Sedak All-Glass Stairs: A Step towards Freedom. Available online: https://www.sedak.com/en/applications/architecture/glass-stairs (accessed on 12 May 2021).
42. Times Colonist. Vehicle Crashes through Window, Damaging Saanich Real Estate Office. 2020. Available online: https://www.timescolonist.com/news/local/vehicle-crashes-through-window-damaging-saanich-real-estate-office-1.24094689 (accessed on 12 May 2021).
43. Lavars, N. Seattle's Iconic Space Needle Tower Gets a Revolving Glass Floor. 2018. Available online: https://newatlas.com/seattle-space-needle-glass-floor/55764/ (accessed on 12 May 2021).
44. Dragun, N. New Flights Mean Chicago Is Calling. 2018. Available online: https://www.escape.com.au/destinations/north-america/usa/new-flights-mean-chicago-is-calling/news-story/36343c5cbf8aca3de0fe6bab4f68f039 (accessed on 12 May 2021).
45. Hello World. 2020. Available online: www.helloworldparis.com (accessed on 12 May 2021).
46. Wellershoff, F.; Sedlacek, G. Glass pavilion rheinbach—stability of glass columns. In Proceedings of the Leading International Glass Conference "Glass Processing Days 2003", Tampere, Finland, 15–18 June 2003; pp. 316–318.
47. Glacier Skywalk. Available online: https://www.gettyimages.fi/photos/glacier-skywalk?family=editorial&phrase=glacier%20skywalk&sort=mostpopular (accessed on 12 May 2021).
48. Bedon, C. Issues on the vibration analysis of in-service laminated glass structures: Analytical, experimental and numerical investigations on delaminated beams. *Appl. Sci.* **2019**, *9*, 3928. [CrossRef]
49. Bedon, C. Diagnostic analysis and dynamic identification of a glass suspension footbridge via on-site vibration experiments and FE numerical modelling. *Compos. Struct.* **2019**, *216*, 366–378. [CrossRef]
50. Wright, P. World's Longest and Highest Glass-Bottomed Bridge Opens in China's "Avatar" Mountains. 2016. Available online: https://weather.com/travel/news/china-glass-bridge-opens (accessed on 12 May 2021).
51. Myers, L. Hayri Atak Envisions a Boutique Hotel Suspended over a Cliff Edge in Norway. 2019. Available online: https://www.designboom.com/architecture/hayri-atak-cliff-concept-boutique-hotel-norway-07-22-2019/ (accessed on 12 May 2021).
52. Frearson, A. MVRDV Replaces Chanel Store's Traditional Facade with Glass Bricks that Are "Stronger than Concrete". 2016. Available online: https://www.dezeen.com/2016/04/20/crystal-houses-amsterdam-chanel-store-mvrdv-glass-facade-technology/ (accessed on 12 May 2021).
53. Tripadvisor. King Power Mahanakhon. Available online: https://www.tripadvisor.it/Attraction_Review-g293916-d14135436-Reviews-King_Power_Mahanakhon-Bangkok.html (accessed on 12 May 2021).
54. Pintos, P. 30 St Mary Axe Tower/Foster + Partners. Available online: https://www.archdaily.com/928285/30-st-mary-axe-tower-foster-plus-partners (accessed on 12 May 2021).
55. Schlaich Bergermann Partner. Glass Bridge Lisbon. 2010. Available online: https://www.sbp.de/fr/projekt/glasbruecke-lissabon/ (accessed on 12 May 2021).
56. MVRDV Markthal. Available online: https://www.mvrdv.nl/projects/115/markthal (accessed on 12 May 2021).

57. A'DAM LOOKOUT + 360° Lunch in Revolving Restaurant Moon. Available online: https://www.adamlookout.com/combi-tickets/#!/lunch (accessed on 12 May 2021).
58. Toronto Star. China Opens Terrifying Tianmen Mountain Glass Walkway. 2016. Available online: https://www.thestar.com/news/world/2016/08/03/china-opens-terrifying-tianmen-mountain-glass-walkway.html (accessed on 12 May 2021).
59. Design & Contract. Referenze: Hotel Hubertus—Valdaora, Bolzano. Available online: https://designandcontract.com/it/referenze/hotel-hubertus-valdaora-bolzano (accessed on 12 May 2021).
60. Crook, L. Snøhetta Completes Europe's First Underwater Restaurant in Norway. 2019. Available online: https://www.dezeen.com/2019/03/20/underwater-restaurant-under-snohetta-baly-norway/ (accessed on 12 May 2021).
61. Vitroplena Structural Glass Solutions—A Transparent Office Floor. 2017. Available online: https://www.vitroplena.be/en/innovative-solutions (accessed on 12 May 2021).
62. Tripster. 360 Chicago Observation Deck in Chicago. Available online: https://www.tripster.com/detail/360-chicago-observation-deck (accessed on 12 May 2021).

Article

Calibrated Numerical Approach for the Dynamic Analysis of Glass Curtain Walls under Spheroconical Bag Impact

Alessia Bez [1], **Chiara Bedon** [1,*], **Giampiero Manara** [2], **Claudio Amadio** [1] and **Guido Lori** [2]

1   Department of Engineering and Architecture, University of Trieste, 34127 Trieste, Italy; alessia.bez@dia.units.it (A.B.); amadio@units.it (C.A.)
2   Permasteelisa S.p.a., 31029 Vittorio Veneto, Italy; g.manara@permasteelisagroup.com (G.M.); g.lori@permasteelisagroup.com (G.L.)
*   Correspondence: chiara.bedon@dia.units.it; Tel.: +39-040-558-3837

Abstract: The structural design of glass curtain walls and facades is a challenging issue, considering that building envelopes can be subjected extreme design loads. Among others, the soft body impact (SBI) test protocol represents a key design step to protect the occupants. While in Europe the standardized protocol based on the pneumatic twin-tire (TT) impactor can be nowadays supported by Finite Element (FE) numerical simulations, cost-time consuming experimental procedures with the spheroconical bag (SB) impactor are still required for facade producers and manufacturers by several technical committees, for the impact assessment of novel systems. At the same time, validated numerical calibrations for SB are still missing in support of designers and manufacturers. In this paper, an enhanced numerical approach is proposed for curtain walls under SB, based on a coupled methodology inclusive of a computationally efficient two Degree of Freedom (2-DOF) and a more geometrically accurate Finite Element (FE) model. As shown, the SB impactor is characterized by stiffness and dissipation properties that hardly match with ideal rigid elastic assumptions, nor with the TT features. Based on a reliable set of experimental investigations and records, the proposed methodology acts on the time history of the imposed load, which is implicitly calibrated to account for the SB impactor features, once the facade features (flexibility and damping parameters) are known. The resulting calibration of the 2-DOF modelling parameters for the derivation of time histories of impact force is achieved with the support of experimental measurements and FE model of the examined facade. The potential and accuracy of the method is emphasized by the collected experimental and numerical comparisons. Successively, the same numerical approach is used to derive a series of iso-damage curves that could support practical design calculations.

Keywords: soft body impact (SBI) test; spheroconical bag (SB) impactor; glass curtain walls; 2-degree of freedom (2-DOF) model; experiments; finite element (FE) numerical models; iso-damage curves

Citation: Bez, A.; Bedon, C.; Manara, G.; Amadio, C.; Lori, G. Calibrated Numerical Approach for the Dynamic Analysis of Glass Curtain Walls under Spheroconical Bag Impact. *Buildings* **2021**, *11*, 154. https://doi.org/10.3390/buildings11040154

Academic Editor: Giuseppina Uva

Received: 17 March 2021
Accepted: 5 April 2021
Published: 7 April 2021

Publisher's Note: MDPI stays neutral with regard to jurisdictional claims in published maps and institutional affiliations.

1. Introduction

Glass curtain walls notoriously represent a challenging issue for designers. Given that they must satisfy specific performance requirements in terms of energy, light, acoustic insulation, etc., curtain walls represent in fact a physical barrier for the occupants. As such, careful consideration is required for their structural assessment, even under extreme events. In this regard, several literature studies have been dedicated to the analysis of glass facades under various loading conditions, including blast events, seismic actions, etc. [1–5]. Among others, the assessment of facade capacities under soft body impact (SBI) still represents one of the most frequent accidental design actions (i.e., due to impact of occupants) and also one of the most severe performance limitations, especially against the potential risk of fall [6], see Figure 1.

(a)   (b)

Figure 1. Example of curtain wall: (**a**) modular unit and (**b**) failure under impact.

So far, several research efforts have been spent on the analysis of the impact performance of glass systems. Relevant outcomes are available in the literature, but they are mostly related to specific design applications like simple glass panels [7], glass balustrades [8,9], load-bearing glass columns [10], existing traditional windows [11] or novel laminated glass windows [12]. In [13], a systemic experimental and numerical analysis has been dedicated to the analysis of glass constructional elements under the pendulum test, with a focus on the numerical description of the impactor features. This study follows the extended investigations reported in [14–16], in which the dynamic performance of glass panels under tire impact has been explored with the support of experimental and numerical tools. Often, within the research community, the SBI performance of glass elements is primarily explored for automotive applications [17], or in terms of danger for people [18,19], while a generalized methodology still lacks for the mechanical characterization of curtain walls.

In this paper, a careful consideration is paid for curtain walls under a spheroconical bag (SB) impactor, with the support of full-scale experiments, Finite Element (FE) numerical models (Code Aster [20]), two Degree of Freedom (2-DOF) modelling techniques and practical design tools that could be useful in support of daily practice (i.e., iso-damage curves). In more detail, Section 2 summarizes the key features of experimental methods in use for curtain walls under SBI, as well as the basic theoretical background of impact. Differing from "ideal" conditions, as shown, the SBI response of curtain walls is affected by the features of the facade (impacted body), by the type of impactor (SB or TT) and by the impact features (impact point, drop height, facade features and impactor properties). A novel approach is thus presented in Section 3, based on the coupled analysis of a simplified 2-DOF model and a more geometrically accurate FE model of the curtain wall unit object of study. The procedural steps that are used for the optimal calibration and minimization of the analysis efforts are hence discussed. Section 4, in more detail, presents an experimental campaign carried out on curtain wall unit specimens under a multitude of impact configurations (15 repetitions and arrangements), while the reliability of the proposed procedure is shown in Section 5. Given that the calibrated 2-DOF parameters are derived from a multitude of impact configurations, the general validity of the procedure is shown. Some further design outcomes are presented in Section 6, where comparative analyses are proposed for the optimal definition of the input impulse. For comparative analysis of the proposed methodology, additional numerical results are also derived from the commercial software SJ MEPLA [21] ("MEPLA", in the following). In conclusion, iso-damage curves are presented in support of design for glass curtain walls under SBI.

2. State-of-Art and Background

2.1. Standardized Experimental Approach

Glass curtain walls and facades are recognized to have a crucial role in buildings, given that they are expected to offer a physical barrier against falling of the occupants, and a physical protection from potential outdoor events. Accordingly, safety must be especially ensured against accidental impact events. The fall of possible debris, in the event of glass breakage, must be prevented with the protection of pedestrians that could be nearby the glass facade. Defenestration events must be also prevented. In this context, the use of laminated safety glass represents an efficient design solution. However, many other performance indicators must be properly assessed, at the component and assembly level.

For a long time, the SBI performance has been assessed through experiments, based on standardized testing procedures. Usually, these tests are carried out in-situ, or in laboratory conditions on full-scale mock-ups. The mostly used impactors, see Figure 2, are the spheroconical bag (SB) and the twin tire (TT). The former is filled with glass spheres and has a total weight M = 50 kg, while the latter consists of two pneumatic tires inflated with 3.5 bar air pressure. A steel mass (M = 50 kg) is included within the tires themselves. In this research paper, the tires consisted of Vredestein V47 pneumatics [22].

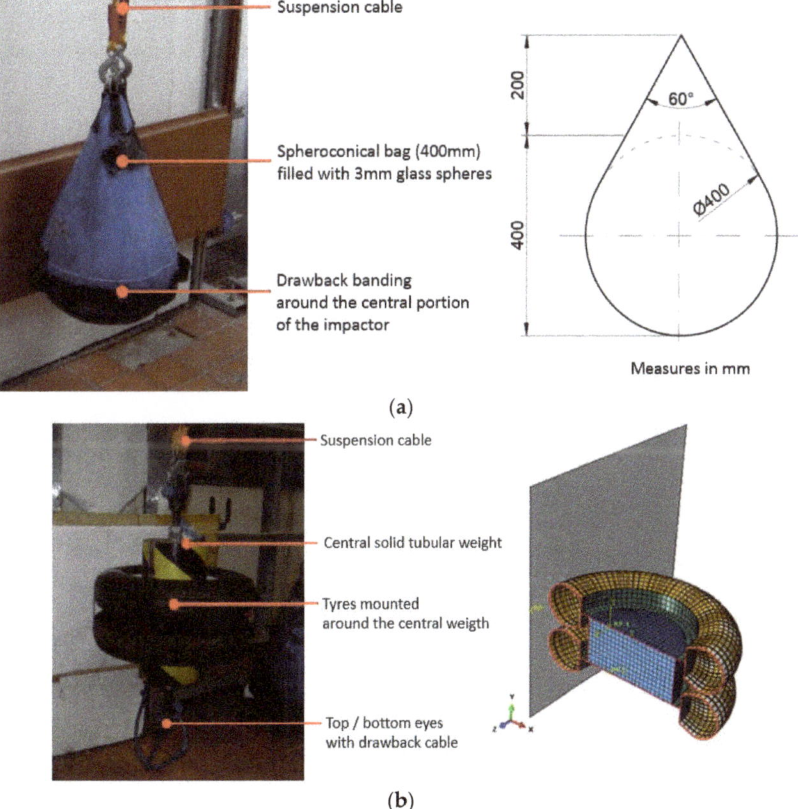

Figure 2. Standardized impactors for glass curtain walls: (**a**) spheroconical bag (SB, with nominal dimensions in mm) or (**b**) twin-tire (TT) impactor (with FE model detail).

Major issues for manufacturers of glass curtain walls under SBI, in this regard, currently derive from the existence of a number of National norms and regulations that follow different procedures.

So far, the International EN 12600 standard [23] introduced the TT pendulum protocol at the European level, with the aim of replacing the SB impactor. The TT is in fact more easily describable in its basic components. Further, the German DIN 18008-4 (Annex A) regulations [24] confirmed the possibility of TT impact FE numerical simulations in place of full-scale experiments. In the years, several research studies highlighted the realistic description of TT features and effects with numerical tools [9,13–16,25]. The MEPLA commercial software, in this regard, also developed a lumped spring-mass model able to capture (with limits) the TT features, and thus offer a geometrically and mechanically simplified numerical model in support of design (see Section 2.2 and [26–28]).

On the other side, a set of standards exists that prescribe a pendulum setup still based on the SB impactor, and includes the French Cahier CSTB 3228 [29] and NF P 8-301 [30] for vertical or horizontal glazing, the CWCT TN 76 [31] (vertical glazing) and ACR[M]001:2005 [32] (horizontal elements) standards in the United Kingdom, but also the American ANSI Z97.1 document [33]. The performance evaluation after impact requires one to assess whether the glass system is able to pass a given setup, and to classify the specimen capacity depending on the lack of breakage, or on the size and features of cracks and fragments (if any). However, given that the cited references represent the primary regulation for various National technical committees, this condition forces facade designers and manufacturers to follow the original SB approach. In this regard, the use of efficient numerical tools could simplify the SB impact assessment for various practical applications. Besides, calibrated parameters for SB simulations are not available in the literature, and cannot be replaced by TT formulations. The Newman's hypotheses discussed in [34] asserts the greater severity of impact tests carried out with the TT impactor, rather than the SB. The latter is in fact recognized to have a behavior that is hardly comparable to an elastic body under impact. Moreover, the SB impactor generally offers a greater dissipative capacity, compared to the TT. Most of the dissipative phenomena are related to the glass spheres inside the bag of Figure 2, which are not bound to reciprocal movements. Accordingly, ideal theoretical assumptions for impact estimates are generally over-conservative for real experimental setup configurations with the SB, thus requiring more advanced characterization and calculation procedures.

In this paper, the proposed methodology aims at exploring the difference of SB and TT parameters, towards the definition and calibration of practical recommendations and practical approaches to equalize the SB effects to the TT, for a given envelope unit under SBI.

2.2. Numerical Analysis of Glass Curtain Walls under Soft Body Impact

2.2.1. Ideal Impact

For a curtain wall under the impact of a pendulum in accordance with Figure 2, the dynamic features and response of the system can be notoriously sensitive to a multitude of parameters. As such, the reliable description of the input time history for the impact force can be challenging. The actual phenomenon still derives from the "ideal" impact configuration, but it should be further characterized. According to the classical theory, the limit condition is represented by an elastic impactor against a rigid body, and the expected force time history is expressed as:

$$F(t) = f(t)\,Ap \tag{1}$$

where the key input parameters are represented by the impacted surface A; the peak pressure p, that depends on the drop height of the impactor; and the time factor $f(t)$, whose typical trend is proposed in Figure 3.

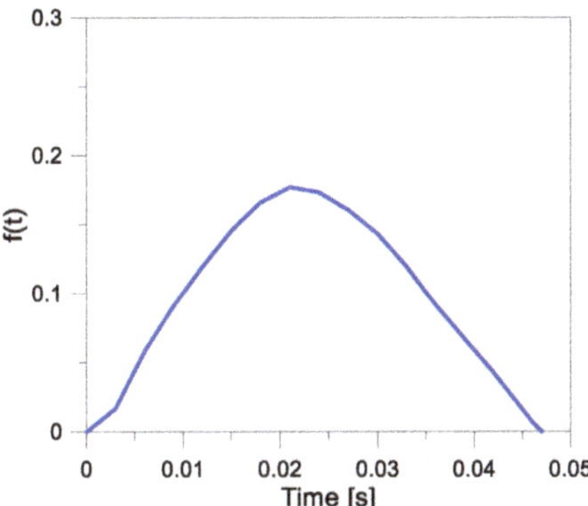

Figure 3. Typical trend for the time factor $f(t)$.

For a rigid impacted surface and elastic impactor, the incoming impulse I_{id} on a given surface A can be rationally calculated as:

$$I_{id} = \int F(t)\, dt = \int [f(t)\, Ap]\, dt = 2\, M_P\, v_P \tag{2}$$

with M_P denoting the mass of the impactor and v_P its velocity, at the time of impact. Assuming that M_P and v_P are the assigned input data, the corresponding impact energy can be hence predicted in:

$$E_{imp} = \frac{1}{2} M_P v_P^2 \tag{3}$$

and thus:

$$I_{id} = 4 \frac{E_{imp}}{v_P} \tag{4}$$

The above assumptions, although theoretically accurate, involve a severely conservative description of input features, both on the side of the impactor and on the side of the impacted glass panel. Equations (2)–(4), more in detail, represent an upper limit (or "ideal") condition, since the real behavior is generally affected by the intrinsic flexibility of the impacted glass panel (i.e., stiffness of facade components, compared to a rigid surface), as well as on the impactor features (i.e., type, size, stiffness, impact energy), which manifest in a series of dissipative phenomena.

2.2.2. Simplified Design Approach

The MEPLA computer software is largely used for structural glass design applications, given that it is specifically tailored to the needs of the civil engineering industry and offers some a useful support for the efficient analysis of standard configurations. For impact purposes, the software proposes a lumped spring-mass model that calculates a time history of impact forces for the so-called "default pendulum impactor" loading case based on EN12600 [23]. This "default pendulum impactor" option, compared to more refined modelling techniques (as for example in [9,13,28]), represents a simplified tool for the description of a test setup with TT impactor. At the same time, the approach assumes the hypothesis of elastic impact on a flexible, undamped facade unit. The final result, compared to an ideal impulse as in Section 2.2.1, is a partially reduced impulse I_{Mepla}:

$$I_{Mepla} = R_{Mepla} \int [f(t)\, Ap]\, dt = R_{Mepla}(2\, M_P\, v_P) < I_{id} \tag{5}$$

in which the reduction factor $R_{Mepla} < 1$ accounts for some intrinsic aspects of the real dynamic phenomenon and bodies involved in the impact event. Research studies in [26,27] and others extensively verified the accuracy of such a schematic representation of the

TT impactor, giving evidence of a good correlation with experimental records of various glazing systems. On the other side, see for example [13,28], the lumped mass-spring model cannot equal a geometrically refined characterization of TT impactor, in terms of component features and pressurized air volume in the tires. As a matter of fact, moreover, it lacks of an accurate geometrical and mechanical characterization of the impactor, as it would be required for the SB. Further effects, in the cited procedure, finally derive from the lack of dissipation features, both for the glazing system and for the impactor. These limitations end in a potential over-prediction of the expected accelerations, displacements and strains for the curtain wall, and thus in a consequent over-design of the load-bearing components against SBI.

3. Design Approach for Glass Facades under Soft Body Impact

3.1. Concept, Assumptions and Goals

In the current investigation, a new calibration procedure in support of design is proposed and validated by means of full-scale experimental tests. Further, the MEPLA software is used for comparative analyses on curtain wall units under variable SB impact. The goal is to predict the impact response of a given glass facade, by taking into account the actual impact force, with the support of analytical and numerical estimates in place of cost- and time-consuming experimental protocols. The flexibility and damping contributions are taken into account both on the side of the curtain wall and the impactor, thus allowing to obtain a generalized solution of the design issue. In this way, the amount of experiments can be reduced to a minimum, or even cancelled. From a practical point of view, see Figure 4, the magnitude of the impulse is expected to be:

$$I_R = R_R \, I < I_{Mepla} < I_{id} \tag{6}$$

with

$$R_R = f\left(E_{imp}, \, K_{imp}, \, impactor\right) < R_{Mepla} < 1 \tag{7}$$

where R_R is a refined reduction coefficient able to account for the boundary conditions and dynamic properties of the curtain wall under SBI. The advantage of Equation (7) is that the maximum theoretical impulse can be minimized to reliable estimates, while keeping mostly constant the total duration of the "ideal" impact event. To achieve this goal, however, some key procedural steps must be taken into account (Section 3.2).

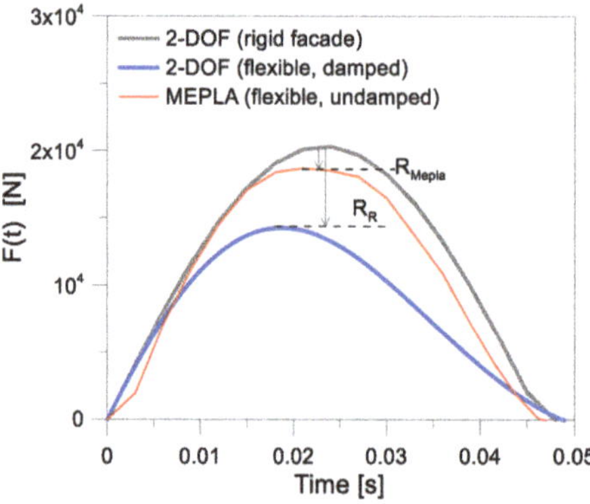

Figure 4. Qualitative comparison of input impact forces for a given glass facade, as obtained under ideal conditions (rigid facade), from MEPLA (flexible, undamped facade) or by accounting for the proposed approach (impactor/facade properties and damping).

3.2. Proposed Procedural Steps

For a given glass curtain wall under SBI, a coupled approach based on a simplified 2-DOF model and a more accurate numerical model (Code Aster) is considered for the modular unit in Figure 5. For the 2-DOF model in Figure 5a, more in detail, components with subscript "F" refer to the facade systems, while the subscripts "P" and "C" are used for the impactor and the contact parameters. The design procedure herein proposed is achieved (and generalized) through the sequence of steps that are schematized in Figure 6. In the STEP 0, the features of the curtain wall and the desired impact energy E_{imp} are defined for the loading condition of interest. Given that $M_P = 50$ kg, the corresponding impact velocity v_P and the related ideal impulse I_{id} are first derived from Equations (3) and (4). At this stage, both the 2-DOF and the FE models of Figure 5 must be properly characterized.

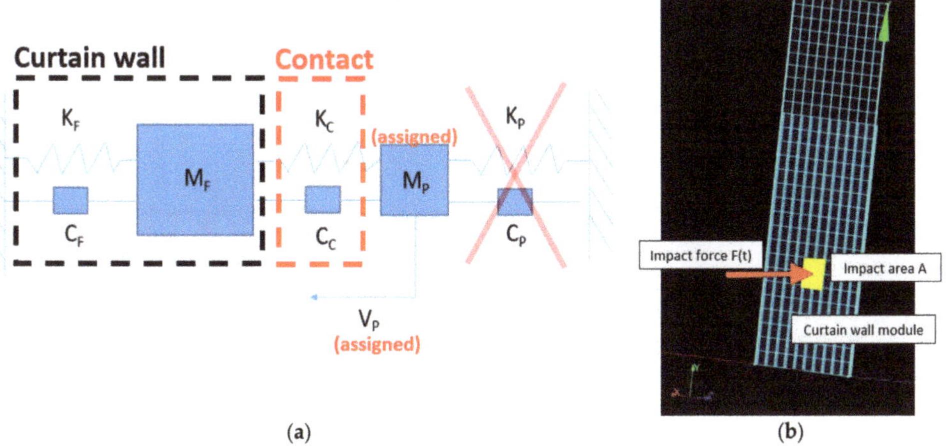

Figure 5. Reference mathematical models for the proposed coupled analysis of glass curtain wall under soft body impact (SBI): (**a**) 2-DOF and (**b**) FE numerical model (Code Aster).

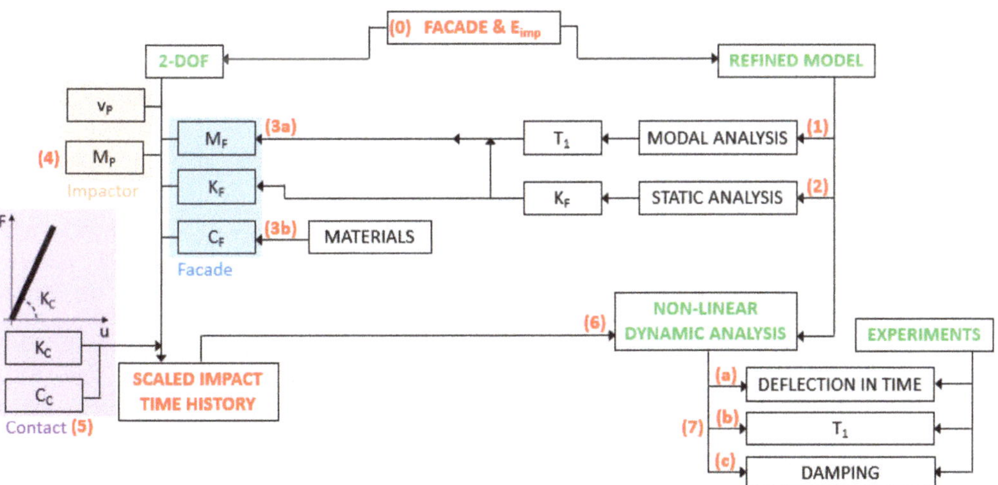

Figure 6. Reference procedural steps for the calibration of input parameters from the herein developed coupled analysis of glass curtain walls under SBI.

Considering the FE model in Figure 5b, the curtain wall module must be described with careful attention for materials, components, boundaries (Section 5). Then, a frequency analysis is carried out to predict the fundamental vibration period of the undamped system (T_1, STEP 1). The equivalent stiffness K_F of the curtain wall represents another key input parameter, given that it is affected by the facade features and also by the point of impact (STEP 2). Such a parameter can be derived from the same FE model, under a unit force that is statically assigned to the desired impact point, and thus based on the corresponding static displacement. The knowledge of T_1 and K_F is the starting point for the 2-DOF characterization, given that the equivalent mass M_F for the modular unit can be calculated as (STEP 3a):

$$M_F = \frac{K_F}{\left(\frac{2\pi}{T_1}\right)^2} \tag{8}$$

Finally, on the side of the curtain wall, the damping term C_F must be also specified (STEP 3b), based on:

$$C_F = \xi \times 2\sqrt{K_F M_F} \tag{9}$$

where ξ is a total damping term (including the effects of materials, possible damage propagation, and aero-elastic effects). A detailed calculation approach would require the experimental calculation of test records (i.e., logarithmic decrement, see Section 5). Given that the proposed approach is intended for efficient early-stage design calculations, however, the accurate total damping ξ may not be available. As a reference, accordingly, a minimum value should be set at least in:

$$\xi_{min} = \xi_{glass} + \xi_{int} + \xi_{silicone} + \xi_{frame} \tag{10}$$

thus accounting for the material properties in use. Based on literature data, for example, the input values in Equation (10) can be expected in the order of 1% for glass, 2% for the soft layers and the metal frame, respectively [35–38].

Depending on the type of impactor in use (see Figure 6) the input parameters must be then properly specified to account for the TT or SB effects (STEP 4), as well as the effect that they transfer to the facade (STEP 5). Given that M_P and v_P are known, in accordance with the 2-DOF model in Figure 5a, such parameters are described in the form of a simple mechanical law for the contact itself (i.e., K_P and C_P are set to 0). The contact parameters K_C and C_C that are required for STEP 5, in other words, are responsible of the actual force (and time history features) that the impactor can transfer to the curtain wall, depending on the time of contact and on the impactor features. In this research study, all these parameters were iteratively calibrated, based on experimental results and numerical fitting (Section 6.2). Once the above parameters are known, the result of the 2-DOF model is a scaled impact force that implicitly accounts for the reduced impulse I_R in Equation (6), and can be thus assigned to the FE model, to perform non-linear dynamic analyses of the system (STEP 6).

In presence of full-scale impact experiments (STEP 7), the FE dynamic analysis can be assessed towards the test records, or give a realistic prediction of the experimental outcomes, in terms of deflection in time (a), period of vibration of the facade (b), damping of the system (c), but also stress peaks, reaction forces, etc. In the present study, STEP 7 is specifically used for the validation. Otherwise, for curtain walls under SBI, the herein proposed calibrated parameters can be used for general applications.

Major efforts are in fact spent for an accurate characterization of the impactor and contact parameters for the 2-DOF model (STEPS 5 and 6), and the reliability of this characterization is supported by the extended experimental program inclusive of various impact configurations (Sections 4 and 5). The developed FE models, in particular, are able to capture the features of the impacted system and also the impactor/contact parameters, and this is proved by the close numerical match of the experimental displacement peaks for all the examined configurations. As far as a similar description of facade components and features is taken into account for a given curtain wall module under variable SBI configurations, the proposed method results in a FE model that can be still used for SB setup investigations. In this way, based on the input facade/impact energy assumptions (STEP 0), the correction coefficient R_R in Equation (6) is implicitly accounted in the 2-DOF procedure and the calculated impact force. For both the SB and TT impactors, simplified mechanical laws are thus presented in Section 6, while the final outcomes of such a calibration procedure are presented in Sections 6.1 and 6.2, giving evidence of the accuracy and potential of the overall procedure.

4. Experimental Analysis of Curtain Walls under Soft Body Impact

The numerical approach herein proposed was validated towards experimental measurements collected from SBI tests carried out on identical configurations of curtain wall panels under various impact conditions. The experiments were carried out in Vittorio Veneto (Italy), during February 2020, with an average ambient temperature of 8–10 °C.

4.1. Materials and Methods

The impact experiments were specifically focused on the dynamic analysis of a single curtain wall specimen configuration as in Figure 7a, with 15 test repetitions. The reference panel was characterized by the presence of:

- a sandwich spandrel section, with a total size of 1500×1300 mm^2, that was positioned on the top of the specimen (for insulation purposes only). The cross-section included an aluminum foil (3 mm in thickness) and a steel foil (2 mm the thickness, S355 the resistance class [39]), while a mineral wool panel was used to fill the gap.
- a visual section (Figure 7b), with a total size of 1500×2700 mm^2, composed of an insulated glass unit (16 mm the cavity thickness) with a 10 mm thick, monolithic layer (on the unexposed side) and a laminated safety panel (two 5 mm thick, annealed panels and a bonding 0.76 mm thick PVB film) on the exposed side. All the glass layers consisted of annealed glass [40].

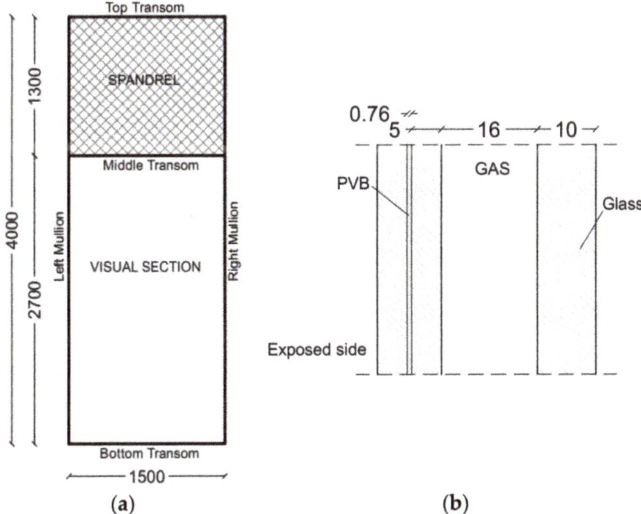

Figure 7. Reference curtain wall specimen: (a) layout and (b) nominal cross-section of the visual section (dimensions in mm).

The specimen included a supporting aluminum frame (alloy type 6063-T6 [41]), with sectional properties reported in Table 1. The mechanical connection between the visual glass panel and the supporting frame was ensured by structural sealant joint (Dowsil^{TM} 993 silicone [42], with a joint thickness of 10 mm). The so-assembled prototype was thus clamped with four steel brackets to the test bench frame (two at the top and two at the bottom corners of the frame), see Figure 8a.

Table 1. Nominal properties of the frame. Key: A= cross-sectional area, I= second moment of area.

		Mullion		Transom		
		Left	**Right**	**Top**	**Middle**	**Bottom**
A	(mm²)	2.12×10^3	2.40×10^3	2.81×10^3	1.77×10^3	1.84×10^3
I	(mm⁴)	1.13×10^7	1.38×10^7	1.61×10^7	0.29×10^7	1.09×10^7

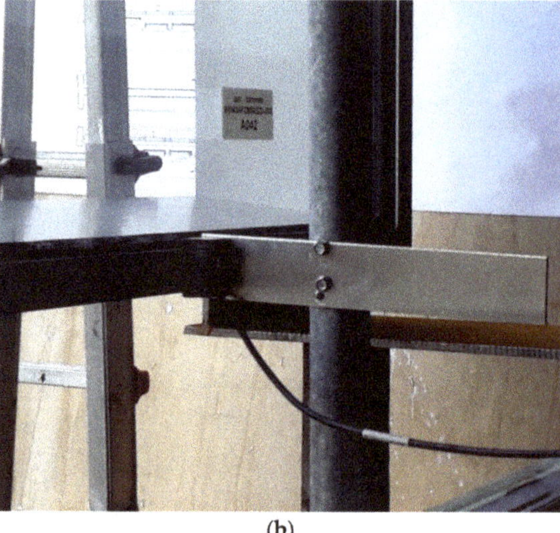

(a)   (b)

Figure 8. Impact experimental test with spheroconical bag: (**a**) tested curtain wall panel and (**b**) detail view of one of the laser sensors in use (bottom transom).

To measure the out-of-plane deformations of the specimen under impact, three laser sensors were used. All the laser sensors were installed on the unexposed side of glass, in some key control points of the curtain wall unit that have been detected as:

- LS1: center of the glass panel;
- LS2: mid-span section of the right-side mullion; and
- LS3: mid-span section of the bottom transom.

To ensure the exhaustiveness and reliability of experimental measurements, the SBI tests were performed on glass specimens under different:

- type of impactor (TT or SB impactor);
- impact point (P#*n*, with a fixed impact energy and impactor); and
- impact energy (E_{imp}, with a fixed impact point and impactor).

To this aim, as also in accordance with Figure 9, five different impact points were selected on the curtain wall specimen, namely represented by:

- P#1: impactor at the glass center, 1 m above the bottom transom;
- P#2: impactor at the glass center;
- P#3: impactor on the middle axis of glass, 0.5 m above the bottom transom;
- P#4: impactor at a distance of 0.35 m from the right mullion, 1 m above the bottom transom;

- P#5: impactor at a distance of 0.35 m from the right mullion, 0.5 m above the bottom transom.

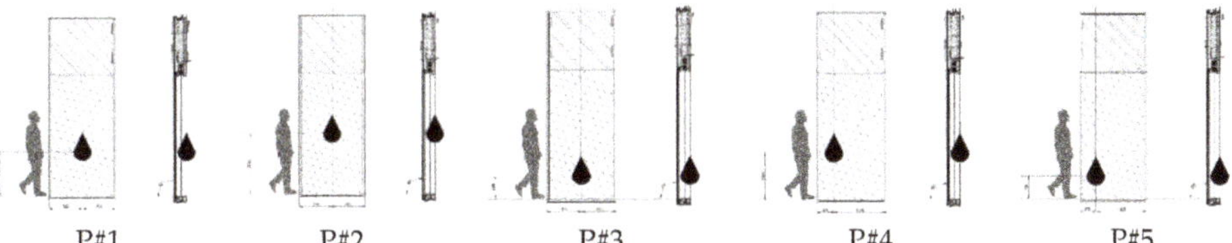

Figure 9. Schematic representation of P#*n* impact points.

The experiments were first carried out for the specimen with spheroconical bag impactor. Successively, the more severe TT impactor was used for additional test repetitions. The final combination of testing configurations is summarized in Table 2.

Table 2. Selected experimental configurations. Key: TT = twin tire impactor; SB = spheroconical bag; x = not available.

	Impact Energy E_{imp} [J]				
Impact Point	100	200	300	450	900
P#1	TT	TT	TT	SB and TT	SB and TT
P#2	x	x	x	SB	SB
P#3	x	x	x	SB	SB
P#4	x	x	x	SB	SB
P#5	x	x	x	SB	SB

4.2. Test Results

The analysis of test results was focused on the measured out-of-plane deformations of selected control points, as well as on the visual detection of possible cracks in glass, so as to evaluate the performance class of the tested specimens, according to the standards requirements. For all the loading configurations in Table 2, it is important to highlight that no glass damage was observed experimentally. In terms of FE numerical investigation herein discussed, accordingly, the lack of failure and crack propagation in the glass panel facilitated a simplified characterization of materials, especially glass, that was assumed to behave linear elastically (Section 5).

5. Finite Element Numerical Analysis

5.1. Solving Approach

The numerical study was carried out with the support of the Code Aster open-source computer software, is integrated into the Salome Meca platform [43]. A series of dynamic non-linear analyses was carried out under various impact configurations. The software choice was intentionally guided, at the time of the research study, by the need of a freely accessible FE software that could be accessible to designers, but at the same time highly refined to support non-linear dynamic simulations.

At the initial stage of the study, to make sure that the reference Code Aster FE model was properly calibrated in its key input parameters, a very high number of benchmarks has been developed under static and dynamic conditions. Based on sensitivity studies, the final FE assembly shown in Figure 4 resulted in brick elements (with 20 nodes and quadratic shape functions) for the glass panel and the sandwich spandrel. In the case of the supporting frame members, mono-dimensional beam elements were used. The mesh size was set in 10 mm. Besides such a series of basic assumptions, the attention was focused on the detailing of several influencing parameters that are summarized in the following sections.

5.2. Reference FE Model for SBI Simulations

The reference FE numerical model was realized to reproduce the actual geometry and dynamic behavior of the tested curtain wall specimens earlier described in Figure 7. Besides the reference mesh pattern that was chosen in Section 5.1, special attention was required for the calibration of some other important parameters.

5.2.1. Geometry

The top spandrel, being composed of a sandwich section with insulation purposes only, was first simplified and numerically represented in the form of an equivalent monolithic section. The material properties were characterized in the form of elasticity and density able to reproduce the actual bending performance of the nominal spandrel section. An equivalent thickness concept was taken into account for the laminated glass section, given that PVB foils under short term and impact events are typically associated to a relatively stiff Young's modulus. Accordingly, a monolithic glass section of 10 mm in thickness was taken into account in place of the 5 + 5/0.76 composite panel.

Regarding the aluminum frame members, the attention was spent for the description of the inertial effects due to the presence of non-metallic thermal barrier components. Compared to inertial values reported in Table 1, the thermal barrier contribution was generally observed to greatly affect the overall period of vibration of the examined curtain wall system. Accordingly, the effective inertial properties for the frame members was herein calculated as prescribed in the Annex C of the EN 14024:2004 document [44]. The final input data are summarized in Table 3.

Table 3. Nominal properties of the frame members. Key: A = cross-sectional area, I = second moment of area; I_t = torsional moment.

		Mullion		Transom		
		Left	Right	Top	Middle	Bottom
A	(mm^2)	2.56×10^3	2.84×10^3	3.08×10^3	1.88×10^3	1.95×10^3
I	(mm^4)	1.40×10^7	1.68×10^7	1.79×10^7	0.33×10^7	1.20×10^7
I_t	(mm^4)	1.00×10^6	1.00×10^6	1.00×10^6	1.00×10^6	1.00×10^6

5.2.2. Gas Cavity

As known, the gas cavity represents a key parameter for insulated glass units, given that it is notoriously responsible of "load sharing" phenomena for the involved glass panels [45–47]. To simulate this interaction for the glass units, various numerical approaches can be found in the literature, and most of them are based on fluid interactions in the cavity, so as to include ideal gas formulations:

$$pV = nRT \tag{11}$$

In the current paper, given that the Code Aster software does not include any fluid interaction for the dynamic non-linear domain, a major simplification was assumed for the gas infill, see Figure 10.

In more detail, an equivalent, linear elastic material was taken into account, with an input modulus of elasticity that was calibrated under the assumption of an isothermal transformation for ideal gases (i.e., a transformation that can often occur under impact test conditions). For a single brick element of the gas cavity, the governing law:

$$pV = cost \tag{12}$$

was thus taken into account.

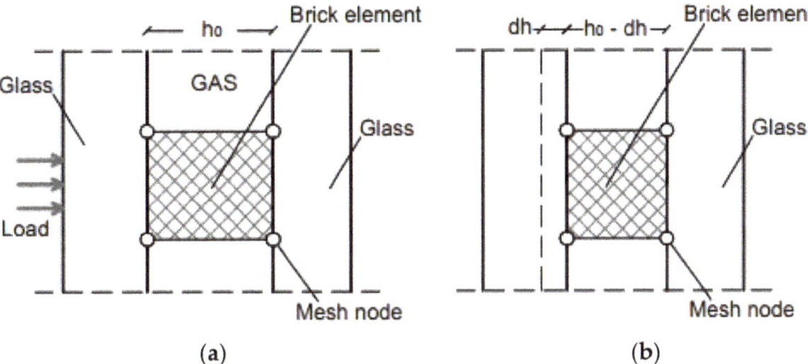

Figure 10. Reference FE numerical model: calculation of the input parameters for the equivalent cavity infill in (**a**) un-deformed and (**b**) deformed configurations.

According to Equation (12), the volume of a single brick element can be expressed as the product of its un-deformed surface area A_{gas} (that was ideally kept constant under deformations) and the un-deformed thickness h_0 (with h_0 = 16 mm the cavity thickness, in this study):

$$p_0 V_{gas} = p_0 A_{gas} h_0 = (p_0 + dp) A_{gas} (h_0 + dh) \tag{13}$$

As far as small thickness variations dh are considered for the cavity subjected to external compressive loads agreeing with Figure 10, the pressure-deformation trend can be derived from measurements of the corresponding brick mesh nodes. Under the simplified hypothesis of a linear elastic material in the small-deformation regime, the so-derived constitutive law allows then to extrapolate an equivalent elastic modulus for the gas infill, that is:

$$E_{eq,gas} = \frac{\Delta p}{dh/h_0} \tag{14}$$

with Δp the pressure variation calculated from Equation (11).

In the present study, the equivalent modulus was estimated in $E_{eq,gas}$ = 0.08 MPa. Besides the intrinsic limits of such an assumption, compared to more accurate calculation approaches, the equivalent modulus $E_{eq,gas}$ can be considered a reliable value, as far as the expected cavity deformations due to impact are small (i.e., in the same order of magnitude of h_0 itself). On the other hand, $E_{eq,gas}$ should be separately estimated under different geometrical configurations (and especially for variations in the un-deformed thickness h_0).

5.2.3. Materials

Given that no visible damage was detected in the reference full-scale experiments (see Section 4.2), linear elastic constitutive laws were used for all the materials in use. The corresponding Young's modulus E, Poisson' ratio ν and density ρ are summarized in Table 4. It is worth noting in Table 4 that no failure stress was explicitly considered for annealed glass (f_t = 45 MPa the characteristic nominal value under quasi-static tensile loading [40]).

Table 4. Input mechanical properties for the materials in use.

	Material			
	Glass	**Aluminum**	**Spandrel ***	**Gas ****
E [MPa]	70,000	70,000	70,000	0.08
ν	0.23	0.34	0.22	0.22
ρ [kg/m^3]	2500	2700	6225	1.225

* = equivalent properties based on the monolithic thickness assumption; ** = equivalent properties based on the linear elastic behavior assumption.

The post-processing analysis of FE data was thus focused on the evolution of maximum deflections for the tested specimens, but also on a necessary analysis of stress peaks in the module components. Major attention was spent for glass, so as to ensure the reliability of a linear elastic material characterization for FE purposes. Among other constructional materials and literature studies, the discussion in [2] and Figure 11 about annealed glass confirms the well-known sensitivity of its compressive and tensile strength to strain rate. This effect typically results in the so called Dynamic Increase Factor (DIF) that can be quantified—for impact and impulsive configurations—in up to 3–4 times the nominal strength value f_t under quasi-static loads. Simple stress estimates can be thus carried out towards the DIF value of strength [3,48,49].

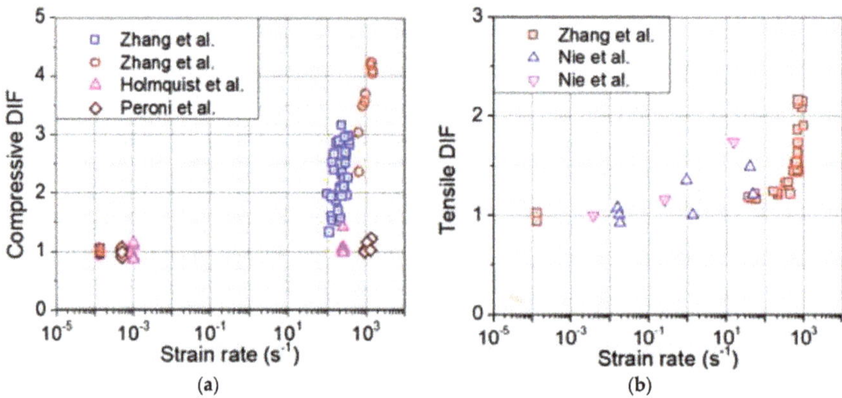

Figure 11. Reference Dynamic Increase Factor (DIF) for glass strength in (**a**) compression or (**b**) tension, as a function of the imposed strain rate. Figure reproduced from [2] with permission from Elsevier®, copyright license agreement n. 5041301127990 (April 2021).

5.2.4. Boundaries and Connections

A special attention was dedicated to the mechanical characterization of the actual supports for the examined curtain wall specimens. At the early design stage, or when no experimental feedback is available, the curtain wall module in Figure 7 should be reasonably described in the form of four point restraints at the corners of the panel. Such a choice would result in reliable calculations for the procedural steps in Figure 6.

A double check and calibration was also carried out for the examined facade modules, based also on the post-process of the available experimental records. On this regard, four axial "discrete" springs were introduced in the Code Aster model, in the region of top and bottom corners, so as to include a certain support flexibility for the restraints of the specimen. The axial stiffness of these springs was calculated with the support of the experimental displacements, given that they proved a certain out-of-plane deformation in the bottom transom. Accordingly, mostly rigid restraints were considered for the top corners ($K_{top} = 10^7$ kN/m), while for the bottom corners the input stiffness was set in $K_{bot} = 1700$ kN/m. Finally, the connection of the glass panel with the supporting frame members was numerically reproduced in the form of a series of equivalent springs with stiffness given by:

$$K_{sil} = \frac{E_{sil} A_{sil}}{t_{sil}} = 1000 \text{ kN/m} \tag{15}$$

where $t_{sil} = 10$ mm represents the nominal thickness of the silicone joint in use, E_{sil} the modulus [42], while A_{sil} is inclusive of the mesh size/spring influence area.

5.2.5. Damping

In the reference FE model, damping was taken into account with the Rayleigh method, as a function of the mass of the curtain wall module. At first, as also in line with Equation (10), the input value was set in $\zeta = 8\%$. The so-calibrated model was thus preliminary taken into account for the 2-DOF characterization in the procedural steps of Figure 6.

Later on, the experimental records were also considered for a post-processing validation of the above assumption (i.e., STEP 7c in Figure 6). Based on the logarithmic decrease of the experimentally measured displacements $u(t)$ for the glass panel:

$$\zeta_{test} = \frac{1}{\sqrt{1 + \left(\frac{2\pi}{\delta}\right)^2}} \tag{16}$$

and

$$\delta = \frac{1}{n} ln \frac{u(t)}{u(t + nT_1)} \tag{17}$$

and being T_1 the fundamental period of vibration of the curtain wall, the damping was experimentally estimated in a mean value of $\zeta_{test} = 12\%$ that is in line with the approximate prediction of Equation (10), but also includes local effects that could derive from the impactor itself or possible local damage (i.e., frame, joints). In this context, it is worth mentioning that the effects of the so-derived damping term ζ have minimum effects on the overall dynamic estimates for the curtain wall analysis. On the side of the 2-DOF impact force, in particular, the curtain wall damping proved to have negligible effects on the corresponding time history of the force. On the side of the curtain wall, similarly, the maximum peak of impact effects (i.e., displacements, stresses, etc.) is notoriously not affected by any damping contribution, given that it can be perceived only in the vibrations that follow the instant of the impact. Accordingly, the overall damping estimation from Equation (10) was still found to be reliable.

5.2.6. Impact Load and Impact Area

For each experimental configuration in Table 2, the impact load was properly considered in the FE model, based on a number of time histories that were derived from the 2-DOF model in Figures 5 and 6. Given that the input time history modifies with the impact features (impactor, impact point, etc.), the corresponding trends were separately collected for the corresponding FE analyses. Figure 12 shows a typical example of variation of the required input time history, as far as the examined curtain wall is subjected to different impact conditions.

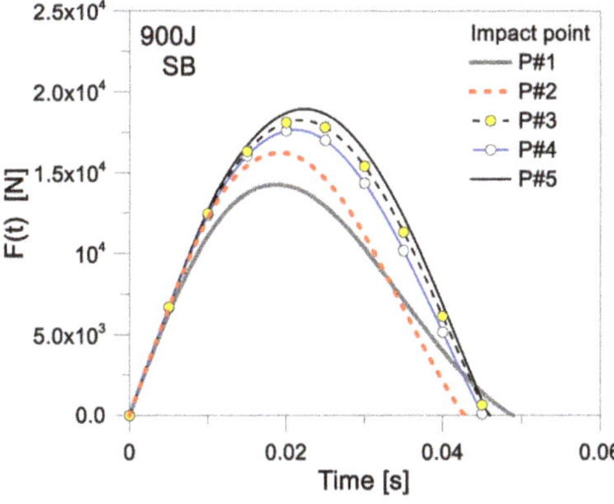

Figure 12. Example of input time histories for the curtain wall under SB, at different impact points (E_{imp} = 900 J).

In order to facilitate the calibration of the key parameters for the 2-DOF and the corresponding FE models, the reference impact area was kept constant for all the Code Aster simulations, with a square shape that was set in $A = 200 \times 200$ mm^2. The latter, see Figure 13a, was experimentally measured in the case of the TT impactor. Similar experimental acquisitions, see Figure 13b, proved that the TT surface A is mostly half the spheroconical bag. For FE purposes, however, the initial A value was kept fix and the input force was scaled accordingly.

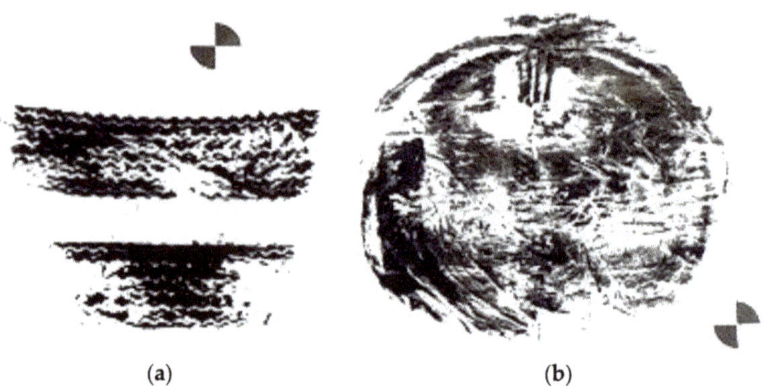

(a)   (b)

Figure 13. Experimental measurement of the reference impact area A corresponding to (a) TT or (b) SB impactors (acquisition for $E_{imp} = 900$ J).

6. Experimental Validation of FE Results

6.1. Impact Response

For the validation of the FE modelling assumptions and the coupled design approach herein proposed (i.e., impactor and contact parameters of the 2-DOF model), the out-of-plane displacements for the three LSn control points were assessed for all the 15 test repetitions. Given that the coupled design approach schematized in Figures 5 and 6 is based on a first vibration mode of the curtain wall object of study, the period was first numerically calculated in $T_1 = 0.1$ s, see Figure 14a.

For the curtain wall under various impact conditions, the typical deformed shape was still found to mostly agree with the global vibration mode of the system, see Figure 14b. The corresponding distribution of stress peaks in glass was calculated as in Figure 14c (exposed panel). For the same modular unit under different impact points, additional local effects were observed in major variations about the localization of stress peaks, and some modifications in the global deformation of the visual section (Figure 14d,e).

In Figures 15 and 16, selected data are shown for two testing configurations only (SB or TT, with P#1 and $E_{imp} = 900$ J). It is of interest that in both the cases, there is a rather good correlation for all the comparative plots, thanks to the reliable input parameters of the coupled approach. For the setup in Figure 15, for example, the maximum response peaks for the visual section under SB impact were quantified in an out-of-plane acceleration of around ≈ 155 m/s^2, with a velocity peak of ≈ 1.7 m/s, after 0.03–0.05 s of analysis. The corresponding strain peak was measured in ≈ 0.0008.

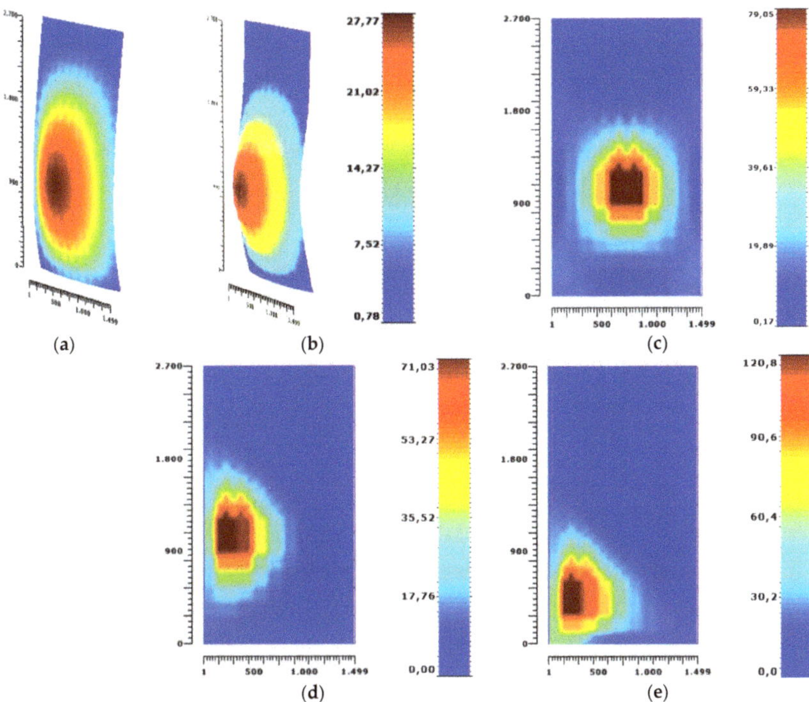

Figure 14. Typical deformed shape of the examined curtain wall (Code Aster): (**a**) fundamental modal shape (frequency analysis) and (**b–e**) example of impact responses (legend values in mm and MPa). (**a**) $T_1 = 0.1$ s. (**b**) Displacement (SB, P#1, $E_{imp} = 900$ J). (**c**) Stress (SB, P#1, $E_{imp} = 900$ J). (**d**) Stress (SB, P#4, $E_{imp} = 450$ J). (**e**) Stress (SB, P#5, $E_{imp} = 900$ J).

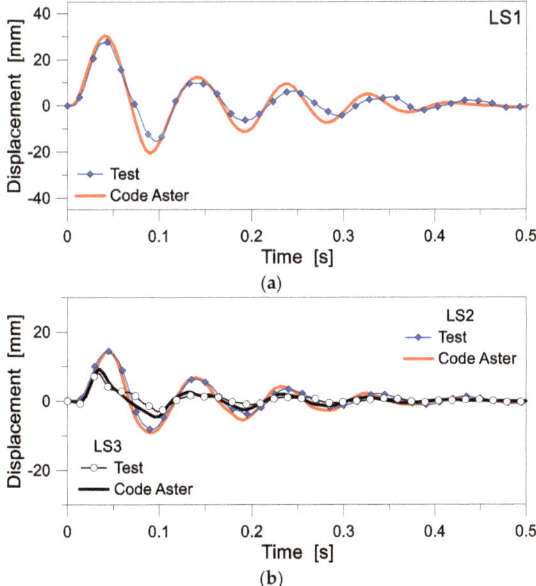

Figure 15. Numerical and experimental out-of-plane displacements for the curtain wall under SB (P#1, $E_{imp} = 900$ J). In evidence, the control points (**a**) LS1 and (**b**) LS2, LS3.

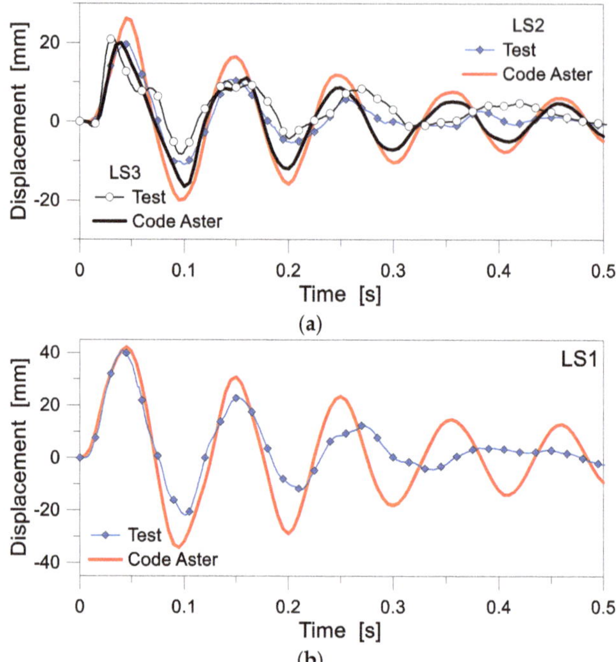

Figure 16. Numerical and experimental out-of-plane displacements for the curtain wall under TT (E_{imp} = 900 J, impact point #1). In evidence, the control points (**a**) LS1 and (**b**) LS2, LS3.

It is worth noting that a general good accuracy was found also for the other testing conditions summarized in Figure 17a, where the numerical peak of out-of-plane displacements at the center of glass (LS1) is proposed under a multitude impact configurations, and compared to the experimental measurements. Given that the drop height, impact point and impactor changes for all the setup configurations, Figure 17a is a further confirmation of the proposal accuracy.

In this regard, it is important to remind that the P#*n* control points in Figure 17a are implicitly associated to different stiffness parameters and dissipation capacities for the impacted modular unit and for the SB or TT impactor in use.

For few impact configurations only, the FE predictions in Figure 17a tend to overestimate the corresponding experimental measurement. This limitation was justified by the sequence of test repetitions that were carried out on a single modular unit. Especially for few TT scenarios, the laser measurement control point gave evidence of minor damage.

Of interest is also the trend of numerically calculated stress peaks in glass as a function of the deflection amplitude, as schematized in Figure 17b for the panel exposed to impact. For the TT configurations (with P#1 impact point but different impact energy values), as expected, the stress peaks in glass reported in Figure 17b linearly increase with E_{imp} and the measured deflection peak. When the SB impactor is used, with different E_{imp} values, a clear correlation still exists between the calculated deflection-stress peaks data, as far as the impact point is kept fix. Moreover, as far as the impact point moves from P#1 towards the edges and the frame of the modular unit, the local stiffness variation results in even higher stress peaks in glass. Regarding the inner glass panel of the IGU visual component, the stress peak for the examined scenarios was numerically estimated in the range from ≈0.3 to ≈0.6 times the values in Figure 17b, depending on the impact configuration.

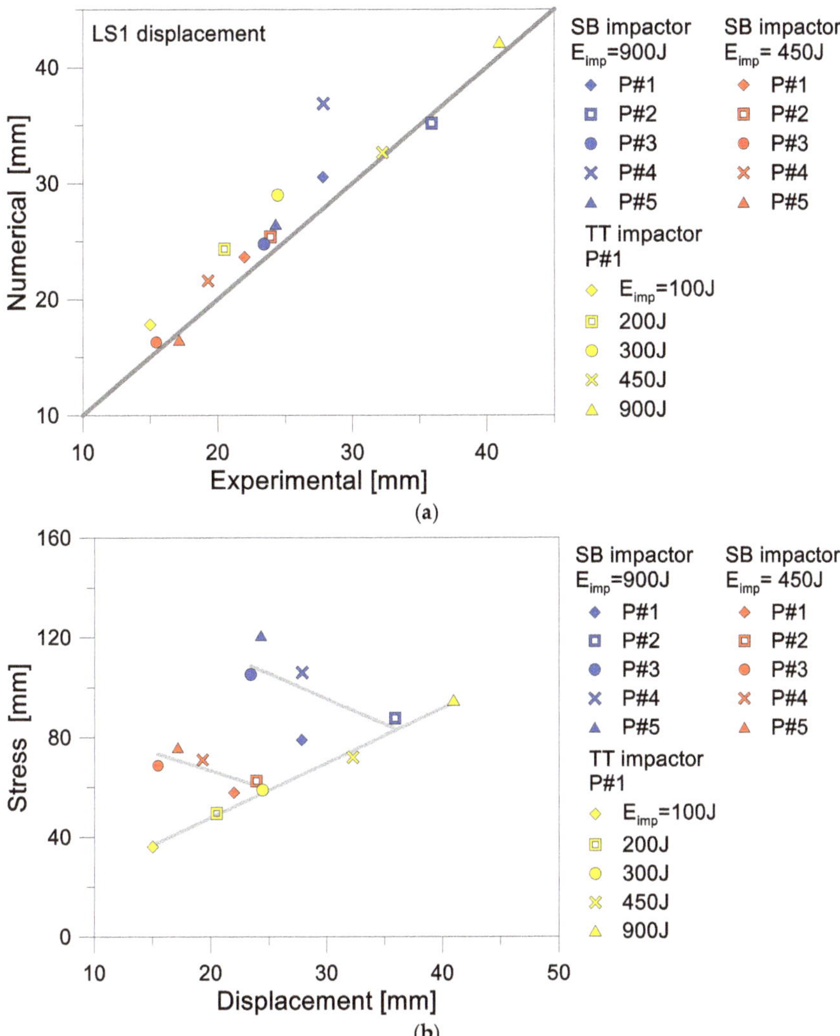

Figure 17. Numerical response analysis (Code Aster): (**a**) displacement peaks (LS1) for the curtain wall under various impact conditions, as a function of the experimental measurements, and (**b**) stress-displacement peaks (glass panel exposed to impact).

Finally, Figure 18 further confirms the general close correlation of FE estimates with the available experimental data, but also Newman's assertion in [34] about the higher severity of the TT impactor, for a given impacted body.

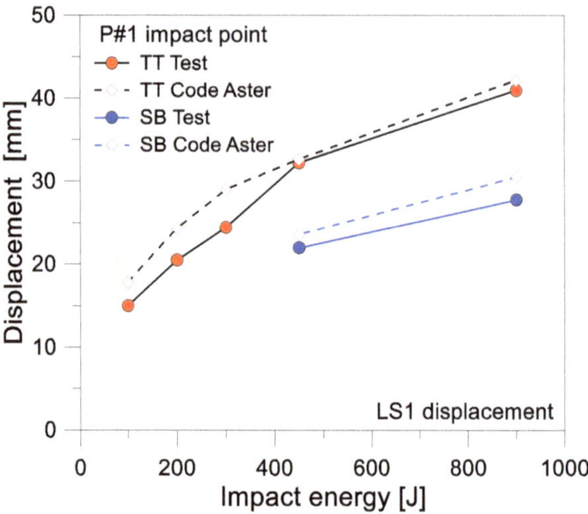

Figure 18. Comparison of numerical and experimental peaks of displacement (LS1) for the curtain wall under various impact conditions, as a function of the impact energy.

6.2. Input Parameters for the Impactor

The key aspect of all the comparisons in Section 6.1 is the use, for the Code Aster numerical model, of input forces that are derived from the 2-DOF model of the system, thus implicitly accounting for the R_R coefficient in Equation (6).

In the current study, the parameters for contact (and thus for the impactor features) where calibrated with the support of experimental measurements, as well as the Code Aster estimates. The detailed procedure is shown in Figure 19, as a sub-section of the general approach in Figure 6. The final goal was obtained by matching the deflection peak of each specimen, for all the 15 test repetitions (STEP 7a in Figure 19), and thus minimizing the % scatter Δ_u for them (STEP 8a), so as to capture a schematic representation of the mechanical behavior for both the TT and the SB impactors (STEP 8b). For the first calibration of impact parameters, major advantage was represented by the available experimental-numerical feedback.

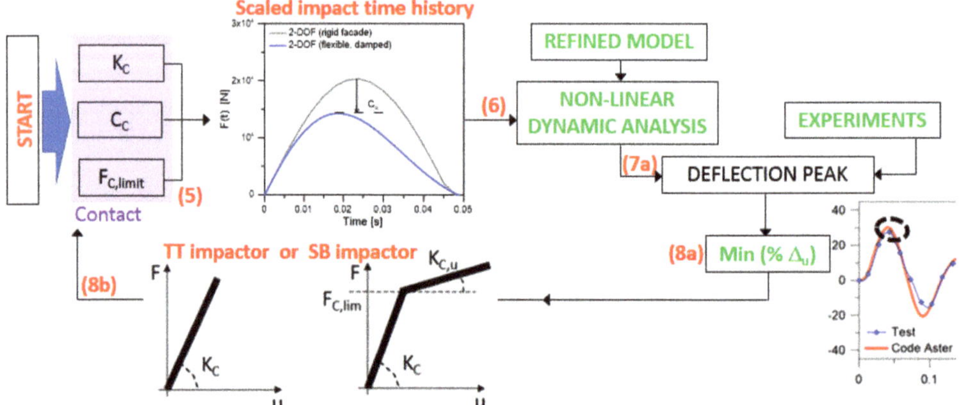

Figure 19. Developed calibration procedure for the 2-DOF input parameters of the impact contact, depending on the impactor in use (linear elastic for TT, elasto-plastic with hardening for SB).

Based on the selected impactors, more in detail, a linear elastic constitutive behavior was taken into account for the tire impactor (thus described in the form of an elastic stiffness K_C). No damping was accounted for the TT ($C_C = 0$), as also in line with past literature documents [9,12]. On the other side, to account for the different behavior of the bag, the latter was schematized in the form of an elasto-plastic law with hardening, with input parameters represented by the initial stiffness K_C, a limit force $F_{C,limit}$ and a residual stiffness $K_{C,u}$ (with $K_{C,u} = 0.9\ K_C$). With the support of benchmark parametric calculations, the limit force $F_{C,limit}$ was introduced in the contact description for the 2-DOF, so as to schematically reproduce the physical impact of the SB pendulum with the curtain wall, and the following separation of the two bodies, due to rebound. Such a plastic force $F_{C,limit}$ was in fact intended to act as a cut-off for the input force that the SB impactor could transfer to the adjacent curtain wall, given that the bag is filled with movable glass spheres.

However, the most influencing parameter for the SB description was represented by the elastic stiffness K_C. Accordingly, the iterative calculations were carried out by setting, in the order, K_C (with $C_C = 0$), thus $F_{C,limit}$ (with $C_C = 0$, and K_C fix) and then C_C (with fix K_C and $F_{C,limit}$ values). The latter, being susceptible to various parameters on the side of the impacting bodies but also on the impact energy, was calculated in a variable magnitude, as a function of E_{imp} and K_F. In any case, its final value for the SB impactor was usually optimized in the range of ≈ 0.1–1 Ns/mm, with limited variations for the overall experimental program.

The overall calibration process was based on the minimization of the percentage scatter of the deflection peaks from the FE model and the corresponding experimental records (LS1), under the 15 available testing configurations. The same iterative approach was thus separately repeated for the specimens with TT or SB impactor, so as to obtain a univocal simplified characterization of the two types of impactors. In Table 5, the so-calibrated values are proposed.

Table 5. 2-DOF input parameters for the curtain wall under SBI (with $M_P = 50$ kg and $\zeta_{test} = 12\%$ *).

		Spheroconical Bag	Twin-Tire
$F_{c,lim}$	[N]	14,000	-
K_c	[N/mm]	230	250
C_c	[Ns/mm]	Variable with E_{imp} and K_F (0.1–1) See Figure 18f	0

* = average logarithmic decrement of experimental displacements (Equation (16)).

In Figure 20, comparative examples are proposed to show the effects of iterative calibrations on the obtained 2-DOF scaled time histories, to further quantify the effects of the various input properties for the SB impactor. As in Figure 20a, the stiffness K_C was found responsible of major variations for the corresponding impact forces, while the cut-off force $F_{C,limit}$ generally resulted in less pronounced variations, compared to a perfectly elastic contact (Figure 20b). In terms of damping, the effects of C_C are proposed in Figure 20c, while Figure 20d confirms the negligible effect of the curtain wall damping C_F. The sensitivity of C_C to the impactor velocity or the facade stiffness is finally shown in Figure 20e, where a regular trend can be noticed with v_P. On the other side, no clear correlation was found for the stiffness K_F (Figure 20f), thus suggesting further explorations.

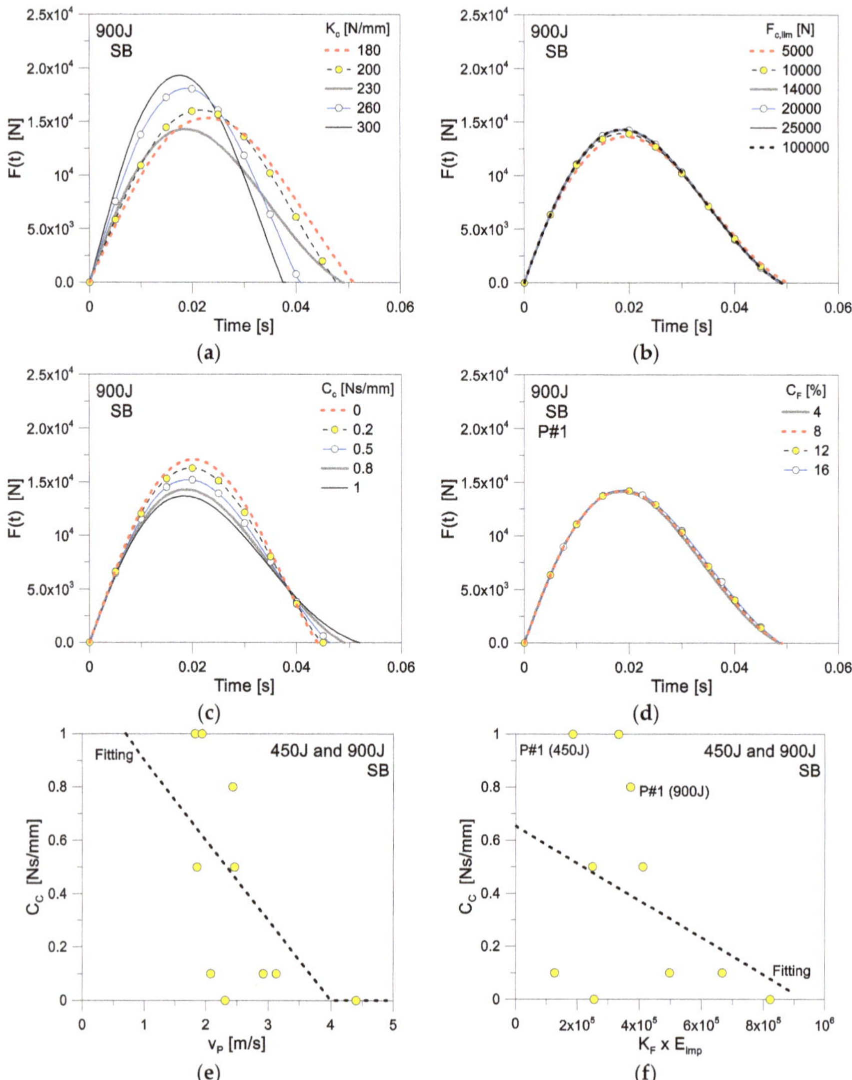

Figure 20. Sensitivity analysis of the 2-DOF impact time histories to selected input parameters: (**a**) contact stiffness, (**b**) contact limit force, (**c**) contact damping and (**d**) facade damping, with (**e,f**) variation of contact damping with impactor velocity, facade stiffness and impact energy.

7. Design Outcomes

7.1. Reduction Coefficient R_R

The reduction coefficient in Equations (6) and (7), as earlier discussed, is implicitly accounted for in the scaled time history of impact forces, but depends on various parameters (impactor and impacted body, thus impact rigidity, impact energy, etc.) and it is a key information for design. Given that a combination of the above parameters can separately affect the corresponding reduction coefficient, some parametric calculations are discussed in this paper, in order to emphasize its sensitivity. In Figure 21, the calculated reduction coefficients are compared for the examined curtain wall system under TT (MEPLA and proposed 2-DOF+Code Aster method) or SB (proposed method). As a reference, the impact conditions of E_{imp} = 100 J, 200 J, 300 J, 450 J or 900 J (P#1) were taken into account. In the case of the SB, values for E_{imp} = 450 J and 900 J only are reported in Figure 20.

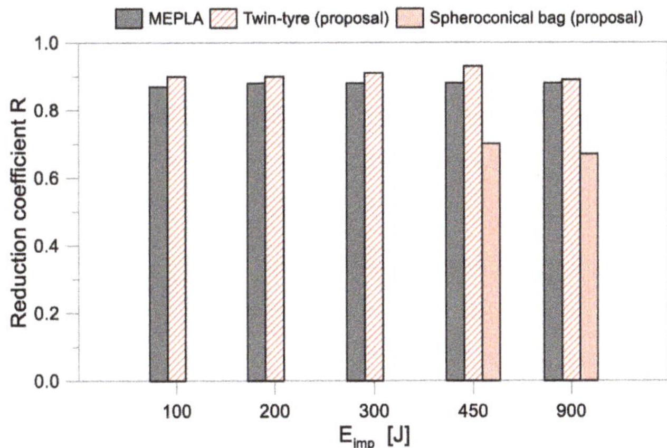

Figure 21. Calculated reduction coefficient R, as a function of impactor type and impact energy E_{imp} (P#1).

As shown, a first important outcome is that the MEPLA implicit coefficient R_{Mepla} is mostly constant under a variable imposed impact energy, and it disregards the impactor features (TT option).

The calculated reduction coefficients R_R from the proposed approach, otherwise, clearly modify with the assigned impact conditions, and also diverge from MEPLA estimates. From the 2-DOF+Code Aster calculations, in particular, it is possible to notice that R_R tends to decrease as far as E_{imp} increases for both the TT and SB impactors. This behavior is in line with expectations, given that dissipative phenomena in the impacted curtain wall qualitatively increase under higher input energies. Moreover, the current 2-DOF+Code Aster data in Figure 21 for the TT are close to the unit, compared to the SB. This means that the TT approaches the "ideal" rigid elastic condition of impact. Moreover, this outcome again confirms Newman's hypothesis about the greater severity of the impact tests carried out with the TT setup (Section 2 and [34]).

As far as the impact rigidity is also introduced in the parametric comparisons, results agreeing with Figure 22 are obtained. The impact energy is set in E_{imp} = 450 J or 900 J respectively, while various impact rigidities are taken into account (P#5). Both the MEPLA and 2-DOF+Code Aster trends show a mostly linear variation of the reduction coefficient with the impact rigidity. Besides, the MEPLA twin-tire values overestimate (up to +24%) the corresponding SB coefficients. Again, it is of interest that the MEPLA coefficients in Figure 22 do not change with the imposed E_{imp}, thus the corresponding calculations can be severely more over-conservative than the reality. Finally, it is possible to notice that the calculated reduction coefficients tend to the unit, as far as the impact stiffness increases. This is because (for relatively stiff impacted bodies) the impact condition is close to an ideal transferred impulse (i.e., on a perfectly rigid surface).

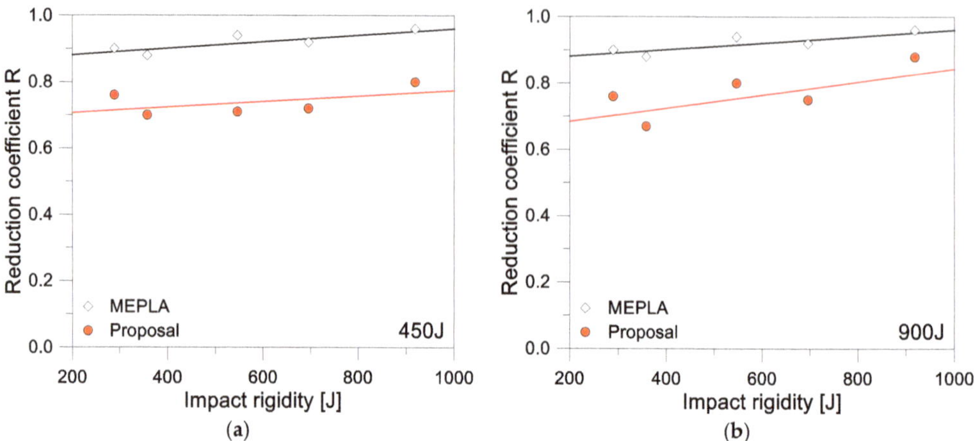

Figure 22. Calculated reduction coefficients R as a function of the impact rigidity, with E_{imp} equal to (**a**) 450 J or (**b**) 900 J.

7.2. Iso-Damage Curves

As is known, iso-damage curves are generally very useful for design, given that they provide a fast and reliable feedback for the structural assessment of a given structural system. This is also the case of glass curtain walls under SBI, as far as the iso-damage curves are properly fitted so as to account for various impact parameters.

In this research study, iso-damage curves were developed on the basis of FE simulations carried out with 25 different impact time histories. Major FE results have been extracted and linearly interpolated, for a more accurate description of maximum displacement, maximum tensile principal stress and probability of breakage parameters. A typical example is proposed in Figure 23, where the "proposal" data denote the result of the coupled calculation approach discussed in this paper. In general, the collected results further confirmed that the expected response for a curtain wall panel is not proportionally dependent on the impulse. In addition, as is also in line with the earlier comparisons, it was noticed further the effect of different impactors (TT or SB) on the examined facade module.

For the facade under SB impactor, it can be noted that the proposed method (already validated by the experimental data) is mostly different from "ideal" and MEPLA assumptions. Given that it proves to properly capture the complex dynamic response of the curtain wall under SBI, it can be used for generalized design problems.

In the case of the curtain wall under TT, the collected comparisons were found to be generally in line with the previous observations. Due to the higher rigidity of the impactor/contact, in the latter case, a close agreement was found also with MEPLA predictions (i.e., due to reduction coefficients tending to the unit, as shown in Figure 19). Besides, the scatter of the so-calculated values from the "ideal" impact conditions were still clearly perceived, thus enforcing the potential of the research approach herein presented, as well as its possible extension to various other configurations of technical interest.

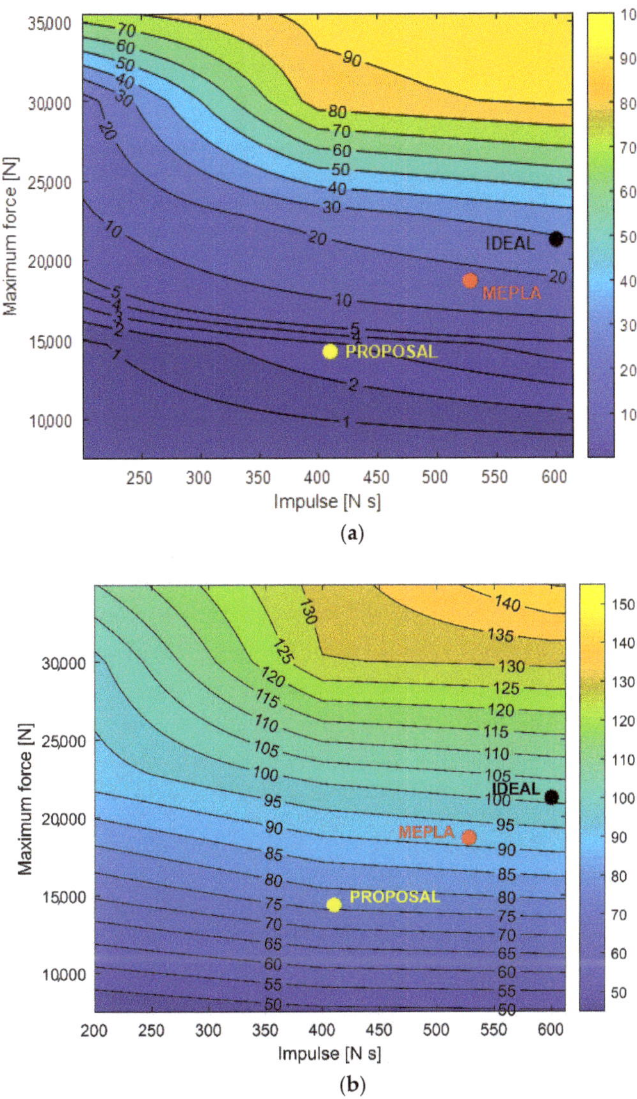

Figure 23. Iso-damage curves based on (**a**) maximum tensile stress or (**b**) probability of failure, based on different calculation methods. In evidence, the proposed approach (SB, $E_{imp} = 900$ J, P#1).

8. Conclusions

The structural design of glass curtain walls is a challenging issue; given that it must satisfy specific performance requirements. Among others, a special care must be paid to the soft body impact (SBI) assessment, where standardized procedures are still required to manufacturers. On one side, the European and German norms introduced the twin-tire (TT) setup and the possibility of numerical simulations in place of experiments. However, several National regulations based on the spheroconical bag (SB) setup are still imposed to producers, and Finite Element (FE) numerical approaches for SB still lack in support of designers. At the same time, the impact response is strongly affected by a multitude of aspects that include the features of the facade (impacted body), the impactor properties and the impact configurations (energy, impact point).

In this paper, the dynamic response of glass curtain walls SBI has been extensively assessed, by taking into account various impact conditions. A special attention has been paid to the SB impactor, towards the calibration of input parameters for the analysis of this demanding setup, to the development of a novel coupled design procedure, by taking advantage of simple full-scale experiments, two Degree of Freedom (2-DOF) calculations and more accurate Finite Element (FE) numerical models for the overall validation of the methodology.

First, the input parameters for the definition of the impact force time history have been proposed. Based on FE numerical model carried out in Code Aster and validated to original experimental results, a refined reduction coefficient R_R has been calculated for curtain walls under SB impact. This coefficient, compared to "ideal" impact conditions (rigid facade) further emphasizes the effects of different impact conditions, and thus the possible consequences on the design detailing of curtain walls. In the case of the SB impactor, in particular, it proved to be less severe than the twin-tire, thus enforcing the need of specific studies in support of reliable design procedures. In conclusion, iso-damage curves have been proposed to represent a robust practical tool for design.

Author Contributions: Conceptualization, G.L. and G.M.; software, A.B. and G.L.; validation, A.B., G.L. and C.B.; formal analysis, A.B. and G.L.; investigation, A.B., G.L. and C.B.; writing—original draft preparation, A.B., C.B. and G.M.; writing—review and editing, A.B., G.L., C.B. and G.M.; supervision, G.M. and C.A. All authors have read and agreed to the published version of the manuscript.

Funding: This research received no external funding.

Institutional Review Board Statement: Not applicable.

Informed Consent Statement: Not applicable.

Data Availability Statement: Additional data will be available upon request.

Acknowledgments: The APCs for the publication of this paper are covered by the "Buildings 2020 Young Investigator Award" granted in January 2021 to the second author. MDPI is gratefully acknowledged. Emanuele Rizzi (University of Trieste) is also gratefully acknowledged for Figure 2b.

Conflicts of Interest: The authors declare no conflict of interest.

References

1. Sivanerupan, S.; Wilson, J.; Gad, E.; Lam, N. Drift Performance of Point Fixed Glass Façade Systems. *Adv. Struct. Eng.* **2014**, *17*, 1481–1495. [CrossRef]
2. Bedon, C.; Zhang, X.; Santos, F.; Honfi, D.; Kozłowski, M.; Arrigoni, M.; Figuli, L.; Lange, D. Performance of structural glass facades under extreme loads—Design methods, existing research, current issues and trends. *Constr. Build. Mater.* **2018**, *163*, 921–937. [CrossRef]
3. Larcher, M.; Arrigoni, M.; Bedon, C.; van Doormaal, J.C.A.M.; Haberacker, C.; Hüsken, G.; Millon, O.; Saarenheimo, A.; Solomos, G.; Thamie, L.; et al. Design of Blast-Loaded Glazing Windows and Facades: A Review of Essential Requirements towards Standardization. *Adv. Civ. Eng.* **2016**, *2016*, 1–14. [CrossRef]
4. Lori, G.; Morison, C.; Larcher, M.; Belis, J. Sustainable façade design for glazed buildings in a blast resilient urban envi-ronment. *Glass Struct. Eng.* **2019**, *4*, 145–173. [CrossRef]
5. Lago, A.; Sullivan, T.J. *A Review of Glass Façade Systems and Research into the Seismic Design of Frameless Glass Facades*; ROSE Research Report 2011/01; Fondazione Eucentre: Pavia, Italy, 2011; ISBN 978-88-6198-059-4.
6. Feldmann, M.; Kasper, R.; Abeln, B.; Cruz, P.; Belis, J.; Beyer, J.; Colvin, J.; Ensslen, F.; Grenier, C.; Zarnic, R.; et al. *Guidance for European Structural Design of Glass Components—Support to the Imple-Mentation, Harmonization and Further Development of the Eurocodes*; Report EUR 26439-Joint Research Centre-Institute for the Protection and Security of the Citizen; Publications Office of the European Union: Copenhagen, Denmark, 2014. [CrossRef]
7. Ramos, A.; Pelayo, F.; Lamela, M.J.; Canteli, A.F.; Aenlle, M.L.; Persson, K. Analysis of structural glass panels under impact loading using operational modal analysis. In Proceedings of the IOMAC'15—6th International Operational Modal Analysis Conference, Gijon, Spain, 12–14 May 2015.
8. Biolzi, L.; Bonati, A.; Cattaneo, S. Laminated Glass Cantilevered Plates under Static and Impact Loading. *Adv. Civ. Eng.* **2018**, *2018*, 7874618. [CrossRef]

9. Kozłowski, M. Experimental and numerical assessment of structural behaviour of glass balustrade subjected to soft body impact. *Compos. Struct.* **2019**, *229*, 111380. [CrossRef]

10. Bedon, C.; Kalamar, R.; Eliášová, M. Low velocity impact performance investigation on square hollow glass columns via full-scale experiments and Finite Element analyses. *Compos. Struct.* **2017**, *182*, 311–325. [CrossRef]

11. Figuli, L.; Papan, D.; Papanova, Z.; Bedon, C. Experimental mechanical analysis of traditional in-service glass windows subjected to dynamic tests and hard body impact. *Smart Struct. Syst.* **2021**, *27*, 365.

12. Mohagheghian, I.; Wang, Y.; Zhou, J.; Yu, L.; Guo, X.; Yan, Y.; Charalambides, M.; Dear, J. Deformation and damage mechanisms of laminated glass windows subjected to high velocity soft impact. *Int. J. Solids Struct.* **2017**, *109*, 46–62. [CrossRef]

13. Pelferne, J.; van Dam, S.; Kuntsche, J.; van Paepegem, W. Numerical simulation of the EN 12600 Pendulum Test for Structural Glass. In Proceedings of the Challenging Glass Conference Proceedings, Ghent, Belgium, 16 June 2016; Volume 5, pp. 429–438.

14. Schneider, J.; Schula, S. Zwei Verfahren zum rechnerischen Nachweis der dynamischen Beanspruchung von Ver-glasungen durch weichen Stoß—Teil 1: Numerische, transiente Simulationsberechnung und vereinfachtes Verfahren mit statischen Ersatzlasten. [Two mechanical design concepts for simulating the soft body impact at glazings—Part 1: Numerical, transient Finite Element simulation and simplified concept with equivalent static loads]. In *Stahlbau Spezial 2011—Glas-bau/Glass in Building*; Ernst & Sohn Verlag: Berlin, Germany, 2011; pp. 81–87.

15. Weller, B.; Reich, S.; Krampe, P. Zwei Verfahren zum rechnerischen Nachweis der dynamischen Beanspruchung von Verglasungen durch weichen Stoß—Teil 2: Numerische Vergleichsberechnungen und experimentelle Verifikation [Two me-chanical design concepts for simulating the soft body impact at glazings—Part 2: Numerical comparison and experimental verification]. In *Stahlbau Spezial 2011—Glasbau/Glass in Building*; Ernst & Sohn Verlag: Berlin, Germany, 2011; pp. 88–92.

16. Schneider, J.; Schula, S. Simulating soft body impact on glass structures. *Proc. Inst. Civ. Eng. Struct. Build.* **2016**, *169*, 416–431. [CrossRef]

17. Chen, S.; Zang, M.; Wang, D.; Yoshimura, S.; Yamada, T. Numerical analysis of impact failure of automotive laminated glass: A review. *Compos. Part B Eng.* **2017**, *122*, 47–60. [CrossRef]

18. Suh, C.M.; Kim, S.H.; Hwang, B.W. Finite Element Analysis of Brain Damage due to Impact Force with a Three Dimensional Head Model. *Key Eng. Mater.* **2005**, *297-300*, 1333–1338. [CrossRef]

19. Sances, J.A.; Carlin, F.H.; Kumaresan, S.; Enz, B. Biomedical Engineering Analysis of Glass Impact Injuries. *Crit. Rev. Biomed. Eng.* **2002**, *30*, 345–378. [CrossRef]

20. Code Aster Software, Électricité de France EDF. Available online: www.code-aster.org/ (accessed on 6 April 2021).

21. *SJ MEPLA Software and User's Manual*; SJ Software GmbH: Aachen, Germany; Available online: https://www.mepla.net/ (accessed on 6 April 2021).

22. Vredstein V47 3.5-8—Technical Data Sheet. Available online: https://www.vredestein.it/ (accessed on 6 April 2021).

23. EN 12600:2002. *Glass in Building—Pendulum Test—Impact Test Method and Classification for Flat Glass*; CEN: Brussels, Belgium, 2002.

24. DIN 18008-4: 2013. *Glas im Bauwesen—Bemessungs- und Konstruktionsregeln—Teil 4: Zusatzanforderungen an ab-sturzsichernde Verglasungen*; Beuth: Berlin, Germany, 2013.

25. de Vries, C.M. Numerical simulation of façade/window glazing fracture under impact loading. *Proc. Challenging Glass 3* **2012**, *3*, 489–500. [CrossRef]

26. Schneider, J.; Bohmann, D. *Glasscheiben unter Stoßbelastung, Experimentelle und Theoretische Untersuchungen für ab-sturzsichernde Verglasungen Bei Weichem Stoß, Bauingenieur*; Springer: Berlin/Heidelberg, Germany, 2002; Volume 77.

27. Breckner, W. *Tragverhalten von Verbundsicherheitsglas unter Stoßbelastung nach DIN EN 12600, Diplomarbeit, Lehrstuhl für Stahlbau*; RWTH: Aachen, Germany, 2001.

28. Brendler, S.; Haufe, A.; Ummenhofer, T. *A Detailed Numerical Investigation of Insulated Glass Subjected to the Standard Pendulum Test*; LS-DYNA Anwenderforum: Bamberg, Germany, 2004.

29. Cahier CSTB 3228. *Resistance Impact on Glass and Glass Roof Infill*; Centre Scientifique et Technique du Bâtiment (CSTB): Marne-la-Vallée, France, 2014.

30. NF P 8-301. *Determination of the Resistance to Soft and Heavy Body Impact*; Association Francaise de Normalisation (AFNOR): Paris, France, 2003.

31. CWCT TN 76. *Technical Note 76: Impact Performance of Building Envelopes: Method for Impact Testing Cladding Panels*; Centre for Window & Cladding Technology (CWCT): Bath, UK, 2012.

32. ACR[M]001:2014. *Test for Non-Fragility of Profiled Sheeted Roof Assemblies*; Advisory Committee for Roofwork (ACR): London, UK, 2014.

33. ANSI Z97.1. *American National Standard for Safety Glazing Materials Used in Buildings—Safety Performance Specifications and Methods of Test*; Accredited Standards Committee (ASC): McLean, VA, USA, 2015.

34. Newman, C.J. Evaluation of an impact standard for curtain walling. *Proc. Inst. Civ. Eng. Struct. Build.* **2004**, *157*, 333–341. [CrossRef]

35. Larcher, M.; Manara, G. *Influence of Air Damping on Structures Especially Glass*; JRC Technical Report, PUBSY JRC 57330; EU Publication Office: Geneva, Switzerland, 2010.

36. Lenk, P.; Coult, G. Damping of Glass Structures and Components. In *Challenging Glass 2—Conference on Architectural and Structural Applications of Glass*; Veer, B.L., Ed.; TU: Delft, Germany, 2010.

37. Barredo, J.; Soriano, M.; Gómez, M.; Hermanns, L. Viscoelastic vibration damping identification methods. Application to laminated glass. *Procedia Eng.* **2011**, *10*, 3208–3213. [CrossRef]
38. Bedon, C.; Fasan, M.; Amadio, C. Vibration Analysis and Dynamic Characterization of Structural Glass Elements with Different Restraints Based on Operational Modal Analysis. *Buildings* **2019**, *9*, 13. [CrossRef]
39. EN 1993-1-1:2005+A1:2014—Eurocode 3. *Design of Steel Structures. General Rules and Rules for Buildings*; CEN: Brussels, Belgium, 2010.
40. EN 572-2:2004. *Glass in Buildings—Basic Soda Lime Silicate Glass Products*; CEN: Brussels, Belgium, 2004.
41. EN 755-2:2016. *Aluminium and Aluminium Alloys—Extruded Rod/Bar, Tube and Profiles—Part 2: Mechanical Properties*; CEN: Brussels, Belgium, 2016.
42. DOWSIL™ 993 Structural Glazing Base and Catalyst—Product Data Sheet. Available online: www.dow.com (accessed on 6 April 2021).
43. Salome-Meca Platform, EDF. Available online: http://www.salome-platform.org/service-and-support/available-programs (accessed on 6 April 2021).
44. EN 14024:2004. *Metal Profiles with Thermal Barrier—Mechanical Performance—Requirements, Proof and Tests for Assessment—Annex C: Effective Momentum of Inertia of Thermal Barrier Profiles*; CEN: Brussels, Belgium, 2004.
45. Morse, S.M.; Norville, H.S. Comparison of methods to determine load sharing of insulating glass units for environmental loads. *Glas. Struct. Eng.* **2016**, *1*, 315–329. [CrossRef]
46. Bedon, C.; Amadio, C. Mechanical analysis and characterization of IGUs with different silicone sealed spacer connections—Part 1: Experiments. *Glas. Struct. Eng.* **2020**, *5*, 301–325. [CrossRef]
47. Bedon, C.; Amadio, C. Mechanical analysis and characterization of IGUs with different silicone sealed spacer connections—Part 2: Modelling. *Glas. Struct. Eng.* **2020**, *5*, 327–346. [CrossRef]
48. Amadio, C.; Bedon, C. Elastoplastic dissipative devices for the mitigation of blast resisting cable-supported glazing fa-çades. *Eng. Struct.* **2012**, *93*, 103–115. [CrossRef]
49. Amadio, C.; Bedon, C. Viscoelastic spider connectors for the mitigation of cable-supported façades subjected to air blast loading. *Eng. Struct.* **2012**, *42*, 190–200. [CrossRef]

MDPI

St. Alban-Anlage 66

4052 Basel

Switzerland

www.mdpi.com

Buildings Editorial Office

E-mail: buildings@mdpi.com

www.mdpi.com/journal/buildings